ASTRONOMY AND ASTROPHYSICS LIBRARY

Series Editors: I. Appenzeller · M. Harwit · R. Kippenhahn
P. A. Strittmatter · V. Trimble

ASTRONOMY AND ASTROPHYSICS LIBRARY

Series Editors: I. Appenzeller · M. Harwit · R. Kippenhahn
P. A. Strittmatter · V. Trimble

The Solar System
By T. Encrenaz and J.-P. Bibring

Astrophysical Concepts 2nd Edition
By M. Harwit

Physics and Chemistry of Comets
Editor: W. F. Huebner

Stellar Structure and Evolution
By R. Kippenhahn and A. Weigert

Modern Astrometry
By J. Kovalevsky

Observational Astrophysics
By P. Léna

Astrophysics of Neutron Stars
By V. M. Lipunov

Supernovae
Editor: A. Petschek

General Relativity, Astrophysics, and Cosmology
By A. K. Raychaudhuri, S. Banerji and A. Banerjee

Tools of Radio Astronomy
By K. Rohlfs

Atoms in Strong Magnetic Fields
Quantum Mechanical Treatment and Applications
in Astrophysics and Quantum Chaos
By H. Ruder, G. Wunner, H. Herold and F. Geyer

The Stars
By E. L. Schatzman and F. Praderie

Physics of the Galaxy and Interstellar Matter
By H. Scheffler and H. Elsässer

Gravitational Lenses
By P. Schneider, J. Ehlers and E. E. Falco

**Relativity in Astrometry, Celestial Mechanics
and Geodesy**
By M. H. Soffel

The Sun An Introduction
By M. Stix

Galactic and Extragalactic Radio Astronomy 2nd Edition
Editors: G. L. Verschuur and K. I. Kellermann

H. Ruder   G. Wunner   H. Herold
F. Geyer

Atoms in Strong Magnetic Fields

Quantum Mechanical Treatment
and Applications
in Astrophysics and Quantum Chaos

With 93 Figures

Springer-Verlag
Berlin Heidelberg New York
London Paris Tokyo
Hong Kong Barcelona
Budapest

Professor Dr. Hanns Ruder
Professor Dr. Heinz Herold
Dr. Florian Geyer
Theoretische Astrophysik, Universität Tübingen, Auf der Morgenstelle 10
D-72076 Tübingen, Germany

Professor Dr. Günter Wunner
Theoretische Physik, Ruhr-Universität Bochum, Lehrstuhl 1
D-44780 Bochum, Germany

Series Editors

Immo Appenzeller
Landessternwarte, Königstuhl
D-69117 Heidelberg, Germany

Martin Harwit
The National Air and Space Museum
Smithsonian Institution
7th St. and Independence Ave. S. W.
Washington, DC 20560, USA

Rudolf Kippenhahn
Rautenbreite 2
D-37077 Göttingen, Germany

Peter A. Strittmatter
Steward Observatory
The University of Arizona
Tucson, AZ 85721, USA

Virginia Trimble
Astronomy Program
University of Maryland
College Park, MD 20742, USA
and Department of Physics
University of California
Irvine, CA 92717, USA

ISBN 3-540-57699-1 Springer-Verlag Berlin Heidelberg New York
ISBN 0-387-57699-1 Springer-Verlag New York Berlin Heidelberg

Library of Congress Cataloging-in-Publication Data. Atoms in strong magnetic fields: quantum mechanical treatment and applications in astrophysics and quantum chaos / H. Ruder ... [et al.]. p. cm. Includes bibliographical references and index. ISBN 3-540-57699-1 (Berlin: alk. paper). − ISBN 0-387-57699-1 (New York: alk. paper) 1. Atomic transition probabilities. 2. Stars−Magnetic fields. 3. White dwarfs. 4. Neutron stars. 5. Quantum chaos. I. Ruder, Hanns. QC454.N8A77 1994 539.7'23−dc20 94-23259

This work is subject to copyright. All rights are reserved, whether the whole or part of the material is concerned, specifically the rights of translation, reprinting, reuse of illustrations, recitation, broadcasting, reproduction on microfilm or in any other way, and storage in data banks. Duplication of this publication or parts thereof is permitted only under the provisions of the German Copyright Law of September 9, 1965, in its current version, and permission for use must always be obtained from Springer-Verlag. Violations are liable for prosecution under the German Copyright Law.

© Springer-Verlag Berlin Heidelberg 1994
Printed in Germany

The use of general descriptive names, registered names, trademarks, etc. in this publication does not imply, even in the absence of a specific statement, that such names are exempt from the relevant protective laws and regulations and therefore free for general use.

SPIN: 10018518   55/3140 - 5 4 3 2 1 0 - Printed on acid-free paper

Preface

In this book we summarize the essential results of our efforts over the years to calculate energies, wave functions, and electromagnetic transitions of atoms as functions of the magnetic field strength from laboratory fields up to neutron star magnetic fields. Motivated by the observational evidence of huge magnetic fields with strengths up to 10^5 T in the vicinity of white dwarf stars and of up to 10^9 T in the vicinity of neutron stars the authors, together with coworkers and candidates for doctor and diploma degrees, have investigated this fascinating quantum mechanical problem more or less continuously since 1978. The extensive tables and figures in the appendices represent the most complete data set to date in this field of research. For practical use all numbers are available by "anonymous ftp" over Internet.

The first direct measurement of a neutron star magnetic field by Trümper and his group, who observed a cyclotron feature at about 50 keV in the spectrum of the X-ray pulsar Hercules X-1 corresponding to a field strength of several 10^8 T, stimulated investigations of atoms within the framework of the adiabatic approximation, which is well justified for such field strengths. This method and its results are discussed in Chaps. 3, 5, and 6.

Angel's idea to explain mysterious lines in the spectra of highly magnetized white dwarf stars in terms of "stationary" Zeeman components in the hydrogen spectrum increased our efforts to also calculate with high accuracy atomic properties in the complex transition regime where Coulomb and Lorentz forces become comparable. In Chap. 7 we summarize the state of the art of this new spectroscopy of stationary lines which became possible using our atomic data and helped to pin down the field strengths of white dwarf stars in the range from 3×10^3 T to 10^5 T.

A further stimulus came from laboratory experiments with highly excited Rydberg atoms where the transition to the magnetically dominated regime can already be performed with magnetic field strengths of just a few Tesla. The almost enthusiastic study of such systems has two reasons: firstly, the transition between quantum and classical physics can be investigated in detail and, secondly, the hydrogen atom is the simplest physically real system which, in the classical limit, changes from regular to chaotic behaviour with increasing magnetic field strength. Therefore, it is an ideal object with which to study phenomena of "quantum chaos" both experimentally and theoretically. A short review of this new and wide field of research is given in Chap. 10.

While our knowledge of hydrogen-like systems in magnetic fields of arbitrary strength is – apart from some minor points – fairly good and complete, the situation for helium-like systems is far less satisfying. In Chap. 9 and in Appendix 4 the results for helium available so far are documented.

This book contains the most comprehensive compilation to date of atomic data for hydrogen, helium, and their isoelectronic sequences in strong magnetic fields. Nevertheless, even more complete and accurate data are urgently needed for detailed analyses of spectra of magnetic white dwarfs, which will be obtained with increasing quality by present and future modern telescopes. Moreover, such data are also prerequisite for quantitative studies of the thermal spectra of isolated neutron stars with strong cosmic magnetic fields. These spectra have been observed by the ROSAT satellite, and will be observed even more extensively and with higher accuracy by future X-ray satellites. Given the progress of numerical methods and the still dramatically increasing power of supercomputers, we feel well equipped to meet these challenges in the next few years.

Finally, we would like to express our thanks to Wolfgang Schweizer, who has done much work in this field together with us as a postdoc, and to our coworkers Moritz Braun, Peter Faßbinder, Hans-Georg Forster, Susanne Friedrich, Roland Niemeier, Heinrich Körbel, Peter Pröschel, Wolfgang Rösner, Wolfgang Strupat, Günther Thurner, Joachim Tworz, and Gudrun Zeller, each of whom has added a more or less extensive mosaic piece to this book in form of a diploma or PhD thesis, and last but not least to an army of students for their help in preparing the figures, the tables, and the references.

Tübingen and Bochum, July 1994   *The authors*

Contents

1. **Introduction** 1
   1.1 Magnetic Fields of Compact Cosmic Objects 1
   1.2 Historical Review 6
   1.3 Notations and Abbreviations 11

2. **Interacting Charged Particles in Uniform Magnetic Fields** 13
   2.1 The N-Body Problem 13
   2.2 The Uncharged Two-Body Problem 14
       2.2.1 Hamiltonian and Wave Functions
             in Centre-of-Mass and Relative Coordinates 15
       2.2.2 Classification, Quantum Numbers and Degeneracy
             in the Non-Interacting Case 15
       2.2.3 Exact Solution for Harmonic Interaction 17
   2.3 Scaling Properties of the Coulomb Problem 19
       2.3.1 The General Case $\boldsymbol{K} \neq 0$ 19
       2.3.2 The Special Case $\boldsymbol{K} = 0$ 20
       2.3.3 Nuclear Charge (Z) Scaling 21

3. **Methods of Solution for the Magnetized Coulomb Problem** 23
   3.1 General Considerations 23
       3.1.1 Characteristic Domains
             of the Magnetic Field Strength 23
       3.1.2 Expansions of the Wave Functions
             and Multiconfiguration Equations 24
       3.1.3 Low-Field High-Field Correspondence 26
   3.2 Numerical Treatment 29
       3.2.1 Hartree-Fock-Like Methods 29
       3.2.2 Coupled-Channels-Like Methods 31
       3.2.3 Diagonalization Methods 32

4. **Results for Low-Lying States** 35
   4.1 Energy Values 35
   4.2 Wavelengths of the Hydrogen Atom 39
   4.3 Wave Functions of the Hydrogen Atom 44
       4.3.1 Graphic Representation of the Spatial
             Probability Distribution of the Electron 46
       4.3.2 Pictorial Representation of the Spatial
             Probability Distribution of the Electron 49

5. Energies for Arbitrarily Excited States in Adiabatic Approximation 57
5.1 Asymptotic Property of the Effective Potentials 57
5.2 Numerical Results and Their Accuracy 58

6. Electromagnetic Transition Probabilities 69
6.1 The General Expressions 69
6.2 Effects of the Finite Proton Mass on the Transition Matrix Element 72
6.3 Results 74
6.4 Expressions for Electromagnetic Transitions in Adiabatic Approximation 78
    6.4.1 Dipole Strengths 78
    6.4.2 Selection Rules 79
    6.4.3 Sum Rules 79
    6.4.4 Asymptotic Formulae 80
    6.4.5 Results and Discussion 82

7. Stationary Lines and White Dwarf Spectra 89
7.1 Stationary Lines 89
7.2 Spectra of Selected Magnetic White Dwarfs 95
    7.2.1 Grw+70°8247 95
    7.2.2 PG 1031+234 99
    7.2.3 SBS 1349+5434 100
    7.2.4 PG 1015+014 101
    7.2.5 G227−35 101
    7.2.6 MR Serpentis 104
    7.2.7 LB 11146 (PG 0945+245) 105
7.3 Table of Magnetic White Dwarfs 106
7.4 Future Work 107

8. Relativistic Effects, Nuclear Mass Effects, and Landau-Excited States 109
8.1 Spin-Orbit Coupling 109
8.2 Effects of the Finite Proton Mass and of Motion Perpendicular to the Magnetic Field 113
8.3 Landau-Excited States 116

9. Helium-Like Atoms in Magnetic Fields of Arbitrary Strengths 123
9.1 Correspondence Diagrams 123
    9.1.1 Short Review of the One-Electron Problem 123
    9.1.2 The Two-Electron System in a Magnetic Field 124
    9.1.3 The Correspondence for $1/Z = 0$ 126
    9.1.4 The Correspondence for the General Case 127

	9.2	Method of Solution	129
		9.2.1 The Hartree-Fock Method for Low to Intermediate Field Strengths ($\beta_Z \lesssim 1$)	130
		9.2.2 The Hartree-Fock Method for High Field Strengths ($\beta_Z \gtrsim 1$)	132
	9.3	Dipole Strengths, Oscillator Strengths, and Transition Probabilities	134
		9.3.1 Selection Rules for the Spherical Ansatz	135
		9.3.2 Selection Rules for the Cylindrical Ansatz	135
	9.4	Results for the Two-Electron Problem	136
		9.4.1 Energy Levels Calculated with a Single-Configuration Ansatz	136
		9.4.2 Influence of Configuration Mixing	137
		9.4.3 Comparison of Energy Values Obtained by Different Methods	140
		9.4.4 Wavelengths, Dipole Strengths, Oscillator Strengths, and Transition Probabilities	143

10. Highly Excited States 153

	10.1	Results	154
		10.1.1 Energy Levels	154
		10.1.2 Transition Probabilities and Comparison with Experiments	155
	10.2	Is There Chaos in Quantum Mechanics?	165
		10.2.1 Introduction	165
		10.2.2 Microwave Ionisation of Rydberg States of the Hydrogen Atom	167
		10.2.3 Statistical Analysis of Energy-Level Sequences	169
		10.2.4 Order and Chaos in the Hydrogen Atom in a Magnetic Field	172
		10.2.5 Level Statistics for the Hydrogen Atom in Magnetic Fields	174
		10.2.6 Resonances in Chaos – the Role of Periodic Orbits	175
		10.2.7 "Scarring" of Wave Functions	177

Outlook 183

A1. Energy Values 185
    A1.1 Tables of the Energy Values 185
    A1.2 Figures of the Energy Values 217

A2. Wavelengths 225
    A2.1 Figures of the Wavelengths 225
    A2.2 Tables of Stationary Wavelengths 231
    A2.3 Figures of Stationary Wavelengths 236

A3. Electromagnetic Transition Probabilities 241
    A3.1 Wavelengths, Dipole Strengths,
         Oscillator Strengths, and Transition Rates 241
    A3.2 Oscillator Strengths and Transition Probabilities
         in Adiabatic Approximation 253
    A3.3 Dipole Strengths of Stationary Transitions 259

A4. Helium and Helium-Like Atoms 263
    A4.1 Tables of the Energy Values 263
    A4.2 Wavelengths, Dipole Strengths,
         Oscillator Strengths, and Transition Rates 283

References 293

Subject Index 307

1 Introduction

1.1 Magnetic Fields of Compact Cosmic Objects

The discovery of huge magnetic fields in the vicinity of white dwarf stars ($B \approx 10^2$–10^5 T) (Kemp et al. 1970; Angel 1978; Angel et al. 1981) and neutron stars ($B \approx 10^7$–10^9 T) (Trümper et al. 1977; Trümper et al. 1978) has opened the possibility of studying the properties of matter under conditions which can never be realized in terrestrial laboratories. (Rapidly time-variable magnetic fields over nuclear dimensions with peak values up to $B \approx 10^{11}$ T are assumed to occur in heavy-ion collisions, cf. Rafelski and Müller 1976). White dwarf stars and neutron stars represent final stages of stellar evolution. Neutron stars are formed from normal stars in a dramatic cosmic event, when the star has consumed its nuclear energy supply and becomes unstable against its own gravitational forces. The catastrophic collapse to a neutron star is usually accompanied by a supernova explosion, in which the star becomes almost as bright as a whole galaxy consisting of a hundred billion suns. Typical values of relevant physical parameters are listed in Table 1.1 in comparison with our sun.

Table 1.1. Physical parameters of compact cosmic objects, compared with the sun

	Sun	Magnetized white dwarf	Pulsar = magnetized rotating neutron star
Mass	2×10^{30} kg $= m_\odot$	$\sim 1\, m_\odot$	$\sim (1$–$2)\, m_\odot$
Radius	7×10^5 km $= R_\odot$	$\sim 10^4$ km $\approx 10^{-2} R_\odot$	~ 10 km $\approx 10^{-5} R_\odot$
Mean density	$1.4\,\text{g/cm}^3$	$\sim 10^6\,\text{g/cm}^3$	$\leq 10^{15}\,\text{g/cm}^3$
Period	27 d	100 s–days ?	10^{-3}–10^3 s
Magnetic field	$\sim 10^{-4}$–10^{-3} T	$\sim 10^2$–10^5 T	$\sim 10^7$–10^9 T

The existence and the physical properties of these compact objects had been predicted theoretically from the knowledge of the equation of state long before they were actually observed. In particular, the expected order of magnitude of the magnetic field strength can easily be estimated by the following consideration. Because of the high conductivity of stellar matter, the magnetic flux is conserved, in good approximation, during the gravitational collapse. As a consequence, the magnetic field is compressed along with the stellar matter and is boosted, taking the radius reduction from Table 1.1, by a factor of 10^4 for white dwarfs and 10^{10} for neutron stars. With possible initial field strengths of about 10^{-2}–10^{-1} T, fields up to 10^3 T or 10^9 T, respectively, are automatically achieved. Physically, these magnetic fields are generated by electric currents in the neutron stars, which are induced – according to the generally accepted theory – during the gravitational collapse and are stable against ohmic dissipation on time scales of 10^7 years. (Note that a neutron star must possess a fraction of electrons and protons of a few percent in order to be stable against beta decay.)

The discovery of these fields was based, for white dwarf stars, on the observation of the circular polarization of the continuum radiation in the optical and the ultraviolet range (Kemp et al. 1970; Angel et al. 1981). The determination of absolute sizes of the magnetic field strengths up to $B \approx 10^3$ T was possible by measuring the splitting of spectral lines in the range of validity of the linear and quadratic Zeeman effect (Kemic 1974a,b; Garstang and Kemic 1974; Garstang 1982). For magnetic field strengths beyond that, it was only on the basis of the results presented in this report that a quantitative interpretation of the spectra, and thus the pinning down of the values of B, could be made.

The evidence of field strengths of the order of 10^8 T in neutron stars is not quite as immediate, and the observations suggest a distinction between pulsars that predominantly emit in the radio or the X-ray range. Radio pulsars can be identified as isolated, rapidly rotating, strongly magnetized neutron stars, whereas X-ray pulsars are rapidly rotating, strongly magnetized neutron stars, which are the compact components in close binary systems pulling matter from the companion star via gravitation onto their own surface. (For comprehensive reviews of the physics of compact objects see, e.g., Shapiro and Teukolsky 1983; Pringle and Wade 1985; Trümper et al. 1986; Helfand and Huang 1987; Nagase 1989).

The determination of the magnetic field strength of radio pulsars is made indirectly by observing the braking of the rotation of the pulsars, which is caused by the emission of extremely intense electromagnetic waves associated with the rotation of the magnetic moment of the pulsar. From classical electrodynamics it is easy to derive the relation $B^2 \propto P\dot{P}$ between the magnetic field strength at the pole, and the period P and its time derivative $\dot{P}$. The values measured for $P\dot{P}$ thus lead, in a more indirect way, to values of B, which are distributed around 2×10^8 T for typical radio pulsars. At these magnetic field strengths, by unipolar induction huge potential differences of up to 10^{17}–10^{18} V occur between the pole and the equator, which correspond, on an atomic length scale of 10^{-10} m, to an electric potential drop of about 3000 V.

The electric forces at the surface thus still outweigh the enormous gravitational forces on electrons – 10^{11} times stronger than on earth – by a factor of 10^{12}, which unavoidably gives rise to field emission. As a consequence, every rapidly rotating highly magnetized neutron star must necessarily be surrounded by a highly relativistic plasma, in which the huge unipolar induction is transformed into kinetic energy. The charges and currents in the plasma change the vacuum fields, and the self-consistent solution of this magnetospheric problem represents an extremely complicated task that has not been solved in a satisfactory way until now. It is evident that, for the whole complex of questions connected with radio pulsars, the accurate knowledge of the properties of matter in superstrong magnetic fields is a prerequisite to a quantitative understanding.

We now turn to a short discussion of the class of X-ray pulsars. To date, approximately 35 objects of this kind, with pulse periods from 69 ms – 835 s, are known. Conclusive evidence was found that these X-ray pulsars are magnetized neutron stars in close binary systems, where, in contrast to radio pulsars, the rotational frequencies usually increase due to accreting gas from their normal companion stars.

Probably the most thoroughly investigated, and understood, system is the X-ray source Hercules X-1, whose story is briefly retold in Fig. 1.1. The figure also includes the scenario in the vicinity of one magnetic polar cap of the neutron star, which has been deduced from the total observed information.

Channelled by the extremely strong magnetic field, ionized matter splashes down with a free-fall velocity of about half the velocity of light on an area of roughly one square kilometer. The matter is stopped by a shock, the kinetic energy is thermalized, and from the observed X-ray spectra one infers temperatures on the order of 10^8 K for this emitting hot spot. From the measured total X-ray luminosity of $\approx 2 \times 10^{30}$ W (about 10^4 times larger than the total radiative power of our sun) of this huge cosmic X-ray tube, it can be concluded that the enormous amount of 10^{11} tons is accreted per second.

Further information can be unravelled by carefully evaluating the spectral distribution of the observed X-ray quanta. Figure 1.1 also shows the X-ray spectrum of Hercules X-1 in the range of 2–150 keV obtained by Trümper et al. (1977) in 1976 using a balloon-borne detector. The spectrum exhibits two line features near 50 keV and 100 keV. In the more pronounced feature near 50 keV, about 1% of the total energy radiated is contained, which still corresponds to 100 times the radiation of our sun over the whole spectrum. Conclusive discussions ruled out an atomic or nuclear origin of these line structures; as the only natural explanation there remained the interpretation in terms of cyclotron transitions between quantized Landau states of electrons in the hot, strongly magnetized plasma above the polar cap. Identifying the energy of the first line feature with the cyclotron energy, $\hbar\omega_\mathrm{c} = \hbar e B/m_\mathrm{e}$, led to $B \approx 5 \times 10^8$ T, which represented the first direct measurement of such a huge magnetic field strength. The other feature can then be explained without difficulty as the second harmonic. (It should be mentioned that, in the meantime, cyclotron features have been observed in the spectra of several other neutron stars.)

To give the reader a more realistic idea of this remarkable astrophysical site in Fig. 1.2 we render the impression of a space trip around the binary system with the X-ray source Hercules X-1. The pictures are the results of computer simulations of the accretion disk, the accretion column and the hot spot.

It is obvious that a quantitative theoretical description of this second class of pulsating sources again requires an accurate knowledge of the properties of matter in superstrong magnetic fields. Of particular interest are, because of the prevailing high temperatures up to 20 keV, cross sections for ionization, recombination, and bremsstrahlung, the propagation of photons in these highly

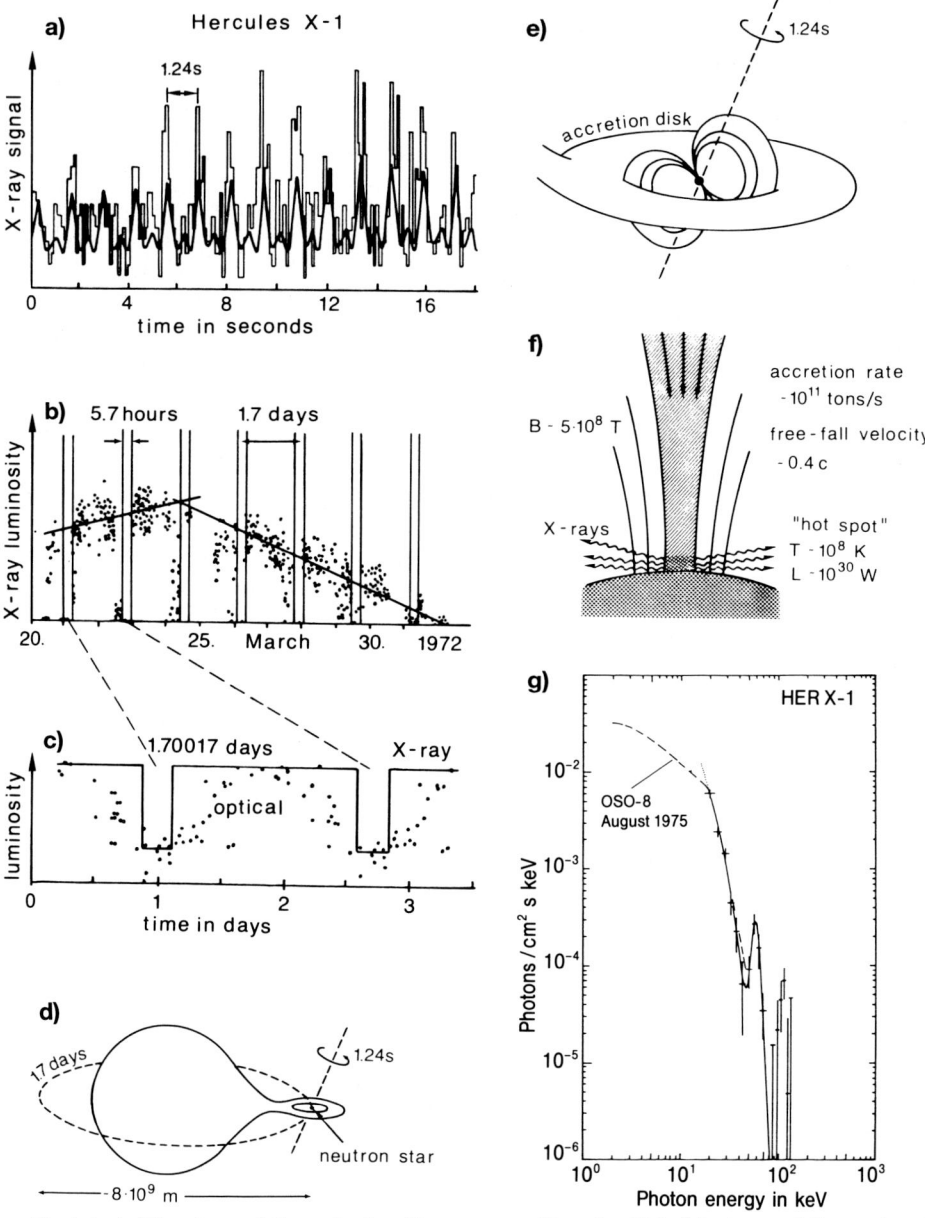

Fig. 1.1a–g. The story of the pulsating X-ray source Hercules X-1: (**a**) observed X-ray signal with a mean pulse period of 1.24 s; (**b**) every 1.70017 days the X-ray source turns off for 5.7 hours; (**c**) correlated luminosity variation between the X-ray source and the optical companion star; (**d**) schematic view of the close binary system, the neutron star pulls over matter from the normal star; (**e**) the material is collected in an accretion disk around the neutron star; (**f**) a close-up of one polar cap of the neutron star, where the observed X-rays are produced; (**g**) the X-ray spectrum of Hercules X-1 (Trümper et al., 1977), note the two cyclotron line features at about 50 keV and 100 keV.

1.1 Magnetic Fields of Compact Cosmic Objects

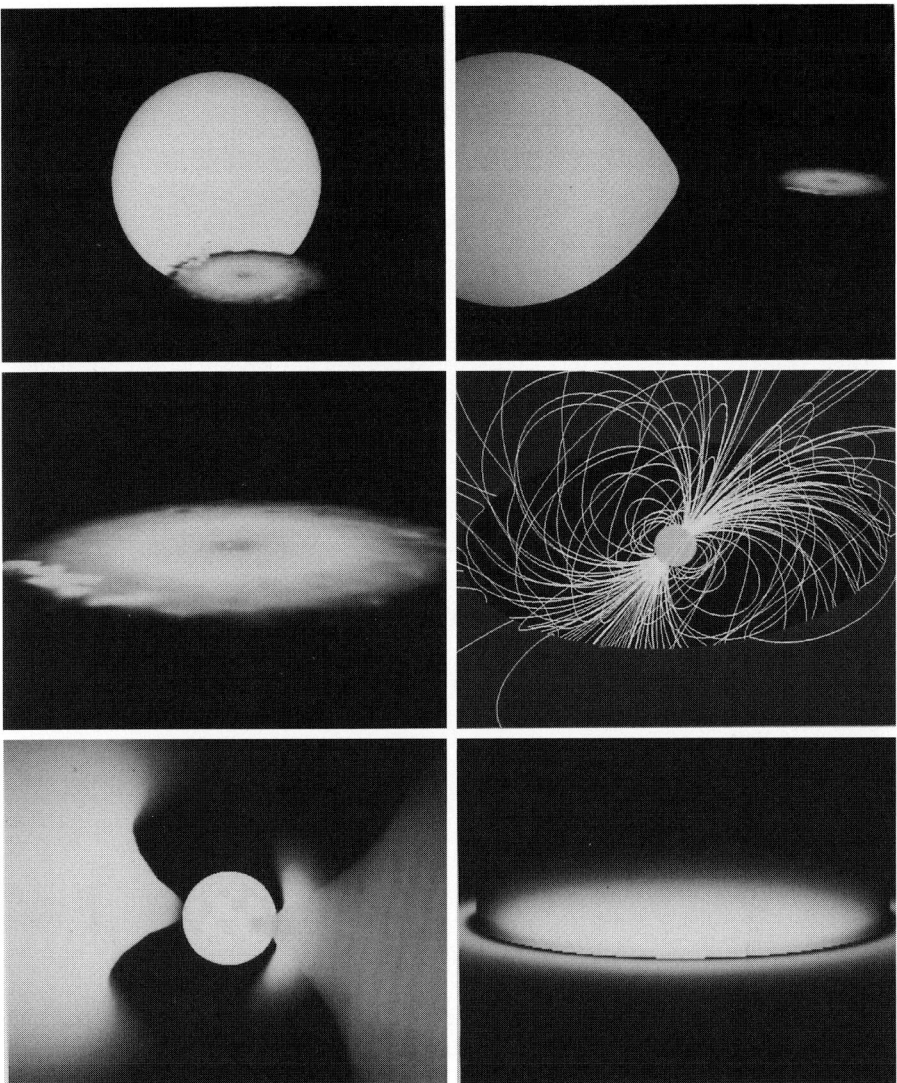

Fig. 1.2. Impression of a space trip around the binary system with the X-ray source Hercules X-1. The pictures are the results of computer simulations of the accretion disk, the accretion column and the hot spot.

magnetized plasmas, and the formation of cyclotron line features. The results of such efforts serve to construct self-consistent models of accretion columns. From the comparison between computed and observed X-ray spectra, reliable conclusions as to the physical conditions in the emitting regions can then be drawn.

In the magnetic field strengths characteristic of neutron stars, the Lorentz forces are larger by 2 to 3 orders of magnitude than the Coulomb binding forces acting on an electron in the ground state of a hydrogen atom. Coulomb binding forces are dominated by the Lorentz forces. Therefore, a recalculation of atomic structure is necessary; this represents, apart from its astrophysical applications, a new and fascinating chapter of quantum mechanics. By complete analogy with the development of field-free quantum mechanics, it is the one-electron problem which here also will serve as a useful guide to working out a new atomic physics in strong magnetic fields. It will be seen that great difficulties occur even in this simplest atomic system if a total understanding of the complete level scheme, wave functions, and electromagnetic transitions in their continuous dependences on the magnetic field is desired. In particular, it is evident that the transition regime from Coulomb dominance to Lorentz dominance causes the greatest problems. It is, however, exactly this region of magnetic field strengths which occurs in strongly magnetized white dwarf stars. Since the atmospheres of these stars mainly consist of hydrogen, the knowledge of the accurate atomic data in this regime is absolutely necessary for the understanding and modeling of the observed spectra. On the other hand, the strongly magnetized white dwarf stars provide almost ideal cosmic strong-field laboratories. The same situation of comparable Lorentz and Coulomb forces can be realized in terrestrial laboratories with magnetic field strengths of several Tesla for highly excited (*Rydberg*) states. These serve as fascinating objects of studies of "quantum" chaos.

1.2 Historical Review

To tackle the general problem of atoms in magnetic fields, all that is required is a knowledge of the fundamentals of quantum mechanics. Therefore, we shall try to formulate this book in such a way that it is understandable to every advanced learner of quantum mechanics, thus saving the reader the tedious labour of constantly having to resort to the original literature. However, in order to give credit to the many researchers who have promoted the field we will list, in the framework of a historical review, some essential contributions.

The need for investigating atoms in strong magnetic fields arose in the physics of excitons (hydrogen-like bound electron-hole pairs) where laboratory fields are already able to dominate the structure of the systems. An influential paper was presented by Loudon (1959), who treated the hydrogen atom in arbitrary field strengths using analytically tractable variational forms. Electromagnetic transitions in magnetized excitonic systems were considered on a similar level by Hasegawa and Howard (1961). Improvements in the accuracy of energy levels and transition probabilities were obtained in the numerical variational calculations of Cabib et al. (1971), Praddaude (1972), Smith et al. (1972), and Garstang and Kemic (1974). A more complete account of the work performed up to 1976 can be found in the review article *Atoms in high magnetic fields* by Garstang (1977).

While these investigations were motivated by the physics of excitons and, in astrophysics, by the observation of magnetic fields with strengths up to several thousand Tesla in white dwarf stars, the direct measurement of a magnetic field strength of 5×10^8 T in the accretion column of the neutron star in the binary system Hercules X-1 by Trümper et al. (1977) provided the decisive stimulus for many theorists to tackle the problem with new drive.

Simola and Virtamo (1978) produced the energy levels of a few low-lying states of hydrogen atoms in strong magnetic fields with an accuracy of about six significant digits and presented the correct correspondence diagrams between the field-free states (without spin) and the states in the high-field limit. A number of papers followed that used different methods (i.e., perturbational expansions, variational ansatzes, Hartree-Fock-like procedures, diagonalization, finite element methods) and dealt with either the low-field or the high-field side of the problem (Simola and Virtamo 1978; Aldrich and Greene 1979; Starace and Webster 1979; Hylton and Rau 1980; Kara and McDowell 1980; Kaschiev et al. 1980; Pavlov-Verevkin and Zhilinskii 1980a; Pavlov-Verevkin and Zhilinskii 1980b; Burdyuzha and Pavlov-Verevkin 1981; Cohen and Hermann 1981; Ruder et al. 1981a; Wunner and Ruder 1981; Wunner et al. 1981a; Wunner et al. 1981b; Wunner et al. 1981c; Bender et al. 1982; Cizek and Vrscay 1982; Friedrich 1982; Wunner and Ruder 1982; Wunner et al. 1982a; Johnson et al. 1983; Le Guillou and Zinn-Justin 1983; Rösner et al. 1983; Baye and Vincke 1984). The first comprehensive list of highly accurate energy values of low-lying states in their dependence of the magnetic field from $0 \leq B \leq 10^9$ T was presented by Rösner et al. (1984).

The most complete list to date consists of the tables given in the appendix of this book. (All tables of the book are available by "anonymous ftp" over Internet. On a network-ready machine you should execute the command *ftp atoms.tat.physik.uni-tuebingen.de* and login as *anonymous*, using your e-mail address as a password. Then execute the commands *cd pub/atoms* and *mget ** to download the relevant TEXfiles of the tables. Should you require values that cannot be extracted from the tables, you can obtain these, if numerically feasible, from the authors on request.)

The data published by Rösner et al. (1984) since then have served as a standard for testing various other mathematical and numerical approaches. Out of a number of papers on this subject, we mention here Williams et al. (1985), who used a trial wave function approach including lowest-order relativistic effects in the context of a neutron star application; Wintgen and Friedrich (1986a), who diagonalized the Hamiltonian in the complete basis of two-dimensional oscillator functions in semi-parabolic coordinates; Rech et al. (1986), who presented four- and five-parameter functions for the energies of low-lying states; Rosen (1986), Handy et al. (1988), who derived rigorous upper or lower bounds on the energy of the ground state; Liu and Starace (1987), who, in two cylindrical adiabatic approximations, determined bounds on the energies of low-lying states; Shertzer (1989), Shertzer et al. (1989), who performed a finite-element analysis for low-lying states; and Miller and Neuhauser (1991), who applied a high-field multiconfigurational Hartree-Fock code to hydrogen and higher elements.

Another line of investigation was devoted to the two-body effects in hydrogenic systems (effects of the finite nuclear mass) in strong magnetic fields (e.g., Avron et al. 1978; O'Connell 1979; Wunner et al. 1980; Herold et al. 1981; Wunner et al. 1981a,b; Baye 1982, 1983; Johnson et al. 1983; Baye and Vincke 1986). It elaborated on the way in which the properties of these systems can be deduced from the results obtained for infinite nuclear mass.

Synchronously with these calculations for low-lying states, which aimed at intense astrophysical magnetic fields, a tremendous development took place in the field of the calculation of energy values and oscillator strengths of highly excited states of the hydrogen atom in strong laboratory magnetic fields. These were instigated by the discovery of regular structures – quasi-Landau resonances – in the optical spectra of highly excited atoms (Rydberg atoms) in fields of a few Tesla (Garton and Tomkins 1969). In addition, experimental work (Zimmermann et al. 1980) had raised the intriguing question as to the existence of an approximate symmetry in the problem (e.g., Delande and Gay 1981; Gay et al. 1983; Delande and Gay 1984; Wintgen and Friedrich 1986b). The first fully quantum mechanical calculation of the hydrogen atom at a laboratory field strength (4.7 T) was produced by Clark and Taylor (1982), who diagonalized the Hamiltonian in a Sturmian basis and presented spectra up to $-40\,\text{cm}^{-1}$ below the field-free ionization threshold. Their calculations were taken up by the Tübingen group (Wunner et al. 1989), which presented results with at least six-digit accuracy for energy values and three-digit accuracy for oscillator strengths for all levels up to the threshold, and established a one-to-one correspondence to experimental spectra determined by the Bielefeld group (Holle et al. 1988). Theoretical spectra of non-hydrogenic Rydberg atoms in strong laboratory magnetic fields were calculated by O'Mahony and Taylor (1986a,b) and O'Mahony (1989), using a combined quantum defect theory R-matrix approach.

Bound-bound transitions between low-lying states of hydrogenic atoms in magnetic fields up to $B = 10^4$ T were first examined in detail by Smith et al (1973a,b) taking a trial wave function approach, and by Kemic (1973, 1974a,b), Garstang and Kemic (1974), Garstang (1977), and Brandi et al (1976), who used wave functions that follow from the diagonalization of the Hamiltonian in a truncated basis of unperturbed hydrogenic states. Kara and McDowell (1980) extended these calculations to $B = 2.35 \times 10^5$ T by diagonalizing in a basis of simple cylindrical functions. Work on transitions of hydrogenic atoms in magnetic fields $B \gg 2.35 \times 10^5$ T was reported in a solid-state context by Hasegawa and Howard as early as 1961; they solved the one-dimensional Schrödinger equation, which results in the "adiabatic approximation" (Schiff and Snyder 1939) in an analytical, though approximate way. Calculations of transition probabilities for bound-bound transitions, for neutron star magnetic fields in the high-field limit were presented by Wunner and Ruder (1980a). These calculations were ultimately refined in such a way that oscillator strengths could be determined with at least a three-digit accuracy for all transitions between low-lying levels as a function of the magnetic field strength, again in the complete range $0 \leq B \leq 10^9$ T (Forster et al. 1984).

All the papers referred to above worked within the framework of the non-relativistic Schrödinger equation. However, bearing in mind that, for a magnetic field strength of 5×10^8 T, the Landau level distance amounts to about 50 keV (i.e., one-tenth of the electron's rest energy), the question of the importance of relativistic effects is a very obvious one. This problem was treated in detail by Lindgren and Virtamo (1979) and by Doman (1980), who arrived at the result that relativistic effects are indeed of the anticipated order of magnitude for the motion of the electron transverse to the magnetic field, but are almost negligible for the bound states of the hydrogen atom. This is understandable, because the energies of the motion parallel to the magnetic field are, at most, on the order of a few hundred eV and thus non-relativistic. Furthermore, the transverse parts of the wave functions in the Landau ground state do not differ relativistically or non-relativistically. A discussion of relativistic effects, such as spin-orbit coupling and its relevance to the correspondence diagram, and the coupling to the electromagnetic vacuum energy (magnetic Lamb shift), was given by Wunner et al. (1985a).

Calculations of bound-free transitions (photoionization), which must take into account the asymptotic Landau structure of the wave functions of the continuum electrons, were first performed in the high-field limit using the adiabatic approximation (Schmitt et al. 1981; Kara and McDowell 1981; Wunner et al. 1982b; Wunner et al. 1983b). Applications to the physics of the surface layers of neutron stars were given by Ventura et al. (1992) and Potekhin and Pavlov (1993), who computed photoabsorption opacities which are needed for the interpretation of X-ray spectra. Methods for calculating the photoionization cross sections became more and more refined (Greene 1983; Friedrich and Chu 1983; Mega et al. 1984; Wintgen and Friedrich 1986c), making possible detailed calculations of photoionization cross sections for lower fields, at first down to 2000 T (Alijah et al. 1990) and most recently down to laboratory field strengths (Delande et al. 1991; Iu et al. 1991; O'Mahony and Mota-Furtado 1991; Watanabe and Komine 1991).

As compared to hydrogen, our knowledge about energy levels and transition strengths of low-lying states in higher atoms and molecules in strong magnetic fields is much more limited and, where available, less accurate. To calculate the binding energy of the hydrogen negative ion for fields of up to 3×10^5 T Henry et al. (1974) expanded the wave function in spherical harmonics and Slater orbitals. Surmelian et al. (1974) used the same basis to calculate variational estimates of the energies of the $1\,^1S$, $2\,^1P$, $2\,^3P$, $2\,^1S$ and $2\,^3S$ states in a strong magnetic field. Mueller et al. (1975) used cylindrical trial wave functions in the adiabatic approximation to derive upper bound variational estimates for the lowest energy levels of H^-, He, and Li^+. Larsen (1979) used different trial functions, which also included correlation between the electrons, to estimate the binding energy of H^- and He in the intermediate regime ($B \sim 5 \times 10^5$ T) and in the high-field limit. Pröschel et al. (1982) applied the Hartree-Fock formalism in cylindrical coordinates to the strong-field triplet ground state of the two-electron problem. Park and Starace (1984) employed an adiabatic approximation in hyperspherical coordinates for the lowest singlet states of

H$^-$ and He. Vincke and Baye (1989) used a simple Slater-determinant basis which allowed for correlation and both transverse and longitudinal mixing to obtain variational estimates of the energies of the lowest singlet and triplet bound states for H$^-$ and He for values of 0 to -2 of the magnetic quantum number for magnetic fields $\geq 1 \times 10^6$ T. Neuhauser et al. (1986) and Miller and Neuhauser (1991) determined wave functions and energies of low-lying states and bound-bound as well as bound-free oscillator strengths for atoms with up to 14 electrons for magnetic fields $B \geq 2 \times 10^7$ T employing the Hartree-Fock formalism with single-particle wave functions in adiabatic approximation. These latter results, however, must be treated with some caution since even the numerics did not seem to be stable. A comprehensive analysis of energy values and transition rates of two-electron systems in arbitrary strong magnetic fields including comparisons with literature values has been given by Thurner et al. (1993).

Only the simplest molecular structures (H$_2^+$, H$_2$) have been investigated so far, within the limit of very high fields by de Melo et al. (1978), Peek and Katriel (1980), Kaschiev et al. (1980), Ozaki and Tomishima (1981), Wunner et al. (1982c), Wille (1987a,b), and for the hydrogen molecule in laboratory fields by Monteiro and Taylor (1990). Density functional formulations have been used to determine approximate values for the binding energies of atoms up to Fe in neutron star fields and to investigate the possible formation of one-dimensional solid state structures (Müller 1984; Jones 1985a,b and 1986; Kössl et al. 1988).

The applications of highly accurate computations of energy levels and transition probabilities of hydrogenic systems was first discussed for neutron star magnetic field strengths in terms of magnetically strongly shifted Ly$_\alpha$ lines of hydrogen-like iron ions (Ruder et al. 1981b; Williams et al. 1985). Applications to white dwarf fields came with the identification of "mysterious" lines in the optical and UV spectra of magnetic degenerates (Angel et al. 1985, Greenstein et al. 1985, Wunner et al. 1985b) with "stationary" lines, i.e. lines whose wavelengths vary only slowly with the field strength, an idea that was originally proposed by Angel (1978). This discovery stimulated even more extensive tabulation of energy levels and transition rates in the pertinent range of magnetic fields (Wunner et al. 1987). A substantial number of magnetized white dwarf spectra has been successfully analyzed meanwhile with the help of these results which at the same time make possible the accurate measurement of magnetic field strengths ranging from some 10^3 to 10^5 T (Angel et al. 1985; Greenstein et al. 1985; Henry and O'Connell 1985; Wunner et al. 1985b; Schmidt et al. 1986b; Latter et al. 1987; Wickramasinghe and Ferrario 1988; Wickramasinghe and Cropper 1988; Schwope et al. 1993; Cohen et al. 1993).

Investigations of magnetic Rydberg atoms received further momentum from the realization that they can serve as a means for testing concepts of "quantum" chaos, and a tide of papers has been addressing this problem. For recent coverage of the topic we refer the reader to the reviews by Friedrich and Wintgen (1989), Hasegawa et al. (1989) and the monographs by Gutzwiller (1990) and Haake (1991).

1.3 Notations and Abbreviations

In this section we will compile, for the reader's convenience, the most important quantities and abbreviations used in this book. For the fundamental constants we adopt the usual notations:

c = speed of light in vacuum
$\hbar$ = Planck's constant
k = Boltzmann's constant
μ_0 = permeability of vacuum
ε_0 = permittivity of vacuum
e = elementary charge
m_e = electron rest mass
m_p = proton rest mass
$\alpha = e^2/(4\pi\varepsilon_0 \hbar c)$ = Sommerfeld's fine structure constant
$E_\infty = \alpha^2 m_e c^2/2$ = Rydberg energy ≈ 13.6058 eV
$a_0 = \hbar/(\alpha m_e c)$ = Bohr radius $\approx 0.529177 \times 10^{-10}$ m
$\mu_B = e\hbar/(2m_e)$ = Bohr magneton

Furthermore we use the special notations:

$E_H = E_\infty/(1 + m_e/m_p)$ = Rydberg energy of the hydrogen atom
$E_{Z\infty} = Z^2 E_\infty$ = Rydberg energy in the Coulomb potential of charge Ze
$a_Z = a_0/Z$ = Bohr radius for nuclear charge Z
$\boldsymbol{B}$ = vector of the magnetic induction
$B_0 = 2\alpha^2 m_e^2 c^2/(e\hbar)$ = reference magnetic field strength $\approx 4.70108 \times 10^5$ T
$B_Z = Z^2 B_0$ = reference magnetic field strength for nuclear charge Z
$\beta = B/B_0$ = magnetic field parameter
$\beta_Z = B/B_Z = \beta/Z^2$ = magnetic field parameter for nuclear charge Z
$B_{\text{crit}} = B_0/(2\alpha^2)$ = reference magnetic field strength for free electrons $\approx 4.41405 \times 10^9$ T
$\omega_c = eB/m_e$ = electron cyclotron frequency
$\Omega_c = eB/m_p$ = proton cyclotron frequency
$a_L = \sqrt{2\hbar/(eB)}$ = Larmor length.

The reference magnetic field strength B_0 is chosen such that for B_0 the Larmor length equals the Bohr radius. More generally one has $a_L = a_0 \beta^{-1/2}$. The cyclotron energy can be expressed in the form $\hbar\omega_c = 4\beta E_\infty$, whence it can be seen that the cut $\beta \approx 1$ marks, for low-lying states, the transition region from the Coulomb-dominated regime to the Lorentz-dominated regime. The strongest laboratory fields reach up to $\beta \approx 10^{-4}$, magnetic white dwarf stars lie between $\beta \approx 10^{-3}$–10^{-1}, and neutron stars between $\beta \approx 10$–10^3. The value of B_{crit} corresponds to that field strength where the quantity $a_{\text{L}}/\sqrt{2}$ equals the Compton wavelength $\lambda_{\text{C}} = \hbar/m_e c = \alpha a_0$, or where the cyclotron energy becomes equal to the electron rest energy.

2 Interacting Charged Particles in Uniform Magnetic Fields

2.1 The N-Body Problem

Let us consider a system of N charged particles with charges e_i and masses m_i ($i = 1, ..., N$) in a uniform magnetic field $\boldsymbol{B}$. The value of the vector potential $\boldsymbol{A}$ at the position $\boldsymbol{r}_i$ of particle i is abbreviated by $\boldsymbol{A}_i = \boldsymbol{A}(\boldsymbol{r}_i)$. In this section no special gauge is adopted. The Hamiltonian of the system is given by

$$H = \sum_{i=1}^{N} \frac{1}{2m_i}(\boldsymbol{p}_i - e_i \boldsymbol{A}_i)^2 + V(\boldsymbol{r}_1, \ldots, \boldsymbol{r}_N) \quad , \tag{2.1}$$

where the potential energy V of the interaction is assumed to be momentum independent and translationally invariant. Generally, the conserved quantity that is associated with the translational invariance of a system, the generalized total momentum, is defined as the generator of infinitesimal translations. In the presence of a magnetic field it must be taken into account that a displacement of the origin changes the vector potential. Therefore, an additional gauge transformation is necessary to preserve the translational invariance of H. Thus we arrive at the following form of the generalized momentum operator:

$$\boldsymbol{P}_{0\mu} = \sum_{i=1}^{N} \left(\frac{\hbar}{i} \frac{\partial}{\partial x_{i\mu}} - e_i \int^{\boldsymbol{r}_i} \frac{\partial \boldsymbol{A}}{\partial x_\mu} \cdot d\boldsymbol{r} \right) \tag{2.2}$$

($\mu = 1, 2, 3$ denotes the Cartesian components).

Introducing $\boldsymbol{B} = \mathrm{rot}\,\boldsymbol{A}$, integrating and neglecting an unimportant constant yields

$$\boldsymbol{P}_0 = \sum_{i=1}^{N} (\boldsymbol{p}_i - e_i \boldsymbol{A}_i + e_i \boldsymbol{B} \times \boldsymbol{r}_i) \quad . \tag{2.3}$$

By construction, $\boldsymbol{P}_0$ commutes with the Hamiltonian (2.1)

$$[\boldsymbol{P}_0, H] = 0 \tag{2.4}$$

and defines a constant of motion. In the special case of the gauge $\boldsymbol{A}(\boldsymbol{r}) = \frac{1}{2} \boldsymbol{B} \times \boldsymbol{r}$, which was used, e.g., by Avron et al. (1978) in their treatment of the problem, $\boldsymbol{P}_0$ is obtained in the form:

$$\boldsymbol{P}_0 = \sum_{i=1}^{N} \left(\boldsymbol{p}_i + \frac{e_i}{2} \boldsymbol{B} \times \boldsymbol{r}_i \right) \ . \tag{2.5}$$

It should be noted that, in general, the components of the generalized momentum operator (2.3) do not commute

$$[P_{0\mu}, P_{0\nu}] = i\hbar \varepsilon_{\mu\nu\lambda} B_\lambda \left(\sum_i e_i \right) \ . \tag{2.6}$$

It is only for a neutral system ($\sum_i e_i = 0$) that the components of $\boldsymbol{P}_0$ can be made sharp simultaneously. $\boldsymbol{P}_0$ should not be confused with the operator of the total kinetic momentum of the system

$$\boldsymbol{P}_{\text{kin}} = \sum_{i=1}^{N} (\boldsymbol{p}_i - e_i \boldsymbol{A}_i) \ , \tag{2.7}$$

which is connected to the motion of the centre of mass

$$\boldsymbol{R} = \frac{1}{M} \sum_i m_i \boldsymbol{r}_i$$

by

$$\dot{\boldsymbol{R}} = \frac{i}{\hbar}[H, \boldsymbol{R}] = \frac{1}{M} \boldsymbol{P}_{\text{kin}} \ . \tag{2.8}$$

Considering

$$\dot{\boldsymbol{P}}_{\text{kin}} = M\ddot{\boldsymbol{R}} = \frac{i}{\hbar} M [H, \dot{\boldsymbol{R}}] = \sum_{i=1}^{N} \frac{e_i}{m_i} (\boldsymbol{p}_i - e_i \boldsymbol{A}_i) \times \boldsymbol{B} \ , \tag{2.9}$$

it can be seen that, in contrast to the field-free case, the total kinetic momentum is by no means conserved. Evidently this is a consequence of the Lorentz forces, which act on the individual particles. The effect of this "motional electric field" has been studied by Blumberg et al. (1979) in connection with the photodetachment of negative ions. With regard to electromagnetic transitions, it is important to note that it is $\boldsymbol{P}_0$ which enters into the law of conservation of momentum. (For a more detailed discussion see Johnson et al. 1983.)

2.2 The Uncharged Two-Body Problem

We now focus on an uncharged system of two particles (charges $e_+ = e$, $e_- = -e$, masses m_+, m_-, $M = m_+ + m_-$) and will use the gauge $\boldsymbol{A}(\boldsymbol{r}) = \frac{1}{2}\boldsymbol{B} \times \boldsymbol{r}$ for the rest of the book.

2.2.1 Hamiltonian and Wave Functions in Centre-of-Mass and Relative Coordinates

Introducing the centre-of-mass coordinate $\boldsymbol{R}$ and the relative coordinate $\boldsymbol{r} = \boldsymbol{r}_- - \boldsymbol{r}_+$, the operator (2.5) reads

$$\boldsymbol{P}_0 = \frac{\hbar}{i}\frac{\partial}{\partial \boldsymbol{R}} - \frac{e}{2}\boldsymbol{B} \times \boldsymbol{r} \quad . \tag{2.10}$$

Because $\sum_i e_i = 0$, it follows from (2.6) that the eigenfunctions Ψ of the Hamiltonian can be chosen in such a way that they are simultaneous eigenfunctions of every component of $\boldsymbol{P}_0$, with eigenvalue $\hbar \boldsymbol{K}$. A simple calculation leads to the form of the wave functions

$$\Psi(\boldsymbol{R},\boldsymbol{r}) = \frac{1}{\sqrt{V}}\exp\left(\frac{i}{\hbar}\left(\hbar\boldsymbol{K} + \frac{e}{2}\boldsymbol{B}\times\boldsymbol{r}\right)\cdot\boldsymbol{R}\right)\Phi(\boldsymbol{r}) \quad , \tag{2.11}$$

where V is the normalization volume. Inserting this into $H\Psi = E\Psi$ yields the eigenvalue equation for $\Phi(\boldsymbol{r})$

$$H_{\text{rel}}\Phi(\boldsymbol{r}) = \left[\frac{\hbar^2\boldsymbol{K}^2}{2M} + \frac{e}{M}(\hbar\boldsymbol{K}\times\boldsymbol{B})\cdot\boldsymbol{r}\right.$$
$$+ \frac{1}{2\mu}\boldsymbol{p}^2 + \frac{e}{2}\left(\frac{1}{m_-} - \frac{1}{m_+}\right)\boldsymbol{B}\cdot(\boldsymbol{r}\times\boldsymbol{p})$$
$$\left. + \frac{e^2}{8\mu}(\boldsymbol{B}\times\boldsymbol{r})^2 + V(\boldsymbol{r})\right]\Phi(\boldsymbol{r}) = E\Phi(\boldsymbol{r}) \quad , \tag{2.12}$$

where $\mu = m_+ m_-/M$ is the reduced mass and $\boldsymbol{p} = (\hbar/i)\partial/\partial\boldsymbol{r}$. It can be seen that, in contrast to the field-free case, the $\boldsymbol{K}$ dependence is not only given by $\hbar^2\boldsymbol{K}^2/(2M)$, but also by the coupling term $(e/M)(\hbar\boldsymbol{K}\times\boldsymbol{B})\cdot\boldsymbol{r}$, so that the wave function $\Phi(\boldsymbol{r})$ actually depends on $\boldsymbol{K}$. We note that a more compact way of writing H_{rel} is

$$H_{\text{rel}} = \frac{1}{2M}(\hbar\boldsymbol{K} + e\boldsymbol{B}\times\boldsymbol{r})^2 + \frac{1}{2\mu}\left(\boldsymbol{p} + \frac{m_+ - m_-}{2M}e\boldsymbol{B}\times\boldsymbol{r}\right)^2 + V(\boldsymbol{r}) \quad . \tag{2.13}$$

2.2.2 Classification, Quantum Numbers and Degeneracy in the Non-Interacting Case

The energy levels of two charged non-interacting particles in a uniform magnetic field $\boldsymbol{B} = B\boldsymbol{e}_z$ are given simply as the sum of the individual Landau energies

$$E(p_{z1}, n_{r1}, m_1, p_{z2}, n_{r2}, m_2)$$
$$= \frac{p_{z1}^2}{2m_+} + \frac{\hbar\omega_c^+}{2}(2n_{r1} + |m_1| - m_1 + 1)$$
$$+ \frac{p_{z2}^2}{2m_-} + \frac{\hbar\omega_c^-}{2}(2n_{r2} + |m_2| + m_2 + 1) \quad , \tag{2.14}$$

where p_{z1} and p_{z2} are the momenta parallel to $\boldsymbol{B}$; $\omega_c^\pm = eB/m_\pm$ are the cyclotron frequencies; $n_{r1}, n_{r2} = 0, 1, \ldots$ are the radial quantum numbers; and m_1 and m_2 are the quantum numbers of the angular momenta l_{z1} and l_{z2}. Of course, the same result for the energies must be obtained from (2.12) with $V = 0$. How this works is also illustrative for the interacting case and is here discussed in some detail. Let us write $\Phi(\boldsymbol{r})$ in the form

$$\Phi(\boldsymbol{r}) = \exp\left(i\frac{m_+ - m_-}{2M}\boldsymbol{K}_\perp \cdot \boldsymbol{r}\right)\Phi_0(\boldsymbol{r}') \quad , \tag{2.15}$$

where we have introduced $\boldsymbol{K}_\perp = K_x\boldsymbol{e}_x + K_y\boldsymbol{e}_y$ and $\boldsymbol{r}' = \boldsymbol{r} - \boldsymbol{r}_0$, with

$$\boldsymbol{r}_0 = -\hbar \boldsymbol{K} \times \boldsymbol{B}/(eB^2) \tag{2.16}$$

(note that $z' = z$). Inserting (2.15) into (2.12) leads to the eigenvalue equation for $\Phi_0(\boldsymbol{r}')$

$$\left[\frac{\hbar^2 K_z^2}{2M} - \frac{\hbar^2}{2\mu}\frac{\partial^2}{\partial z^2} - \frac{\hbar^2}{2\mu}\left(\frac{\partial^2}{\partial x'^2} + \frac{\partial^2}{\partial y'^2}\right) \right.$$
$$\left. + \frac{eB}{2}\left(\frac{1}{m_-} - \frac{1}{m_+}\right)l'_z + \frac{e^2 B^2}{8\mu}(x'^2 + y'^2)\right]\Phi_0(\boldsymbol{r}') = E\Phi_0(\boldsymbol{r}') \quad . \tag{2.17}$$

The eigenvalues are obtained as

$$E(K_x, K_y, K_z, p_z, n_r, m) = \frac{\hbar^2 K_z^2}{2M} + \frac{p_z^2}{2\mu}$$
$$+ \frac{1}{2}\hbar\omega_c^+(2n_r + |m| - m + 1) + \frac{1}{2}\hbar\omega_c^-(2n_r + |m| + m + 1) \quad , \tag{2.18}$$

where $n_r = 0, 1, \ldots$ is the radial quantum number of the relative motion and $\hbar m$ is the eigenvalue of the relative angular momentum l'_z. The eigenfunctions of the transverse part of (2.17) are given by the well-known Landau states of a charged particle in a uniform magnetic field, $\phi_{nm}^{\text{Lan}}(\boldsymbol{r}'_\perp)$ (cf. Canuto and Ventura 1977), where n is related to n_r and m by $n = n_r + (|m| + m)/2$. It should be pointed out that, with respect to the system of the original coordinates $\boldsymbol{r}$, the origin of the Landau states is displaced to the point $\boldsymbol{r} = \boldsymbol{r}_0$ (see (2.16)).

As far as the motion parallel to the magnetic field is concerned, the connection between (2.14) and (2.18) is given simply by the usual transition to centre-of-mass and relative coordinates. A certain energy of the transverse motion corresponds to a definite value of $n_+ \equiv n_{r1} + \frac{1}{2}(|m_1| - m_1)$ and $n_- \equiv n_{r2} + \frac{1}{2}(|m_2| + m_2)$, where there is a twofold infinite degeneracy, since the quantum numbers m_1 and m_2 may assume the values $m_1 = -n_+, -n_+ + 1, \ldots, +\infty$ and $m_2 = n_-, n_- - 1, \ldots, -\infty$. For fixed n_+ and n_-, the value of n_r and of m in (2.18) is well defined, and the twofold indefinite degeneracy is represented by the quantum numbers K_x and K_y. Thus the equivalence of (2.14) and (2.18), with respect to the energy value and to the degree of degeneracy, is demonstrated. For three special mass ratios ($m_-/m_+ = 0, 0.1, 1$), the level scheme for the non-interacting case is shown in Fig. 2.1. The levels are classified according to (2.18). This classification provides a suitable starting point for the treatment of the interacting case.

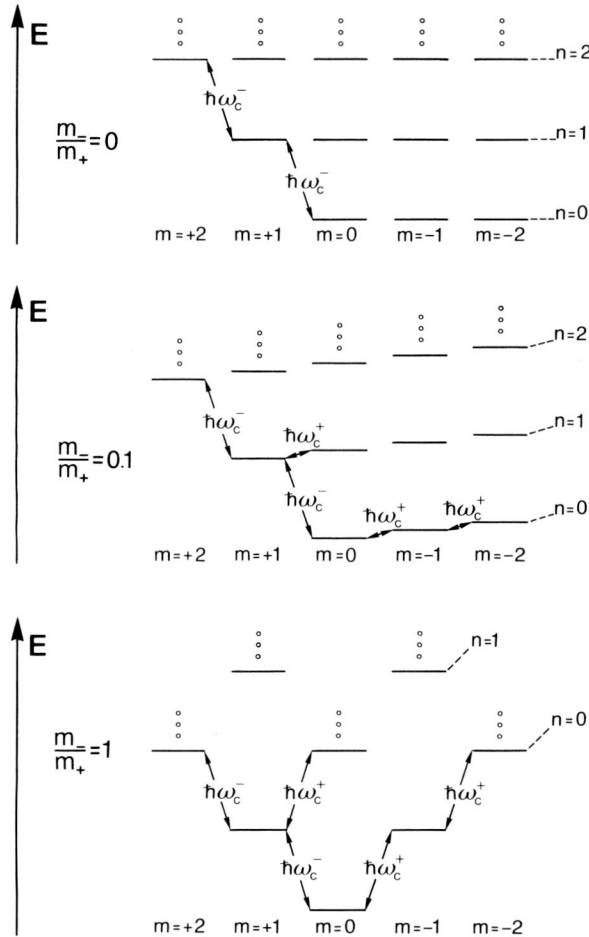

Fig. 2.1. Level scheme of two non-interacting charged particles with masses m_+ and m_- for three mass ratios. The quantum number m is the eigenvalue of the z component of the relative angular momentum, n denotes the Landau quantum number, and $\hbar\omega_c^+$, $\hbar\omega_c^-$ are the cyclotron energies. Each level is drawn for the case of no longitudinal motion and is twofold infinitely degenerate with respect to $\boldsymbol{K}_\perp$. It can be seen that the m degeneracy of the states with the same n, which occurs for $m_-/m_+ = 0$, is removed more and more as the mass ratio increases.

2.2.3 Exact Solution for Harmonic Interaction

In general, the two-body problem with interaction no longer permits analytical solutions. However, if we adopt a harmonic interaction between the two particles, the energies and wave functions still can be determined analytically. For equal masses this is mentioned by Avron et al. (1978), but this statement is also correct for unequal masses, as will be shown later (cf. Herold et al. 1981).

Furthermore, the solution for harmonic interaction is very instructive in gaining insight into the properties of the spectrum for a more general interaction. Let us therefore consider the eigenvalue equation (2.12) for $V(r) = \frac{1}{2}\mu\omega_0^2 r^2$. For $\Phi(r)$ we make the ansatz

$$\Phi(r) = \exp\left(i\frac{m_+ - m_-}{2M}\alpha \boldsymbol{K}_\perp \cdot \boldsymbol{r}\right)\Phi_0(r') , \qquad (2.19)$$

with $r' = r - \alpha r_0$ and

$$\alpha = \left(1 + \frac{M\mu\omega_0^2}{e^2 B^2}\right)^{-1} = \left(1 + \frac{\omega_0^2}{\omega_c^+ \omega_c^-}\right)^{-1}. \qquad (2.20)$$

Note that for $\omega_0 = 0$, (2.19) reduces to (2.15). Inserting (2.19) into (2.12) leads to the eigenvalue equation for $\Phi_0(r')$

$$\left[\frac{\hbar^2 K_z^2}{2M} - \frac{\hbar^2}{2\mu}\frac{\partial^2}{\partial z^2} + \frac{1}{2}\mu\omega_0^2 z^2 - \frac{\hbar^2}{2\mu}\left(\frac{\partial^2}{\partial x'^2} + \frac{\partial^2}{\partial y'^2}\right) + \frac{eB}{2}\left(\frac{1}{m_-} - \frac{1}{m_+}\right)l_z'\right.$$
$$\left. + \left(\frac{e^2 B^2}{8\mu} + \frac{\mu\omega_0^2}{2}\right)(x'^2 + y'^2) + \frac{\hbar^2 \boldsymbol{K}_\perp^2}{2M}(1-\alpha)\right]\Phi_0(r') = E\Phi_0(r'). \qquad (2.21)$$

It can be seen that the relative motion parallel to the field is harmonic with the unperturbed oscillator frequency ω_0 (quantum number $n_z = 0, 1, \ldots$). The motion perpendicular to the field is that of a two-dimensional harmonic oscillator around $r' = 0$ with frequency $\omega_\perp = (\omega_0^2 + e^2 B^2/4\mu^2)^{1/2}$ (quantum numbers n_r and m). The energy levels are thus obtained immediately as

$$E(K_x, K_y, K_z, n_z, n_r, m)$$
$$= \frac{\hbar^2 K_z^2}{2M} + \hbar\omega_0(n_z + \tfrac{1}{2}) + \hbar\omega_\perp(2n_r + |m| + 1)$$
$$+ \frac{\hbar eB}{2}\left(\frac{1}{m_-} - \frac{1}{m_+}\right)m + \frac{\hbar^2 K_\perp^2}{2M}\left(1 + \frac{\omega_c^+ \omega_c^-}{\omega_0^2}\right)^{-1}. \qquad (2.22)$$

In addition to an increase in the frequency of the transverse oscillations, the $\boldsymbol{K}_\perp$ degeneracy of the energy levels is removed. It is certainly a speciality of the harmonic interaction that the dependence of the energy on $\boldsymbol{K}_\perp$ is quadratic for all values of $\boldsymbol{K}_\perp$. For more general interactions, the $\boldsymbol{K}_\perp$ degeneracy is also removed and the energy depends on $\boldsymbol{K}_\perp$. However, as long as the interaction is axially symmetric with respect to the direction of the magnetic field, the dependence is only on $|\boldsymbol{K}_\perp|$ and certainly not quadratic for all values of $\boldsymbol{K}_\perp$.

In order to gain some insight into the physical meaning of the quantum number $\boldsymbol{K}_\perp$, it is appropriate to calculate the expectation value of $\boldsymbol{P}_{\text{kin}}$. From (2.3) and (2.7) we have the connection

$$\boldsymbol{P}_{\text{kin}} = \boldsymbol{P}_0 + e\boldsymbol{B} \times \boldsymbol{r} . \qquad (2.23)$$

For the eigenfunctions $\Phi(r)$ of the harmonic interaction, the expectation value of $\boldsymbol{P}_{\text{kin}}$ can be evaluated in closed form, yielding

$$\langle (\boldsymbol{P}_{\text{kin}})_\perp \rangle = \hbar \boldsymbol{K}_\perp (1-\alpha) = \hbar \boldsymbol{K}_\perp \left(1 + \frac{\omega_c^+ \omega_c^-}{\omega_0^2}\right)^{-1} \quad (2.24\text{a})$$

$$\langle (\boldsymbol{P}_{\text{kin}})_z \rangle = \hbar \boldsymbol{K}_z \quad . \quad (2.24\text{b})$$

We see from (2.24a) that the quantum number $\boldsymbol{K}_\perp$ is indeed related to an average motion of the system perpendicular to the magnetic field. The absolute size of the mean kinetic momentum of this transverse motion is, however, scaled by the factor $(1-\alpha)$. In the limiting case $\omega_0^2/(\omega_c^+\omega_c^-) \to 0$, i.e., where the interaction is small compared with the effect of the magnetic field, the expectation value of $(\boldsymbol{P}_{\text{kin}})_\perp$ tends to zero, and no net transverse motion is possible, although $\hbar \boldsymbol{K}_\perp$ may be very different from zero. The intuitive physical meaning of this result is that only an uncharged system with an attractive interaction can perform a transverse motion through the magnetic field. For a small interaction, this motion is strongly reduced with respect to the longitudinal motion. In the other limiting case in which the interaction dominates, $\omega_0^2/(\omega_c^+\omega_c^-) \to \infty$, $\boldsymbol{K}_\perp$ becomes, of course, asymptotically the usual centre-of-mass momentum.

2.3 Scaling Properties of the Coulomb Problem

We now turn to a discussion of the scaling properties of (2.12) if $V(\boldsymbol{r})$ is taken to be the Coulomb interaction $V_\text{C}(\boldsymbol{r}) = -e^2/(4\pi\varepsilon_0 r)$.

2.3.1 The General Case $\boldsymbol{K} \neq 0$

The eigenfunctions and the eigenvalues of (2.12) depend on $\boldsymbol{K}, B, M$, and μ, i.e., $\Phi = \Phi(\boldsymbol{K}, B, M, \mu; \boldsymbol{r})$ and $E = E(\boldsymbol{K}, B, M, \mu)$. A closer inspection of the structure of the Schrödinger equation (2.12) yields the scaling laws ($\lambda = m_+/(m_+ + m_-)$)

$$\Phi(\lambda \boldsymbol{K}, B, M, \mu; \boldsymbol{r}) = \lambda^{3/2} \Phi(\boldsymbol{K}, B/\lambda^2, M/\lambda, \mu/\lambda; \lambda \boldsymbol{r}) \quad , \quad (2.25\text{a})$$

$$E(\lambda \boldsymbol{K}, B, M, \mu) = \lambda\, E(\boldsymbol{K}, B/\lambda^2, M/\lambda, \mu/\lambda) \quad . \quad (2.25\text{b})$$

Thus, in order to exploit the scaling freedom in the quantum number $\boldsymbol{K}$, the solutions of (2.12) must be known, for some given value of $\boldsymbol{K} \neq 0$, as *continuous* functions of the magnetic field strength and of the masses involved.

Since for strong magnetic fields these solutions could only be obtained numerically, the practical treatment of the two-body problem in the general case $\boldsymbol{K} \neq 0$ constitutes a laborious task. For the physically interesting case of very strong magnetic fields and the lowest Landau level in Sect. 8.2 numerical results for the hydrogen atom will be presented. In the following section, we shall concentrate on the case $\boldsymbol{K} = 0$, which represents, also in magnetic fields, the physically reasonable starting point for investigating the internal properties of two-body systems.

2.3.2 The Special Case $K = 0$

Here the Hamiltonian in (2.12) becomes axially symmetric, and Φ can always be chosen as an eigenfunction of $l_z = (\mathbf{r} \times \mathbf{p})_z$ with eigenvalue $\hbar m$. The M-dependence of the energy then is contained simply in a diagonal term, and the scaling laws can be formulated most conveniently in terms of the solutions of the problem for $M \to \infty$ (without loss of generality we take $m_+ \to \infty$)

$$\Phi_m(m_-, m_+, B; \mathbf{r}) = \lambda^{3/2} \Phi_m(m_-, m_+ \to \infty, B/\lambda^2; \lambda \mathbf{r}) \quad , \tag{2.26a}$$

$$E_m(m_-, m_+, B) = \lambda E_m(m_-, m_+ \to \infty, B/\lambda^2) - \hbar m eB/m_+ \tag{2.26b}$$

(cf. Pavlov-Verevkin and Zhilinskii 1980a,b). It can be seen that, for a given m_- and $K = 0$, once the problem has been solved in the limit $m_+ \to \infty$ for every value of B, the solutions of the two-body problem with mass m_- and arbitrary finite values of m_+ are also known. From (2.26), the scaling law for the dipole strengths $d_{\tau'\tau}^{(q)} = |\langle \tau'|r^{(q)}/a_0|\tau\rangle|^2$, where τ, τ' denote sets of quantum numbers characterizing the initial and the final state, respectively, and $r^{(q)}$ ($q = 0, \pm 1$) are the spherical components of the relative vector, can be derived:

$$d_{\tau'\tau}^{(q)}(m_-, m_+, B) = \lambda^{-2} d_{\tau'\tau}^{(q)}(m_-, m_+ \to \infty, B/\lambda^2) \quad . \tag{2.27}$$

From (2.26b) and (2.27) the scaling behaviour of oscillator strengths

$$f_{\tau'\tau}^{(q)} = \frac{2\mu}{\hbar^2}(E_{\tau'} - E_\tau) a_0^2 d_{\tau'\tau}^{(q)}$$

and transition probabilities

$$w_{\tau'\tau}^{(q)} = \frac{4}{3} \frac{e^2}{4\pi\varepsilon_0} \frac{1}{\hbar} \left(\frac{E_{\tau'} - E_\tau}{\hbar c}\right)^3 a_0^2 d_{\tau'\tau}^{(q)}$$

can directly be calculated, with the result

$$f_{\tau'\tau}^{(q)}(m_-, m_+, B) = f_{\tau'\tau}^{(q)}(m_-, m_+ \to \infty, B/\lambda^2) \tag{2.28}$$
$$- \frac{1}{\lambda} \frac{\hbar e B q}{m_+ E_\infty} d_{\tau'\tau}^{(q)}(m_-, m_+ \to \infty, B/\lambda^2) \quad ,$$

and

$$w_{\tau'\tau}^{(q)}(m_-, m_+, B) = \frac{1}{3} \lambda w_0 \, d_{\tau'\tau}^{(q)}(m_-, m_+ \to \infty, B/\lambda^2) \tag{2.29}$$
$$\times \left(\left[E_{\tau'}(m_-, m_+ \to \infty, B/\lambda^2) - E_\tau(m_-, m_+ \to \infty, B/\lambda^2) - \frac{1}{\lambda} \frac{\hbar e B q}{m_+ E_\infty} \right] / E_\infty \right)^3 \quad ,$$

where $w_0 = \alpha^5 m_- c^2 / 2\hbar$. In particular, it is only in the case where the magnetic quantum number is not changed in the transition ($q = 0$) that these expressions reduce to the simple forms

$$f_{\tau'\tau}^{(0)}(m_-, m_+, B) = f_{\tau'\tau}^{(0)}(m_-, m_+ \to \infty, B/\lambda^2) \quad , \tag{2.30}$$

and

$$w^{(0)}_{\tau'\tau}(m_-, m_+, B) = \lambda\, w^{(0)}_{\tau'\tau}(m_-, m_+ \to \infty, B/\lambda^2) \quad , \tag{2.31}$$

Important applications of the mass scaling laws are the hydrogen atom with finite proton mass and positronium (cf. Wunner et al. 1981a,b). It should be noted that, contrary to widely held opinion (cf. O'Connell 1979), the effect of the finite proton mass is not always of the anticipated order of magnitude m_e/m_p. The last term in (2.26b) causes a shift between states with different magnetic quantum numbers m by (with the proton cyclotron frequency $\Omega_c = eB/m_p$)

$$\Delta m\, \hbar \Omega_c \approx \Delta m \times 29.6\,\text{eV} \times B/(4.7 \times 10^8\,\text{T}) \quad , \tag{2.32}$$

which, in high fields, is comparable to the Coulomb binding energies, and thus is by no means negligible (cf. Wunner et al. 1980). This term also gives rise to the nontrivial modifications in (2.28) and (2.29). The mass scaling formulae will find their application in Chap. 6.

2.3.3 Nuclear Charge (Z) Scaling

In contrast to the field-free case, the well-known Z-scaling is possible, in the presence of a magnetic field, only in the limit $m_+ \to \infty$. The scaling laws for the wave functions and energies read:

$$\Phi_m(Z, B; \mathbf{r}) = Z^{3/2} \Phi_m(Z=1, B/Z^2; Z\mathbf{r}) \quad , \tag{2.33a}$$
$$E_m(Z, B) = Z^2 E_m(Z=1, B/Z^2) \tag{2.33b}$$

(cf. Surmelian and O'Connell 1974). From these we can derive the scaling laws

$$d^{(q)}_{\tau'\tau}(Z, B) = Z^{-2}\, d^{(q)}_{\tau'\tau}(Z=1, B/Z^2) \quad , \tag{2.34a}$$
$$f^{(q)}_{\tau'\tau}(Z, B) = Z^0\, f^{(q)}_{\tau'\tau}(Z=1, B/Z^2) \quad , \tag{2.34b}$$
$$w^{(q)}_{\tau'\tau}(Z, B) = Z^4\, w^{(q)}_{\tau'\tau}(Z=1, B/Z^2) \quad , \tag{2.34c}$$

for the dipole strengths, oscillator strengths and rates of transitions in hydrogenic ions with nuclear charge Z.

Obviously the practical application of the scaling laws derived in this section requires the knowledge of the continuous B-dependence of energies and dipole strengths for $m_+ \to \infty$. A closed-form analytical solution to this problem is impossible, and one therefore has to resort to numerical methods.

3 Methods of Solution for the Magnetized Coulomb Problem

3.1 General Considerations

Measuring energies in units of the Rydberg energy E_∞, lengths in units of the Bohr radius a_0, and the magnetic field strength in units of B_0, the Hamiltonian of an electron in a static Coulomb potential and in a uniform magnetic field then reads for spin-down states

$$H = -\Delta - \frac{2}{r} + 2\beta l_z + \beta^2 \varrho^2 - 2\beta \quad , \tag{3.1}$$

where the magnetic field is assumed to point in the z-direction and $\varrho^2 = x^2 + y^2$. The energies of the corresponding spin-up states are obtained by simply adding 4β. The eigenstates of (3.1) can be classified according to z-parity π and the z-component l_z of orbital angular momentum, which are exact symmetries of H, but in general no further separation of the two-dimensional problem is possible.

In spite of this complication, the problem of hydrogen atoms in magnetic fields of arbitrary strength initially appears to be an elementary exercise in quantum mechanics, which – in principle – could be solved with any desired accuracy, in the worst case by numerical means. Motivated by applications to magnetized semiconductors in solid-state physics, magnetic Rydberg states in atomic physics, and, in particular, to strongly magnetized compact objects in astrophysics (see Sect. 1.1), a large number of theoreticians have tackled this fundamental quantum mechanical problem (see Sect. 1.2).

3.1.1 Characteristic Domains of the Magnetic Field Strength

Energies and wave functions calculated from the infinite-nuclear-mass Hamiltonian (3.1) can be connected to those belonging to finite nuclear mass by use of the appropriate mass scaling laws discussed in Sect. 2.3.

The basic difficulty in solving the Schrödinger equation belonging to the Hamiltonian (3.1) for *arbitrary* field strength lies in the fact that the spherical symmetry of the Coulomb potential, on the one hand, and the cylindrical symmetry of the magnetic field, on the other, prevent a separation of variables, so that closed-form analytical solutions are not possible in general. Because of the different symmetries, three domains of the magnetic field strength can be distinguished in a natural way:

i) The weak-field regime, in which the Coulomb potential predominates and the magnetic field can be considered as a small perturbation. This range, the regime of the linear and quadratic Zeeman effect, in which the magnetic field mixes the angular momentum states of a given n manifold, was clarified in the early days of quantum mechanics. It should be pointed out, however, that even in this regime, progress has been made in recent years in understanding the separability of the Zeeman wave functions in momentum space, the existence of an additional approximate constant of motion for given n, and, related to this, the symmetry classification of the states of a diamagnetic band in terms of a rotational-vibrational scheme (see, e.g., Herrick 1982; Delande and Gay 1984; Wintgen and Friedrich 1986a,b; Wunner 1986).

ii) The intense-field regime, in which the magnetic field predominates, and the Coulomb potential can be treated as a perturbation which influences the (slow) motion of the electron parallel to the magnetic field but does not affect the (rapid) gyrations – described quantum mechanically by Landau states – in the plane perpendicular to the field. This situation is adequately accounted for by the well-known *adiabatic approximation* introduced by Schiff and Snyder as early as 1939.

iii) The intermediate-field regime (often called "strong-field"), in which the electron experiences electric and magnetic forces of comparable strength. It is obvious that the elaboration of energies and wave functions in this regime causes the greatest difficulties and is incomplete even today. Moreover, the absolute sizes of the field strengths worked with in this regime evidently depend on the state of excitation of the electron. By considering the equality of Coulomb and Lorentz forces for an electron in a circular Bohr orbit with principal quantum number $n_{\rm p}$, $B_{n_{\rm p}} \approx B_0/(2n_{\rm p}^3) \approx 8.7 \times (30/n_{\rm p})^3$ T is obtained as a rough measure. Thus for white dwarf and neutron star magnetic fields, low-lying states are found to be subject to a strong-field, or even intense-field, situation, while at laboratory field strengths studies of the strong-field regime must concentrate on Rydberg states.

3.1.2 Expansion of the Wave Functions and Multiconfiguration Equations

The nonintegrability of the Hamiltonian (3.1) implies that we are forced, in the complete quantum theoretical treatment of the problem, to resort to numerical methods. To obtain "exact" (as opposed to variational) solutions, it is suggestive to invoke basis function expansions that are inspired by the symmetries of the limiting cases $B \to 0$ and $B \to \infty$. These are, in the case $B \to 0$, expansions in terms of oscillator functions in semiparabolic coordinates (adapted to the SO(2,2) = SO(2,1) $\oplus$ SO(2,1) dynamical symmetry of the Coulomb problem (cf. Englefield 1971)), or expansions in terms of spherical harmonics and a set of complete radial functions (ordinary SO(3) symmetry), whereas the case $B \to \infty$ suggests expansions in terms of Landau functions with a complete longitudinal basis. The different expansions are employed according to the approach to the intermediate-field regime, whether from the side of low fields or intense fields, and overlapping results are expected in the transitional regime.

Our procedure for determining as accurately as possible eigenvalues and eigenfunctions of (3.1) begins with the expansion of the wave functions in terms of spherical harmonics for the range in which the Coulomb forces still outweigh the Lorentz forces, i.e., $\beta \lesssim 1$ for low-lying states,

$$\Phi_{m\pi}(\boldsymbol{r}) = \sum_l \frac{1}{r} f_{lm}(r) Y_{lm}(\vartheta, \varphi) \quad , \tag{3.2}$$

while for magnetic fields where the spherical symmetry of the Coulomb potential is perturbed to an ever-increasing extent ($\beta \gtrsim 1$ for low-lying states), the expansion in terms of Landau states $\phi_{nm}^{\text{Lan}}(\varrho, \varphi)$ (n is the Landau quantum number, m is the magnetic quantum number; see, e.g., Canuto and Ventura 1977) is more appropriate:

$$\Phi_{m\pi}(\boldsymbol{r}) = \sum_n g_{nm}(z) \phi_{nm}^{\text{Lan}}(\varrho, \varphi) \quad . \tag{3.3}$$

Since the z-parity of $Y_{lm}(\vartheta, \varphi)$ is $(-1)^{l-m}$, for given m and π the series (3.2) runs over only even or odd angular momenta, respectively. The z-parity of the expansion (3.3) is carried by the longitudinal wave functions $g_{nm}(z)$, and no restriction occurs with respect to the Landau quantum numbers n.

Inserting the wave functions (3.2) or (3.3) together with the Hamiltonian (3.1) into the Schrödinger extremum principle of ordinary non-relativistic quantum mechanics leads to a set of coupled differential equations for simultaneously determining the expansion functions and the energy value E of the state, namely:

Spherical expansion

$$-E f_{lm}(r) - \frac{d^2}{dr^2} f_{lm}(r) + \sum_{l'} V_{\text{eff}}^{ll'm}(r) f_{l'm}(r) = 0 \quad ,$$

$$(l = 0, 2, 4, \ldots \text{ or } l = 1, 3, 5, \ldots) \quad . \tag{3.4}$$

Cylindrical expansion

$$-E g_{nm}(z) - \frac{d^2}{dz^2} g_{nm}(z) + \sum_{n'} V_{\text{eff}}^{nn'm}(z) g_{n'm}(z) = 0 \quad ,$$

$$(n = 0, 1, 2, \ldots) \quad . \tag{3.5}$$

The effective potentials $V_{\text{eff}}^{ll'm}(r)$ in (3.4) are given by

$$V_{\text{eff}}^{ll'm}(r)$$
$$= \left\langle Y_{lm}(\vartheta, \varphi) \left| \frac{l(l+1)}{r^2} - \frac{2}{r} + 2\beta(m-1) + \beta^2 r^2 \sin^2\vartheta \right| Y_{l'm}(\vartheta, \varphi) \right\rangle_{\vartheta, \varphi}$$
$$= \delta_{ll'} \left(\frac{l(l+1)}{r^2} - \frac{2}{r} + 2\beta(m-1) + \frac{2}{3}\beta^2 r^2 \right)$$
$$- \frac{2}{3}\beta^2 r^2 \left(\frac{2l+1}{2l'+1} \right)^{1/2} (l\,0, 2\,0\,|\,l'\,0)\,(l\,m, 2\,0\,|\,l'\,m) \quad . \tag{3.6}$$

The effective potentials $V_{\text{eff}}^{nn'm}(z)$ are defined by

$$V_{\text{eff}}^{nn'm}(z) = \left\langle \phi_{nm}^{\text{Lan}}(\varrho,\varphi) \left| \frac{-2}{(\varrho^2 + z^2)^{1/2}} \right| \phi_{n'm}^{\text{Lan}}(\varrho,\varphi) \right\rangle_{\varrho,\varphi} . \qquad (3.7)$$

It can be seen that, trivially, for $\beta = 0$, the system (3.4) reduces to the equations for determining the radial wave functions $R_{n_p l}$ (n_p = principal quantum number) of the field-free hydrogenic problem. The system (3.5), on the other hand, also decouples if the terms $V_{\text{eff}}^{nn'm}(z)$ for $n \neq n'$ can be omitted; this *adiabatic approximation*, which amounts to taking the wave function as a simple product of one Landau state ϕ_{nm}^{Lan} and a longitudinal function $g_{nm\nu}$ (with ν nodes on the z-axis, i.e., $\pi = (-1)^\nu$), becomes increasingly accurate with increasing β (see, e.g., Friedrich 1982; Rösner et al. 1983), and is even exact in the limit $\beta \to \infty$.

3.1.3 Low-Field High-Field Correspondence

The problem of the correspondence between the quantum numbers in the limits $\beta \to 0$ and $\beta \to \infty$ was solved conclusively by Simola and Virtamo (1978), who applied the non-crossing rule of quantum mechanics and, in this way, ruled out previous incorrect level correspondences based, for example, on nodal surface conservation or computations in truncated spaces. The procedure is a simple counting rhyme in which states with given (conserved) quantum numbers m and π are related to each other beginning with the state with lowest energy. This correspondence between the low-field and the high-field states of the non-relativistic Coulomb problem is listed in Table 3.1 for the lowest states with $|m| \leq 3$ and is schematically shown in Fig. 3.1 for z-parity ± 1 and $|m| \leq 2$. The prime at the field free quantum numbers indicates that the energetically degenerate angular momentum states are linearly combined so as to be adjusted to the symmetry of the magnetic field. The coefficients of the linear combinations for some of these states can be found in Table 4.1.

Modifications of the correspondence in the physical hydrogen atom caused by spin-orbit coupling will be discussed in Sect. 8.1.

3.1 General Considerations 27

Table 3.1. Correspondence for the lowest states of the nonrelativistic Coulomb problem with $|m| \leq 3$

$m=0,\ \pi=+1$		$m=0,\ \pi=-1$		$m=-1,\ \pi=+1$		$m=-1,\ \pi=-1$	
n_p, l, m	n, m, ν	n_p, l, m	n, m, ν	n_p, l, m	n, m, ν	n_p, l, m	n, m, ν
$1s_0$	$0,0,0$	$2p_0$	$0,0,1$	$2p_{-1}$	$0,-1,0$	$3d_{-1}$	$0,-1,1$
$2s_0$	$0,0,2$	$3p_0$	$0,0,3$	$3p_{-1}$	$0,-1,2$	$4d_{-1}$	$0,-1,3$
$3d'_0$	$0,0,4$	$4f'_0$	$0,0,5$	$4f'_{-1}$	$0,-1,4$	$5g'_{-1}$	$0,-1,5$
$3s'_0$	$0,0,6$	$4p'_0$	$0,0,7$	$4p'_{-1}$	$0,-1,6$	$5d'_{-1}$	$0,-1,7$
$4d'_0$	$0,0,8$	$5f'_0$	$0,0,9$	$5f'_{-1}$	$0,-1,8$	$6g'_{-1}$	$0,-1,9$
$4s'_0$	$0,0,10$	$5p'_0$	$0,0,11$	$5p'_{-1}$	$0,-1,10$	$6d'_{-1}$	$0,-1,11$
$5g'_0$	$0,0,12$	$6h'_0$	$0,0,13$	$6h'_{-1}$	$0,-1,12$		
$5d'_0$	$0,0,14$	$6f'_0$	$0,0,15$	$6f'_{-1}$	$0,-1,14$		
$5s'_0$	$0,0,16$	$6p'_0$	$0,0,17$	$6p'_{-1}$	$0,-1,16$		

$m=-2,\ \pi=+1$		$m=-2,\ \pi=-1$		$m=-3,\ \pi=+1$		$m=-3,\ \pi=-1$	
n_p, l, m	n, m, ν	n_p, l, m	n, m, ν	n_p, l, m	n, m, ν	n_p, l, m	n, m, ν
$3d_{-2}$	$0,-2,0$	$4f_{-2}$	$0,-2,1$	$4f_{-3}$	$0,-3,0$	$5g_{-3}$	$0,-3,1$
$4d_{-2}$	$0,-2,2$	$5f_{-2}$	$0,-2,3$	$5f_{-3}$	$0,-3,2$	$6g_{-3}$	$0,-3,3$
$5g'_{-2}$	$0,-2,4$	$6h'_{-2}$	$0,-2,5$	$6h'_{-3}$	$0,-3,4$		
$5d'_{-2}$	$0,-2,6$	$6f'_{-2}$	$0,-2,7$	$6f'_{-3}$	$0,-3,6$		
$6g'_{-2}$	$0,-2,8$						
$6d'_{-2}$	$0,-2,10$						

$m=+1,\ \pi=+1$		$m=+1,\ \pi=-1$		$m=+2,\ \pi=+1$		$m=+2,\ \pi=-1$	
n_p, l, m	n, m, ν	n_p, l, m	n, m, ν	n_p, l, m	n, m, ν	n_p, l, m	n, m, ν
$2p_{+1}$	$1,+1,0$	$3d_{+1}$	$1,+1,1$	$3d_{+2}$	$2,+2,0$	$4f_{+2}$	$2,+2,1$
$3p_{+1}$	$1,+1,2$	$4d_{+1}$	$1,+1,3$	$4d_{+2}$	$2,+2,2$	$5f_{+2}$	$2,+2,3$
$4f'_{+1}$	$1,+1,4$	$5g'_{+1}$	$1,+1,5$	$5g'_{+2}$	$2,+2,4$	$6h'_{+2}$	$2,+2,5$
$4p'_{+1}$	$1,+1,6$	$5d'_{+1}$	$1,+1,7$	$5d'_{+2}$	$2,+2,6$	$6f'_{+2}$	$2,+2,7$
$5f'_{+1}$	$1,+1,8$	$6g'_{+1}$	$1,+1,9$	$6g'_{+2}$	$2,+2,8$		
$5p'_{+1}$	$1,+1,10$	$6d'_{+1}$	$1,+1,11$	$6d'_{+2}$	$2,+2,10$		
$6h'_{+1}$	$1,+1,12$						
$6f'_{+1}$	$1,+1,14$						
$6p'_{+1}$	$1,+1,16$						

$m=+3,\ \pi=+1$		$m=+3,\ \pi=-1$	
n_p, l, m	n, m, ν	n_p, l, m	n, m, ν
$4f_{+3}$	$0,+3,0$	$5g_{+3}$	$0,+3,1$
$5f_{+3}$	$0,+3,2$	$6g_{+3}$	$0,+3,3$
$6h'_{+3}$	$0,+3,4$		
$6f'_{+3}$	$0,+3,6$		

28   3 Methods of Solution for the Magnetized Coulomb Problem

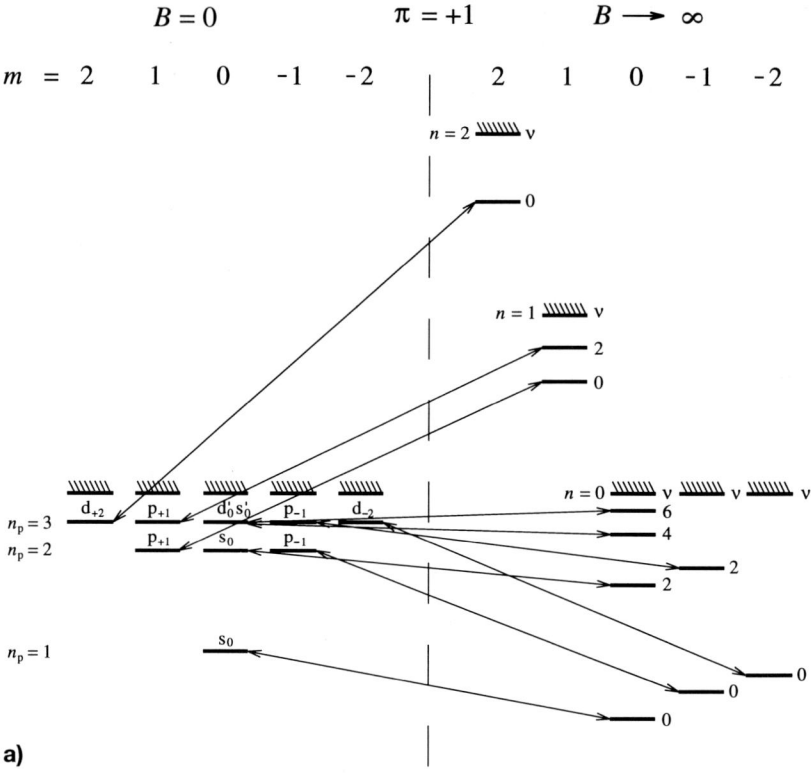

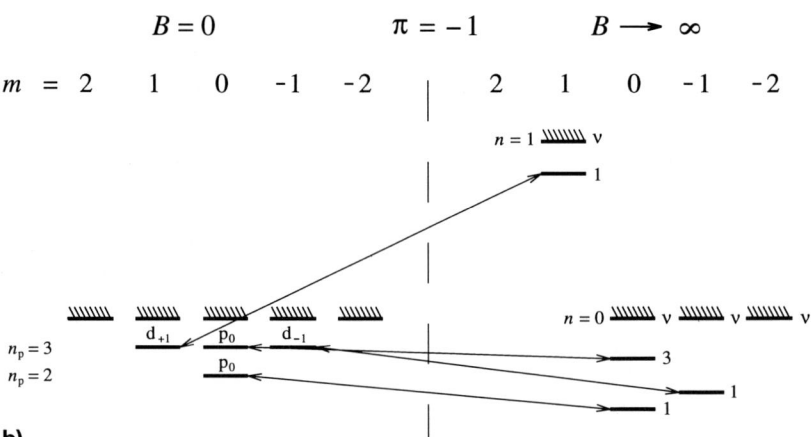

Fig. 3.1a,b. Schematic correspondence diagram between the low-field and the high-field states of the nonrelativistic Coulomb problem with $|m| \leq 2$ for z-parity $+1$ (**a**) and z-parity -1 (**b**)

3.2 Numerical Treatment

3.2.1 Hartree-Fock-Like Methods

The systems of coupled differential equations (3.4) and (3.5) possess a structure similar to that of the well-known Hartree-Fock equations for determining the *best* single-particle wave functions of a Slater determinant, with the simplification that here the effective potentials are independent of the unknown functions f_{lm} or g_{nm}. It is therefore straightforward, in solving our equations, to employ the numerical integration procedures implemented in the successful multiconfiguration Hartree-Fock code of Froese-Fischer (1977, 1978), although the presence of the magnetic field, of course, requires changes at several points in the code. The modifications necessary in the strong-field case (3.5) were described by Pröschel (1982) and Pröschel et al. (1982). The method used in computing the effective potentials (3.7) was adopted from Friedrich (1982) and slightly modified. For the spherical expansion (3.4), the tail procedure in particular had to be adjusted to the r^2 contributions of the effective potentials (3.6).

To obtain starting functions for the computation, a *one-configuration* calculation was carried out first, that is, the eigenvalue equation was solved numerically, taking into account solely that angular momentum in (3.2) or that Landau level in (3.3) that is substantially contained in the corresponding state in the limit $\beta \to 0$ or $\beta \to \infty$, respectively. In those cases where for $\beta \to 0$ the wave function in zeroth-order perturbation theory becomes a superposition of energetically degenerate angular momentum states, independent one-configuration calculations were performed for these angular momenta, and subsequently the Hamiltonian was diagonalized in this restricted basis to yield the proper starting functions. The systems of equations (3.4) or (3.5) were then solved by consecutively adding one more component to the forms (3.2) or (3.3) and taking the results of the $(n-1)$-configuration calculation as a start for the n-configuration calculation. Alternatively, in actual computations, once a multicomponent wave function had been determined for some value of β, it could be used as an input to solve (3.4) or (3.5) at a slightly altered value of β.

Typical examples for the convergence of this procedure with an increasing number n_c of expansion terms in the two forms of wave functions are shown in Fig. 3.2 for the ground state (asymptotic quantum numbers $n_p l m / n m \nu \cong 1s_0/0\ 0\ 0$) and for the excited state $3d_{-1}/0\ -1\ 1$ for intermediate values of the magnetic field strength, where both methods reasonably can be applied. Fig. 3.2 shows that the spherical expansion generally exhibits a more pronounced saturation behaviour with respect to n_c than does the cylindrical expansion.

As expected, when the magnetic field strength is increased, a higher number of components is needed in the spherical expansion to reach saturation, while, in the cylindrical computation, the $E(n_c)$ curves flatten. Comparing the two states considered in Fig. 3.2, it can be seen that the spherical expansion yields results of similar quality at values of β that are larger by roughly a factor of 50 for the $1s_0/0\ 0\ 0$ state than for the $3d_{-1}/0\ -1\ 1$ state. Evidently this

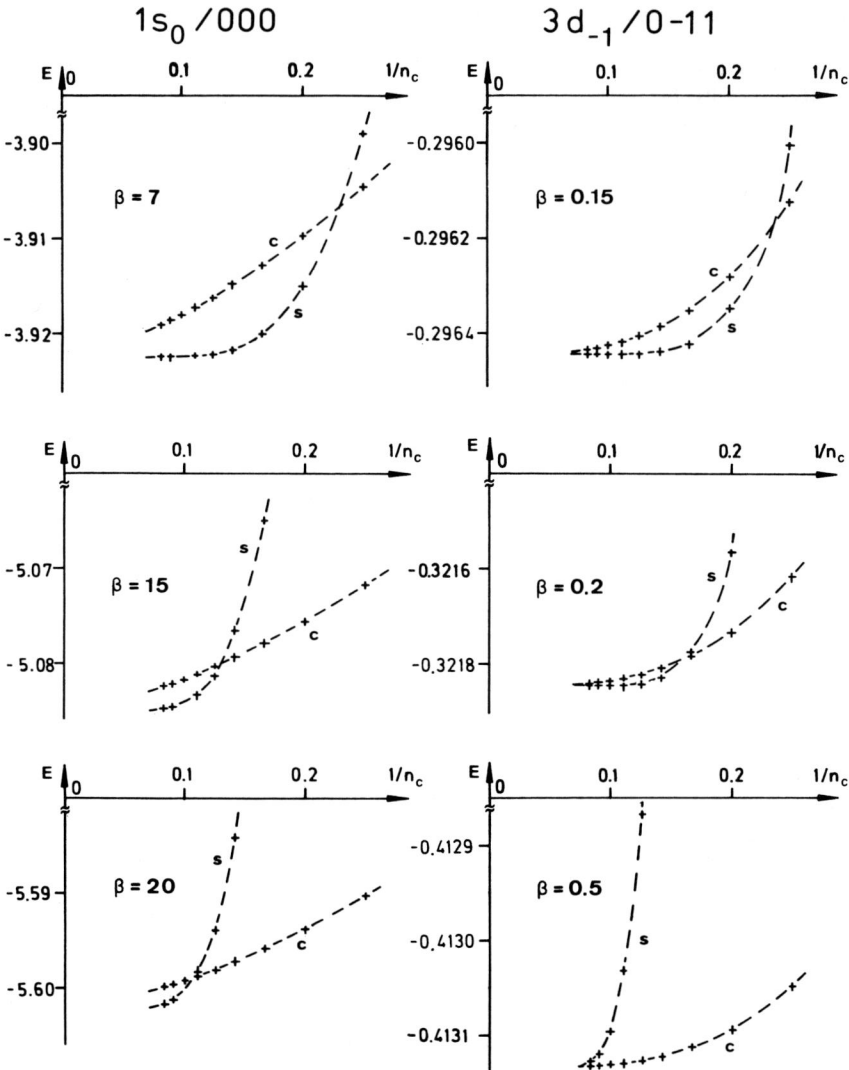

Fig. 3.2. Two examples, the ground state $1s_0/0\,0\,0$ and an excited state $3d_{-1}/0\,-1\,1$, illustrating the convergence behaviour of the energy values E (in units of E_∞) as a function of $1/n_c$ (n_c = number of terms included in the spherical (s) and cylindrical (c) expansions (3.2) and (3.3)) for values of β in the transition regime from Coulomb force to Lorentz force dominance

is because the $1s_0/0\,0\,0$ wave function is concentrated more closely to the nucleus and, therefore, higher values of β are necessary to render the magnetic field dominant. Furthermore, it can be observed that, in spite of values of $\beta \approx 10-20$, the convergence of the cylindrical computation for the $1s_0/0\,0\,0$ state is relatively poor. This can be explained by the fact that the singularity of the

Coulomb potential causes a cusp at the origin in all wave functions containing a Y_{00} contribution (i.e., all $m = 0$ and positive z-parity wave functions), a behaviour that can be approximated only by the superposition of a sufficiently large number of (smooth) cylindrical expansion functions. Indeed, it can be seen that, for the $3d_{-1}/0 -1\ 1$ state, whose wave function vanishes at the origin, the cylindrical computation converges much faster.

3.2.2 Coupled-Channels-Like Methods

In the region of intermediate field strengths where neither the spherical symmetry nor the cylindrical symmetry is dominant, it was found that it becomes increasingly difficult, particularly for principal quantum numbers $n_p \geq 4$, to obtain convergence with the methods described above. Therefore, we employed two further methods for this region and for higher excited states. One of them is a coupled-channels-like method starting from the Landau state expansion (3.3), the other is a diagonalization method working with the spherical expansion (3.2).

The set of differential equations (3.5) possesses, formally, the structure of coupled channels (with the Landau excitation energies $E_{nm} = 2\beta(2n+|m|+1)$ acting as the threshold energies of open channels) known from other problems of (field-free) atomic physics. Solving, for a given number n_c of Landau channels, the coupled differential equations (3.5), with their long-range potentials, and for many, partly highly excited, channel states, constitutes a non-trivial numerical problem. The essence of the method is (Wintgen 1985, Geyer 1987) that first, for given energy and z-parity, n_c linearly independent solutions are determined by choosing n_c independent initial conditions at the origin compatible with the definite z-parity, and integrating outward the set of n_c equations for these n_c independent solutions. The exponentially increasing contributions can be eliminated by forming new linear combinations, while integrating outward in such a way that the divergent contributions, first of the highest, then of the next lowest, Landau states, mutually cancel; the remaining exponentially decreasing contributions of these states can be neglected beyond that point. In this way the number of contributing channels, and of the solution functions to be integrated, is successively decreased until the lowest Landau channel, that is one function, remains, which is then used for determining the energy eigenvalue. In our calculations it was possible to include up to $n_c = 96$ Landau channels. Results were obtained in the magnetic field range $\beta = 3 \times 10^{-2}$ to $\beta = 10^3$ for states with principal quantum numbers $n_p = 1$ to 5. The method described also lends itself to calculating continuum wave functions in the positive-energy domain, where one or several Landau channels are open, and to determining oscillator strengths of bound-bound and bound-free transitions involving the states.

3.2.3 Diagonalization Methods

To determine energies and wave functions of the Hamiltonian (3.1) for low-lying states for values of $\beta \lesssim 1$, basis function expansions with subsequent diagonalization proved most efficient. In our own computations we set up the Hamiltonian matrix in the *complete* and *orthonormal* basis formed by the products of spherical harmonics $Y_{lm}(\vartheta, \varphi)$ and the Laguerre functions (with fixed exponent)

$$G_{nl}^{(\zeta)}(r) = \zeta^{3/2} \left[\frac{n!}{(n+2l+2)!}\right]^{1/2} \exp\left(-\zeta r/2\right) (\zeta r)^l L_n^{(2l+2)}(\zeta r) \quad . \quad (3.8)$$

Here ζ denotes an inverse-length parameter, and the $L_n^{(2l+2)}$ are generalized Laguerre polynomials as defined by Abramowitz and Stegun (1972). Matrix elements with respect to this basis can be evaluated in closed analytical form (see, e.g., Wunner et al. 1986a), and because of the $l = 0 \oplus l = 2, m = 0$ tensorial form of the diamagnetic term, only angular momentum components which are equal or differ by ± 2 are coupled. This gives rise to a banded, sparse form of the Hamiltonian matrix, which can be diagonalized by efficient algorithms. The basis functions (3.8) resemble the Sturmian basis used previously by Edmonds (1973) and Clark and Taylor (1980, 1982), the only difference being the superscript of the Laguerre polynomials ($2l+2$ in (3.8) instead of $2l+1$ for Sturmians) which makes our basis orthogonal, in contrast to the Sturmian basis, which has a non-diagonal overlap matrix. In this way, we only have to deal with an ordinary, not a general, eigenvalue problem, which allows the use of faster algorithms and makes larger matrix dimensions tractable.

For the calculation of the few lowest (≤ 9) eigenvalues for given m and parity, we used the Davidson (1975) procedure (Wunner et al. 1986a). This is an iterative method that requires keeping only two columns of the matrix in the memory of the computer at a time, but leads to unacceptably long computing times if one goes beyond the few lowest eigenvalues. The method of choice turned out to be the spectral transformation Lanczos method (STLM) as developed and implemented by Ericsson and Ruhe (1980). In this method, appropriate matrix manipulations lead to a modified eigenvalue problem whose eigenvalue sequence $\{\nu_i\}$ is inversely related to the sequence of the original problem, $\{\lambda_i\}$, shifted by some real number μ, viz. $\nu_i = (\lambda_i - \mu)^{-1}$. Thus the largest and smallest eigenvalues of the transformed problem correspond to those eigenvalues λ_i that lie in the immediate neighbourhood of μ; this is the ideal situation for using the Lanczos algorithm, which has the property of converging first, and most rapidly, for the extremum eigenvalues ν_i of a matrix. Therefore, the eigenvalues λ_i in a given interval $[\alpha, \beta]$ can be obtained by moving from the lower to the upper end of the interval with a monotonously increasing sequence of shifts $\mu_o, \mu_1, \mu_2, \ldots$, and by determining the extremum eigenvalues via the Lanczos algorithm. Details of the procedure are described by Zeller (1990).

Eigenvectors are then found by inverse iteration. Convergence was ascertained by varying the size of the basis and the scale parameter ζ. For states up to principal quantum numbers of 6 and field strengths $\beta \sim 10^{-1} - 10^1$, matrix sizes of 3600 were sufficient to provide energies with a relative accuracy of 10^{-6} and oscillators strengths with a relative accuracy of 10^{-3}.

One advantage of this method lies in the fact that it can be extended, in a straightforward way, to highly excited ("Rydberg") states in strong laboratory field strengths ($\beta \sim 10^{-5}$), where a similar competition between Coulomb and magnetic interaction takes place, as it does for low-lying states in astrophysical fields (see the detailed discussion in Chap. 9). The handle is the inverse length parameter ζ in (3.8): the size of ζ determines the radial extension of the wave function and can be adjusted to the energy range under consideration. A rule of thumb for the choice of ζ, borrowed from the field-free hydrogenic case, is $\zeta \sim 2\sqrt{E}$. Thus, for low-lying states, discussed in this chapter, ζ was chosen in the range 0.2 to 1, while for highly excited states, ($E \sim 10^{-4}$) values of $\zeta \sim 1/50$ were appropriate.

In this way the energies and oscillator strengths of highly excited states also could be determined with the above-mentioned accuracies up to energies of $E \sim 0$ in laboratory magnetic fields. Up to energies of $E \approx -20\,\text{cm}^{-1}$ ($\sim 2 \times 10^{-4}$ Rydberg) basis sizes of 6400 were well sufficient to provide the desired accuracies ($\Delta E/E \leq 10^{-6}$, $\Delta f/f \leq 10^{-3}$). Typically, about 50 successive states could be described by a single value of ζ in this range.

Above that energy, the basis sizes had to be steadily increased until, at energies around the field-free ionization threshold, a matrix dimension of 303 601 (with a half band width of 552) was reached. In this extreme situation, where almost all of the user-addressable memory of a four-processor Cray 2 machine (224 Megawords) was exhausted, the number of non-zero matrix elements amounted to $\sim 10^8$. A detailed discussion of highly excited states in strong magnetic fields will be presented in Chap. 10.

4 Results for Low-Lying States

4.1 Energy Values

The results of our extensive numerical calculations are presented in the form of tables and figures in Appendix A1. The energy values (in units of $-E_\infty$) are given for the 38 lowest-lying states of the static Coulomb problem – corresponding to the H atom with infinite proton mass – and for magnetic field strengths in the range $10^{-4} \leq \beta \leq 10^3$. The tables are ordered according to the azimuthal quantum number m and the z-parity π. The states are labelled by their asymptotic ($\beta = 0$ and $\beta \to \infty$) quantum numbers. Those states that, in the limit $\beta \to 0$, remain linear combinations of different angular momentum states are additionally marked by a prime, and the angular momentum given indicates the dominant l-component if two angular momentum states are involved. For linear combinations with more than two angular momentum states, this procedure is somewhat arbitrary, as can be seen from the $5g'_0$ and $5d'_0$ states, where the ($l = 4$)-component is dominant. Therefore, we choose the energy as the ordering parameter. For a given degenerate (n_p, m, π)-multiplet, we denote the state with the highest l' which is lowered most when the magnetic field is switched on (cf. Garstang and Kemic 1974; Garstang 1977). The coefficients of the linear combinations are obtained by diagonalizing the matrices following from the standard perturbation treatment of degenerate states. In Table 4.1 these coefficients are listed for the 38 states considered. (Table 4.1a contains the states with positive parity, Table 4.1b those with negative parity, ordered according to increasing energy.)

In Tables A1.2a–h we cover the whole range of β-values using a logarithmic mesh with 6 points per decade at 1, 1.5, 2, 3, 5, and 7. In the region of the transition from Coulomb force to Lorentz force dominance, the energy values and, to a yet larger extent, the wave functions and the electromagnetic transition probabilities strongly vary with the magnetic field strength. This behaviour is especially pronounced in the vicinities of close anticrossings. Therefore, for values of β between 10^{-3} and 10^{-1}, we calculated the energy values on a much finer mesh of the β-scale. These results are presented in Tables A1.3a–e.

We mention as one application that the numbers presented in these tables can serve as a basis for the quantitative interpretation of the features in the spectra of highly magnetized white dwarf stars possessing polar magnetic field strengths of up to several 10^4 T. For the reader's convenience, and to facilitate the use of the tables and figures, we have listed, in Table 4.2 (and additonally

Table 4.1a. Coefficients of linear combinations in the limit $\beta \to 0$, z-parity $\pi = +1$

state	$n_p s_0$	$n_p d_0$	$n_p g_0$	$n_p p_{-1}$	$n_p f_{-1}$	$n_p d_{-2}$	$n_p g_{-2}$
$3d'_0$	0.402024	0.915629					
$3s'_0$	0.915629	−0.402024					
$4d'_0$	0.492699	0.870200					
$4s'_0$	0.870200	−0.492699					
$5g'_0$	0.328915	0.647320	0.687598				
$5d'_0$	0.447214	0.534522	−0.717137				
$5s'_0$	0.831754	−0.543381	0.113678				
$4f'_{-1}$				0.246668	0.969100		
$4p'_{-1}$				0.969100	−0.246668		
$5f'_{-1}$				0.335711	0.941965		
$5p'_{-1}$				0.941965	−0.335711		
$5g'_{-2}$						0.177339	0.984150
$5d'_{-2}$						0.984150	−0.177339
$6g'_{-2}$						0.255597	0.966784

Table 4.1b. Coefficients of linear combinations in the limit $\beta \to 0$, z-parity $\pi = -1$

state	$n_p p_0$	$n_p f_0$	$n_p d_{-1}$	$n_p g_{-1}$	$n_p f_{-2}$	$n_p h_{-2}$
$4f'_0$	0.584710	0.811242				
$4p'_0$	0.811242	−0.584710				
$5f'_0$	0.677109	0.735882				
$5p'_0$	0.735882	−0.677109				
$5g'_{-1}$			0.387392	0.921915		
$5d'_{-1}$			0.921915	−0.387392		
$6h'_{-2}$					0.286789	0.957994
$6f'_{-2}$					0.957994	−0.286789

in Table A1.1), the relations between states, range of β, tables, and figures. In the column "range of β", (s) denotes the standard and (f) the finer mesh.

Tables A1.2a–h and A1.3a–e represent the most complete and accurate compilation of the energy values of the one-electron problem in a static Coulomb potential and a uniform magnetic field of arbitrary strength published to date. The energy values of states with positive m can be calculated by simply adding $4\beta m$ to the corresponding energies with $-m$.

Wherever a comparison with the literature is possible, our values either reproduce the earlier results to within the number of digits given by us, or they improve on these data. It should be emphasized that the true energy values of the hydrogen atom and its isoelectronic sequence can be gained in their dependence on the magnetic field strength with the help of the scaling laws for the nuclear charge and for arbitrary finite mass of the positive charge discussed in Sect. 2.3.

With increasing principal quantum number n_p, the transition from the spherical to the cylindrical expansion occurs more and more at smaller val-

4.1 Energy Values   37

Table 4.2. Relations between states, range of β, tables, and figures

m	π	states $n_\mathrm{p}lm/nm\nu$			range of β	Table	Figure
0	+1	$1s_0/0\,0\,0$,	$2s_0/0\,0\,2$,	$3d'_0/0\,0\,4$	$10^{-4}-10^3$, (s)	A1.2a	A1.2a
0	+1	$3s'_0/0\,0\,6$,	$4d'_0/0\,0\,8$,	$4s'_0/0\,0\,10$	$10^{-4}-10^3$, (s)	A1.2a	A1.2a
0	+1	$5g'_0/0\,0\,12$,	$5d'_0/0\,0\,14$,	$5s'_0/0\,0\,16$	$10^{-4}-10^3$, (s)	A1.2a	A1.2a
0	−1	$2p_0/0\,0\,1$,	$3p_0/0\,0\,3$,	$4f'_0/0\,0\,5$	$10^{-4}-10^3$, (s)	A1.2b	A1.2b
0	−1	$4p'_0/0\,0\,7$,	$5f'_0/0\,0\,9$,	$5p'_0/0\,0\,11$	$10^{-4}-10^3$, (s)	A1.2b	A1.2b
−1	+1	$2p_{-1}/0\,-1\,0$,	$3p_{-1}/0\,-1\,2$,	$4f'_{-1}/0\,-1\,4$	$10^{-4}-10^3$, (s)	A1.2c	A1.2c
−1	+1	$4p'_{-1}/0\,-1\,6$,	$5f'_{-1}/0\,-1\,8$,	$5p'_{-1}/0\,-1\,10$	$10^{-4}-10^3$, (s)	A1.2c	A1.2c
−1	−1	$3d_{-1}/0\,-1\,1$,	$4d_{-1}/0\,-1\,3$,	$5g'_{-1}/0\,-1\,5$	$10^{-4}-10^3$, (s)	A1.2d	A1.2d
−1	−1	$5d'_{-1}/0\,-1\,7$			$10^{-4}-10^3$, (s)	A1.2d	A1.2d
−2	+1	$3d_{-2}/0\,-2\,0$,	$4d_{-2}/0\,-2\,2$,	$5g'_{-2}/0\,-2\,4$	$10^{-4}-10^3$, (s)	A1.2e	A1.2e
−2	+1	$5d'_{-2}/0\,-2\,6$,	$6g'_{-2}/0\,-2\,8$		$10^{-4}-10^3$, (s)	A1.2e	A1.2e
−2	−1	$4f_{-2}/0\,-2\,1$			$10^{-4}-10^3$, (s)	A1.2e	A1.2f
−2	−1	$5f_{-2}/0\,-2\,3$,	$6h'_{-2}/0\,-2\,5$,	$6f'_{-2}/0\,-2\,7$	$10^{-4}-10^3$, (s)	A1.2f	A1.2f
−3	+1	$4f_{-3}/0\,-3\,0$,	$5f_{-3}/0\,-3\,2$		$10^{-4}-10^3$, (s)	A1.2g	A1.2g
−3	−1	$5g_{-3}/0\,-3\,1$			$10^{-4}-10^3$, (s)	A1.2g	A1.2h
−4	+1	$5g_{-4}/0\,-4\,0$			$10^{-4}-10^3$, (s)	A1.2h	A1.2i
0	+1	$2s_0/0\,0\,2$,	$3d'_0/0\,0\,4$,	$3s'_0/0\,0\,6$	$10^{-3}-10^{-1}$, (f)	A1.3a	A1.2a
0	+1	$4d'_0/0\,0\,8$,	$4s'_0/0\,0\,10$,	$5g'_0/0\,0\,12$	$10^{-3}-10^{-1}$, (f)	A1.3a	A1.2a
0	+1	$5d'_0/0\,0\,14$,	$5s'_0/0\,0\,16$		$10^{-3}-10^{-1}$, (f)	A1.3a	A1.2a
0	−1	$2p_0/0\,0\,1$,	$3p_0/0\,0\,3$,	$4f'_0/0\,0\,5$	$10^{-3}-10^{-1}$, (f)	A1.3b	A1.2b
0	−1	$4p'_0/0\,0\,7$,	$5f'_0/0\,0\,9$,	$5p'_0/0\,0\,11$	$10^{-3}-10^{-1}$, (f)	A1.3b	A1.2b
−1	+1	$2p_{-1}/0\,-1\,0$,	$3p_{-1}/0\,-1\,2$,	$4f'_{-1}/0\,-1\,4$	$10^{-3}-10^{-1}$, (f)	A1.3c	A1.2c
−1	+1	$4p'_{-1}/0\,-1\,6$,	$5f'_{-1}/0\,-1\,8$,	$5p'_{-1}/0\,-1\,10$	$10^{-3}-10^{-1}$, (f)	A1.3c	A1.2c
−1	−1	$3d_{-1}/0\,-1\,1$,	$4d_{-1}/0\,-1\,3$,	$5g'_{-1}/0\,-1\,5$	$10^{-3}-10^{-1}$, (f)	A1.3d	A1.2d
−1	−1	$5d'_{-1}/0\,-1\,7$			$10^{-3}-10^{-1}$, (f)	A1.3d	A1.2d
−2	+1	$3d_{-2}/0\,-2\,0$,	$4d_{-2}/0\,-2\,2$,	$5g'_{-2}/0\,-2\,4$	$10^{-3}-10^{-1}$, (f)	A1.3e	A1.2e
−2	+1	$5d'_{-2}/0\,-2\,6$			$10^{-3}-10^{-1}$, (f)	A1.3e	A1.2e

ues of β. This effect is easily understood when we compare the level distances in the unperturbed one-electron problem ($\propto 1/n_\mathrm{p}^3$) with the perturbation energy associated with the diamagnetic term ($\propto \beta^2 n_\mathrm{p}^4$), from which immediately follows that the transition to the magnetic-field dominated region is shifted to smaller values of β proportional to $n_\mathrm{p}^{-7/2}$. This means at the same time that, for given β, the number of spherical harmonics necessary to produce satisfactory accuracy rapidly increases with n_p, as can be clearly seen from the results shown in Tables A1.2a–h.

For illustrational purposes, the continuous dependence of the energy values on the magnetic field strength is depicted in various figures. In Fig. 4.1 (and Fig. A1.1) all energy levels with $n_\mathrm{p} \leq 5$ are plotted.

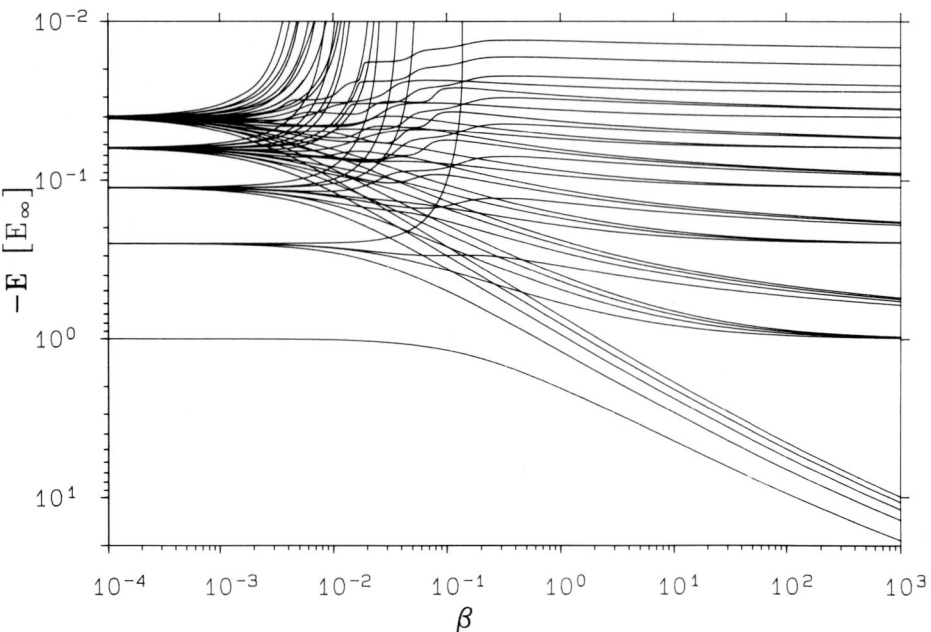

Fig. 4.1. Energy values (in units of E_∞) of all states of an electron in a static Coulomb potential with principal quantum numbers $n_p \leq 5$ as continuous functions of the magnetic field parameter $\beta = B/(4.701 \times 10^5 \text{ T})$. The Rydberg series, corrected by the splitting of the linear and quadratic Zeeman effect, is easily recognized for small values of β. At β-values around 10^{-2} a complete rearrangement of the energy levels occurs. States with positive m are shifted to higher Landau levels, and for $\beta \geq 10$ the level structure of the strong-field regime is reached, where states with $\nu > 0$ converge towards the Rydberg series again (*hydrogen-like states*), while the z-nodeless ground states in every $m \leq 0$ band are strongly lowered (*tightly bound states*) and diverge logarithmically in the limit $\beta \to \infty$.

Additionally, in the Figs. A1.2a–i, all states of the Tables A1.2a–h with definite m and π are grouped in one figure, whereas in the Figs. A1.3a–e, all states which, in the field-free limit, originate from one principal quantum number are combined. These figures show, for small values of β, the Rydberg structure of the level scheme corrected by the splitting of the linear and quadratic Zeeman effect, while at values of β around 10^{-2} (for the low-lying states under consideration) a complete rearrangement of the level structure occurs (the breakpoint is shifted left with increasing principal quantum number, in accordance with the $n_p^{-7/2}$ dependence discussed above). It is only for β greater than, say, 10 that the graphs become well-behaved again, and the level order characteristic of the strong-field regime is reached. In this latter case the energies of states with $\nu > 0$ converge towards the Rydberg series $1/\overline{\nu}^2$ again, with $\overline{\nu} = \nu/2$ for ν even, and $\overline{\nu} = (\nu + 1)/2$ for ν odd (*hydrogen-like states*), while the energies of the z-nodeless ground states in every $m \leq 0$ band are strongly lowered (*tightly bound states*), and in fact diverge logarithmically in the limit $\beta \to \infty$ (Loudon 1959). We note that, by using an asymptotic property of the potential functions, the energy values of excited states with $m \leq -3$ and arbitrary ν can be determined within an accuracy of $\lesssim 1\%$ for magnetic field strengths $\beta \gtrsim 1$, with the accuracy improving rapidly as β, $|m|$, or ν are increased (Rösner et al. 1983). These results ideally complement those listed in Tables A1.2a–h. On the other hand, the energies of states with positive m, which correspond to states in excited Landau levels, become less and less negative with β increasing from zero and eventually merge with the continuum at $\beta < 0.15$, and even sooner for higher principal quantum numbers. Finally, it should be noted that, at intermediate values of β, the monotonous behaviour of the energy graphs of $m \leq 0$ states is more and more destroyed as β is increased; a slight dip is already observed in the state $2s_0/0\ 0\ 2$, a stronger one in $3d'_0/0\ 0\ 4$, and the wiggles become more pronounced in higher states. This phenomenon is related to the existence of an approximate symmetry in the magnetized hydrogenic problem, revealed for higher n_p states by exponentially diminishing level anticrossings in the computations by Zimmerman et al. (1980), and Delande and Gay (1981), and already clearly indicated in the $(m = 0)$-levels of Fig. A1.2a.

4.2 Wavelengths of the Hydrogen Atom

With the accurate energy values for the static Coulomb problem at hand it is an easy task to compute the wavelengths of transitions as functions of the magnetic field strength. The wavelengths of the physical hydrogen atom are obtained from those of the static Coulomb problem by employing the scaling laws from Sect. 2.3 for taking into account the finiteness of the proton mass ($\lambda = m_p/(m_p + m_e) \approx 0.9994557$). While the corrections in the first term on the right-hand side of (2.26b) are of the expected order of magnitude, the second term (multiples of the proton cyclotron energy $\hbar\Omega_c \approx 29.6\,\text{meV} \times (B/4.7 \times 10^5\,\text{T})$) can produce sizeable shifts if the magnetic field is large enough (cf. Wunner et al. 1980 and Sect. 8.2). For wavelengths below 1000 nm the contributions in $|\Delta m| = 1$ dipole transitions are non-negligible for $B \geq 5 \times 10^4\,\text{T}$ and become of special importance in the high-field regime.

In Figs. A2.1–A2.4 of Appendix A2 and in Figs. 4.2a,b the wavelengths of all dipole transitions possible between states with (field-free) principal quantum numbers $n_p \leq 5$ are plotted in their dependence on the magnetic field strength for $10^{-4} \leq \beta \leq 10^3$. The correspondence between data and figures (including the stationary components; see Chap. 7) is summarized in Table 4.3 (and Table A2.1).

To convey to the reader a general impression of the complexity of the spectrum, even of the simplest atomic system, in Figs. 4.2 the wavelength spectrum of the hydrogen atom is shown from the soft X-ray range up to the far infrared as a function of the magnetic field strength.

We distinguish, in a very natural way, two types of electromagnetic transitions. Of first type are transitions that evolve from the field-free Lyman, Balmer, Paschen, and Brackett transitions ($n'_p \neq n_p$). These transitions are

Table 4.3. Relations between transitions, number of components, tables, and figures

Transition	$n'_p \to n_p$	number of components	Figure	Stationary components	Table	Figure
Lyman α	$2 \to 1$	3	A2.1	1	A2.2a	A2.5
Lyman β	$3 \to 1$	3	A2.1	—	—	—
Lyman γ	$4 \to 1$	6	A2.1	—	—	—
Lyman δ	$5 \to 1$	6	A2.1	—	—	—
Balmer α	$3 \to 2$	15	A2.2a	5	A2.2a–b	A2.6a
Balmer β	$4 \to 2$	18	A2.2a	7	A2.2b–d	A2.6b
Balmer γ	$5 \to 2$	27	A2.2b	7	A2.2d–e	A2.6c
Balmer δ	$6 \to 2$	30	A2.2b	*	—	—
Balmer ε	$7 \to 2$	39	A2.2b	*	—	—
Paschen α	$4 \to 3$	42	A2.3	2	A2.2f	A2.7a
Paschen β	$5 \to 3$	51	A2.3	7	A2.2f–h	A2.7b
Brackett α	$5 \to 4$	90	A2.4	3	A2.2h	A2.8
$n'_p \leq 5 \to n_p \leq 5$	$n'_p \neq n_p$	261	4.2a	—	—	—
$n'_p \leq 5 \to n_p \leq 5$	$n'_p = n_p$	105	4.2b	—	—	—

* There are numerous stationary points, which, however, are not very pronounced, for details see the figures.

dominant in the low and medium field regime ($\beta \lesssim 0.1$) and are plotted in great detail in Figs. A2.1–A2.4. The other type represents transitions between states with the same (field-free) principle quantum numbers ($n'_p = n_p$). The wavelengths of these transitions (Fig. 4.2b) tend, of course, to infinity if β goes to zero. Furthermore, the electromagnetic transition probabilities vanish, at least proportional to β^3, due to the vanishing photon energies. Therefore, these transitions are unimportant in the infrared and optical spectral range for low magnetic field strengths. At high field strengths, however, they complement the total spectrum of the magnetized hydrogen atom as an essential part.

We will now discuss the first type of transitions beginning with the Lyman lines, which exhibit a rather smooth behaviour and are continuously shifted to shorter wavelengths, with the exception of the transition originating in $2p_{-1} \to 1s_0$, which first runs through a broad maximum of $\lambda = 134.1$ nm at $\beta \approx 0.12$ before starting the monotonous descent essentially caused by the strong energetic lowering of the $1s_0/0\ 0\ 0$ final state. Turning to Balmer and Paschen transitions, we clearly recognize, in the region of small magnetic field strengths ($\beta \approx 10^{-3}$), the splitting of the unperturbed lines into three equidistant Zeeman components. For larger β these components continue splitting by the quadratic Zeeman effect. The onset of the quadratic Zeeman effect is shifted to smaller β-values with increasing wavelengths. Beyond this region ($\beta \approx 10^{-2}$), where the perturbation theory treatment breaks down, the lines are completely torn apart by the magnetic field within one β-decade and the spectrum becomes totally distorted. Since the energy levels of states with different m and different z-parity are allowed to cross (cf. Fig. 4.1), the wavelengths of corresponding transitions go to infinity at certain values of β.

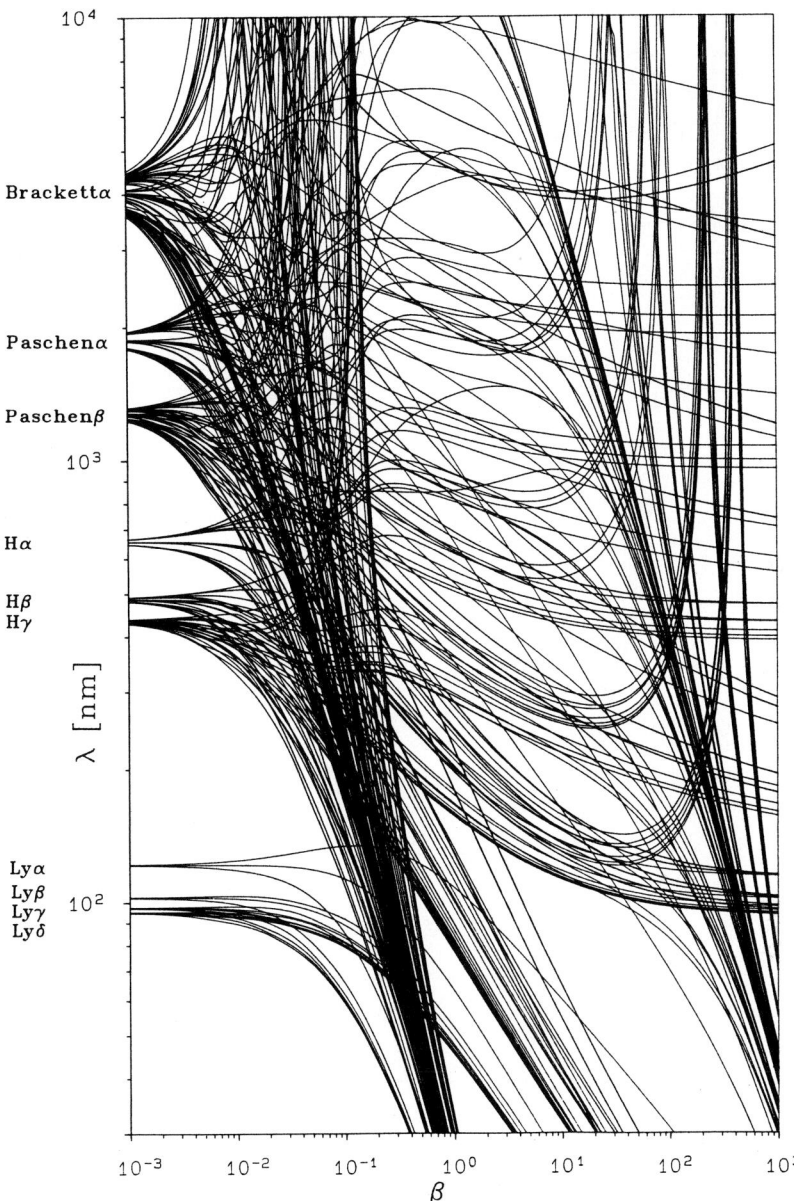

Fig. 4.2a. The wavelength spectrum of the hydrogen atom from the soft X-ray range (30 nm) up to the far infrared (10000 nm) as a function of the magnetic field strength in the interval 470 T to 4.7×10^8 T on a doubly logarithmic scale. All possible transitions between states with (field-free) principal quantum numbers $n_p \leq 5$ and $n'_p \neq n_p$ are included. Effects of the finiteness of the proton mass are taken into account. The two rapidly declining bunches of lines correspond to cyclotron-like transitions of electrons (*left-hand bunch*) and protons (*right-hand bunch*), respectively. The *stationary* lines in the intermediate region are particularly well recognizable if the figure is viewed sideways at flat angles.

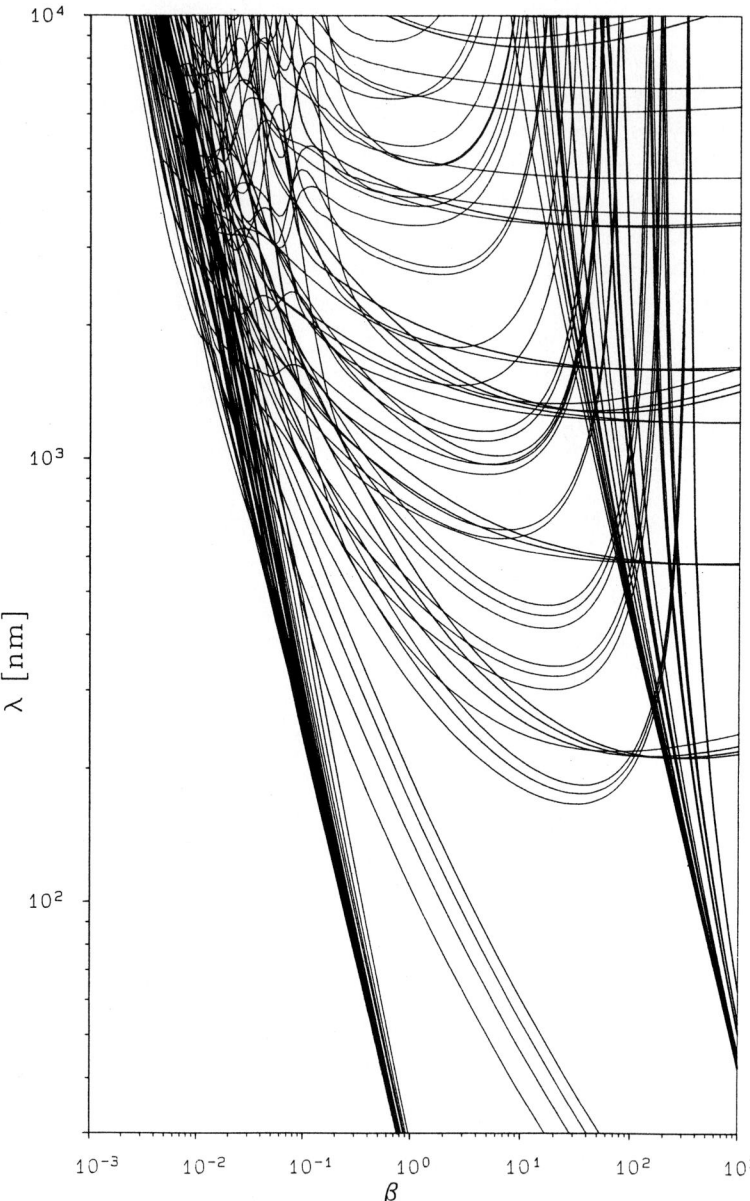

Fig. 4.2b. The wavelengths (from the soft X-ray range (30 nm) up to the far infrared (10000 nm)) of all possible transitions between states with (field-free) principal quantum numbers $n_p \leq 5$ and $n'_p = n_p$ as a function of the magnetic field strength in the interval 470 T to 4.7×10^8 T on a doubly logarithmic scale. Effects of the finiteness of the proton mass are taken into account. The two rapidly declining bunches of lines again correspond to cyclotron-like transitions of electrons (*left-hand bunch*) and protons (*right-hand bunch*), respectively. Due to the vanishing energy level distances, all wavelengths tend to infinity if β goes to zero.

Reordering appears only in intense fields, indicative of the fact that, in the limit $B \to \infty$, the level scheme approaches that of the one-dimensional Coulomb problem, which consists of tightly bound levels and levels whose energies equal those of the field-free hydrogen atom. As a consequence, numerous lines tend toward the wavelengths of the unperturbed hydrogen series on the right-hand side of the figures. Further ordering is evident from the clustering of many lines into two conspicuous bunches (clearly seen in Fig. 4.2a, where all lines with $n'_p \neq n_p$ are shown), which decline proportionally to $1/\beta$. The left-hand bunch comprises all electron cyclotron transitions with $\lambda^{(e)}_{\text{cycl}} \approx 22.8\,\text{nm}/\beta$. The shortest wavelengths then correspond to cyclotron transitions of the electron from the first-excited to the ground-state Landau level. For neutron star magnetic fields ($\beta = 10^3$), the appropriate photon energies lie at $54\,\text{keV}$, and thus in the X-ray region. For photon energies like these, which amount to more than 10 percent of the rest energy of the electron, naturally relativistic effects of the same order are to be expected. This is indeed confirmed by the rigorous relativistic treatment (cf. Daugherty and Ventura 1977; Herold et al. 1982a,b) for cyclotron transitions in which the transverse state of excitation of the electron changes; for the Coulomb binding energies, however, in which the longitudinal motion plays the essential role, relativistic effects remain well below 0.1 percent even at magnetic fields of 10^9 T (Lindgren and Virtamo 1979), which is understandable because of the smallness of the ratio between the Coulomb binding energy and the rest energy of the electron.

The bunch on the right-hand side is due to the finiteness of the proton mass, by which levels with adjacent azimuthal quantum numbers are shifted with respect to each other by the proton cyclotron energy, giving rise to transitions with wavelengths $\lambda^{(p)}_{\text{cycl}} = (m_p/m_e)\lambda^{(e)}_{\text{cycl}} \approx 4.18 \times 10^4\,\text{nm}/\beta$ (cf. Sect. 8.2). All lines in the proton cyclotron bunch would either tend to wavelengths of the unperturbed hydrogen atom or even go to infinity (in the case of energy levels coinciding in the $B \to \infty$ limit of the one-dimensional Coulomb problem), if the finiteness of the proton mass were neglected. These two bunches can also be clearly seen in the second type of transitions ($n'_p = n_p$, Fig. 4.2b), reflecting the fact that, at high field strengths, cyclotron transitions become dominant and the effect of the Coulomb interaction between electron and proton can be regarded as a perturbation The lines below $\sim 100\,\text{nm}$ between the two cyclotron bunches correspond to transitions to tightly bound states. The four lines of this type in Fig. 4.2b with $n'_p = n_p$ are the transitions between the states $2s_0/0\;0\;2 \to 2p_{-1}/0\;-1\;0$, $3p_{-1}/0\;-1\;2 \to 3d_{-2}/0\;-2\;0$, $4d_{-2}/0\;-2\;2 \to 4f_{-3}/0\;-3\;0$, $5f_{-3}/0\;-3\;2 \to 5g_{-4}/0\;-4\;0$.

The behaviour of the wavelengths in the intermediate regime appears to be fully disordered, which indicates that, in this region, the classical problem, too, is of chaotic type. Clearly, any attempt to observe, and resolve, a line spectrum of hydrogen at a given magnetic field strength in the intermediate regime is doomed to failure. An element of order, however, is brought in even in this domain by several transitions whose wavelengths go through minima and maxima in certain intervals of the magnetic field strength, that is, they are less sensitive to variations of the magnetic field than the many fast-running components. An inhomogeneous field with a variation of, say, a factor of two (as is the case for a dipole field) around extrema of wavelengths will therefore filter out exactly these stationary components, and thus it is possible to observe, in this instance, a clearly arranged spectrum with a few well-resolved features. Speculative as this may sound, nature has indeed provided us with cosmic laboratories to test this hypothesis, namely strongly magnetized white dwarf stars (see Chap. 7).

4.3 Wave Functions of the Hydrogen Atom

To gain some idea of the spatial structure of the wave functions of hydrogen atoms in strong magnetic fields, for the states $1s_0/0\ 0\ 0$ (at $\beta = 1$) and $3d_{-1}/0\ -1\ 1$ (at $\beta = 0.1$), the four lowest normalized expansion functions $\frac{1}{r}f_{lm}(r)$ and $g_{nm}(z)$ of equations (3.2) and (3.3), following from computations using 12 configurations each, are represented in Fig. 4.3 in their dependence on r or z, respectively. The coefficients given in Fig. 4.3 denote the overlaps of the normalized expansion functions with the total wave function. Evidently, in every instance it is the lowest expansion coefficient (i.e. $l = 0$ or $l = 2$ in the spherical calculation, and $n = 0$ in the cylindrical one) which is maximum, assuming values close to unity.

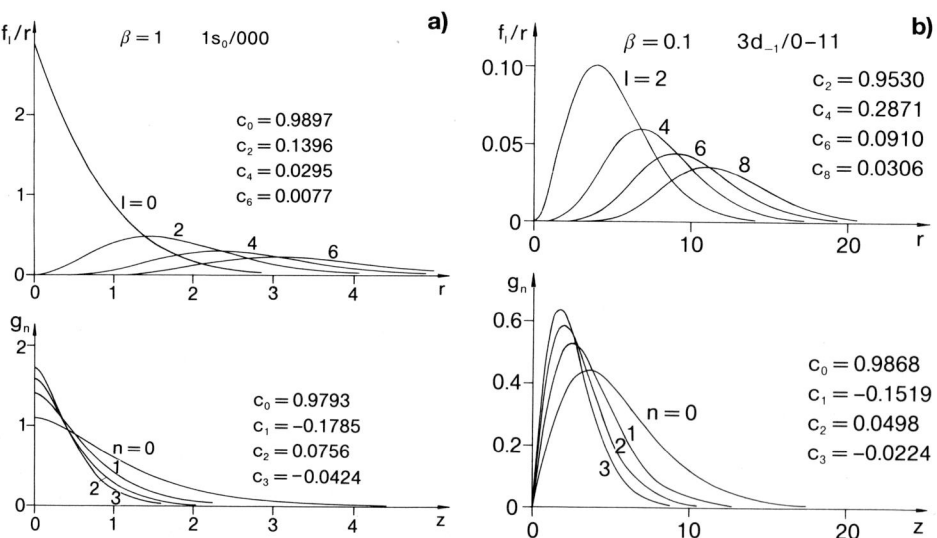

Fig. 4.3a,b. The spatial structure of the four lowest normalized expansion functions $\frac{1}{r}f_{lm}(r)$ and $g_{nm}(z)$ for the state $1s_0/0\ 0\ 0$ (a) and the state $3d_{-1}/0\ -1\ 1$ (b), following from computations using 12 angular momentum or Landau components each. The coefficients given denote the amplitudes with which the respective components contribute to the total wave function.

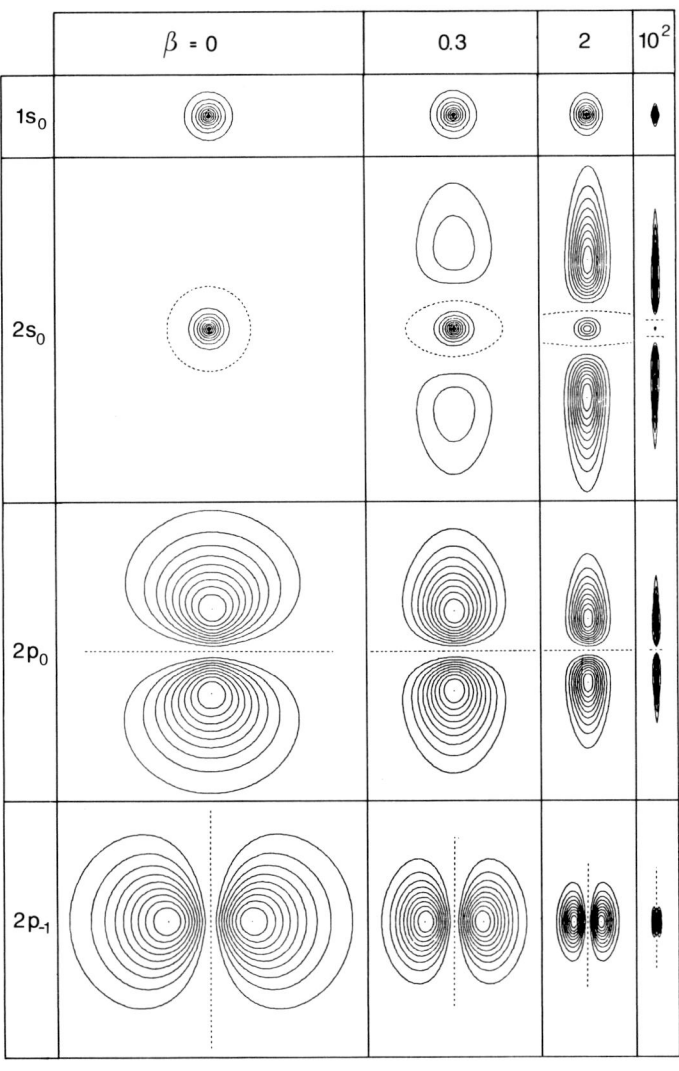

Fig. 4.4. Contour plots of the spatial probability distribution of the electron in the xz-plane (decreasing from maximum to zero in increments of 0.1 of the maximum value) at four different values of the magnetic field strength. Nodal surfaces are indicated by broken curves. As β increases, the atomic states are squeezed together in the direction transverse to the field; in tightly bound states ($1s_0, 2p_{-1}$) the extension parallel to the field also shrinks, while *hydrogen-like* states become strongly elongated, particularly for higher quantum numbers, due to the necessary orthogonality of the longitudinal wave functions in the high-field limit.

The graphs show that, in the spherical expansion, the lowest expansion function is dominant at small distances from the nucleus, where the spherical symmetry of the Coulomb potential still prevails, while further out, the superposition of the higher angular momentum contributions is apparently requisite to account for the cylindrical symmetry imposed by the magnetic field. Correspondingly, in the cylindrical expansion the lowest expansion term dominates at sufficiently large distances from the nucleus, whereas in the vicinity of the nucleus the contributions of many Landau levels are necessary to reproduce the features in the wave function that are caused by the singularity of the Coulomb potential. Next we visualize the drastic changes in the shapes of hydrogenic wave functions brought about by strong magnetic fields. Because of the remaining rotational symmetry about the direction of the field, it suffices to consider intersections of $\Phi^*\Phi$ with the xz-plane. For four low-lying states, $1s_0/0\ 0\ 0$, $2s_0/0\ 0\ 2$, $2p_0/0\ 0\ 1$, $2p_{-1}/0\ -1\ 0$, Fig. 4.4 provides curves of equal spatial probability distribution of the electron (decreasing from maximum to zero in increments of 0.1 of the maximum value) at four different values of the magnetic field strength, $B = 0\,\mathrm{T}$, $1.4 \times 10^5\,\mathrm{T}$, $9.4 \times 10^5\,\mathrm{T}$, and $4.7 \times 10^7\,\mathrm{T}$.

The most striking feature is the strong constriction of the atomic states (roughly proportional to $\beta^{-\frac{1}{2}}$) perpendicular to the magnetic field as the field strength is increased – a consequence of the Lorentz forces becoming larger and larger. Two of the states shown ($1s_0/0\ 0\ 0$ and $2p_{-1}/0\ -1\ 0$) additionally experience a sizeable reduction in their linear extension parallel to the field, which causes them to appreciably gain energy and become *tightly bound* states. As demonstrated by the state $2s_0/0\ 0\ 2$, however, other states may even enlarge their extension parallel to the field. This effect is enhanced even more in states with higher numbers of nodes in the z-direction, due to the "lack of space" in the quantized "tubes" of radius a_L prescribed by the magnetic field.

4.3.1 Graphic Representation of the Spatial Probability Distribution of the Electron

For a denser sequence of magnetic field values, in Figs. 4.5 the spatial probability distribution of the electron in the xz-plane is presented in three-dimensional plots for the states $3d'_0/0\ 0\ 4$ and $4f'_0/0\ 0\ 5$. We remind the reader that, in the limit $B \to 0$, these states are linear combinations of the field-free states, $|3s_0\rangle$, $|3d_0\rangle$ and $|4p_0\rangle$, $|4f_0\rangle$, respectively, adjusted to the symmetry of the magnetic field (see Table 4.1). In the plots of the $3d'_0/0\ 0\ 4$ state, the s-admixture is clearly recognizable by the peak at the origin. The figures illustrate again how the states are compressed towards the z-axis as the magnetic field is rising, with new nodal surfaces being formed. They also show that the states possess essentially only two outward appearances, which suddenly change from one into another in a rather narrow range of magnetic field strength, but remain largely unaltered otherwise.

Next, we will follow this transition by means of the three complex and strongly interacting states $5g'_0/0\ 0\ 12$, $5d'_0/0\ 0\ 14$, and $5s'_0/0\ 0\ 16$. For these states the transition region lies around $\beta \approx 10^{-2}$. In this region these states

4.3 Wave Functions of the Hydrogen Atom 47

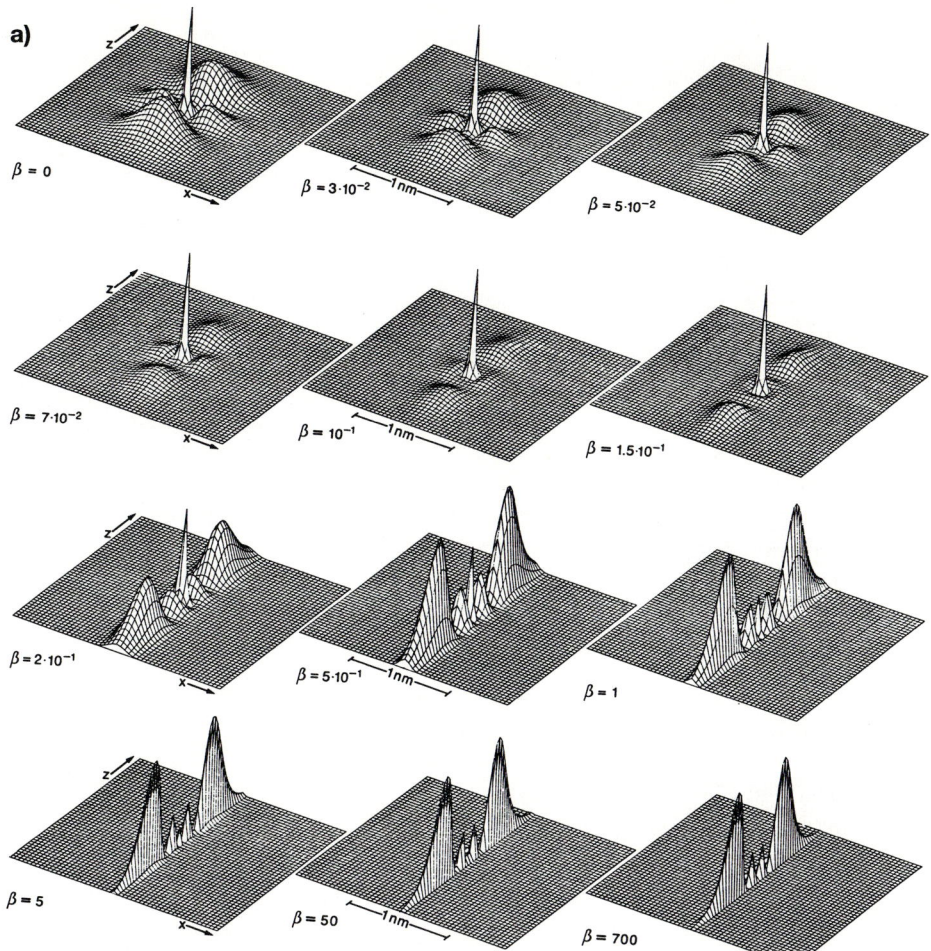

Fig. 4.5a. Three-dimensional plot of the spatial probability distribution of the electron (normalized to the maximum value) in the xz-plane of the state $3d'_0/0\ 0\ 4$ evolving from $|3d'_0\rangle = 0.916\,|3d_0\rangle + 0.402\,|3s_0\rangle$ (for $\beta = 0$) for various magnetic field strengths. The cusp at the origin, recognizable in weaker fields, is caused by the s-admixture. Evidently the state essentially exists in two outward appearances, which suddenly change from one into another in a rather narrow range of magnetic field strength, but remain largely unaltered otherwise.

undergo extremely small anticrossings (see Fig. 4.1 and Fig. A1.2a). A sequence of three-dimensional plots of the spatial probability distribution on the inside backcover demonstrates the drastic changes in the atomic structure in this region. An enlargement of the energies around $\beta \approx 10^{-2}$ of these states is also shown on that sheet; crossed circles mark the β-values where the wave functions are plotted. The accurate calculation of the wave functions of these states is very expensive and demands a large amount of computing facility. To guarantee convergence, up to 6400 expansion functions (80 spherical harmonics with l up to 160 and 80 expansion terms for each radial function) must be used. From this

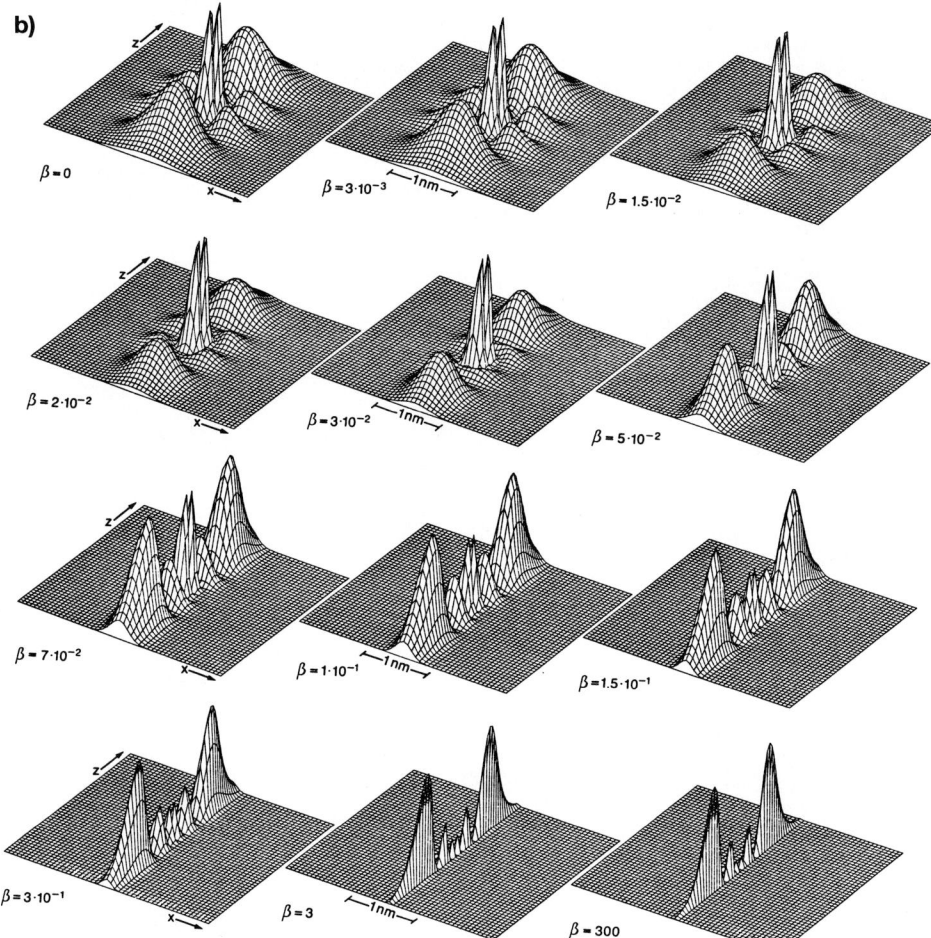

Fig. 4.5b. Three-dimensional plot of the spatial probability distribution of the electron (normalized to the maximum value) in the xz-plane of the state $4f'_0/0\ 0\ 5$ evolving from $|4f'_0\rangle = 0.811\,|4f_0\rangle + 0.585\,|4p_0\rangle$ (for $\beta = 0$) for various magnetic field strengths. Evidently the state essentially exists in two outward appearances, which suddenly change from one into another in a rather narrow range of magnetic field strength, but remain largely unaltered otherwise.

sequence one can again clearly recognize the transition from the spherical to the cylindrical structure with increasing values of β. Furthermore, the increasing extension of the higher excited states at $\beta = 0.1$ is obvious. Of special interest is the behaviour of the wave functions in the vicinity of the narrow anticrossings. As is well-known, the structure of the states is exchanged. As a consequence, the $4s'_0$ state runs almost unaltered in its structure through the $5g'_0$, $5d'_0$, and $5s'_0$ states. These rapid changes in the structures give rise to drastic variations in the electromagnetic transition probabilities (see Chaps. 6 and 7).

4.3.2 Pictorial Representation of the Spatial Probability Distribution of the Electron

The idea of visualizing the electron cloud of the hydrogen atom is very old. Synthetic photographs for 23 low-lying (field-free) states up to $n_\mathrm{p} = 5$ were produced by White as early as 1931. (These pictures are reproduced in many textbooks on quantum mechanics, e.g., Finkelnburg (1967).) For this purpose he constructed a mechanical device, which is depicted in Fig. 4.6 and described as follows:

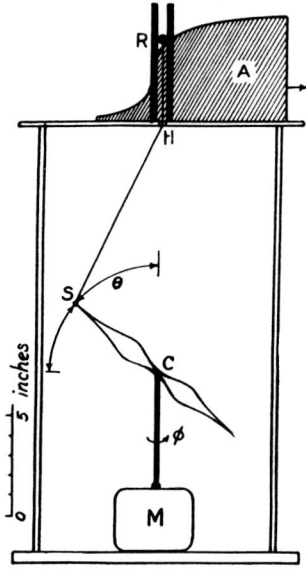

Fig. 4.6. A mechanical device, which, when set in motion and photographed, represents the electron cloud for the various states of the hydrogen-like atoms. The model shown in the figure is for a $3d$ electron (from White 1931).

"Spindles like the one shown in the center of the figure are turned out on a lathe so that in projection they give the density distribution. Such a spindle is pivoted at its center by a small pin at C, and set in rotation about the vertical axis by means of a motor M. This motion gives the required symmetry about ϕ. At the same time that rotation about the ϕ axis is taking place, the angle θ is changed slowly from $\theta = 0$ to $\theta = \pi/2$ by means of a swivel S and a double cord SHR, which passes through a hole in the table top to a roller R, the motion of which is confined to the slot as shown. Curves of thin wood, e.g., A in Fig. 4.6, are cut so that when they are moved slowly but with uniform speed along the table top in the direction indicated by the arrow the angular velocity $d\theta/dt$ is changed in the proper manner to give, as nearly as possible, the correct density when photographed. A time exposure started when $\theta = 0$ and stopped when $\theta = \pi/2$ for the model shown yields the figure for the $3d$ state $m = 0$.

In constructing the various curves to be used at A, the changing angle of the cord HR was taken into account, and also the fact that, as θ increases, each point on the spindle moves in larger and larger circles and therefore faster. A small slot milled in the lower half of each spindle allows the angle $\theta = 0$ to be reached."

Though ingenious in its time, nowadays, in the era of computers and high resolution graphics stations, White's method seems medieval. We have developed a method that produces pictures of the spatial probability distribution on a graphics display screen (Ruder et al. 1989). The basic idea is that the three-dimensional scalar field $\Psi^*\Psi$ of the spatial probability distribution of the electron emits "light" with an intensity that is proportional to the local numerical values of the scalar field and that can penetrate it without absorption. The electron cloud is therefore equivalent to an "optically thin self-radiating gaseous nebula". The intensity that a distant observer receives out of a specific direction from such an object is just the integral along the line of sight of the intensities emitted towards the observer. By scanning the whole object, with the line-of-sight moving across it in a fine grid, a raster image can be computed, which is a good approximation of what the observer actually sees. In Fig. 4.7 the procedure is schematically illustrated and the notation is explained.

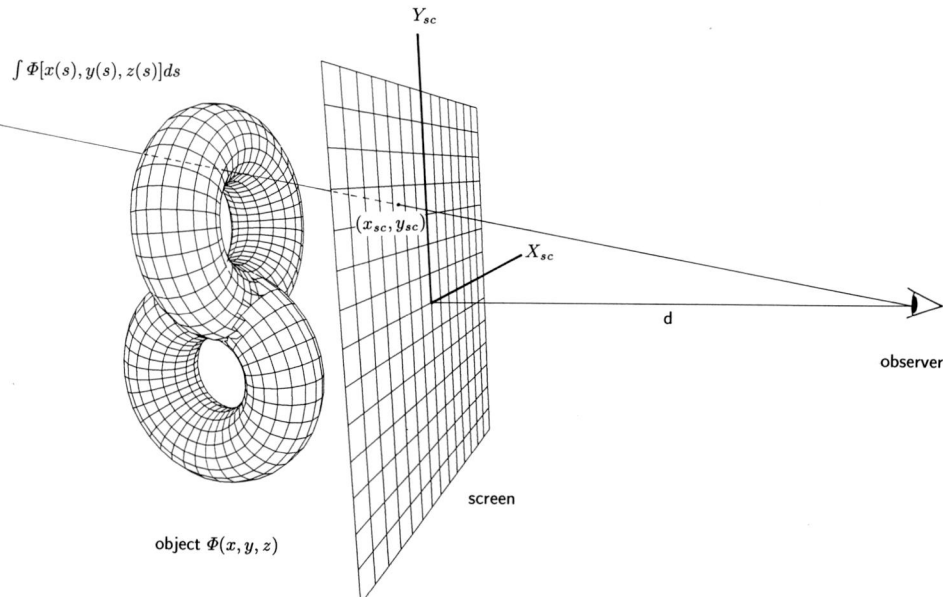

Fig. 4.7. Schematic representation of the method of line-of-sight integration. For each pixel on the screen the intensity is set proportional to the integral through the object along the line-of-sight.

4.3 Wave Functions of the Hydrogen Atom

For a realistic impression it is useful to place the observer at a finite distance d (central projection). The intensity I as a function of the coordinates x_{sc}, y_{sc} on the screen is given by

$$I(x_{sc}, y_{sc}) = \int_{s_{min}}^{s_{max}} \Phi[x(s), y(s), z(s)] ds \quad , \tag{4.1}$$

where s is the arc length of the line-of-sight. The two-dimensional field $I(x_{sc}, y_{sc})$ can be mapped immediately onto the bit planes of a raster graphics display using either an intensity- or a colour-coding scheme and yields a true picture of the self-radiating object.

We have employed this method with a raster of 575 × 575 lines-of-sight and 256 intensity levels to visualize the wave functions of states with $n_p = 5$. In Fig. 4.8 all 15 states with (field-free) principal quantum number $n_p = 5$ are shown for five values of $\beta = 0.001, 0.01, 0.05, 0.1, 0.5$ corresponding to values of the magnetic field of $B = 470, 4700, 2.35 \times 10^4, 4.7 \times 10^4, 2.35 \times 10^5$ T for the hydrogen atom.

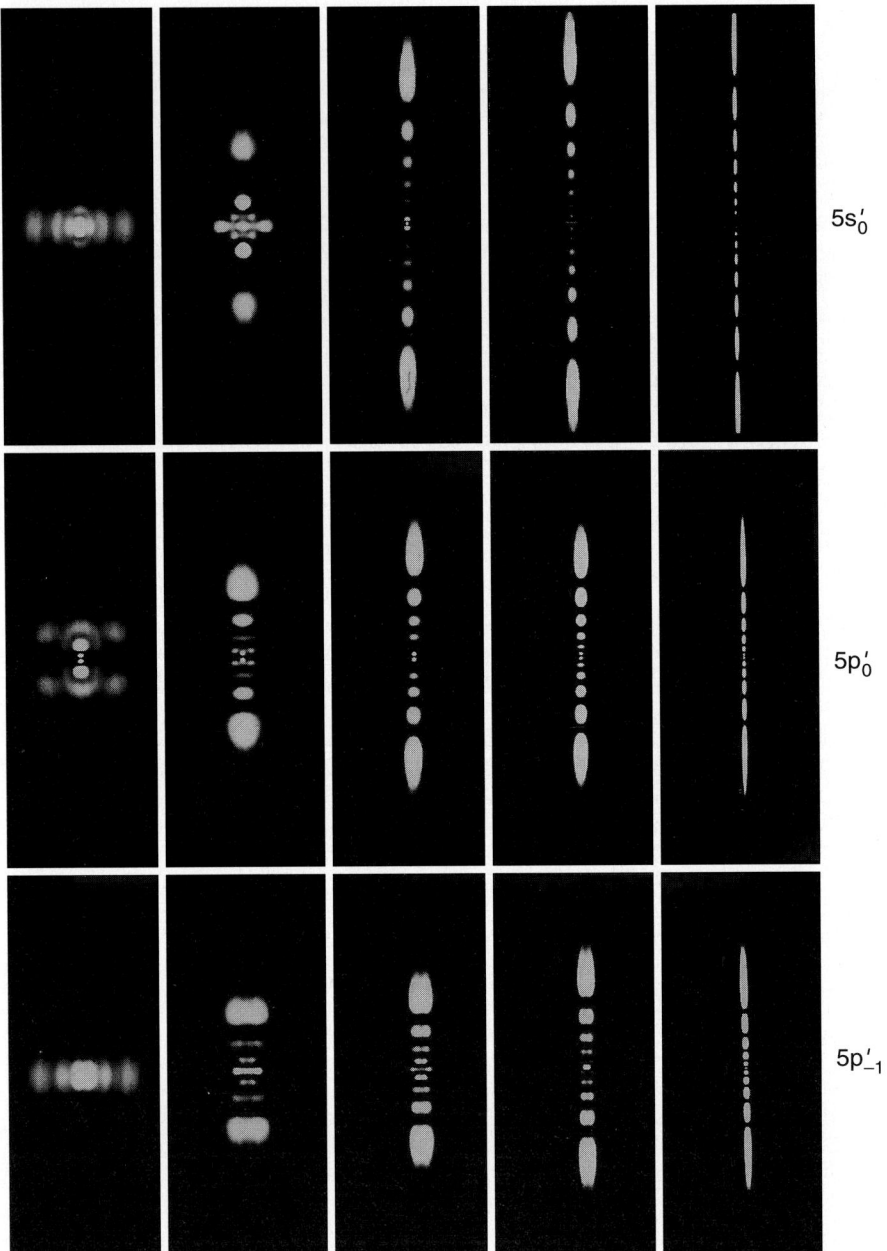

Fig. 4.8a. Pictures of the electron cloud obtained by the method of line-of-sight integration for all states with the (field-free) principal quantum number $n_p = 5$. The magnetic field strength increases from left to right (*Column 1:* $\beta = 0.001$, $B = 470$ T; *Column 2:* $\beta = 0.01$, $B = 4700$ T; *Column 3:* $\beta = 0.05$, $B = 2.3 \times 10^4$ T; *Column 4:* $\beta = 0.1$, $B = 4.7 \times 10^4$ T; *Column 5:* $\beta = 0.5$, $B = 2.3 \times 10^5$ T). The states are from top to bottom $5s'_0$, $5p'_0$, and $5p'_{-1}$.

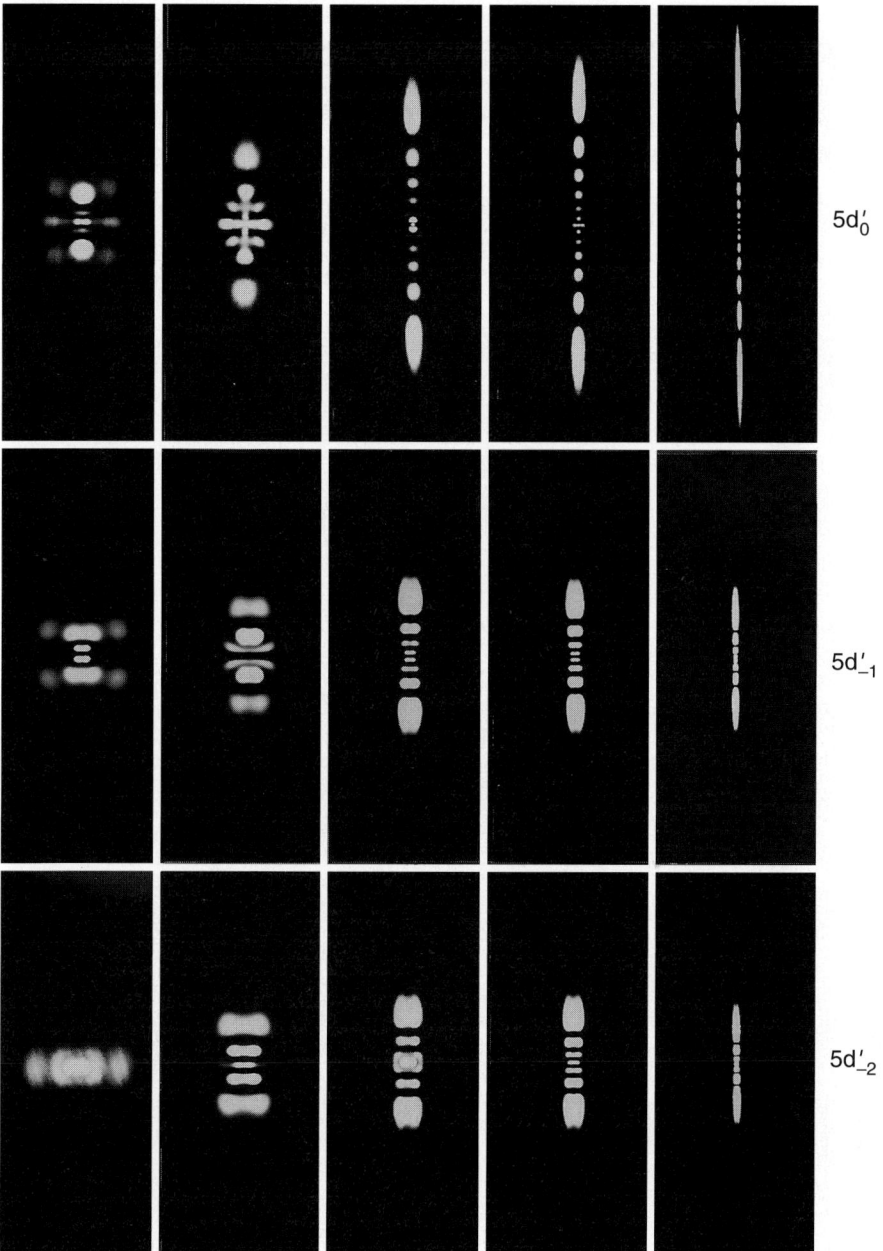

Fig. 4.8b. Pictures of the electron cloud obtained by the method of line-of-sight integration for all states with the (field-free) principal quantum number $n_\mathrm{p} = 5$. The magnetic field strength increases from left to right (*Column 1:* $\beta = 0.001$, $B = 470\,\mathrm{T}$; *Column 2:* $\beta = 0.01$, $B = 4700\,\mathrm{T}$; *Column 3:* $\beta = 0.05$, $B = 2.3 \times 10^4\,\mathrm{T}$; *Column 4:* $\beta = 0.1$, $B = 4.7 \times 10^4\,\mathrm{T}$; *Column 5:* $\beta = 0.5$, $B = 2.3 \times 10^5\,\mathrm{T}$). The states are from top to bottom $5d'_0$, $5d'_{-1}$, and $5d'_{-2}$.

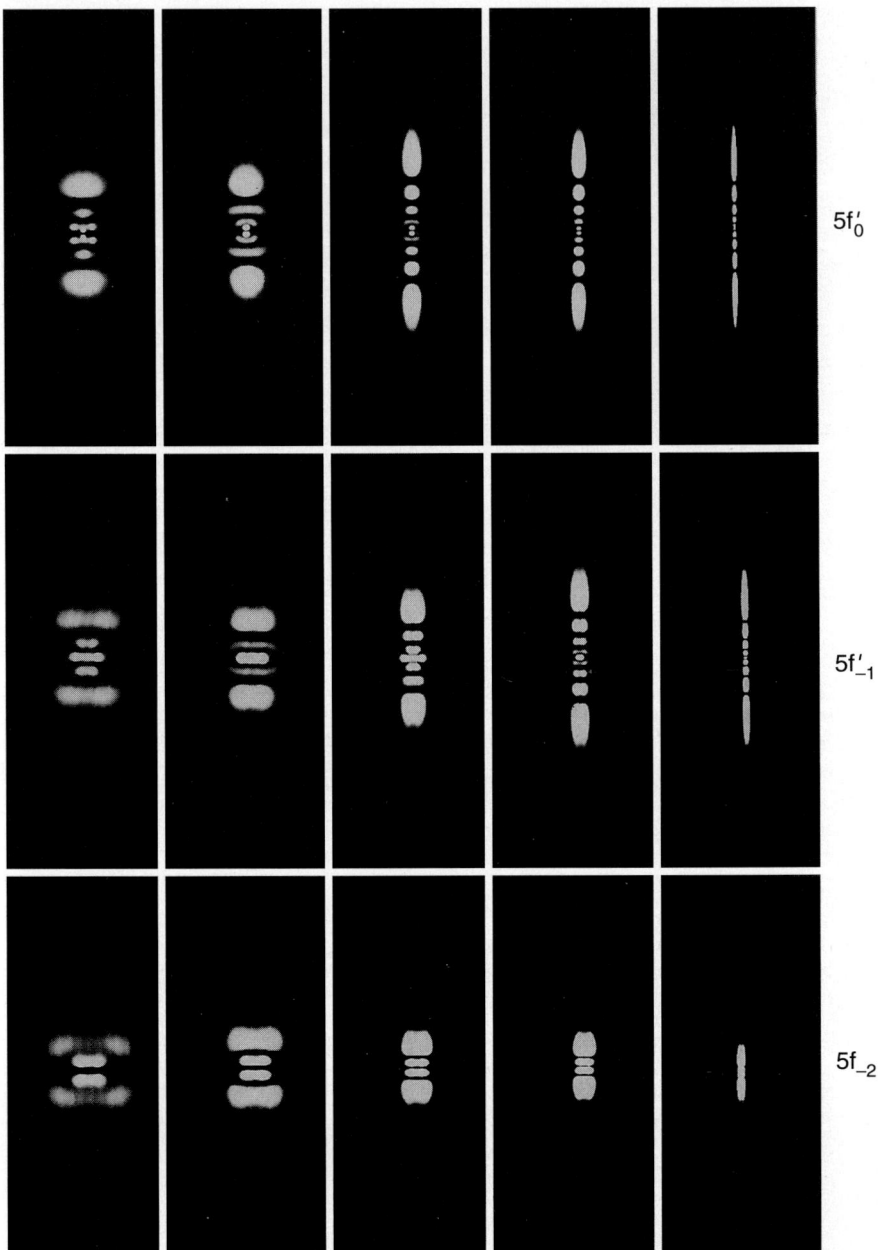

Fig. 4.8c. Pictures of the electron cloud obtained by the method of line-of-sight integration for all states with the (field-free) principal quantum number $n_p = 5$. The magnetic field strength increases from left to right (*Column 1:* $\beta = 0.001$, $B = 470$ T; *Column 2:* $\beta = 0.01$, $B = 4700$ T; *Column 3:* $\beta = 0.05$, $B = 2.3 \times 10^4$ T; *Column 4:* $\beta = 0.1$, $B = 4.7 \times 10^4$ T; *Column 5:* $\beta = 0.5$, $B = 2.3 \times 10^5$ T). The states are from top to bottom $5f'_0$, $5f'_{-1}$, and $5f_{-2}$.

4.3 Wave Functions of the Hydrogen Atom   55

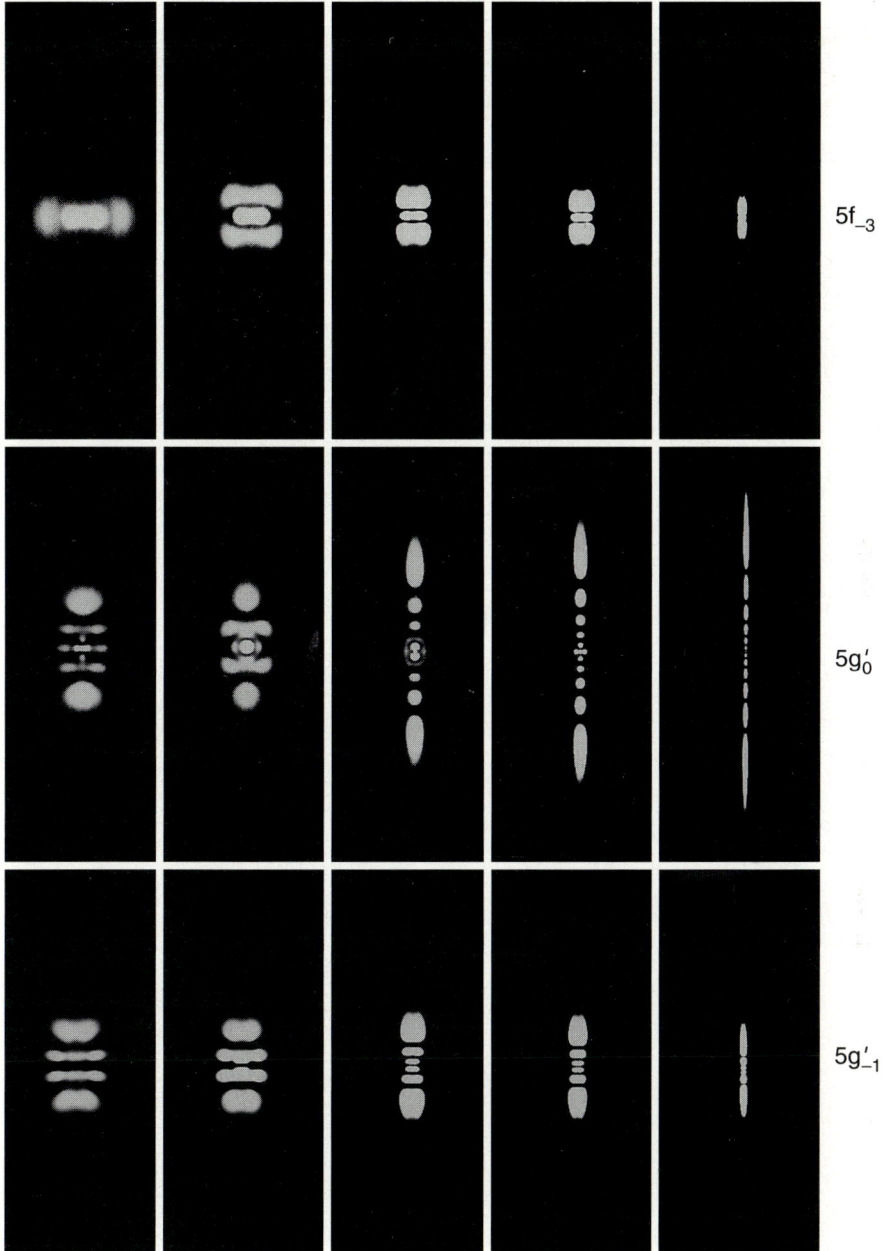

Fig. 4.8d. Pictures of the electron cloud obtained by the method of line-of-sight integration for all states with the (field-free) principal quantum number $n_p = 5$. The magnetic field strength increases from left to right (*Column 1*: $\beta = 0.001$, $B = 470$ T; *Column 2*: $\beta = 0.01$, $B = 4700$ T; *Column 3*: $\beta = 0.05$, $B = 2.3 \times 10^4$ T; *Column 4*: $\beta = 0.1$, $B = 4.7 \times 10^4$ T; *Column 5*: $\beta = 0.5$, $B = 2.3 \times 10^5$ T). The states are from top to bottom $5f_{-3}$, $5g'_0$, and $5g'_{-1}$.

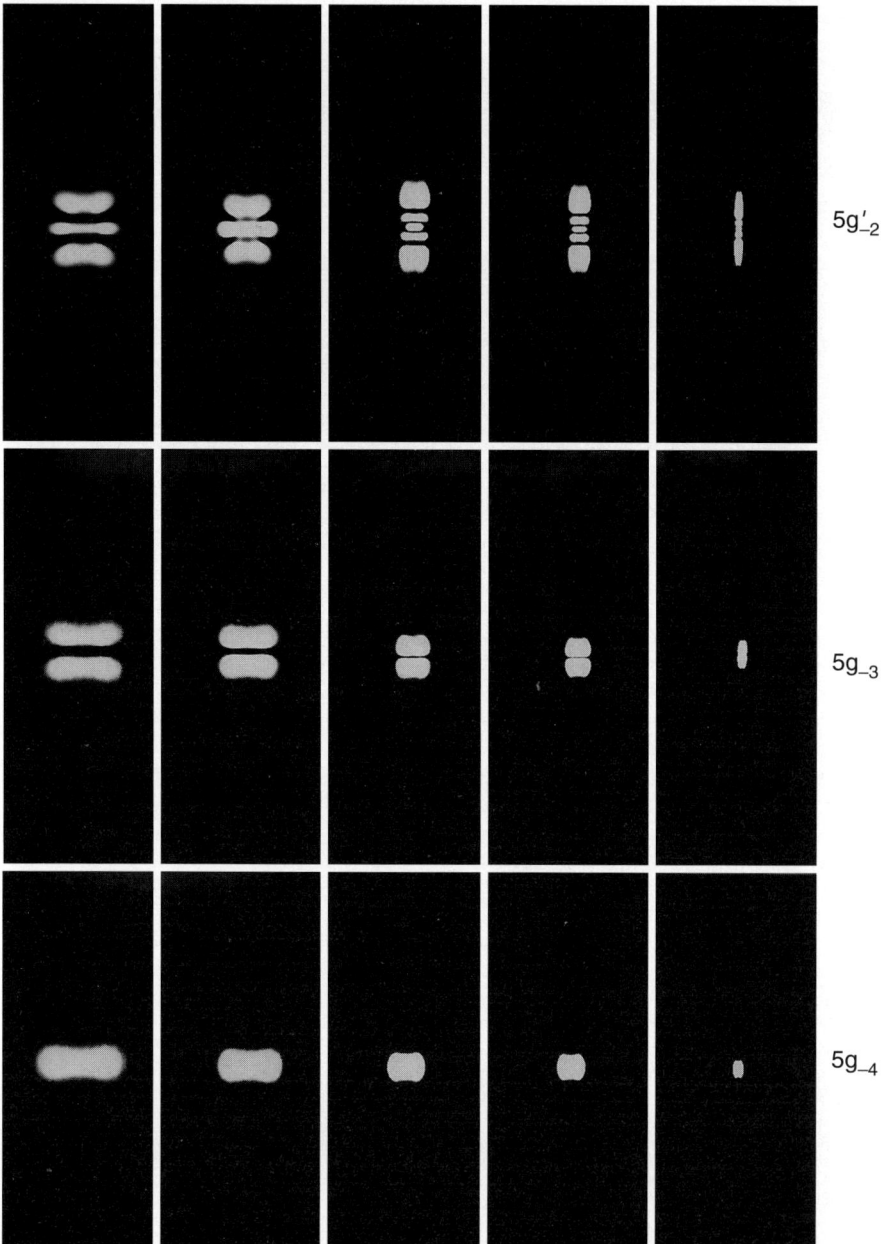

Fig. 4.8e. Pictures of the electron cloud obtained by the method of line-of-sight integration for all states with the (field-free) principal quantum number $n_\mathrm{p} = 5$. The magnetic field strength increases from left to right (*Column 1:* $\beta = 0.001$, $B = 470\,\mathrm{T}$; *Column 2:* $\beta = 0.01$, $B = 4700\,\mathrm{T}$; *Column 3:* $\beta = 0.05$, $B = 2.3 \times 10^4\,\mathrm{T}$; *Column 4:* $\beta = 0.1$, $B = 4.7 \times 10^4\,\mathrm{T}$; *Column 5:* $\beta = 0.5$, $B = 2.3 \times 10^5\,\mathrm{T}$). The states are from top to bottom $5g'_{-2}$, $5g_{-3}$, and $5g_{-4}$.

5 Energies for Arbitrarily Excited States in Adiabatic Approximation

5.1 Asymptotic Property of the Effective Potentials

In the previous chapter we presented the complete list of results for principal quantum numbers up to $n_p = 5$. It is evident that the expenditure necessary to cover arbitrary n_p does not, in general, justify the effort. However, in the limit of intense magnetic fields, where the adiabatic approximation is applicable – that is, the factorization of the wave function in a single product of a Landau wave function and a longitudinal part –, an asymptotic property of the effective potentials together with a quantum defect description allows us to determine the energy values of states with an *arbitrary* degree of excitation for *any* magnetic quantum number in intense magnetic fields from a single table. Explaining the method and the use of the table is the subject of this section.

An attempt to get a grasp of the energy values for $m = 0, -1$ and -2 and arbitrary ν was undertaken by Friedrich (1982), who took the energy values in the form

$$E(B, n = 0, m, \nu)/E_\infty = -\frac{1}{(\bar{\nu} + \delta(B, m, \nu))^2} \quad , \tag{5.1}$$

where $\bar{\nu} = \nu/2$ for ν even and $\bar{\nu} = (\nu+1)/2$ for ν odd, and $\delta(B, m, \nu)$ denotes a formal quantum excess. The advantage of this representation lies in the fact that, for ν odd or ν even and for given B and m, δ tends to some constant value as $\bar{\nu}$ increases. Furthermore, it is already well-known (Loudon 1959) that, for $B \to \infty$, the quantum excess converges to zero independently of m.

We now show in which way, for given ν, the remaining B-m-manifold can be reduced to a one-dimensional manifold. In particular, for sufficiently large magnetic fields, this allows us to determine the energies of both the tightly bound (i.e., $\nu = 0$) and the hydrogen-like (i.e., $\nu \neq 0$) states for *arbitrary* m (and B), which is of prime importance, e.g., in calculating thermodynamic properties in terms of partition functions. The natural starting point is the asymptotic ($B \to \infty$) form of the wave function, in which the motions perpendicular and parallel to the field are decoupled (adiabatic approximation)

$$\Phi_{0m\nu}(\varrho, \varphi, z) = g_{0m\nu}(z)\phi_{nm}^{\text{Lan}}(\varrho, \varphi) \quad , \tag{5.2}$$

where ϕ_{0m}^{Lan} is the wave function of the lowest Landau state. The longitudinal part $g_{0m\nu}$ of the wave function and the energy eigenvalue are obtained by solving one-dimensional Schrödinger equations with effective potentials (expressed in units of E_∞):

$$V_{\text{eff}}^{(\beta,s)}(z) = \frac{4}{s!} \int_0^\infty \frac{e^{-\varrho^2} \varrho^{2s+1}}{(\varrho^2/\beta + z^2)^{1/2}} d\varrho \quad . \tag{5.3}$$

The numerator of the integrand is bell-shaped, with a maximum at $\varrho = \sqrt{s+\frac{1}{2}}$ and turning points at $\varrho = \left(s+\frac{3}{4} \pm \sqrt{s+\left(\frac{3}{4}\right)^2}\right)^{\frac{1}{2}}$ (for $s=0$: $\varrho = \sqrt{\frac{3}{2}}$). This suggests an expansion of the denominator about $\varrho = \sqrt{\frac{s+1}{2}}$, and integrating by terms we arrive at

$$V_{\text{eff}}^{(\beta,s)}(z) = -\frac{2}{\left((s+1/2)/\beta + z^2\right)^{1/2}}$$
$$\times \left(1 - \frac{1}{4(s+1/2+\beta z^2)} + \frac{3(s+5/4)}{(s+1/2+\beta z^2)^2} - \ldots\right) \quad . \tag{5.4}$$

Asymptotically, i.e., for large s, the value of the large parentheses tends toward unity, and thus the effective potentials can be well approximated by the model potential

$$V_{\text{as}}^{(p)}(z) = -\frac{2}{(1/p + z^2)^{1/2}} \quad , \tag{5.5}$$

which no longer depends on the two quantities β and s, but solely on the ratio $p = \beta/(s+\frac{1}{2})$.

5.2 Numerical Results and Their Accuracy

Solving the Schrödinger equation

$$-\frac{d^2}{dz^2} g_{p\nu} + V_{\text{as}}^{(p)}(z) g_{p\nu} = E_{\text{as}}(p,\nu) g_{p\nu} \tag{5.6}$$

for fixed ν and all values of the parameter p yields the energy values and wave functions of the model potential for all values of β. Of course, these results will be useful to the problem under consideration only when the conditions of the underlying approximations are justified. Table 5.1 provides the results obtained by numerical integration of (5.6) for $10^{-3} \leq p \leq 10^3$ and $0 \leq \nu \leq 20$. The mesh of p points is chosen such that in a quadratic interpolation of $\log(-E_{\text{as}})$ versus $\log(p)$ for intermediate p values, the relative error of the energies is about 10^{-4}.

Table 5.1a. Asymptotic energies $E_{as}(p,\nu)$ in units of $-E_\infty$ for positive z-parity

p	0	2	4	6	8	10
1×10^{-3}	0.0581200	0.0412718	0.0301061	0.0226026	0.0174411	0.0137937
2×10^{-3}	0.0809609	0.0540693	0.0375643	0.0271733	0.0203880	0.0157834
5×10^{-3}	0.1249810	0.0757740	0.0490188	0.0337085	0.0243939	0.0183922
1×10^{-2}	0.1729717	0.0961880	0.0587096	0.0388612	0.0274079	0.0202930
2×10^{-2}	0.2385599	0.1201618	0.0690410	0.0440437	0.0303317	0.0220935
5×10^{-2}	0.3626671	0.1570339	0.0832195	0.0507290	0.0339684	0.0242811
1×10^{-1}	0.4952507	0.1882728	0.0939800	0.0555319	0.0365009	0.0257744
2×10^{-1}	0.6728368	0.2216422	0.1044965	0.0600387	0.0388244	0.0271250
5×10^{-1}	0.9999998	0.2676577	0.1177147	0.0654818	0.0415697	0.0286986
1	1.339554	0.3029093	0.1270534	0.0691994	0.0434099	0.0297407
2	1.781865	0.3378302	0.1357690	0.0725849	0.0450630	0.0306685
5	2.568065	0.3827154	0.1463303	0.0765884	0.0469911	0.0317411
1×10^1	3.353997	0.4153230	0.1536203	0.0792929	0.0482777	0.0324511
2×10^1	4.342679	0.4465683	0.1603412	0.0817456	0.0494337	0.0330852
5×10^1	6.026028	0.4855983	0.1684070	0.0846391	0.0507842	0.0338213
1×10^2	7.637232	0.5133183	0.1739293	0.0865893	0.0516865	0.0343103
2×10^2	9.587212	0.5394446	0.1789867	0.0883535	0.0524972	0.0347478
5×10^2	12.75946	0.5715242	0.1850100	0.0904282	0.0534439	0.0352564
1×10^3	15.66458	0.5939561	0.1891051	0.0918224	0.0540760	0.0355946

Table 5.1a. (Continued)

p	12	14	16	18	20
1×10^{-3}	0.0111461	0.0091751	0.0076745	0.0065082	0.0055863
2×10^{-3}	0.0125443	0.0101918	0.0084351	0.0070920	0.0060433
5×10^{-3}	0.0143297	0.0114644	0.0093729	0.0078025	0.0065938
1×10^{-2}	0.0156011	0.0123552	0.0100210	0.0082880	0.0069676
2×10^{-2}	0.0167855	0.0131747	0.0106113	0.0087276	0.0073030
5×10^{-2}	0.0182012	0.0141429	0.0113024	0.0092379	0.0076909
1×10^{-1}	0.0191544	0.0147882	0.0117592	0.0095730	0.0079442
2×10^{-1}	0.0200077	0.0153614	0.0121628	0.0098681	0.0081662
5×10^{-1}	0.0209924	0.0160180	0.0126224	0.0102023	0.0084166
1	0.0216389	0.0164465	0.0129210	0.0104184	0.0085781
2	0.0222109	0.0168238	0.0131827	0.0106074	0.0087194
5	0.0228679	0.0172551	0.0134810	0.0108221	0.0088790
1×10^1	0.0233004	0.0175377	0.0136758	0.0109621	0.0089829
2×10^1	0.0236850	0.0177883	0.0138481	0.0110855	0.0090744
5×10^1	0.0241296	0.0180771	0.0140459	0.0112273	0.0091791
1×10^2	0.0244237	0.0182675	0.0141764	0.0113203	0.0092479
2×10^2	0.0246861	0.0184371	0.0142922	0.0114029	0.0093086
5×10^2	0.0249901	0.0186331	0.0144259	0.0114981	0.0093788
1×10^3	0.0251917	0.0187628	0.0145142	0.0115609	0.0094251

Table 5.1b. Asymptotic energies $E_{\text{as}}(p,\nu)$ in units of $-E_\infty$ for negative z-parity

p	1	3	5	7	9
1×10^{-3}	0.0488272	0.0351236	0.0259931	0.0197900	0.0154659
2×10^{-3}	0.0658606	0.0448357	0.0317939	0.0234377	0.0178744
5×10^{-3}	0.0965563	0.0604514	0.0403634	0.0285120	0.0210850
1×10^{-2}	0.1274778	0.0743088	0.0473365	0.0324111	0.0234588
2×10^{-2}	0.1663432	0.0897172	0.0545299	0.0362543	0.0257313
5×10^{-2}	0.2315793	0.1118806	0.0640506	0.0411081	0.0285197
1×10^{-1}	0.2920695	0.1294165	0.0710313	0.0445256	0.0304355
2×10^{-1}	0.3618428	0.1470380	0.0776507	0.0476723	0.0321687
5×10^{-1}	0.4658067	0.1695558	0.0856373	0.0513628	0.0341672
1	0.5497827	0.1853586	0.0909779	0.0537727	0.0354537
2	0.6340706	0.1996239	0.0956386	0.0558413	0.0365468
5	0.7384878	0.2155881	0.1006951	0.0580511	0.0377037
1×10^{1}	0.8070849	0.2252601	0.1036845	0.0593417	0.0383742
2×10^{1}	0.8638234	0.2328619	0.1059981	0.0603326	0.0388866
5×10^{1}	0.9195172	0.2400290	0.1081524	0.0612494	0.0393587
1×10^{2}	0.9483434	0.2436418	0.1092290	0.0617056	0.0395930
2×10^{2}	0.9679674	0.2460701	0.1099495	0.0620101	0.0397492
5×10^{2}	0.9837092	0.2480037	0.1105216	0.0622515	0.0398728
1×10^{3}	0.9904966	0.2488349	0.1107671	0.0623550	0.0399258

Table 5.1b. (Continued)

p	11	13	15	17	19
1×10^{-3}	0.0123685	0.0100910	0.0083757	0.0070562	0.0060214
2×10^{-3}	0.0140287	0.0112786	0.0092523	0.0077206	0.0065361
5×10^{-3}	0.0161749	0.0127791	0.0103406	0.0085340	0.0071604
1×10^{-2}	0.0177194	0.0138378	0.0110971	0.0090933	0.0075848
2×10^{-2}	0.0191691	0.0148175	0.0117896	0.0096005	0.0079679
5×10^{-2}	0.0209136	0.0159801	0.0126028	0.0101918	0.0084107
1×10^{-1}	0.0220928	0.0167568	0.0131414	0.0105802	0.0087005
2×10^{-1}	0.0231470	0.0174452	0.0136156	0.0109207	0.0089532
5×10^{-1}	0.0243488	0.0182235	0.0141493	0.0113014	0.0092344
1	0.0251149	0.0187163	0.0144838	0.0115399	0.0094100
2	0.0257614	0.0191300	0.0147643	0.0117388	0.0095561
5	0.0264413	0.0195630	0.0150570	0.0119457	0.0097078
1×10^{1}	0.0268333	0.0198117	0.0152246	0.0120639	0.0097943
2×10^{1}	0.0271319	0.0200006	0.0153516	0.0121535	0.0098600
5×10^{1}	0.0274063	0.0201740	0.0154680	0.0122354	0.0099198
1×10^{2}	0.0275421	0.0202597	0.0155254	0.0122758	0.0099493
2×10^{2}	0.0276326	0.0203167	0.0155630	0.0123027	0.0099688
5×10^{2}	0.0277042	0.0203618	0.0155940	0.0123239	0.0099843
1×10^{3}	0.0277349	0.0203811	0.0156069	0.0123330	0.0099910

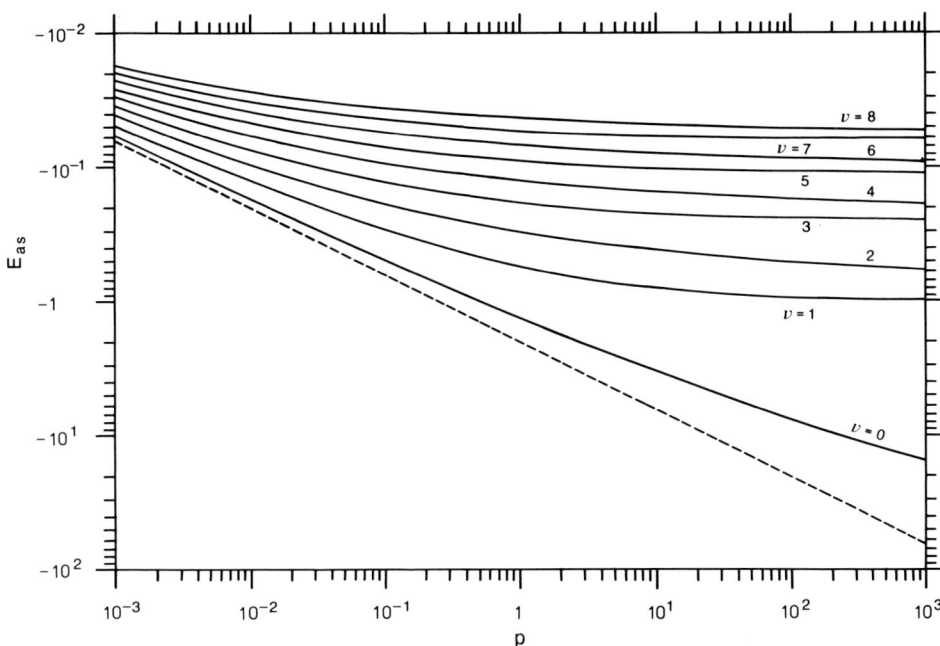

Fig. 5.1. Asymptotic energies $E_{\text{as}}(p,\nu)$ in units of the Rydberg energy as continuous functions of p. The dashed line marks the depth of the asymptotic potential V_{as}^p at $z=0$.

For illustrational purposes, in Fig. 5.1 the energies are shown as continuous functions of p. By construction, the energies of the model potential are intended to substitute the two-dimensional manifold of the adiabatic approximation energies. Ultimately, however, they should serve as a simple asymptotic approximation to the exact energy values of the hydrogen atom in strong magnetic fields. Therefore, a few words must be said beforehand about the accuracy of the adiabatic approximation, as a function of the magnetic field strength, in dependence on the quantum numbers s and ν. In Fig. 5.2, the relative error of the adiabatic energies E_{ad} with respect to the accurate energies E_{ac} is plotted on a doubly logarithmic scale as a function of β for $10^{-2} \leq \beta \leq 10^3$ for various values of s and ν.

It can be seen in Fig. 5.2 that the accuracy of the adiabatic approximation rapidly increases with increasing β, s and ν, the error falling below about 1 percent for all ν if $s > 2$ and $\beta \leq 1$. With this in mind, we are now able to quantify the range of applicability of our asymptotic expansion. We first consider the tightly bound states, which exhibit the largest deviation. In Table 5.2, taking $p = \beta/(s+\frac{1}{2})$ in accordance with (5.4) and (5.5), the asymptotic energies $E_{\text{as}}(p, \nu = 0)$ are compared with the results of the adiabatic approximation for $10^{-3} \leq \beta/(s+\frac{1}{2}) \leq 10^3$, and s from 0 up to 20.

It is found that, although originally devised for large s, the asymptotic expansion yields fairly good approximations to the adiabatic energies also for small values of s and even for $s = 0$. The convergence of the adiabatic energies

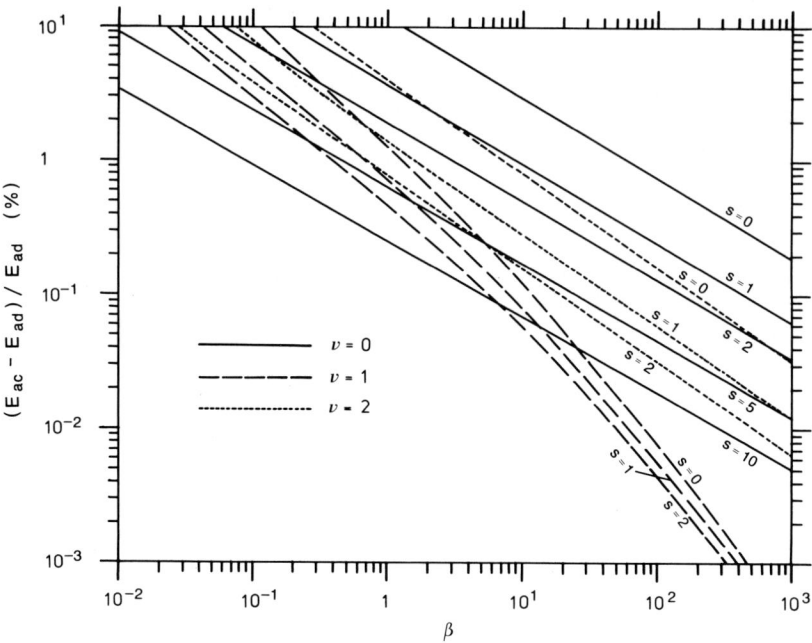

Fig. 5.2. Relative error (in percent) of the adiabatic approximation energies E_{ad} with respect to the accurate energies E_{ac} as a function of β for different values of s and ν

toward the asymptotic values with increasing s and increasing β is illustrated in Fig. 5.3, where the ratios $E_{\mathrm{ad}}/E_{\mathrm{as}}$ taken from Table 5.2 are plotted as functions of $\beta/(s+\frac{1}{2})$ for different values of s. We see that, for $s \geq 3$ and $\beta/(s+\frac{1}{2}) \geq 1$, the deviations of E_{as} from E_{ad} fall below 1 percent. The solid parts of the lines correspond to those ranges of β where the accuracy of the adiabatic approximation in turn becomes better than 1 percent (cf. Fig. 5.2). Thus, Fig. 5.3 reveals the nice property that the asymptotic expansion approaches the adiabatic approximation in the same degree as the adiabatic approximation approaches the exact solution.

Table 5.2. Comparison of the adiabatic approximation energies $E_{\text{ad}}(\beta, s, \nu = 0)$ and the asymptotic energy values $E_{\text{as}}(p = \beta/(s+\frac{1}{2}), \nu = 0)$

$\beta/(s+\frac{1}{2})$	0	1	2	3	4
1×10^{-3}	0.0648396	0.0613950	0.0602044	0.0596413	0.0593161
2×10^{-3}	0.0895110	0.0852584	0.0837135	0.0829754	0.0825471
5×10^{-3}	0.1364873	0.1310395	0.1288975	0.1278586	0.1272518
1×10^{-2}	0.1871016	0.1807221	0.1780208	0.1766936	0.1759141
2×10^{-2}	0.2555705	0.2483450	0.2449890	0.2433160	0.2423273
5×10^{-2}	0.3836087	0.3756830	0.3713291	0.3691092	0.3677848
1×10^{-1}	0.5188984	0.5110894	0.5059101	0.5032146	0.5015930
2×10^{-1}	0.6984760	0.6917463	0.6857290	0.6825189	0.6805689
5×10^{-1}	1.026083	1.023117	1.016093	1.012185	1.009773
1	1.363217	1.365699	1.358118	1.353714	1.350956
2	1.799554	1.810587	1.802771	1.797956	1.794881
5	2.570276	2.598904	2.591534	2.586409	2.583020
1×10^{1}	3.337029	3.384952	3.378713	3.373617	3.370115
2×10^{1}	4.298623	4.372032	4.367795	4.363033	4.359566
5×10^{1}	5.932046	6.050035	6.050031	6.046279	6.043143
1×10^{2}	7.493688	7.654424	7.658963	7.656462	7.653820
2×10^{2}	9.383438	9.595069	9.605363	9.604555	9.602630
5×10^{2}	12.45976	12.75117	12.77091	12.77301	12.77237
1×10^{3}	15.28061	15.64144	15.66962	15.67438	15.67494

Table 5.2. (Continued)

$\beta/(s+\frac{1}{2})$	5	7	10	20	$-E_{\text{as}}$
1×10^{-3}	0.0591050	0.0588475	0.0586424	0.0583892	0.0581200
2×10^{-3}	0.0822683	0.0819277	0.0816557	0.0813192	0.0809609
5×10^{-3}	0.1268552	0.1263692	0.1259799	0.1254970	0.1249810
1×10^{-2}	0.1754030	0.1747750	0.1742707	0.1736436	0.1729717
2×10^{-2}	0.2416768	0.248750	0.2402295	0.2394247	0.2385599
5×10^{-2}	0.3669089	0.3658246	0.3649481	0.3638509	0.3626671
1×10^{-1}	0.5005156	0.4991771	0.4980912	0.4967274	0.4952507
2×10^{-1}	0.6792667	0.6776418	0.6763184	0.6746502	0.6728368
5×10^{-1}	1.008148	1.006108	1.004437	1.002317	0.9999998
1	1.349084	1.346718	1.344768	1.342285	1.339554
2	1.792775	1.790093	1.787869	1.785018	1.781865
5	2.580659	2.577615	2.575062	2.571757	2.568065
1×10^{1}	3.367633	3.364390	3.361641	3.358048	3.353997
2×10^{1}	4.357048	4.353703	4.350826	4.347021	4.342679
5×10^{1}	6.040746	6.037451	6.034541	6.030609	6.026028
1×10^{2}	7.651659	7.648567	7.645753	7.641861	7.637232
2×10^{2}	9.600839	9.598100	9.595493	9.591767	9.587212
5×10^{2}	12.77127	12.76922	12.76705	12.76373	12.75946
1×10^{3}	15.67450	15.67313	15.67140	15.66852	15.66458

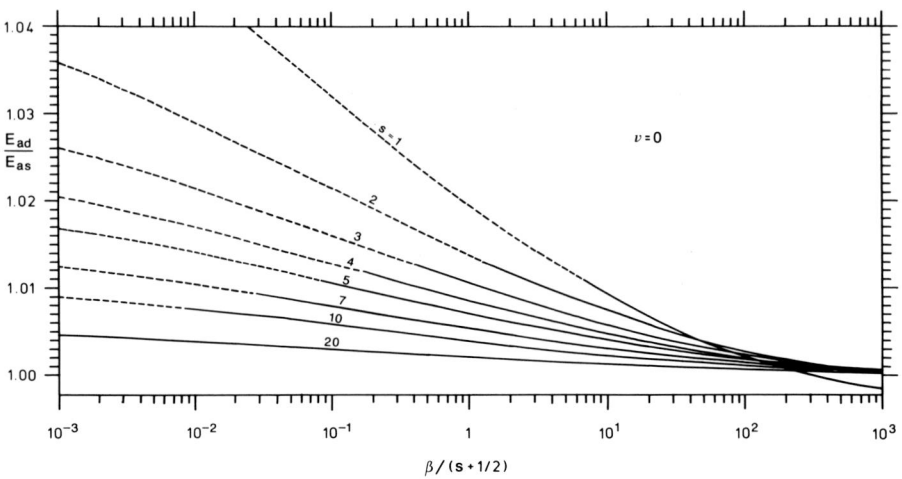

Fig. 5.3. Ratio of adiabatic and asymptotic energy values of the tightly bound state as a function of $\beta/(s+1/2)$ for various values of s

We next consider the convergence of E_{as} towards E_{ad} with increasing ν. For this purpose, in Fig. 5.4a ($s = 3$) and Fig. 5.4b ($s = 20$) the ratios E_{ad}/E_{as} are shown as functions of $\beta/(s+\frac{1}{2})$ for $1 \leq \nu \leq 20$. The shape of the curves evidently depends on ν, but is rather insensitive to s. It should be noted, however, that the global acccuracy increases, from $s = 3$ to $s = 20$, by roughly a factor of 6. The conclusion that can be drawn from Figs. 5.2–5.4 is that, for magnetic fields such that $\beta \geq 1$, the set of asymptotic energies $E_{as}(p, \nu)$ provides the energy values of hydrogenic systems with an accuracy of better than $\simeq 1$ percent for fixed ν and *arbitrary* $s \geq 3$, with the accuracy improving rapidly as β, s, or ν are increased. In addition, fair approximations are obtained even in the cases $s = 0, 1, 2$. To be in a position to extrapolate the energies $E_{as}(p, \nu)$, for given p, to values of ν not contained in Table 5.1, use can be made of the convergence of the quantum excess $\delta_{as}(p, \nu)$, defined by

$$E_{as}(p,\nu) = -\frac{1}{(\bar{\nu} + \delta_{as}(p,\nu))^2} \quad , \tag{5.7}$$

with increasing ν. The quantum excesses for $\nu = 1$ (negative z-parity) and $\nu = 2$ (positive z-parity), which can be extracted from Table 5.1, are given in Table 5.3 in the range $10^{-3} \leq p \leq 10^3$, together with those for $\nu = 19$ and $\nu = 20$, which can be considered as approximations to the limiting ($\nu \to \infty$) values of the quantum excesses.

Computing the quantum excess differences $\epsilon_{as}^-(p,\nu) = \delta_{as}(p,\nu) - \delta_{as}(p,1)$ (ν odd), $\epsilon_{as}^+(p,\nu) = \delta_{as}(p,\nu) - \delta_{as}(p,2)$ (ν even), it turns out that, for $p \geq 0.5$, these lie very close to zero, and a graphical representation of $\epsilon_{as}^{\mp}(p,\nu)$ as a function of p (Fig. 5.5a and Fig. 5.5b), in connection with the tabulated values of $\delta_a(p,1)$, $\delta_{as}(p,2)$, allows the determination of the quantum excesses with high accuracy.

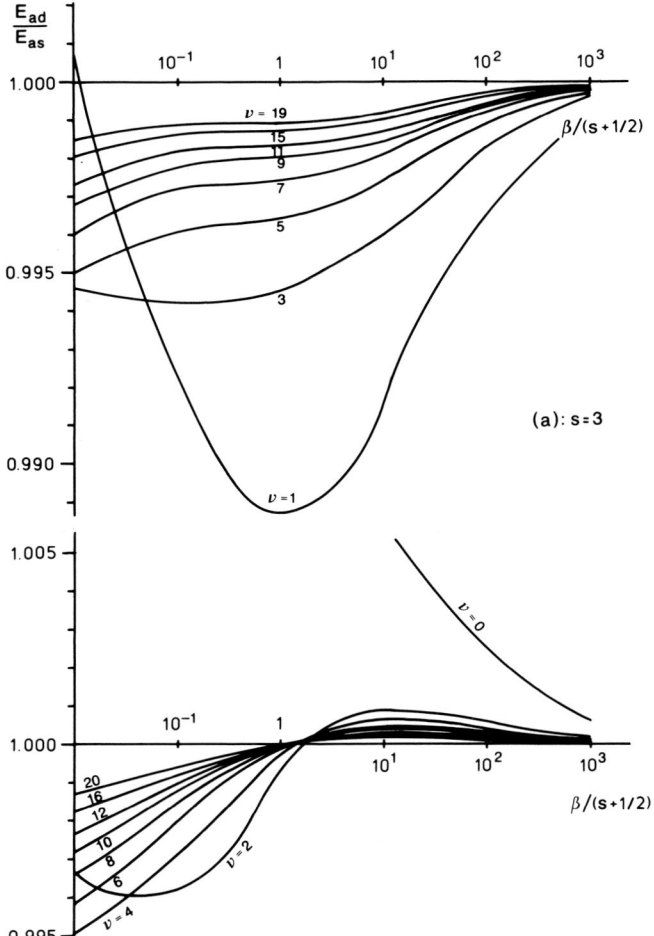

Fig. 5.4a. Ratio of adiabatic and asymptotic energy values for $s = 3$ as a function of $\beta/(s+1/2)$ for different values of ν (*upper part*, negative z-parity; *lower part*, positive z-parity)

A remarkable feature in the positive z-parity subspace is (see Fig. 5.5b) that at $p_0 \approx 12$ (i.e. $\beta \approx 12(s+\frac{1}{2})$), all quantum excesses of the asymptotic energies are practically equal; a similar behaviour was observed by Friedrich (1982), who, in the accurate energies, found values of $p_0 \approx 18$ for $s = 0$, $p_0 \approx 13.7$ for $s = 1$, and $p_0 \approx 12.8$ for $s = 2$. The rapid convergence to our asymptotic value is evident from these numbers.

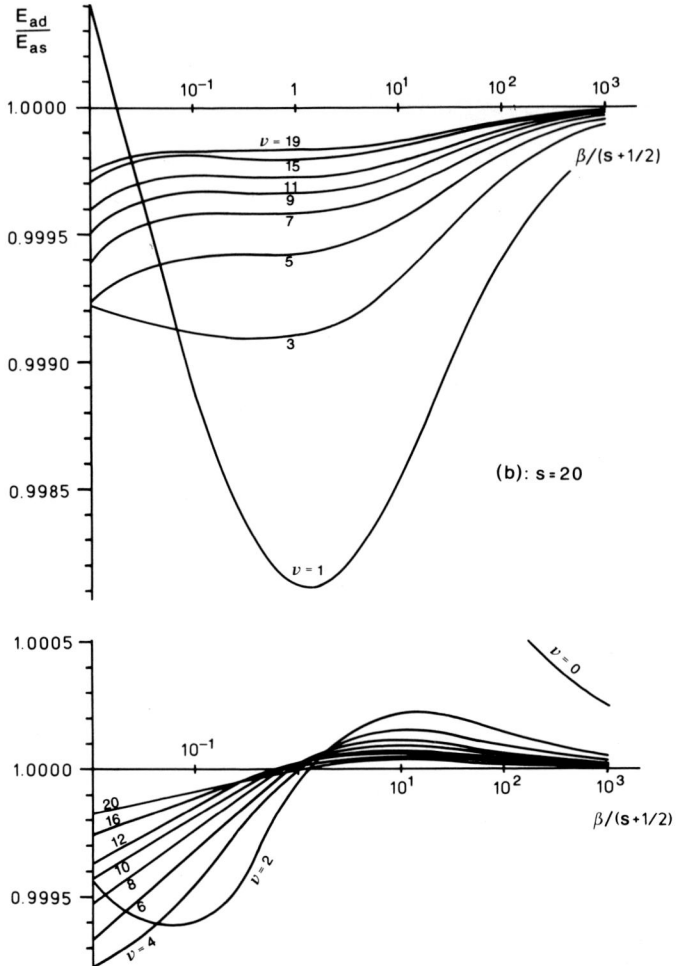

Fig. 5.4b. Ratio of adiabatic and asymptotic energy values for $s = 20$ as a function of $\beta/(s + 1/2)$ for different values of ν (upper part, negative z-parity; lower part, positive z-parity). Note the change of the vertical scale with respect to Fig. 5.4a.

Surprisingly, in special cases, solutions to the Schrödinger equation (5.6) can be obtained analytically. For this purpose we choose the wave function in the form

$$g_{p\nu}(z) = z^k \sum_i c_i^{(\nu)} x^i \exp(-\sqrt{-E_{as}}\, x) \tag{5.8}$$

with $x = (1/p + z^2)^{1/2}$ and $k = 0, 1$ for ν even or ν odd, respectively. This leads to a four-termed recurrence formula for the coefficients $c_i^{(\nu)}$. The condition for this series to break off at a finite $i = i_{max}^{(\nu)}$ requires that the energy assumes the values

$$E_{as}(p,\nu) = -\frac{1}{(i_{max}^{(\nu)}+k)^2} \quad \text{with} \quad i_{max}^{(\nu)} \geq 1 \quad , \tag{5.9}$$

where the corresponding p values follow, together with the coefficients $c_i^{(\nu)}$, from the evaluation of the recurrence formula with given k and $i_{max}^{(\nu)}$. Graphically, in Fig. 5.1, these analytical solutions represent the points of intersection of the $E_{as}(p,\nu)$-functions with horizontal lines $E = -1/n^2$, $n = 1, 2, 3, \ldots$ In Table 5.4, we present the solutions with $i_{max}^{(\nu)} \leq 2$, the coefficients of which are determined from equations of second degree at most.

Table 5.3. Quantum exesses $\delta_{as}(p,\nu)$ (defined by (5.7)) for $\nu = 1, 19$ (negative z-parity) and for $\nu = 2, 20$ (positive z-parity)

p	$\delta_{as}(p,1)$	$\delta_{as}(p,19)$	$\delta_{as}(p,2)$	$\delta_{as}(p,20)$
1×10^{-3}	3.5255	2.887	3.9224	3.379
2×10^{-3}	2.8966	2.369	3.3006	2.864
5×10^{-3}	2.2182	1.818	2.6328	2.315
1×10^{-2}	1.8008	1.482	2.2243	1.980
2×10^{-2}	1.4519	1.203	1.8848	1.702
5×10^{-2}	1.0780	0.9039	1.5235	1.403
1×10^{-1}	0.85036	0.7208	1.3047	1.220
2×10^{-1}	0.66242	0.5685	1.1241	1.066
5×10^{-1}	0.46520	0.4063	0.93290	0.9001
1	0.34867	0.3087	0.81695	0.7970
2	0.25583	0.2296	0.72048	0.7092
5	0.16367	0.1494	0.61645	0.6125
1×10^{1}	0.11312	0.1044	0.55170	0.5510
2×10^{1}	0.075939	0.0708	0.49643	0.4976
5×10^{1}	0.042846	0.0404	0.43503	0.4376
1×10^{2}	0.026874	0.0255	0.39575	0.3987
2×10^{2}	0.016412	0.0156	0.36153	0.3647
5×10^{2}	0.0082463	0.0078	0.32276	0.3259
1×10^{3}	0.0047859	0.0045	0.29755	0.3005

As is seen from the first line of Table 5.4, and the monotonic behaviour of the energies as functions of p (cf. Fig. 5.1), the binding energy of the tightly bound state is larger than the Rydberg energy E_∞ for $p > \frac{1}{2}$, which means that in the real physical problem of the hydrogen atom with a given magnetic field all tightly bound states with $-m = s < 2\beta - 1/2$ lie below $-E_\infty$, and thus below all hydrogen-like states. The large number of these states, essential e.g. to thermodynamical considerations, clearly shows the importance of the simple and accurate asymptotic expansions presented here.

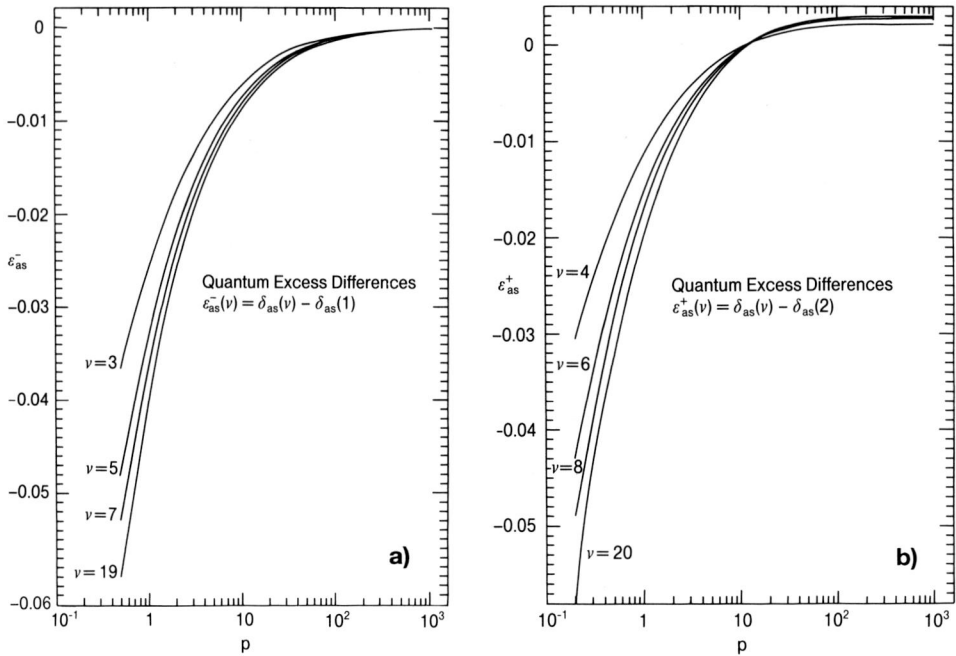

Fig. 5.5a,b. (a) Quantum excess differences for negative z-parity states as a function of p; (b) Quantum excess differences for positive z-parity states as a function of p

Table 5.4. Analytical solutions of the form (5.8) with $i_{max}^\nu \leq 2$.

k	i_{max}	p	E_{as}	c_1/c_0	c_2/c_0	ν
0	1	$1/2$	-1	1	0	0
0	2	$(3-\sqrt{7})/16$	$-1/4$	$1/2$	$-(2-\sqrt{7})/12$	0
0	2	$(3+\sqrt{7})/16$	$-1/4$	$1/2$	$-(2+\sqrt{7})/12$	2
1	1	$1/16$	$-1/4$	$1/2$	0	1
1	2	$(3-\sqrt{5})/108$	$-1/9$	$1/3$	$-(1-\sqrt{5})/54$	1
1	2	$(3+\sqrt{5})/108$	$-1/9$	$1/3$	$-(1+\sqrt{5})/54$	3

6 Electromagnetic Transition Probabilities

The calculation of electromagnetic transitions is a standard chapter of quantum mechanics, the essential formula being the famous "golden rule" (see, e.g., Heitler 1953, Messiah 1973). However, the resulting final expressions for field-free transitions or for Zeeman transitions which may be found in the literature cannot simply be applied to atoms in strong magnetic fields. This is because in the derivation of these expressions *averaging* processes over m are involved, whereas in strong magnetic fields the m-degeneracy is completely destroyed. Let us, therefore, begin by briefly reviewing the considerations, and the formulae, which are valid also for strong magnetic fields.

6.1 The General Expressions

The crucial quantity entering into the calculation of a $\Delta m = q$ dipole transition from an initial state $|\tau\rangle$ to a final state $|\tau'\rangle$ is the dipole matrix element

$$p_{\tau'\tau}^{(q)} = \left\langle \tau' \left| \frac{r^{(q)}}{a_0} \right| \tau \right\rangle \quad , \quad q = 0, \pm 1 \quad , \tag{6.1}$$

where $r^{(q)} = \sqrt{(4\pi/3)}\, r Y_{1,q}(\vartheta, \varphi)$ stands for the spherical component of the position vector $\boldsymbol{r}$, and a_0 is the Bohr radius. Important quantities derived from the dipole matrix element are the dipole strength

$$d_{\tau'\tau}^{(q)} = \left| p_{\tau'\tau}^{(q)} \right|^2 \quad , \tag{6.2}$$

the oscillator strength

$$f_{\tau'\tau}^{(q)} = \left(\frac{\hbar\omega}{E_\infty} \right) \left| p_{\tau'\tau}^{(q)} \right|^2 \quad , \tag{6.3}$$

($\hbar\omega = E_{\tau'} - E_\tau$, and E_∞ is the Rydberg energy), and the transition probability

$$w_{\tau'\tau}^{(q)} = \frac{1}{3} w_0 \left(\frac{\hbar\omega}{E_\infty} \right)^3 \left| p_{\tau'\tau}^{(q)} \right|^2 \quad , \tag{6.4}$$

with $w_0 = \alpha^5 m_e c^2 / 2\hbar \approx 8.03 \times 10^9\,\mathrm{s}^{-1}$.

The polarization behaviour and the angular distributions are the same as for the σ- and π-components of the normal Zeeman effect. This means that the radiation is linearly polarized in $\Delta m = 0$ transitions (angular distribution proportional to $\sin^2 \vartheta$), and circularly polarized in $\Delta m = \pm 1$ transitions (angular distribution proportional to $(1 + \cos^2 \vartheta)$).

We remind the reader that by virtue of the scaling laws valid in a magnetic field discussed in detail in Sect. 2.3 the dipole strengths, oscillator strengths and transition rates in hydrogenic ions with nuclear charge Z and finite core mass m_+ can be related to those computed for $Z = 1$ and $m_+ \to \infty$.

The evaluation of the dipole matrix element (6.1) for the *spherical case*, where the initial and the final states are described by wave functions of the form

$$\Phi_{m\pi}(\boldsymbol{r}) = \sum_l \frac{1}{r} f_{lm}(r) Y_{lm}(\vartheta, \varphi) \quad , \tag{6.5}$$

is easily carried out applying the well-known integral properties of products of spherical harmonics. The results read (with primed quantities referring to the final state, the fixed quantum number m is omitted)

$$p^{(q)}_{\tau'\tau} = \sum_l \left[\left(\frac{2l+1}{2l+3} \right)^{1/2} (l,0;1,0|l+1,0)(l,m;1,q|l+1,m+q) \right.$$

$$\times \int_0^\infty dr\, f'_{l+1}(r) r f_l(r) + \left(\frac{2l+1}{2l-1} \right)^{1/2} (l,0;1,0|l-1,0)$$

$$\left. \times (l,m;1,q|l-1,m+q) \int_0^\infty dr\, f'_{l-1}(r) r f_l(r) \right] \quad , \quad q = 0, \pm 1 \quad . \tag{6.6}$$

In the *cylindrical case*

$$\Phi_{m\pi}(\boldsymbol{r}) = \sum_n g_{nm}(z) \phi^{\mathrm{Lan}}_{nm}(\varrho, \varphi) \quad . \tag{6.7}$$

use can be made of the relations

$$r^{(+1)} \phi^{\mathrm{Lan}}_{nm} = -(2\beta)^{-1/2} \left[(n+1)^{1/2} \phi^{\mathrm{Lan}}_{n+1,m+1} + (n-m)^{1/2} \phi^{\mathrm{Lan}}_{n,m+1} \right]$$

$$r^{(-1)} \phi^{\mathrm{Lan}}_{nm} = -(2\beta)^{-1/2} \left[n^{1/2} \phi^{\mathrm{Lan}}_{n-1,m-1} + (n-m+1)^{1/2} \phi^{\mathrm{Lan}}_{n,m-1} \right] \tag{6.8}$$

(which can be proved, for example, with the help of Appendix A in Canuto and Ventura 1977), and of the orthogonality of the Landau wave functions to yield:

$$p^{(0)}_{\tau'\tau} = \sum_n \int_{-\infty}^{+\infty} dz\, g'_n(z) z g_n(z) \quad , \tag{6.9a}$$

$$p^{(+1)}_{\tau'\tau} = -(2\beta)^{-1/2} \sum_n \left[(n+1)^{1/2} \int_{-\infty}^{+\infty} dz\, g'_{n+1}(z) g_n(z) \right.$$

$$\left. + (n-m)^{1/2} \int_{-\infty}^{+\infty} dz\, g'_n(z) g_n(z) \right] \quad , \tag{6.9b}$$

$$p_{\tau'\tau}^{(-1)} = -(2\beta)^{-1/2} \sum_n \left[n^{1/2} \int_{-\infty}^{+\infty} dz\, g'_{n-1}(z) g_n(z) \right.$$
$$\left. + (n - m + 1)^{1/2} \int_{-\infty}^{+\infty} dz\, g'_n(z) g_n(z) \right] \quad . \quad (6.9c)$$

In actual calculations, the integrations over r or z remaining in (6.6) or (6.9) were performed through a Simpson algorithm inserting into (6.6) and (6.9) the numerically gained wave functions $f_{lm}(r)$ of (6.5) and $g_{nm}(z)$ of (6.7).

When one works with the diagonalization method, the radial functions $f_l(r)$ in the spherical expansion are obtained as linear combinations of the Laguerre functions (3.8),

$$\frac{1}{r} f_l(r) = \sum_n a_{nl} G_{nl}^{(\zeta)}(r) \quad . \tag{6.10}$$

The radial integration, in (6.5), over the basis functions can be carried out in closed analytical form,

$$\int_0^\infty G_{n'l'}^{(\zeta)}(r) r G_{nl}^{(\zeta)}(r) r^2 dr =$$

$$\begin{cases} (-1)^{n+n'} \frac{1}{\zeta} \binom{2}{n'-n} \sqrt{\frac{n'!(2l+n+2)}{n!(2l+n')!}} & n'-2 \le n \le n'\, ,\ l' = l-1 \\ (-1)^{n+n'} \frac{1}{\zeta} \binom{2}{n-n'} \sqrt{\frac{n!(2l+n'+4)}{n'!(2l+n+2)!}} & n' \le n \le n'+2\, ,\ l' = l+1 \\ 0 & \text{otherwise} \end{cases} \tag{6.11}$$

so that the dipole strengths are ultimately only algebraic (quadratic) forms of the eigenvector components a_{nl} of the initial and final state, respectively.

Note that (6.11) requires both initial and final states to be described by basis functions with the *same* value of the length scale parameter ζ. In the numerical calculations this is the case when the transitions are from a Rydberg state to another Rydberg state or from a low-lying state to another low-lying state. The situation is more involved when, in laboratory field strengths, transitions from *low-lying* to *Rydberg* states are considered, as the numerical treatment requires, for feasible basis sizes, the use of ζ values that differ by almost two orders of magnitude. However, since the low-lying states remain practically unaffected by the field at these field strengths one can work with the *field-free* hydrogenic functions and expand them analytically, for arbitrary values of ζ, in terms of the basis (3.8). We list the eigenvector components of the field-free $1s$, $2s$ and $2p$ states for arbitrary ζ, which are necessary for evaluating the dipole strengths of Lyman and Balmer transitions to Rydberg states (with given ζ) in laboratory field strengths:

$R_{1s}(r) = 2e^{-r}$:

$$\left\langle G_{n0}^{(\zeta)} \middle| R_{1s} \right\rangle_r = 2\zeta^{-\frac{3}{2}} \left(\frac{2\zeta}{2+\zeta} \right)^3 \sqrt{(n+2)(n+1)} \left(\frac{2-\zeta}{2+\zeta} \right)^n \quad , \tag{6.12a}$$

$R_{2s}(r) = \dfrac{1}{\sqrt{2}} e^{-\frac{1}{2}r} \left(1 - \dfrac{r}{2}\right)$:

$$\left\langle G_{n0}^{(\zeta)} \middle| R_{2s} \right\rangle_r = \qquad (6.12b)$$

$$\dfrac{1}{\sqrt{2}} \zeta^{-\frac{3}{2}} \dfrac{(2\zeta)^3}{(1+\zeta)^5} \sqrt{(n+2)(n+1)} \left(\dfrac{1-\zeta}{1+\zeta}\right)^{n-1} [(1-\zeta)(\zeta-2) + 2n\zeta] \quad ,$$

$R_{2p}(r) = \dfrac{1}{2\sqrt{6}} r e^{-\frac{1}{2}r}$:

$$\left\langle G_{n1}^{(\zeta)} \middle| R_{2p} \right\rangle_r = \qquad (6.12c)$$

$$\dfrac{1}{2\sqrt{6}} \zeta^{-\frac{5}{2}} \left(\dfrac{2\zeta}{1+\zeta}\right)^5 \sqrt{(n+4)(n+3)(n+2)(n+1)} \left(\dfrac{1-\zeta}{1+\zeta}\right)^n \quad .$$

6.2 Effects of the Finite Proton Mass on the Transition Matrix Element

We now discuss the modifications arising in the electromagnetic transition probabilities due to the finiteness of the proton mass. In contrast to the field-free situation these modifications are by no means obvious in the presence of a magnetic field on account of the coupling between the collective and the internal motion.

Starting point is the Hamiltonian for the two-body problem in a uniform external magnetic field extended by the interaction with the quantized radiation field described by the vector potential operator

$$\boldsymbol{A}_{\text{rad}}(\boldsymbol{r}) = \sum_{\boldsymbol{k},s} \sqrt{\dfrac{\hbar}{2\varepsilon_0 \omega_k V}} \left(\boldsymbol{e}_s e^{i\boldsymbol{k}\cdot\boldsymbol{r}} a_{\boldsymbol{k},s} + \boldsymbol{e}_s^* e^{-i\boldsymbol{k}\cdot\boldsymbol{r}} a_{\boldsymbol{k},s}^\dagger\right) \quad , \qquad (6.13)$$

(where V is the normalization volume, $\boldsymbol{e}_s$ ($s = 1, 2$) are the two orthogonal polarization unit vectors and $\boldsymbol{k}$ is the wave number vector of the photon, $a_{\boldsymbol{k},s}$ and $a_{\boldsymbol{k},s}^\dagger$ are the photon annihilation and creation operators, respectively) which has the form

$$H = \dfrac{1}{2m_+} \left[\boldsymbol{p}_+ - \dfrac{1}{2} e\boldsymbol{B} \times \boldsymbol{r}_+ - e\boldsymbol{A}_{\text{rad}}(\boldsymbol{r}_+)\right]^2$$

$$+ \dfrac{1}{2m_-} \left[\boldsymbol{p}_- + \dfrac{1}{2} e\boldsymbol{B} \times \boldsymbol{r}_- + e\boldsymbol{A}_{\text{rad}}(\boldsymbol{r}_-)\right]^2 + V(r) \quad . \qquad (6.14)$$

Introducing, as in Sect. 2.2.1, the center-of-mass coordinate $\boldsymbol{R}$ and the relative coordinate $\boldsymbol{r} = \boldsymbol{r}_- - \boldsymbol{r}_+$ the wave functions Ψ_i and Ψ_f for the initial and the final state of the atom are given by (2.11), i.e. the atomic states are characterized by

6.2 Effects of the Finite Proton Mass on the Transition Matrix Element

the generalized momenta K_i and K_f and the internal wave functions Φ_i and Φ_f. Considering the process of the emission of a photon with quantum numbers k, s, the Hamiltionian of the matter-radiation interaction sandwiched with the initial and final photon state yields

$$H_{\text{int}}^{k,s} = \sqrt{\frac{\hbar}{2\varepsilon_0 \omega V}} \left[\frac{e}{m_-} e_s^* \cdot (p_- + \frac{1}{2} eB \times r_-) e^{-ik\cdot r_-} \right.$$
$$\left. - \frac{e}{m_+} e^{-ik\cdot r_+} e_s^* \cdot (p_+ - \frac{1}{2} eB \times r_+) \right] . \quad (6.15)$$

To obtain the final form of the transition matrix element the expression (6.15) is further sandwiched with the atomic wave functions Ψ_i and Ψ_f. For the evaluation it is necessary to insert in (6.15) the coordinates R and r leading to

$$H_{\text{fi}}^{k,s} = e\sqrt{\frac{\hbar}{2\varepsilon_0 \omega V}} \int \frac{1}{V} e^{i(K_i - K_f - k)\cdot R}$$
$$\Phi_f^*(r) e_s^* \cdot \left\{ e^{-i\frac{m_+}{M} k\cdot r} \left[\frac{1}{M} \hbar K_i + \frac{1}{m_-} p + \frac{e}{2m_-}(B \times r) \right] \right. \quad (6.16)$$
$$\left. + e^{i\frac{m_-}{M} k\cdot r} \left[-\frac{1}{M} \hbar K_i + \frac{1}{m_+} p - \frac{e}{2m_+}(B \times r) \right] \right\} \Phi_i(r) d^3 R\, d^3 r .$$

From the integration over R one sees immediately that it is the generalized momentum which enters into the law of momentum conservation, as mentioned already in Sect. 2.1. Expression (6.16) simplifies essentially for the dipole approximation ($k \cdot r \ll 1$) since then the two different exponential factors can be replaced by 1:

$$H_{\text{fi}}^{k,s} = e\sqrt{\frac{\hbar}{2\varepsilon_0 \omega V}} \delta_{K_i, K_f + k} \quad (6.17)$$
$$\left\langle \Phi_f \left| e_s^* \cdot \left[\frac{1}{m_-}(p + \frac{1}{2} eB \times r) + \frac{1}{m_+}(p - \frac{1}{2} eB \times r) \right] \right| \Phi_i \right\rangle .$$

Using the commutator relation

$$\frac{i}{\hbar} \left[H_{\text{rel}}^{(K)}, r \right] = \frac{1}{m_-}(p + \frac{1}{2} eB \times r) + \frac{1}{m_+}(p - \frac{1}{2} eB \times r) \quad (6.18)$$

for the internal Hamiltonian $H_{\text{rel}}^{(K)}$ of (2.13) we arrive at the very simple formula

$$H_{\text{fi}}^{k,s} = e\sqrt{\frac{\hbar}{2\varepsilon_0 \omega V}} \frac{i}{\hbar}(E_f - E_i)\langle \Phi_f | e_s^* \cdot r | \Phi_i \rangle , \quad (6.19)$$

which is formally identical to the field-free expression. In the last step a further approximation had to be used, namely that the wave functions Φ_i and Φ_f are eigenfunctions of $H_{\text{rel}}^{(K)}$ with the same value of K, i.e. we have assumed that $K_f \approx K_i$, which is well justified in atomic transitions. Finally, with $E_f - E_i = \hbar\omega$ the transition probability then contains the usual length formula of the dipole matrix element.

Apart from small effects due to rescaling with the reduced mass as discussed in Sect. 2.3 the essential effect of the finite proton mass arises for $\Delta m \neq 0$ transitions due to the changes of the energies caused by the proton cyclotron motion, already discussed in Sect. 2.3.2. (This can quantitatively be seen in Fig. 6.2, p.76.)

The derivation given above for dipole transitions can be performed, in principle, for higher multipole transitions, too. The resulting formulae, however, do not possess the simple structure of the field-free case. We omit a detailed discussion since, in practice, such transitions are not important.

6.3 Results

It is evident that our approach to determining the wave functions would ultimately provide the exact solutions of the full Schrödinger equation, if infinitely many expansion terms could be included in (6.5) or (6.7). However, the rapid convergence of the results with increasing number n_c of expansion terms considered permits us to gain accuracies that are entirely sufficient for all practical purposes using numbers of configurations which are still numerically feasible. As a typical example, Fig. 6.1 shows the convergence behaviour of the dipole matrix elements of the transition $2p_0/0\ 0\ 1 \rightarrow 1s_0/0\ 0\ 0$ as a function of $1/n_c$ in different ranges of the magnetic field strength.

It is recognized that even in the transitional region ($\beta = 1$), where both methods can reasonably be applied, the results of the spherical computation rapidly approach the asymptotic value using only a few configurations, while the cylindrical computation does not exhibit a similar saturation behaviour yet, and thus here a genuine extrapolation to $1/n_c \rightarrow 0$ is still requisite. The curves belonging to the cylindrical computation, however, evidently run more flatly once β is increased well above unity. (Note, in this context, the change of the vertical scale of the $\beta = 1$ diagram in Fig. 6.1 by a factor of 200). All in all the analysis of the convergence behaviour suggests that a four-digit accuracy can be guaranteed for the dipole matrix elements, and thus for all the derived quantities, with the exception, perhaps, of the transitional region, where in a few cases the fourth digit may become uncertain, at worst.

In Table A3.1 of the Appendix 3 we have listed the wavelengths, dipole strengths, oscillator strengths, and transition rates of all transitions possible between states with the field-free principal quantum numbers $n_p \leq 3$ for the hydrogen atom in magnetic fields spanning the range from $\beta = 10^{-4}$ to 10^3. In the table, the states are characterized by their asymptotic ($\beta \rightarrow 0$ and $\beta \rightarrow \infty$) quantum numbers and every block entry gives, from top to bottom, the wave length λ (in Ångstroms), the dipole strength $d_{\tau'\tau}$ (in units of a_0^2), the oscillator strength $f_{\tau'\tau}$, and the rate $w_{\tau'\tau}$ (in units of w_0) of the respective transition. The wavelength is easily converted into the photon energy of the transition by aid of the relation $\hbar\omega/E_\infty = (\lambda/911.27\text{Å})^{-1}$ (alternatively, $\hbar\omega$ can be extracted directly from the energy values of the states). It must be emphasized that in

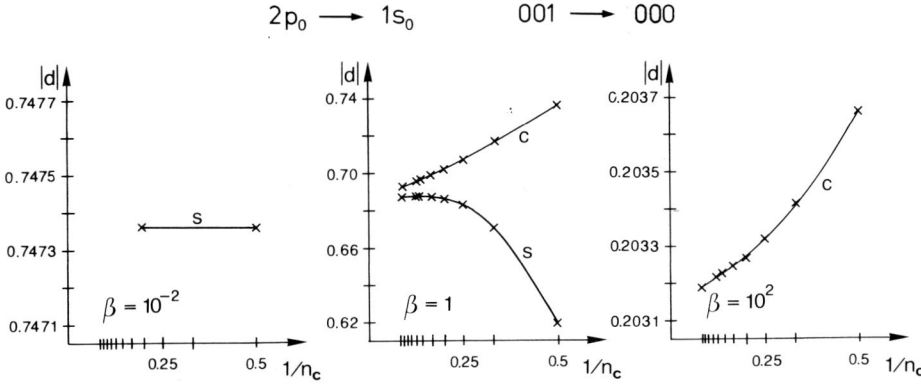

Fig. 6.1. A typical example illustrating the convergence behaviour of the dipole matrix elements of a transition as a function of the number n_c of terms included in the spherical (S) or cylindrical (C) expansion of the wave function in three different ranges of the magnetic field strength. The states are characterized by their asymptotic ($\beta = 0$ and $\beta \to \infty$) quantum numbers n_p, l, m and n, m, ν, respectively.

calculating the wavelengths, oscillator strengths and transition rates presented in Table A3.1 account has been taken of the finiteness of the proton mass, which causes relative shifts between states of different m by multiples of the proton cyclotron energy ($\hbar\Omega_c \approx 29.6\,\text{eV}\,\beta/10^3$) that become sizeable, in particular, in the high-field regime (cf. Sect. 2.3.2 and Sect. 6.2). As a consequence, even states with $m < 0$ belonging to the lowest Landau level may be raised into the range of positive energy values for sufficiently large β; accordingly transitions in Table A3.1 have been labelled by a star at the numerical value of the wavelength once at least one of the two states involved has reached the continuum. The same labelling applies to the transitions where states are raised to positive energies primarily by transverse excitation into higher Landau levels. Table A3.1 confirms the argument given by Wunner et al. (1983a) that for $\beta \gg 1$ the oscillator strengths of $\Delta m = -1$ transitions between neighbouring Landau levels should behave like $f^{-1}_{n-1\,m-1\,\nu'\,nm\nu} \approx -2n\delta_{\nu'\nu}$, whence the "cyclotron" transition $nm\nu \to n-1\,m-1\,\nu$ should almost completely exhaust the $\Delta m = -1$ sum rule of oscillator strengths; the numerical values in Table A3.1 indeed exhibit this behaviour.

The results presented in Table A3.1 of Appendix 3 are the most accurate and complete ones published so far in the literature. They are the basis for the calculation of line spectra of hydrogenic systems in arbitrary magnetic field strengths. Furthermore they can serve to test the accuracy of calculations that used approximation schemes of one sort or another.

In order to gain a pictorial impression of the global behaviour of electromagnetic transitions of the hydrogen atom over the complete range of magnetic field strengths, in Fig. 6.2 oscillator strengths for transitions between low-lying states of the hydrogen atom are presented as continuous functions of the magnetic field strength.

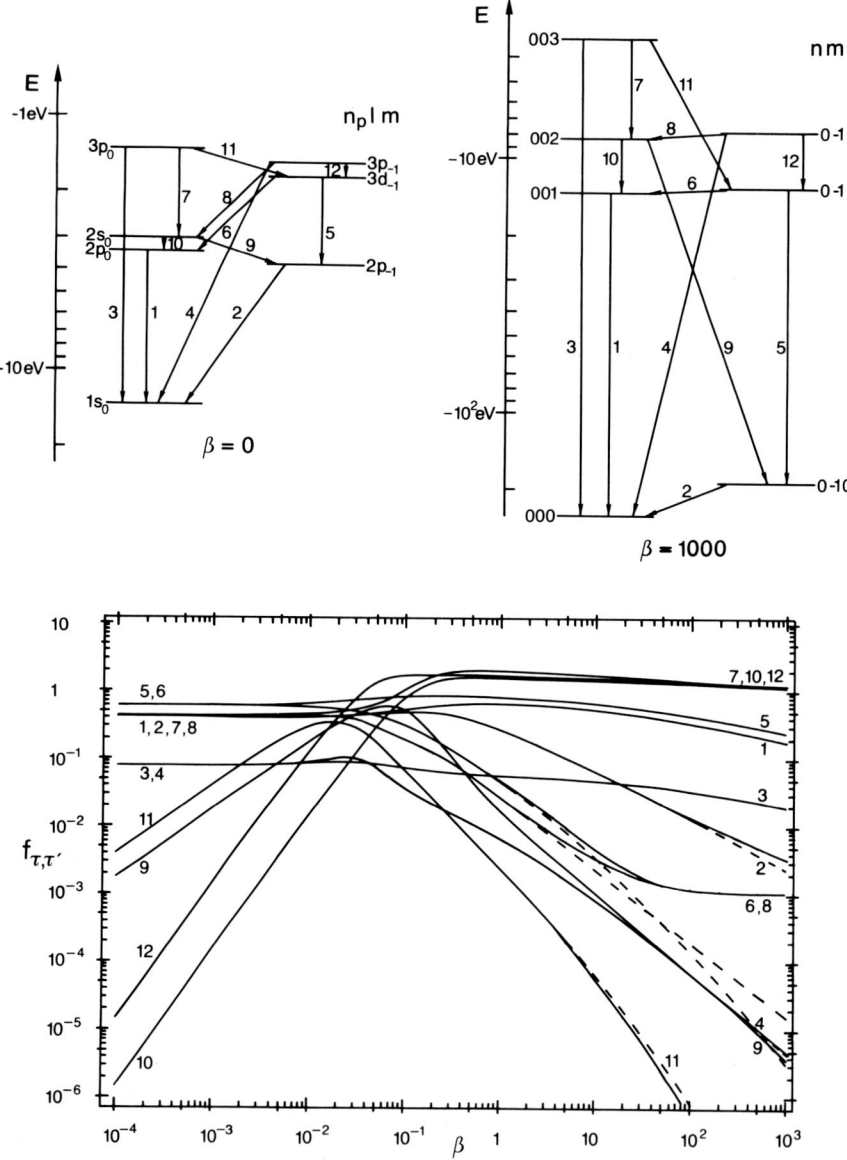

Fig. 6.2. Oscillator strengths for transitions between low-lying states of the hydrogen atom (with finite proton mass) as continuous functions of the magnetic field strength. Broken curves represent the behaviour for infinite proton mass. All transitions have been taken into account which are allowed between the levels shown to scale in the upper half of the figure for $\beta = 0$ and $\beta = 10^3$. Levels degenerate at $\beta = 0$ have been split weakly for the sake of clarity. The level scheme for $\beta = 10^3$ has been drawn for infinite proton mass; to obtain the correct scheme for the hydrogen atom, the $m = -1$ ladder has to be shifted upward by the proton cyclotron energy ($\approx 29.6\text{eV}$ at $\beta = 10^3$). It is recognized that the behaviour of the oscillator strengths greatly changes in the range $10^{-2} \leq \beta \leq 10^{-1}$, resulting in a complete rearrangement in the high-field regime.

6.3 Results

We restrict ourselves to the states with $m = 0$ and $m = -1$ shown for $\beta = 0$ and $\beta = 10^3$ in the upper half of Fig. 6.2 in level schemes that correspond to each other, and to all dipole transitions possible between them. In the lower half of Fig. 6.2 the oscillator strengths of these 12 transitions are plotted as functions of the magnetic field strength. The figure shows that in the weak-field regime the oscillator strengths of those transitions which are also allowed at $B = 0$ (transitions 1–8) deviate only slightly from the field-free values, while the oscillator strengths of the other transitions (9–12) increase with simple power laws ($\propto \beta$ for the $\Delta m = -1$ transitions 9 and 11, and $\propto \beta^2$ for the $\Delta m = 0$ transitions 10 and 12), in correspondence to the removal of the energy degeneracy of these states. In the high-field regime one first notices the $\Delta m = 0$ transitions (7,10,12) between hydrogen-like states, which possess oscillator strengths greater than one; the f-values of transitions *within* an m-band to the respective tightly bound state (1,3,5) are, compared with these, smaller, they also decrease, however, only weakly with the magnetic field strength. In contrast to that, the oscillator strengths of transitions *between* m-bands exhibit, from the transitional region on, a relatively steep descent, which is understandable in terms of the shrinking linear extensions of the states perpendicular to the field expressed by the $\beta^{-1/2}$-dependence in the $\Delta m = \pm 1$ dipole matrix elements (6.9b,c). Superimposed on this is, of course, the β-dependence of the overlaps entering into (6.9b,c) and, in proceeding to the oscillator strengths, that of the energy differences. As regards the overlaps, one finds a quickly improving quasi-orthogonality of longitudinal wave functions belonging to different m (whence, in adiabatic approximation $|p^{(\pm 1)}_{0\,m\pm 1,\nu';0,m,\nu}|^2 \simeq \delta_{\nu\nu'}/2\beta \cdot \{^{|m|}_{|m-1|}\}$, see Sect. 6.4), which is ultimately responsible for the rapid suppression of ν-non-conserving $\Delta m \neq 0$ transitions in the high-field regime (transitions 4, 9, 11 in Fig. 6.2). The energy differences, on the other hand, are eventually ($\beta \gg 1$) dominated by the raise of the $m = -1$ relative to the $m = 0$ band by the proton cyclotron energy, $\hbar\Omega_c = 4\beta E_H m_e/m_p$, whence the oscillator strengths of ν-conserving $\Delta m \neq 0$ transitions asymptotically tend to

$$f^{\pm 1}_{0,m\pm 1,\nu;0,m,\nu}(\infty) = 2m_e/m_p \cdot \{^{|m|}_{|m-1|}\} \simeq 1.089 \times 10^{-3} \{^{|m|}_{|m-1|}\} \quad . \tag{6.20}$$

Fig. 6.2 shows that in the $\Delta m = +1$, $\Delta \nu = 0$ transitions 6 and 8 between hydrogen-like states deviations from the case $m_p \to \infty$ (dashed lines) become important at $\beta \geq 10$, and that the asymptotic behaviour is already assumed for $\beta \geq 500$, while in the transition 2 evidently larger values of β are required for the proton cyclotron energy to dominate the energetic lowering of the tightly bound states ($\propto (\ln \beta)^2$), and for the f-value of the transition to converge to the asymptotic value. It is also seen that not even the increase of the proton cyclotron energy linear with β is capable of preventing the evanescence of the $\Delta m \neq 0$ transitions 4, 9, and 11 discussed before. As a final remarkable feature of Fig. 6.2 we note that the switch-over from the Coulomb-dominated to the magnetic-field-dominated regime occurs in a rather narrow region around $\beta = 0.1$, where all oscillator strengths differ from each other only by roughly one order of magnitude.

78   6 Electromagnetic Transition Probabilities

With the values tabulated in the Appendices every reader has the means at his disposal to determine, by quadratic interpolation, the energies and dipole strengths of transitions between low-lying states of the hydrogen atom for any value of the magnetic field strength up to 4.7×10^9 T, with an accuracy that will be sufficient for all practical purposes. In addition, by virtue of the Z scaling laws it is thus possible to also cover quantitatively the complete isoelectronic sequence of hydrogen in intense magnetic fields. If values for other magnetic field strengths or further states are needed the reader is recommended to contact the authors.

6.4 Expressions for Electromagnetic Transitions in the Adiabatic Approximation

For magnetic fields with $B \geq 10^7$ T, which are important in neutron star physics, the ansatz for the wave functions in the form of a product of a Landau state and a longitudinal wave function is an excellent approximation. Therefore, it is worthwhile presenting the relevant expressions for this adiabatic approximation (cf. Wunner and Ruder 1980a).

6.4.1 Dipole Strengths

One easily checks that, in spite of photon energies up to the order of 200 eV, in transitions between states with the same Landau quantum number n the dipole approximation is still justified. For transitions between the hydrogen-like states the argumentation is the same as in the field-free case, and in transitions to the m ground states the typical inverse wave numbers $k^{-1} = \lambda/2\pi$ are of the order of 20 a_0, which is still much larger than the linear extensions of the ground state wave functions. We note that the latter conclusion is no longer true in the infinite-field limit, where the ratio of the z-extension of the ground state and the photon wavelength diverges as $\ln \beta$ (the ground state energies increase as $(\ln \beta)^2$, the z-extension of the wave functions decreases only as $(\ln \beta)^{-1}$; cf. Loudon 1959). Because of the large Landau level distances (of the order of some keV) for the magnetic field strengths considered here, the dipole approximation possesses only limited validity for transitions between states with different n (the error being of the order of 1% for $\beta = 50$, and of the order of 20% for $\beta = 1000$ in transitions with $\Delta n = \pm 1$).

In adiabatic approximation the wave functions (6.7) are simple products of a Landau state $\phi^{\text{Lan}}_{nm}(\varrho, \varphi)$ and a longitudinal wave function $g_{nm\nu}(z)$ with ν nodes and the general expressions (6.9) for the dipole strengths $|p^{(q)}_{\tau'\tau}|^2$ reduce to

$$\left|p^{(0)}_{n'm'\nu',nm\nu}\right|^2 = \delta_{m'm}\delta_{n'n}\left|\langle g_{nm\nu'}|z/a_0|g_{nm\nu}\rangle\right|^2 \;, \tag{6.21a}$$

$$\left|p^{(+1)}_{n'm'\nu',nm\nu}\right|^2 = \delta_{m'm+1}\left[\delta_{n'n}\left|\langle g_{nm+1\nu'}|g_{nm\nu}\rangle\right|^2(n-m)\right.$$
$$\left.+\delta_{n'n+1}\left|\langle g_{n+1m+1\nu'}|g_{nm\nu}\rangle\right|^2(n+1)\right]/2\beta \quad , \tag{6.21b}$$

$$\left|p^{(-1)}_{n'm'\nu',nm\nu}\right|^2 = \delta_{m'm-1}\left[\delta_{n'n}\left|\langle g_{nm-1\nu'}|g_{nm\nu}\rangle\right|^2(n+1-m)\right.$$
$$\left.+\delta_{n'n-1}\left|\langle g_{n-1m-1\nu'}|g_{nm\nu}\rangle\right|^2 n\right]/2\beta \quad . \tag{6.21c}$$

Although the dipole approximation may no longer be valid for $\Delta n \neq 0$ transitions, we have listed the general relations (6.21a–c), because in discussing sum rules the formal contributions of these transitions will have to be taken into account.

6.4.2 Selection Rules

From (6.21a–c) the usual selection rules for dipole transitions may immediately be read off:

$$\Delta m = 0, \pm 1 \quad . \tag{6.22}$$

Longitudinal states $g_{nm\nu}$ with ν even have positive z-parity (i.e., they are symmetric functions of z), those with ν odd have negative z-parity. Hence it also follows from (6.21a–c) that in $\Delta m = 0$ transitions the initial and the final state must have opposite z-parity, whereas in $\Delta m = \pm 1$ transitions they must have the same z-parity. The possible $\Delta n = 0$ transitions of the lowest states of a given m-band which are compatible with these selection rules are schematically shown in Fig. 6.3.

6.4.3 Sum Rules

The sum rule for $\Delta m = 0$ transitions has the same form as without magnetic field:

$$\sum_{\tau'} f^{(0)}_{\tau'\tau} = 1 \quad , \tag{6.23}$$

whereas for $\Delta m = \pm 1$ transitions the sum rules are modified by the presence of the magnetic field (Hasegawa and Howard 1961),

$$\sum_{\tau'} f^{(\pm 1)}_{\tau'\tau} = 1 \pm \langle\tau|\frac{1}{\hbar}L_z|\tau\rangle \pm \langle\tau|(\varrho/a_{\rm L})^2|\tau\rangle \tag{6.24}$$

(L_z is the z-component of the angular momentum operator). The third term on the right-hand side of (6.24), which is negligible in weak magnetic fields, must be fully taken account of for the magnetic field strengths considered here. Evaluating (6.23) and (6.24) with the wave functions in adiabatic approximation $g_{nm\nu}(z)\phi^{\rm Lan}_{nm}(\varrho,\varphi)$, it follows from the definition (6.3) and from (6.21a–c) that we have contributions to (6.23) only from $n' = n$, and to (6.24) from $n' = n, n \pm 1$. The second and third term on the right-hand side of (6.24) can be determined analytically. Thus one finally obtains

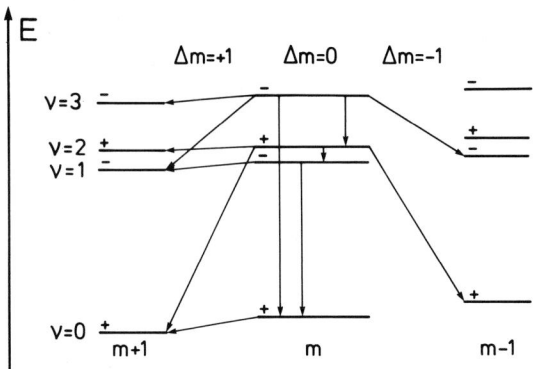

Fig. 6.3. Allowed transitions of the lowest states within a given m-band. States with even z-parity are labeled by $+$, those with odd z-parity by $-$. Transitions leading to the m-band are not shown. To obtain all allowed transitions the picture has to be repeated for every value of m.

$$\sum_{n'\nu'} f^{(q)}_{n'm+q\nu',nm\nu} = \begin{cases} 1 & \text{for } q = 0 \quad (n' = n) & (6.25a) \\ 2n+2 & \text{for } q = +1 \quad (n' = n, n+1) & (6.25b) \\ -2n & \text{for } q = -1 \quad (n' = n, n-1) \end{cases} . \quad (6.25c)$$

It must be noted, however, that the relations (6.25) hold only approximately (the error being of the same order as the error of the adiabatic approximation), since the manipulations which enter into the derivation of (6.23) and (6.24) are valid only for the exact wave functions of the problem. The evaluation of (6.25) therefore provides a useful test on the quality of the adiabatic approximation.

For the dipole strengths, sum rules can be formulated as well. As an immediate consequence of the completeness of the states we have

$$\sum_{n'\nu'} |p^{(q)}_{n'm+q\nu',nm\nu}|^2$$

$$= \begin{cases} \langle g_{nm\nu}|(z/a_0)^2|g_{nm\nu}\rangle & \text{for } q = 0 \quad (n' = n) & (6.26a) \\ (2n+1-m)/2\beta & \text{for } q = \pm 1 \quad (n' = n, n+q) \end{cases}, \quad (6.26b)$$

where in deriving (6.26b) again the properties of the Landau wave functions have been used. In contrast to (6.25), the relations (6.26) are exact, because only the completeness of the states is needed, and therefore (6.26) can be employed to test the accuracy of the numerical wave functions.

6.4.4 Asymptotic Formulae

For $\beta \gg 1$ and sufficiently large (odd) ν the longitudinal wave functions $g_{nm\nu}$ become independent of n and m and are given in very good approximation by

$$g_{nm\nu}(z) = \sqrt{2}z R_{n_p 0}(|z|) \quad (\beta \gg 1, \ \nu \gg 1, \text{odd}, \ n_p = (\nu+1)/2) \quad , \quad (6.27)$$

where

6.4 Electromagnetic Transitions in the Adiabatic Approximation

$$R_{n_p 0}(r) = 2n_p^{-3/2} \exp(-r/n_p) \, F(-n_p + 1, 2, 2r/n_p) \tag{6.28}$$

is the radial function of the s-state with principal quantum number n_p of the field-free hydrogen atom and F is the confluent hypergeometric series. As was mentioned already, (6.27) becomes exact for $\beta \to \infty$.

Let us consider the matrix element $\langle g_{nm\nu}|z|g_{nm0}\rangle$. Since for $\beta \gg 1$ the ground states g_{nm0} are strongly concentrated about the origin, it is only the behavior of $g_{nm\nu}$ in the corresponding neighbourhood of $z = 0$ that is essential to the calculation of this matrix element. Expanding (6.28) in powers of r/n_p, we find

$$R_{n_p 0}(r) \approx 2n_p^{-3/2}\left[\left(1 - \frac{2r}{1!2!} + \frac{(2r)^2}{2!3!} \mp \ldots\right) + O\left(\frac{r}{n_p}\right)\right] \;, \tag{6.29}$$

which means that for $n_p \gg 1$ the r-dependence of the states $R_{n_p 0}$ is practically independent of n_p for $r \ll n_p$. This is understandable because the energies of these states are almost the same, and therefore the wave functions differ from each other only for large r. Using (6.27), (6.28), and (6.29), we obtain

$$\langle g_{nm\nu}|z|g_{nm0}\rangle \approx (\nu + 1)^{-3/2} c_{nm} \quad (\beta \gg 1, \; \nu \gg 1, \text{odd}) \;, \tag{6.30}$$

with

$$c_{nm} = 4 \int_{-\infty}^{+\infty} g_{nm0}(z)\left(1 - \frac{2|z|}{1!2!} + \frac{(2|z|)^2}{2!3!} \mp \ldots\right) z^2 dz \;. \tag{6.31}$$

The constants c_{nm} can be calculated with the numerical wave functions g_{nm0}.

From (6.30) we immediately gain the following asymptotic formulae for $\Delta m = 0$ transitions from the ground state of an m-band for $\beta \gg 1$, and $\nu \gg 1$, odd:

dipole strengths:

$$d^{(0)}_{nm\nu,nm0} \approx c_{nm}^2 (\nu + 1)^{-3} \;, \tag{6.32}$$

oscillator strengths:

$$f^{(0)}_{nm\nu,nm0} \approx \left|\frac{E_{nm0}}{E_\infty}\right| c_{nm}^2 (\nu + 1)^{-3} \;, \tag{6.33}$$

transition rates:

$$w^{(0)}_{nm\nu,nm0} \approx \frac{1}{3} w_0 \left|\frac{E_{nm0}}{E_\infty}\right| c_{nm}^2 (\nu + 1)^{-3} \;. \tag{6.34}$$

6.4.5 Results and Discussion

Here we provide tabular material for the electromagnetic transitions of the hydrogen atom in strong magnetic fields, which we have obtained using numerical wave functions in the adiabatic approximation, and discuss the results. In the composition of the tables we closely follow the manner of presentation chosen by Bethe (1933) in his classical review article about the field-free hydrogen atom (see also Bethe and Salpeter 1957).

Table 6.1 shows the dipole strengths of the transitions with $\Delta m = 0$ to the ground states of the m-bands ($n = 0$, $m = 0, -1, \ldots, -4$, $\nu = 0$) for a magnetic field strength of 2.35×10^7 T ($\beta = 50$). It will turn out that these are the predominant transitions in strong magnetic fields. Because of the z-parity selection rules only final states with odd ν occur in Table 6.1.

Table 6.1. Dipole strengths for $\Delta m = 0$ transitions from the tightly bound states ($\nu = 0$) of the H atom for $\beta = 50$ ($n = 0$)

initial state final state	$m = 0, \nu = 0$ $m' = 0, \nu'$	$m = -1, \nu = 0$ $m' = -1, \nu'$	$m = -2, \nu = 0$ $m' = -2, \nu'$	$m = -3, \nu = 0$ $m' = -3, \nu'$	$m = -4, \nu = 0$ $m' = -4, \nu'$
$\nu' = 1$	0.06366	0.12314	0.17011	0.21042	0.24646
$\nu' = 3$	0.00554	0.00931	0.01182	0.01374	0.01531
$\nu' = 5$	0.00155	0.00255	0.00319	0.00367	0.00406
$\nu' = 7$	0.00064	0.00105	0.00131	0.00150	0.00166
$\nu' = 9$	0.00033	0.00053	0.00066	0.00076	0.00084
$\nu' = 11$	0.00019	0.00031	0.00038	0.00044	0.00048
$\nu' = 13$	0.00012	0.00019	0.00024	0.00028	0.00030
$\nu' = 15 - \infty$	0.00036	0.00059	0.00073	0.00084	0.00092
asympt. formula	$\dfrac{0.33}{(\nu'+1)^3}$	$\dfrac{0.53}{(\nu'+1)^3}$	$\dfrac{0.66}{(\nu'+1)^3}$	$\dfrac{0.76}{(\nu'+1)^3}$	$\dfrac{0.84}{(\nu'+1)^3}$
discr. spectr.	0.07239	0.13767	0.18844	0.23165	0.27003
cont. spectr.	0.04759	0.05222	0.05317	0.05322	0.05298
total	0.11998	0.18989	0.24161	0.28487	0.32301

For $\nu = 1$–13, the dipole strengths have been calculated using the numerical wave functions. The factors in the asymptotic formulae have been obtained from (6.31) with the corresponding ground state wave functions. Comparing with the values calculated numerically for $\nu = 9, 11, 13$, it is seen that the asymptotic $(\nu + 1)^{-3}$ behavior is already reached at $\nu = 9$ for the m-values considered here. Table 6.1 also provides the sum of the dipole strengths of the $\Delta m = 0$ transitions into the continuous spectrum. The latter has been obtained as the difference between the total sum of the dipole strengths according to (6.26a), and the contribution from the discrete spectrum. The right-hand side of the sum rule (6.26a) has been evaluated with the numerical ground state wave functions.

The accuracy of the calculated matrix elements can be estimated by looking at the sum rule for $\Delta m = -1$ transitions from the m ground state, which are given for $\beta = 50$ in Table 6.2. Evidently the sum rule (6.26b) is almost completely exhausted ($\approx$ 99%) by the transition $(m, \nu = 0) \to (m - 1, \nu = 0)$. Since all further contributions to the sum (6.26b) are nonnegative and negligibly small, this indicates that the essential matrix elements are accurate at least within 1%.

Table 6.2. Dipole strengths for $\Delta m = -1$ transitions from the tightly bound states ($\nu = 0$) of the H atom for $\beta = 50$ ($n = 0$)

initial state final state	$m = 0, \nu = 0$ $m' = -1, \nu'$	$m = -1, \nu = 0$ $m' = -2, \nu'$	$m = -2, \nu = 0$ $m' = -3, \nu'$	$m = -3, \nu = 0$ $m' = -4, \nu'$
$\nu' = 0$	9.820×10^{-3}	1.990×10^{-2}	2.993×10^{-2}	3.994×10^{-2}
$\nu' = 2$	2.2×10^{-5}	1.7×10^{-5}	1.4×10^{-5}	1.2×10^{-5}
$\nu' = 4$	4.1×10^{-6}	3.2×10^{-6}	2.6×10^{-6}	2.3×10^{-6}
$\nu' = 0 - 4$	9.846×10^{-3}	1.992×10^{-2}	2.994×10^{-2}	3.996×10^{-2}
total	1.0×10^{-2}	2.0×10^{-2}	3.0×10^{-2}	4.0×10^{-2}

Comparing the orders of magnitudes of $\Delta m = 0$ dipole strengths with the ones of $\Delta m \neq 0$ transitions (for $\Delta m = +1$ transitions, all the dipole strengths between states with $n = 0$ are smaller than 3×10^{-5} and are therefore not explicitly given in a table), it is seen that in a strong magnetic field the dipole strengths for $\Delta m = 0$ are considerably larger. The physical reason for this is that in $\Delta m = \pm 1$ transitions the transverse atomic dipole contributes, i.e., the linear extension of the atom perpendicular to the magnetic field is important, which is reduced with respect to the longitudinal extension by the appropriate factor. For a magnetic field strength of 4.70×10^8 T ($\beta = 1000$) the dipole strengths are shown in Tables 6.3 and 6.4.

Table 6.3. Dipole strengths for $\Delta m = 0$ transitions from the tightly bound states ($\nu = 0$) of the H atom for $\beta = 1000$ ($n = 0$)

initial state final state	$m = 0, \nu = 0$ $m' = 0, \nu'$	$m = -1, \nu = 0$ $m' = -1, \nu'$	$m = -2, \nu = 0$ $m' = -2, \nu'$	$m = -3, \nu = 0$ $m' = -3, \nu'$	$m = -4, \nu = 0$ $m' = -4, \nu'$
$\nu' = 1$	0.01061	0.01958	0.02676	0.03300	0.03863
$\nu' = 3$	0.00112	0.00196	0.00259	0.00311	0.00356
$\nu' = 5$	0.00032	0.00056	0.00074	0.00088	0.00100
$\nu' = 7$	0.00014	0.00023	0.00031	0.00036	0.00042
$\nu' = 9$	0.00007	0.00012	0.00016	0.00019	0.00021
$\nu' = 11 - \infty$	0.00014	0.00024	0.00031	0.00037	0.00042
asympt. formula	$\dfrac{0.07}{(\nu' + 1)^3}$	$\dfrac{0.12}{(\nu' + 1)^3}$	$\dfrac{0.15}{(\nu' + 1)^3}$	$\dfrac{0.18}{(\nu' + 1)^3}$	$\dfrac{0.21}{(\nu' + 1)^3}$
discr. spectr.	0.01226	0.02269	0.03085	0.03790	0.04423
cont. spectr.	0.02757	0.03381	0.03721	0.03949	0.04117
total	0.03983	0.05650	0.06806	0.07739	0.08540

Table 6.4. Dipole strengths for $\Delta m = -1$ transitions from the tightly bound states ($\nu = 0$) of the H atom for $\beta = 1000$ ($n = 0$)

initial state final state	$m=0, \nu=0$ $m'=-1, \nu'$	$m=-1, \nu=0$ $m'=-2, \nu'$	$m=-2, \nu=0$ $m'=-3, \nu'$	$m=-3, \nu=0$ $m'=-4, \nu'$
$\nu'=0$	4.952×10^{-4}	9.972×10^{-3}	1.498×10^{-3}	1.998×10^{-3}
$\nu'=2$	2.6×10^{-7}	2.0×10^{-7}	1.7×10^{-7}	1.4×10^{-7}
$\nu'=4$	4.6×10^{-8}	3.5×10^{-8}	2.9×10^{-8}	2.6×10^{-8}
$\nu'=0-4$	4.955×10^{-4}	9.974×10^{-4}	1.498×10^{-3}	1.998×10^{-3}
total	5.0×10^{-4}	1.0×10^{-3}	1.5×10^{-3}	2.0×10^{-3}

The features discussed for $\beta = 50$, especially the dominance of $\Delta m = 0$ transitions, are found to be even more pronounced. Comparing Tables 6.1 and 6.3, it is seen that, in going from $\beta = 50$ to $\beta = 1000$, the dipole strengths of the $\Delta m = 0$ transitions to the ground states of the m-bands decrease by roughly a factor of 4 to 6. This is mainly due to the fact that the atomic z-dipole, which is important in $\Delta m = 0$ transitions, is scaled down by the stronger concentration of the $\nu = 0$ wave functions about the origin for increasing β. As a consequence, the relative contribution of the discrete spectrum to the $\Delta m = 0$ dipole sum rule also decreases with the magnetic field strength.

Turning to a discussion of the magnetic field dependence of the dipole strengths for $\Delta m = -1$ transitions, we note that, according to (6.21c),

$$|p^{(-1)}_{n'=0, m-1, \nu'; n=0, m, \nu}|^2 = \langle g_{0,m-1,\nu'} | g_{0,m,\nu} \rangle|^2 (|m|+1)/(2\beta) \quad .$$

For $\Delta \nu = 0$, the overlap of the wave functions is almost unity due to the similarity of the effective potentials for m and $m-1$, and therefore the magnetic field dependence is practically given by the factor $1/\beta$ alone. The comparison of Tables 6.2 and 6.4 shows that this asymptotic behavior is already reached fairly well for the magnetic field strengths considered here. For $\Delta \nu \neq 0$, the overlap is very small, and vanishes in the limit $\beta \to \infty$. Because of this β dependence of the overlap, the dipole strengths of $\Delta \nu \neq 0$ transitions decrease more strongly than $1/\beta$.

The magnetic field dependence of the dipole strengths enters of course into the β dependences of both the oscillator strengths and the transition rates. There, however, also the magnetic field dependence of the energy differences contributes. In particular, the increasing binding energies of the $\nu = 0$ states can counteract the decrease of the dipole strengths in such a way that there is an increase in the rates of $\Delta m = 0$ transitions to the tightly bound states for the range of β values considered here.

In Tables A3.2a–c and A3.3a–c, the oscillator strengths of $\Delta m = 0, \pm 1$ transitions are presented for $\beta = 50$ and $\beta = 1000$. We find the same characteristic properties as for the dipole strengths. Because of the energy weighting, the continuous spectrum contributes to the $\Delta m = 0$ sum rule to a larger extent. As mentioned, the sum rules (6.26) can be used to test the quality of the adiabatic approximation. Especially suitable for this purpose is the $\Delta m = -1$

sum rule. Looking at Tables A3.2b and A3.3b, it is found that the employed adiabatic approximation wave functions fulfill the sum rule with an accuracy of the order of 10^{-2} for $\beta = 50$, and 10^{-3} for $\beta = 1000$. In Tables A3.2c and A3.3c for the $\Delta m = +1$ transitions the large discrepancy between the summed oscillator strengths and the value of the sum rule (6.25b) is striking. This is easily explained by the fact that, in contrast to the $\Delta m = 0, -1$ transitions, here transitions to states with $n' = 1$ are involved (cf. (6.21b)). The dipole strengths for these transitions are $|\langle g_{1m+1\nu'}|g_{0m\nu}\rangle|^2/(2\beta)$, which for $\nu = \nu'$ approximately yields $1/(2\beta)$. Since, on the other hand, the Landau level distance is 4β, obviously it is the transition $(n = 0, m, \nu) \to (n' = 1, m+1, \nu)$ which exhausts the oscillator sum rule almost completely.

Finally in Tables A3.4a–d ($\beta = 50$) and Tables A3.5a–d ($\beta = 1000$) we present the transition probabilities for dipole emission of the states ($n = 0$, $m = 0, \ldots, -4$, $\nu = 0, \ldots, 5$) to all lower states ($n' = 0, m', \nu'$) in units of 10^8 s^{-1}. Entries which are left blank in Tables A3.4 and A3.5 refer to transitions which are not allowed in emission for energetic reasons or are forbidden by the selection rules discussed above. Absorption rates can be obtained from the tabulated values in the usual way by simply interchanging initial and final states. In Tables A3.4 and A3.5, in the column "sum" the summation over ν', in the column "total" over all final states ($n' = 0, m', \nu'$), has been performed. The latter is the total decay constant, the reciprocal value of which is equal to the lifetime (last column) of the initial state ($n = 0, m, \nu$) with respect to dipole emission. It turns out that states with odd z-parity ($\nu = 1, 3, \ldots$) are much more short-lived than states with even z-parity ($\nu = 0, 2, \ldots$), which possess lifetimes comparable with the ones of the field-free hydrogen atom. This is because states with odd z-parity can perform a fast $\Delta m = 0$ transition to the corresponding ground state, whereas states with odd z-parity can only perform transitions either with low energy to hydrogen-like states or with $\Delta m = \pm 1$ to the neighbouring ground states for which the dipole strengths are strongly reduced. Again, the effect is even more enhanced for $\beta = 1000$. Also it should be noted that the m ground states (except $m = 0$) have finite lifetimes because of the $\Delta m = +1$ transition to the neighbouring $(m + 1)$ ground state. With the help of the tables, the decay of a given excited state ($n = 0, m, \nu$) can be traced in a very simple way. States with odd z-parity almost exclusively perform a $\Delta m = 0$ transition to the m ground state, and then decay successively via $\Delta m = +1$ transitions to the $m = 0$ ground state. States with even z-parity first perform a hydrogen-like $\Delta m = 0$ transition with z-parity change, and then decay as described above.

For the purposes of illustration all transitions for $n = 0$; $m = 0, -1, -2$; $\nu \leq 5$ with rates larger than 10^8 s^{-1} are drawn in Fig. 6.4 for $\beta = 1000$. Figure 6.5 shows, again for $\beta = 1000$, the photon spectrum associated with the transitions.

The intensities have been obtained assuming for every state an average occupation number of one electron. Of course, the actual intensities in the photon spectrum in a given physical situation will depend on the manner in which the excited states are populated. One recognizes in Fig. 6.5 in the range

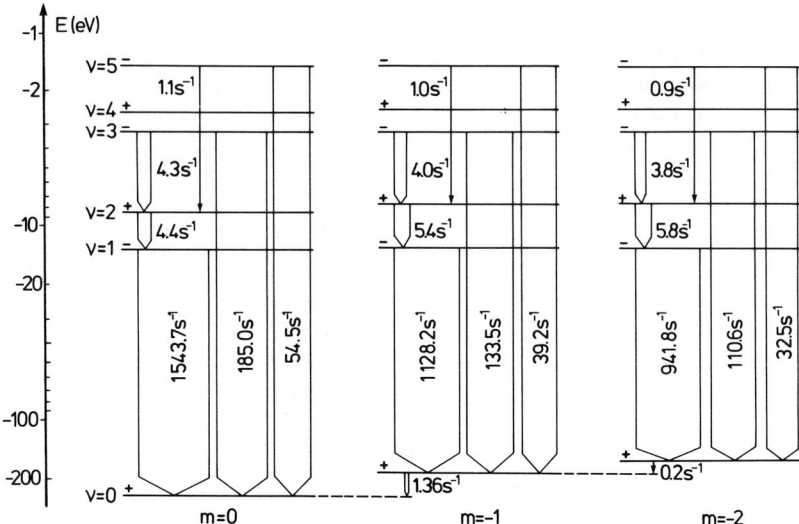

Fig. 6.4. Energy levels and transition probabilities for the hydrogen atom for $\beta = 1000$ ($n = 0$; $m = 0, -1, -2$; $\nu = 0 - 5$). The signs $+$ and $-$ refer to the even and odd z-parity of the states, respectively. All transitions with rates $w \gtrsim 10^8\,\mathrm{s}^{-1}$ are shown. The numbers give the values of w in units of $10^8\,\mathrm{s}^{-1}$; the arrow widths are proportional to the logarithms of these numbers.

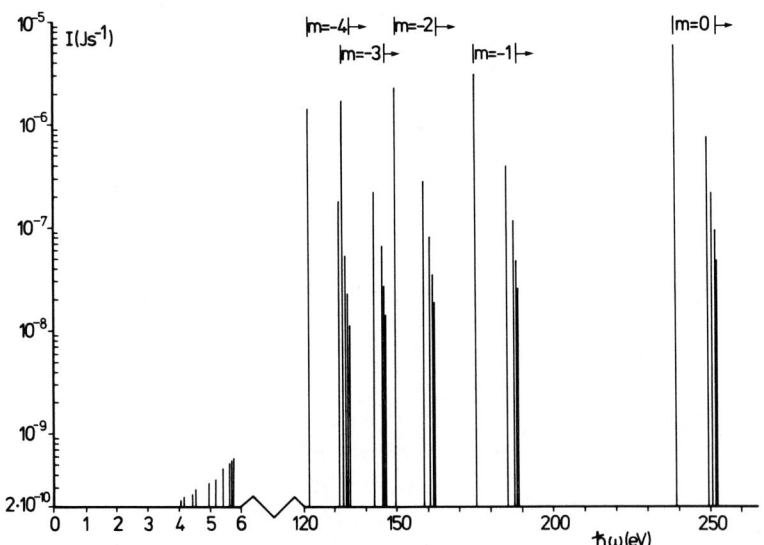

Fig. 6.5. Photon spectrum for the hydrogen atom for $\beta = 1000$ ($n = 0$). The intensities are obtained assuming an averaged occupation number of one electron for every state. All transitions shown are $\Delta m = 0$ transitions. Lines at high energies correspond to transitions to the respective m-band ground state, the arrows indicate the onset of the continua. Lines at low energies correspond to transitions between hydrogen-like states. The only transition with $\Delta m \neq 0$ and $I > 2 \times 10^{10}\,\mathrm{Js}^{-1}$ is from $(m = -1, \nu = 0) \to (m' = 0, \nu' = 0)$ ($\hbar\omega = 63.7\,\mathrm{eV}, I = 13.8 \times 10^{-10}\,\mathrm{Js}^{-1}$), and is not explicitly shown.

from 120 to 250 eV the $\Delta m = 0$ transitions to the m ground states grouped according to their different m-values. The position of the continuous spectra are indicated by arrows. In the low energy range (of a few eV) the lines of $\Delta m = 0$ transitions (drawn for $m = 0, \ldots, -4$) between hydrogen-like states occur. The only transition with $\Delta m \neq 0$ which exceeds the threshold of 2×10^{-10} Js^{-1}, which we have set for Fig. 6.5, is the one from $(m' = -1, \nu' = 0) \to (m = 0, \nu = 0)$ (not shown in Fig. 6.5). Its energy is 63.7 eV, its intensity is 13.8×10^{-10} Js^{-1}, and thus is only somewhat larger than the intensities of the hydrogen-like lines.

Concluding this section, it should be remarked that we have discussed in detail transitions to the tightly bound states, since their existence constitutes the major modification with respect to the field-free case. On the other hand, the transitions between hydrogen-like states may also be of astrophysical importance, because they possess the structure of redshifted and blueshifted hydrogen series. A quantitative account of these transitions is given in Wunner and Ruder (1980b). Electromagnetic transitions for helium-like systems in magnetic fields of arbitrary strength will be treated in Chap. 9.

7 Stationary Lines and White Dwarf Spectra

7.1 Stationary Lines

One of the most spectacular applications of the calculations presented in the foregoing chapters was the identification of absorption features in the optical spectra of magnetized white dwarf stars, which had defied interpretation for almost 50 years, in terms of *stationary* components of hydrogen lines in magnetic fields of several 10^5 T (Angel et al. 1985; Greenstein et al. 1985; Wunner et al. 1985b; Schmidt et al., 1986a,b). By *stationary* components we mean those transitions whose wavelengths go through maxima or minima as functions of the magnetic field strength. These lines, between 300 nm and 1000 nm, are particularly well recognized if Fig. 4.2a is viewed sideways at flat angles. The fact that these transitions can produce sharp absorption features in white dwarf spectra is obvious when one considers that the magnetic field strength varies (in a dipolar geometry) by a factor of two across the white dwarf and thus all fast moving wavelengths are smeared out.

To do justice to the importance of stationary components for the quantitative interpretation of the spectra of strongly magnetized white dwarf stars, we present in the tables of Appendix A2.2 and the figures of Appendix A2.3 in detail the wavelengths and in the tables of Appendix A3.3 the dipole strengths of all important transitions which become stationary in the region of white dwarf magnetic field strengths at EUV, UV, optical, and IR wavelengths. In the tables the effects of the finiteness of the proton mass are taken into account for the values of the magnetic field strengths and the wavelengths (cf. the scaling laws discussed in Sect. 2.3). All states denoted by a prime are linear combinations of field-free degenerate states with different angular momenta. Therefore, all these states are no pure l-states and can perform, even in an arbitrarily small magnetic field, transitions with $|\Delta l| > 1$ such as $4f'_0 \rightarrow 2s_0$, which is forbidden for the pure $4f_0$ state. From the accuracy of the energy values computed with the spherical basis expansion (7–9 significant digits) we can conclude that the corresponding wavelengths can be guaranteed to within the numbers of digits quoted in the tables. Slight drops in accuracy can occur for $B \gtrsim 7 \times 10^4$ T when in the computation of the energies of the states involved in a transition the switch occurs from the spherical to the cylindrical basis set. From the last energy values actually computed in the cylindrical basis, and the extrapolated values, we can, however, derive rigorous upper and lower bounds on the wavelengths in these cases, which are both given in the tables at the appropriate places. Precision is quickly regained after the switch, but at any rate the accuracy achieved is sufficient to predict the position of stationary points also beyond 7×10^4 T. Since we concentrate on stationary features here, no values for wavelengths are given for magnetic fields in which the lines have become fast running.

Additionally in Table 7.1, for all corresponding Lyman, Balmer, Paschen, and Brackett transitions, the accurate values of the magnetic field strengths and the wavelengths as well as the type of the extrema are summarized. Paschen and Brackett transitions have been included since transitions from the Paschen and Brackett series are "blue"-shifted into the near infrared or even into the optical region in certain ranges of the magnetic field strength.

Table 7.1. Values of the magnetic fields and wavelengths of the extrema of the stationary components of Lyman α, Balmer α, Balmer β, Balmer γ, Paschenα, Paschen β and Brackett α

Transition	B in 10^2 T	β	λ in nm	Type
Lyman α				
$2p_{-1} \to 1s_0$	560	0.12	134.26	max
Balmer α				
$3p_{-1} \to 2s_0$	59.7	0.0127	708.8	max
	200	0.042	677.4	min
	$\sim 10^4$	~ 3	1460.0	max
$3p_0 \to 2s_0$	232	0.0494	583.0	min
	1200	0.26	676.3	max
$3d_{-2} \to 2p_{-1}$	117	0.0249	744.94	max
$3d_{-1} \to 2p_0$	616	0.131	853.23	max
	2420	0.515	844.47	min
	42500	9.05	947.7	max
$3s'_0 \to 2p_{+1}$	46	0.00979	699.28	max
	140	0.029	664.3	min
Balmer β				
$4p'_{-1} \to 2s_0$	11	0.0023	492.05	max
	447	0.095	376.15	min
	790	0.168	380.4	max
$4f'_{-1} \to 2s_0$	61	0.013	511.96	max
	370	0.079	440.8	min
	997	0.212	460.7	max
$4p'_0 \to 2s_0$	340	0.073	363.72	min
	757	0.161	369.0	max
$4f'_0 \to 2s_0$	310	0.066	412.37	min
	837	0.178	424.7	max
$4d_{-2} \to 2p_{-1}$	12.8	0.00272	493.05	max
$4d_{-1} \to 2p_0$	21.8	0.00464	497.06	max
$4s'_0 \to 2p_{+1}$	13	0.00216	491.71	max
	55.9	0.0119	454.9	min
Balmer γ				
$5p'_{-1} \to 2s_0$	3.9	0.00082	435.85	max
	465	0.099	335.3	min
	696	0.148	336.8	max
$5f'_{-1} \to 2s_0$	12.8	0.00272	439.09	max
	456	0.097	349.1	min
	733	0.156	351.2	max

Table 7.1. (Continued)

Transition	B in 10^2 T	β	λ in nm	Type
Balmer γ				
$5p'_0 \to 2s_0$	350	0.075	331.2	min
	710	0.151	334.06	max
$5f'_0 \to 2s_0$	350	0.074	342.5	min
	724	0.154	346.0	max
$5s'_0 \to 2p_{+1}$	26	0.0055	417.5	min
	34	0.0073	418.5	max
	92	0.0196	408.5	min
$5d'_0 \to 2p_{+1}$	92	0.0197	419.5	min
$5g'_0 \to 2p_{+1}$	53	0.0113	451.0	max
	70	0.0149	430.5	min
Paschen α				
$4p'_{-1} \to 3d'_0$	11.2	0.00238	1967.7	max
	379	0.0806	989.1	min
	1880	0.40	1324	max
$4p'_0 \to 3d'_0$	280	0.060	886.7	min
	1700	0.36	1170	max
Paschen β				
$5p'_{-1} \to 3d'_0$	4.0	0.00084	1297.1	max
	400	0.085	760.0	min
	1500	0.31	884.7	max
$5f'_{-1} \to 3d'_0$	13.7	0.00292	1327.99	max
	410	0.087	836.0	min
	1600	0.33	1000.7	max
$5p'_0 \to 3d'_0$	290	0.061	715.87	min
	1500	0.31	865.5	max
$5f'_0 \to 3d'_0$	290	0.061	773.0	min
	1500	0.32	955.3	max
Brackett α				
$5s'_0 \to 4f'_{-1}$	149	0.0317	1056.9	min
	362	0.0770	1184	max
$5d'_0 \to 4f'_{-1}$	150.0	0.03191	1132.18	min
	358	0.0762	1254.8	max
$5g'_0 \to 4f'_{-1}$	154.6	0.0329	1272.57	min
	354.6	0.0754	1378.79	max

To help visualize the behaviour of the wavelengths, the results for the stationary components of Lyman α, Balmer α, Balmer β, Balmer γ, Paschen α, Paschen β, and Brackett α are graphically summarized in Figs. A2.5–A2.8 of Appendix A2.3. A compilation of all these stationary lines in the optical and the near-infrared region is shown in Fig. 7.1. Out of all Lyman components it is only the transition between the states with the field-free quantum numbers $2p_{-1}$ and $1s_0$ which, for $B > 3 \times 10^3$ T, runs through a maximum (Fig. A2.5). Extrema below 3×10^3 T, which are far less pronounced, can still be calculated with high accuracy in the framework of the quadratic Zeeman effect.

Out of the 15 Balmer α components, 5 transitions exhibit stationary behaviour (Fig. 7.1 and Fig. A2.6a). Similar to Lyman α, the transition $3d_{-2} \to 2p_{-1}$ is one between states which, in the high-field limit, become *tightly bound*,

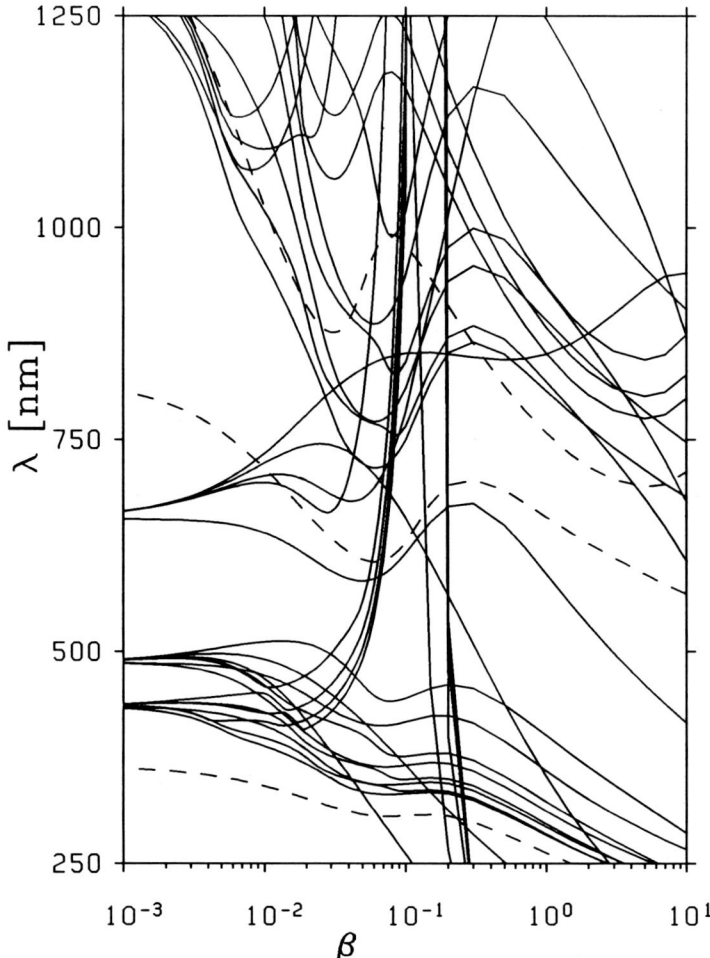

Fig. 7.1. The wavelengths of all transitions of Balmer α, β, γ, Paschen α, β, and Brackett α which become stationary in the optical and the near-infrared region as continuous functions of the magnetic field strength in the interval 4.7×10^2 T to 4.7×10^6 T. The continuum edges of the transitions are plotted as dashed curves. The second minima at $\beta \approx 5$ ($B \approx 2.35 \times 10^6$ T) in the Brackett α lines are caused by effects of the finite proton mass. For a detailed description of all lines see Table A2.2 of Appendix A2.2 and Figs. A2.5–A2.8 of Appendix A2.3.

and possesses only a maximum. The undulating behaviour in the wavelengths of the transitions $3p_0 \to 2s_0$ and $3p_{-1} \to 2s_0$ is essentially caused by slight wiggles in the magnetic field dependence of the energy of the $2s_0$ state at $\beta \approx 0.08$ and $\beta \approx 0.15$ ($B \approx 3.6 \times 10^4$ T and $B \approx 6.9 \times 10^4$ T) (see Table A1.2a, Table A1.3a, Fig. A1.1).

The second maximum of the transition $3p_{-1} \to 2s_0$ at infrared wavelengths (not shown in the figure) is due to effects of the finite proton mass. These are also the cause for the structure shown by the transition $3d_{-1} \to 2p_0$ at high fields. The transition $3s_0' \to 2p_{+1}$ does not possess a finite second maximum, since the energy levels of the two states intersect at $B \approx 6\times 10^4$ T. The situation is quite similar with Balmer β (Fig. A2.6b), where the undulating shape of the transitions $4p_{-1}' \to 2s_0$, $4f_{-1}' \to 2s_0$, $4p_0' \to 2s_0$, and $4f_0' \to 2s_0$ is again caused by the features in the energy of the $2s_0$ state. Compared with these, the maxima of the other three transitions lie at considerably smaller magnetic field strengths, and only the transition $4s_0' \to 2p_{+1}$ runs also through a minimum, before the wavelength diverges as a consequence of the intersecting energy levels (at $B \approx 8\times 10^4$ T). The behaviour of the energy of the $2s_0$ level determines to an ever increasing extent the stationary Balmer transitions from higher principal quantum numbers, as well as the form of the $2s_0$ Balmer jump (dashed line). This is illustrated by Fig. A2.6b for Balmer β and even more pronounced by Fig. A2.6c for Balmer γ. The wavelengths of all higher Balmer transitions lie between the wavelengths of the corresponding stationary components of Balmer γ and the $2s_0$ Balmer jump. The additional three stationary Balmer γ transitions $5g_0' \to 2p_{+1}$, $5d_0' \to 2p_{+1}$, and $5s_0' \to 2p_{+1}$ with extrema around $B \approx 5 \times 10^3$ T reflect the structure of the near anticrossings of the $5g_0'$, $5d_0'$, and $5s_0'$ states since the energy of the $2p_{+1}$ state is approximately constant in this region of the magnetic field strength. At higher magnetic fields the $2p_{+1}$ state intersects the $5l_0'$ states, the wavelengths of the three transitions go to infinity, and reappear as fast running lines.

Within the range of magnetic field strengths that is of importance to strongly magnetized white dwarfs, the $3d_0'$ state also exhibits wiggles in the energy dependence similar to those of $2s_0$. The corresponding $3d_0'$ Paschen jump (dashed line in Figs. A2.7a and A2.7b) extends far into the red region (minimum of 605 nm around 3×10^4 T), with the consequence that stationary points occur for Paschen α transitions in the near infrared, and for higher Paschen transitions to $3d_0'$ in the optical region. This situation is demonstrated explicitly for Paschen α in Fig. A2.7a and for Paschen β in Fig. A2.7b. To be complete, the three Paschen β transitions $5p_{+1}' \to 3d_{+2}$, $5p_0' \to 3s_0$, and $5p_{-1}' \to 3s_0$ with minima around $B \approx 5 \times 10^3$ T have been included in Fig. A2.7b. For $\beta \lesssim 1$ the stationary components of Brackett α in Fig. A2.8 can be interpreted in a completely analogous manner, the underlying state is here the $4p_{-1}'$ state. The minima at $\beta \approx 5$ are caused by the influence of the finite proton mass.

All the figures show that in the neighbourhood of most of the extrema the curves remain flat enough that even with a variation of the magnetic field strength by a factor of two across a white dwarf star distinct spectral features can be expected. The Tables A2.2 can be used to determine, for example by polynomial approximations, the exact values of the magnetic fields and the wavelenghts at the stationary points of the transitions shown in Figs. A2.5–A2.8. The numbers and the types of the extrema are given in Table 7.1. We note in this context that the values computed by Angel et al. (1985) without taking into account the effects of the finite proton mass become inaccurate for $B \geq 5 \times 10^4$ T and $|\Delta m| = 1$ in a degree that they are no longer useful.

Of course, a complete description of the spectra of hydrogen and hydrogenic ions in arbitrary magnetic fields not only requires the knowledge of the wavelengths but also of the strengths of the respective electromagnetic transitions. The necessary data go beyond what we have tabulated in Appendix A3.1 for transitions between states with $n \leq 3$, as higher principal quantum numbers are involved in Balmer β, γ, etc. and in the Paschen transitions. Therefore in Appendix A3.3 we also present separate tables for dipole strengths of important stationary lines of the Lyman, Balmer, and Paschen series computed from our accurate strong-magnetic-field wave functions. As an example, Fig. 7.2 shows the magnetic field dependences of energy levels and dipole matrix elements of stationary σ_- components of H_β and H_γ transitions. In Fig. 7.2 and in the rest of this chapter we use the "astronomical" unit of 1 Megagauss = 1 MG = 10^2 T for the magnetic field strength. One recognizes strong oscillations of the transition strengths in the range 30–500 MG, associated with avoided crossings and related changes in the character of the wave functions of the final states. This is, already at low principal quantum numbers, an indication of the well-known, extremely small, anticrossings which occur in Rydberg states in laboratory fields (Zimmerman et al., 1980). It is evident that even if the magnetic field of a white dwarf comes to lie in the neighbourhood of a stationary point there may yet be no observable feature if the dipole strength of the transition happens to pass through a zero around this magnetic field value.

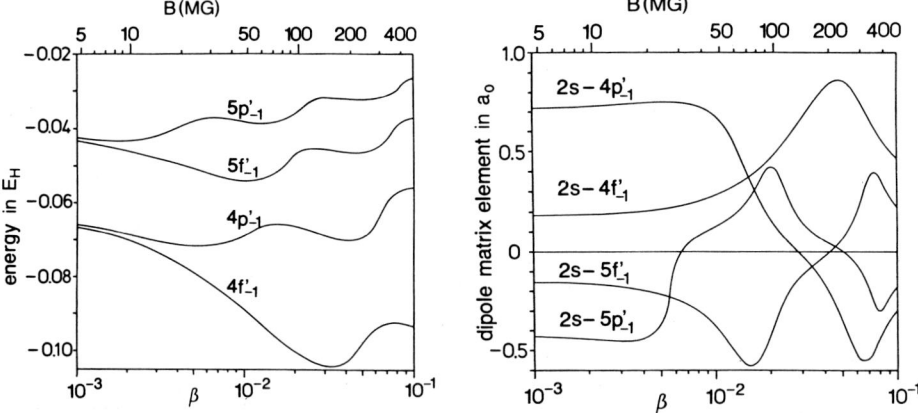

Fig. 7.2. Dipole matrix elements of stationary σ_- components of H_β and H_γ and energy levels of the H atom for the final states involved in the transitions in the range of magnetic field strength 4.7 to 470 MG (1 MG = 10^2 T). The avoided crossings of the energy levels are reflected in strong oscillations of the transition strengths.

7.2 Spectra of Selected Magnetic White Dwarfs

In this section we will elaborate on the progress made recently in analyses of the spectra of magnetic white dwarf stars owing to the existence of stationary lines in the hydrogenic spectrum. One particular white dwarf has played the rôle of a "Rosetta stone" in this development, and it is therefore appropriate to begin by retelling the story of this object.

7.2.1 Grw+70°8247

42 light years away in the stellar configuration Draco lies a 13 magnitude star long suspected of having a strong magnetic field. Its spectrum had consistently defied interpretation ever since the first shallow absorption features were discovered in the spectrum of this white dwarf star, known as Grw+70°8247, almost 50 years ago by Minkowski (1938). The features appeared at wavelengths that were completely inexplicable, and so Minkowski came to the conclusion that this was a most a unique object. The photographic spectrum taken by Greenstein and Matthews (1957), shown in Fig. 7.3a, displays two distinct features at $\lambda\lambda$ 3650 and 4135 Å, with sharp cut-offs at the blue side and wide extensions to the red side, while the CCD spectrum taken in 1974 by Angel et al. (1985), shown in Fig. 7.3b, exhibits a similarly shaped feature at 5850 Å, and two broad absorption features around 7000 Å and 8450 Å.

It had been speculated for a long time whether an interpretation was possible by molecular bands of C_2 or by exotic metallic lines, and when all else had failed Angel (1978) proposed an identification in terms of atomic hydrogen lines in strong magnetic fields using variational energy values fragmentarily available at that time. The circular polarization of its optical continuum (Kemp et al. (1970)), had in fact given a clue to the existence of a strong magnetic field in the vicinity of this object, but the lack of reliable quantum mechanical calculations for atoms in fields above $\sim$ 30 MG (the perturbation theory treatment of the quadratic Zeeman effect fails beyond 5–10 MG, and extensions by Kemic (1974a) are valid only up to $\sim$ 30 MG) presented a major obstacle to checking the identification in terms of magnetically strongly shifted atomic lines and pinning down the prevailing field strengths. This obstacle was only removed when the quantum mechanical problem of hydrogen atoms in magnetic fields of arbitrary strength could be solved conclusively.

Clearly, any attempt to observe, and resolve, a line spectrum of hydrogen at a given magnetic field strength in the intermediate regime is doomed to failure. As already mentioned in Sect. 4.2 an element of order is brought in even in this domain by the "stationary" transitions whose wavelengths go through maxima and minima in certain intervals of the magnetic field strength and thus approximately produce features around the maximum or minimum values of the wavelengths when the magnetic field varies across these intervals. Since the magnetic field of a white dwarf with a dipolar field distribution has a variation by a factor of two from the pole to the equator, this opens the possibility of well resolved "stationary" line features being produced even in these field

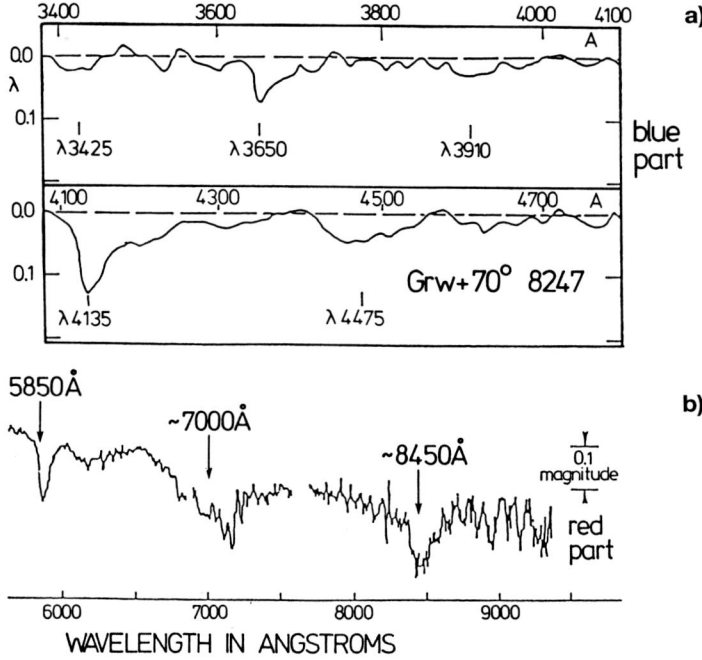

Fig. 7.3a,b. Optical spectrum of the white dwarf star Grw+70°8247 (a) blue part (Greenstein and Matthews (1957)), (b) red part (Angel et al. (1985)). The features – occurring at strange wavelengths for a white dwarf – had defied identification ever since their first discovery by Minkowski in 1938. Note in particular the band-head like features at 3650, 4135 and 5850 Å.

strengths. This in fact finally provided the key to the correct explanation of the shallow absorption features observed in Grw+70°8247, and it opened a new era of stellar atomic spectroscopy, the "spectroscopy of stationary lines".

In Fig. 7.4 and Fig. 7.5 we demonstrate the excellent agreement between the wavelength positions of the extrema of stationary components of H_α, H_β, H_γ and absorption features in the red and blue part of the spectrum of Grw+70°8247. In particular, the sharp blue edges and "red-shaded" extensions of the features at 3650 Å, 4135 Å, and 5800 Å are well accounted for by the minimum character of the corresponding stationary components. More complicated structures of other features are produced by blends of stationary components, such as the broad feature around 8500 Å to which a stationary transition of H_α and one of Paschen β contribute. Thus we have the result that – in terms of wavelengths and qualitative line shapes – all spectral features can consistently be explained by *stationary* lines of atomic hydrogen in a dipolar magnetic field with a polar field strength of $\sim$ 320 MG (Angel et al. 1985; Greenstein et al. 1985; Wunner et al. 1985b; Henry and O'Connell 1985). No other previously known white dwarf star had a magnetic field even one tenth this value.

Our interpretation so far relied solely on the behaviour of the wavelengths as a function of the field. But astronomers are more ambitious and want to

7.2 Spectra of Selected Magnetic White Dwarfs   97

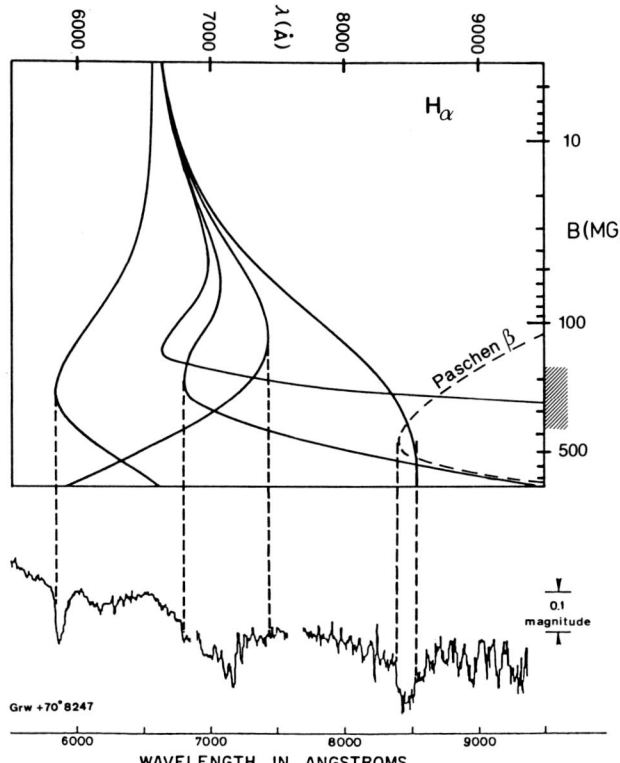

Fig. 7.4. Stationary H$_\alpha$ transitions of the hydrogen atom in magnetic fields from 4 to 700 MG in comparison with the red part of the optical spectrum of Grw+70°8247 (Angel et al. (1985)). The sharp blue edge of the feature at 5850 Å coincides with the wavelength minimum of a single stationary H$_\alpha$ transition, as indicated by the dashed line, and the red extension of the feature is explained by the variation of the wavelength around the minimum in an extended magnetic field whose strength varies from $\sim$ 160 to $\sim$ 320 MG (see the corresponding hatching along the B coordinate). Such a variation is present, e.g., in a dipolar field. The broader feature around 7000 Å is accounted for by a blend of two stationary H$_\alpha$ components in the same range of field, while even a strongly blue-shifted stationary Paschen β component contributes to the feature around 8450 Å.

actually calculate synthetic spectra, to be compared to the observed spectra in order to explore the precise emission and absorption conditions in these strongly magnetized objects. Of course, transitions rates have to be incorporated in the calculations of synthetic spectra, and it is evident from the strong dependence of the transition rates on the field strength (cf. Fig. 7.2) that these calculations provide a very sensitive tool indeed to probe the physical conditions prevailing in these objects. Figure 7.6 shows a synthetic spectrum of Grw+70°8247 calculated by Wickramasinghe and Ferrario (1988) assuming a centred dipolar field with a polar field strength of 320 MG, in comparison with the observed spectrum. The agreement is excellent and confirms the interpretation given above. However, it must be noted that in the calculations of line shapes parametrized

98   7 Stationary Lines and White Dwarf Spectra

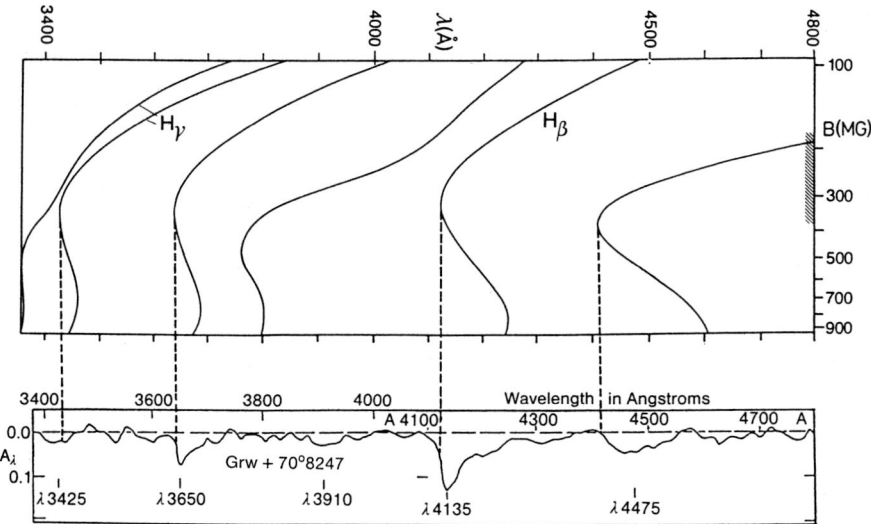

Fig. 7.5. Stationary H_β and H_γ components in magnetic fields from 100 to 900 MG in comparison with the short-wavelength part of the spectrum of Grw+70°8247 (Greenstein and Matthews (1957)). Again all the features are explained in a consistent way in terms of stationary transitions of hydrogen in the range of magnetic field found in Fig. 7.4 (compare hatching along the B-coordinate).

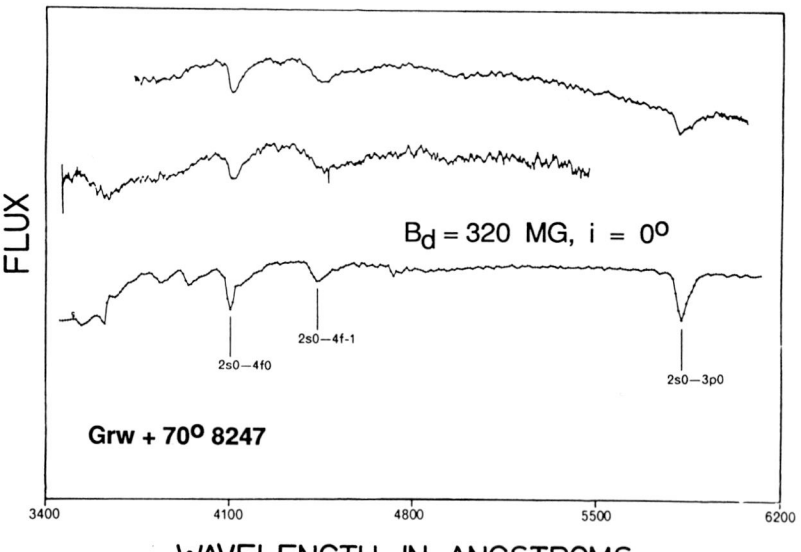

Fig. 7.6. Synthetic spectrum of Grw+70°8247 (*bottom curve*) calculated by Wickramasinghe and Ferrario (1988) in a dipole field model using the wavelength and bound-bound transition information of Appendix A2.2 and A3.3 in comparison with observed spectra (*two upper curves*). A polar field strength of 320 MG and an angle of inclination of 0° between the magnetic axis and the line of sight was assumed in the calculation.

Stark-broadened profiles are assumed lacking a more detailed theory for broadening of nondegenerate lines. This theory will require the knowledge of the shifts in energy of the lines split by the magnetic field as functions of an arbitrarily oriented electric field caused by the ions and electrons in the atmosphere. This problem is still far from being solved.

7.2.2 PG 1031+234

Unlike Grw+70°8247, whose spectrum shows no time variation, the magnetic degenerate PG 1031+234 (Schmidt et al. 1986a,b; Latter et al. 1987) rotates with a period of 3.4 hours and exhibits strongly asymmetric changes in the spectrum and the polarization during one period. Figure 7.7 shows the spectrum of this white dwarf star together with the computed behaviour of the wavelengths of the H atom as a function of field strength.

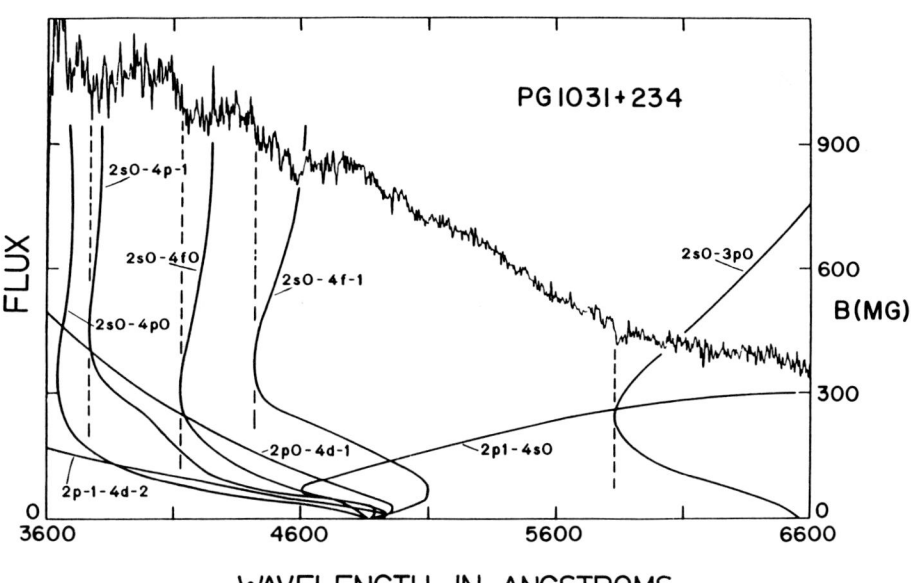

Fig. 7.7. Observed spectrum of the high field magnetic white dwarf PG 1031+234 (Schmidt et al. 1986b) at a given rotational phase, together with the computed behaviour of wavelengths as a function of field strength. A field strength ranging up to ~ 500 MG is deduced from the data for this object.

The comparison with Grw+70°8247 shows that the features attributed to the $2s_0 \rightarrow 4f_0$, $2s_0 \rightarrow 4f_{-1}$ and $2s_0 \rightarrow 3p_0$ transitions are slightly red-shifted indicating a somewhat higher field strength. From the positions of the stationary components and the observed phase modulations Schmidt et al. (1986a,b) concluded that this object possesses a nonaxisymmetric field morphology with an equivalent polar field strength of ~ 500 MG and a high-field "spot" with

~ 10^3 MG. This value currently represents the "world record" for the magnetic field strength of a white dwarf star. At the same it is the highest magnetic field in which hydrogen lines have ever been seen in nature.

7.2.3 SBS 1349+5434

This faint object with an apparent visual magnitude of $m_v = 16.4$, was discovered recently by Liebert et al. (1994) using the Multiple Mirror Telescope (MMT) of the University of Arizona. The combined circular polarization and MMT flux spectra are shown in Fig. 7.8 together with the calculated behaviour of wavelengths of prominent transitions of hydrogen as a function of field strength.

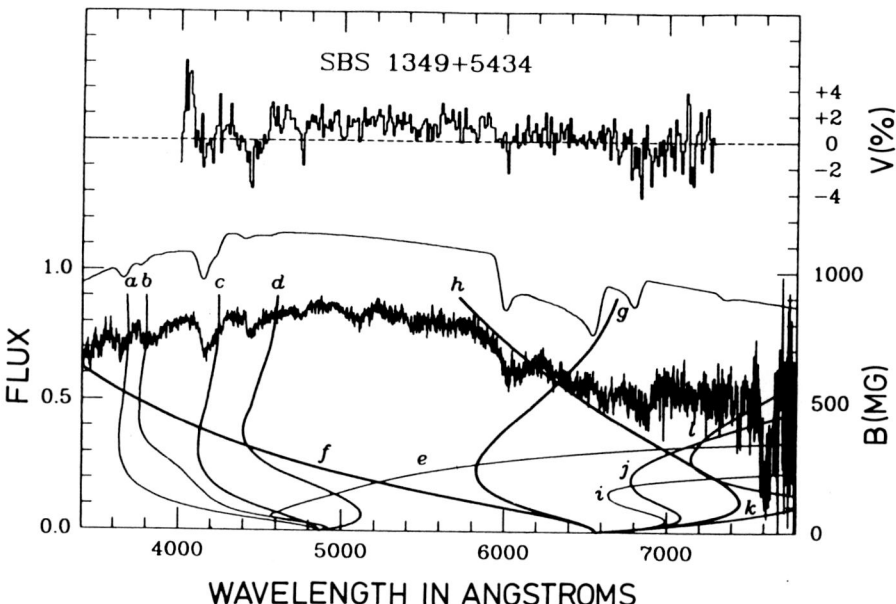

Fig. 7.8. Flux spectrum for SBS 1349+5434 and Stokes parameter V of the circular polarization compared to the calculated wavelengths of prominent hydrogen transitions in fields from 0–900 MG (H$_\beta$: (a) $2s_0 \to 4p_0$, (b) $2s_0 \to 4p_{-1}$, (c) $2s_0 \to 4f_0$, (d) $2s_0 \to 4f_{-1}$, (e) $2s_0 \to 4f_{-1}$; H$_\alpha$: (f) $2p_{\pm 1} \to 3d_{\pm 1}$, (g) $2s_0 \to 3p_0$, (h) $2p_{-1} \to 3d_{-2}$, (i) $2p_{+1} \to 3s_0$, (j) $2s_0 \to 3p_{-1}$, (k) $2p_0 \to 3d_{-1}$; P$_\beta$: $3d_0 \to 5p_0$ (l)). The smooth curve results from a simple spectral model for a dipolar field distribution with polar field strength 760 MG. The flux is given in units of 10^{-26} erg/(cm^2 s Hz). (From Liebert et al. 1994)

Obvious in the flux spectrum are the number of diffuse absorption features at unusual wavelengths, which at once implicate a magnetic degenerate object. From the presence of a sharp blue edge to the $2s_0 \to 4f_0$ feature (curve c), one can infer a considerable portion of the star with $200 \lesssim B \lesssim 500$ MG. The absence of a sharp *red* edge to that profile implies that the stellar field is generally weaker than 800 MG over the observed hemisphere, while the lack of any evidence of the red turnaround of $2s_0 \to 3p_{-1}$ (curve j) near 7050 Å confirms that the field strength certainly exceeds 50 MG. In a first model calculation Liebert et al. (1994) constructed synthetic flux spectra using the atomic data presented in this book. The model shown as the smooth spectrum in Fig. 7.8 is an example of an acceptable fit, with polar field strength $B = 760$ MG and an angle $i = 20°$ to the line of sight. The simple model already nicely reproduces both the locations and the profiles of features, especially $2s_0 \to 4f_0$ (c) and $2p_{-1} \to 3d_{-2}$ (h). In spite of the high polar field strength of 760 MG (a value exceeded among known white dwarfs by only the high-field "spot" on PG 1031+234), the circular polarization is unexpectedly weak, which can be interpreted as evidence for a substantially more complicated field pattern. Obviously comprehensive spectroscopic studies throughout the rotational cycle of the star will be required – in addition to the atomic data base presented in this book – to reveal this field pattern.

7.2.4 PG 1015+014

The inspection of Fig. 4.2a for the wavelengths shows that for fields in the range $\sim$ 50–100 MG, apart from the occurrence of a few stationary components, the optical region is sprayed with slow and moderately fast moving components of Balmer α and β which can have an impact on the spectrum even in the presence of a field spread. PG 1015+014 is an example of a white dwarf with a field in this range. Spectropolarimetric observations by Wickramasinghe and Cropper (1988) show that its spectrum and broad-band circular polarization varies with a period of 98.75 minutes. The spectrum at two rotational phases is shown in Fig. 7.9 together with the suggested identifications. The synthetic spectra shown were obtained by Wickramasinghe and Cropper (1988) for a centered dipole field distribution with a polar field strength of 100–120 MG viewed at angle $i = 75°$ to the dipole axis. Again the overall agreement is very good and confirms the presence of a magnetic field of this strength. More refined calculations of synthetic spectra may have to be extended to nonaxisymmetric field structures, for which there are evidences in the observed spectral variations during one period.

7.2.5 G227–35

This star is an example of a magnetic white dwarf in which the modulations of the flux spectrum are too weak to make possible an identification of stationary lines, but where in polarization spectra features can be detected that are due to stationary lines. Figure 7.10 shows the total unpolarized flux F_λ and the normalized Stokes parameter of circular polarization, V, as functions of the wavelength, measured for G227–35 by Cohen et al. (1993), in comparison with the behaviour of H_α and H_β components which have stationary points between 20 and 200 MG, or are changing slowly. The position of the prominent feature in the circular polarization near 7450 Å is in excellent agreement with the calculated stationary point of the $2p_{-1} \to 3d_{-2}$ component of H_α at 7449.4 Å, which occurs at 117 MG.

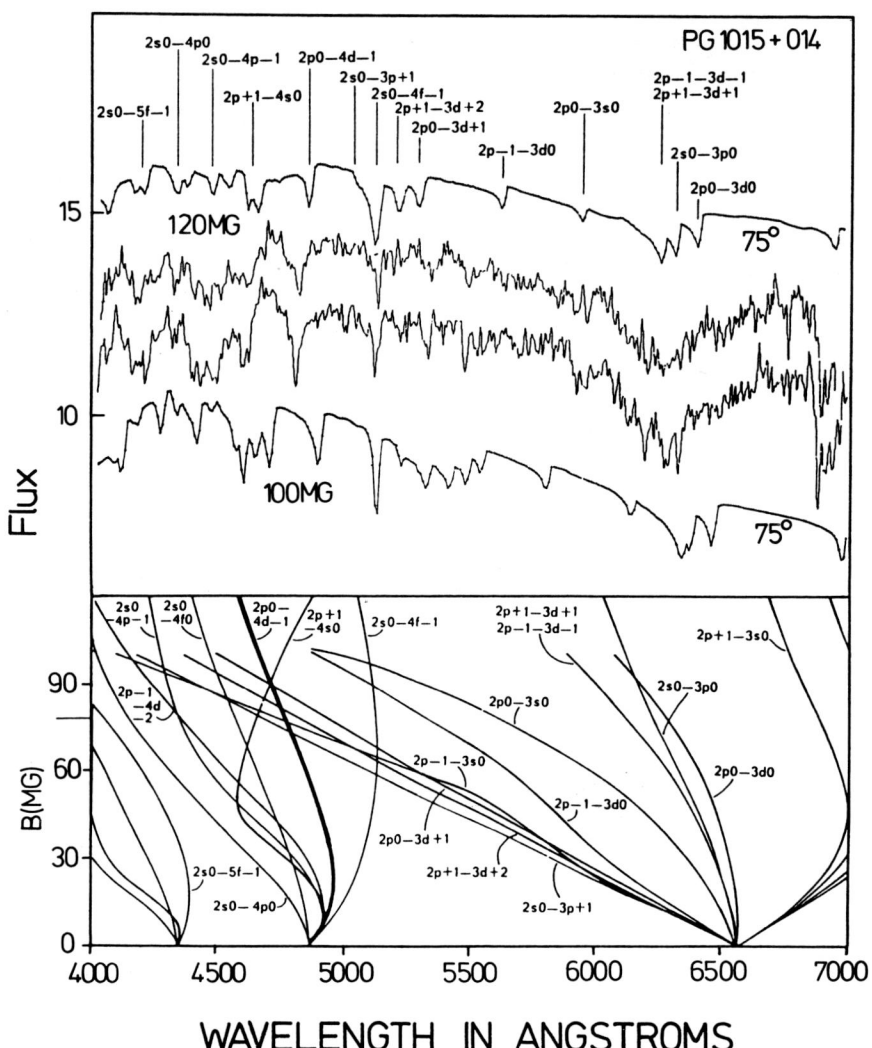

Fig. 7.9. Spectra of the intermediate field magnetic white dwarf PG 1015+014 at two different rotational phases (*two curves in the middle*) compared with synthetic spectra (*top and bottom curve*) using centered dipole models. The magnetic field dependence of wavelengths is shown in the bottom panel. The magnetic field of this object lies in the range 100–120 MG. (From Wickramasinghe and Cropper 1988)

Further confirmation of this identification is seen, in the expanded view of the feature in Fig. 7.11, in the existence of an extended blue wing, which is expected because the stationary point is a maximum of wavelength. The wing extends to about 7400 Å, which corresponds to fields from 80 to 160 MG. A simple centered dipole model for the polarization spectrum leads to a value of the polar field strength of about 130 MG. By contrast, the total flux shows a

7.2 Spectra of Selected Magnetic White Dwarfs   103

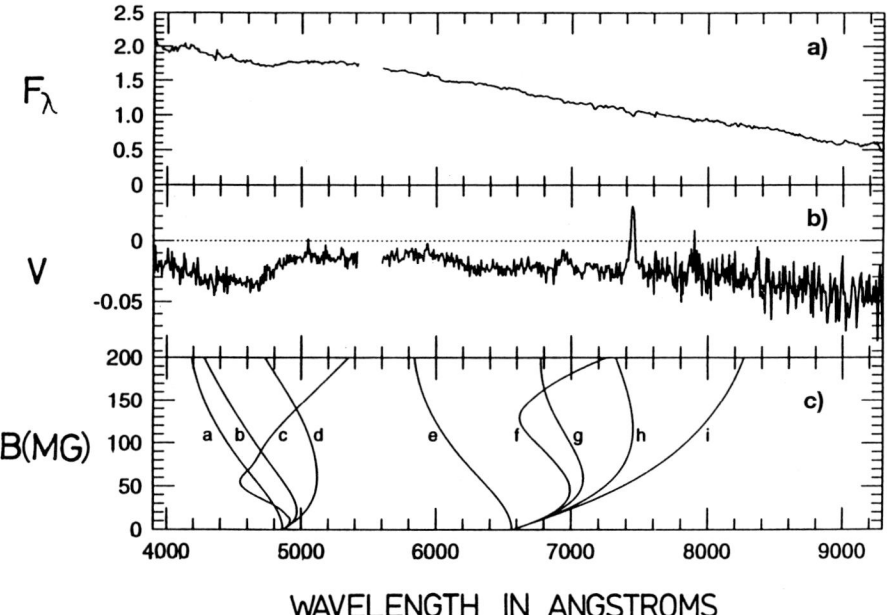

Fig. 7.10a–c. (a) Flux spectrum F_λ for G227–35, (b) normalized Stokes parameter V of the circular polarization, and (c) wavelength vs. magnetic field in MG for some of the components of H$_\alpha$ and H$_\beta$ ((a) $2s_0 \to 4f_0$, (b) $2p_0 \to 4d_{-1}$, (c) $2p_{+1} \to 4s_0$, (d) $2s_0 \to 4f_{-1}$, (e) $2s_0 \to 3p_0$, (f) $2p_{+1} \to 3s_0$, (g) $2s_0 \to 3p_{-1}$, (h) $2p_{-1} \to 3d_{-2}$, (i) $2p_0 \to 3d{-1}$). The feature in the circular polarization near 7450 Å is due to the transition h, which has its stationary point at 7449.4 Å and 117 MG. (From Cohen et al. 1993)

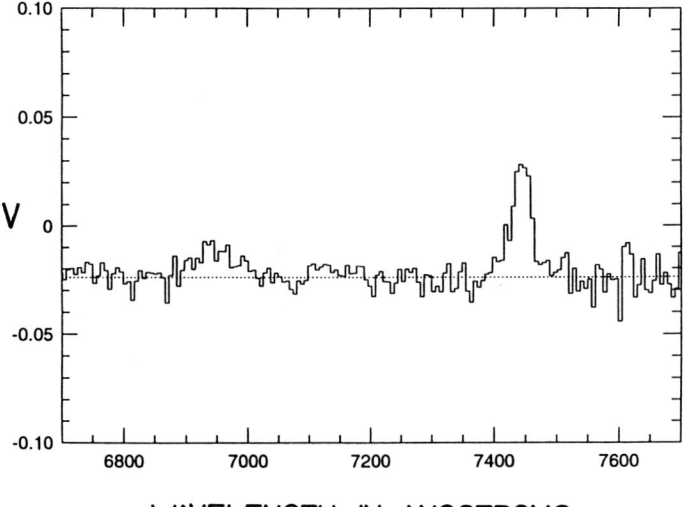

Fig. 7.11. Expanded view of the circular polarization spectrum around the feature at 7450 Å. Note the blue-shaded wing of the feature, in accordance with the maximum character of the stationary point. (From Cohen et al. 1993)

real though inconspicuous absorption feature at 7450 Å. It is clear from this example that polarimetry provides an additional powerful tool for detecting stationary lines and thus measuring the field strengths of magnetic degenerates.

7.2.6 MR Serpentis

We now present an example of a system where the magnetic white dwarf is the compact component in a close binary system, with a normal star accreting matter onto the white dwarf (AM Herculis systems in astronomical terminology). In contrast to the isolated white dwarfs discussed above, the remarkable feature in these systems is that the spectrum is produced in a localized region in the vicinity of the white dwarf with almost no spread in the magnetic field strength. In these regions the field strengths are not as high as those observed in isolated white dwarfs but high enough as to produce a complete splitting of Balmer components into individual lines. Figure 7.12 shows the spectrum of the AM Herculis system MR Serpentis taken by Schwope et al. (1993). The magnetic field in the absorbing region is found to be about 27 MG.

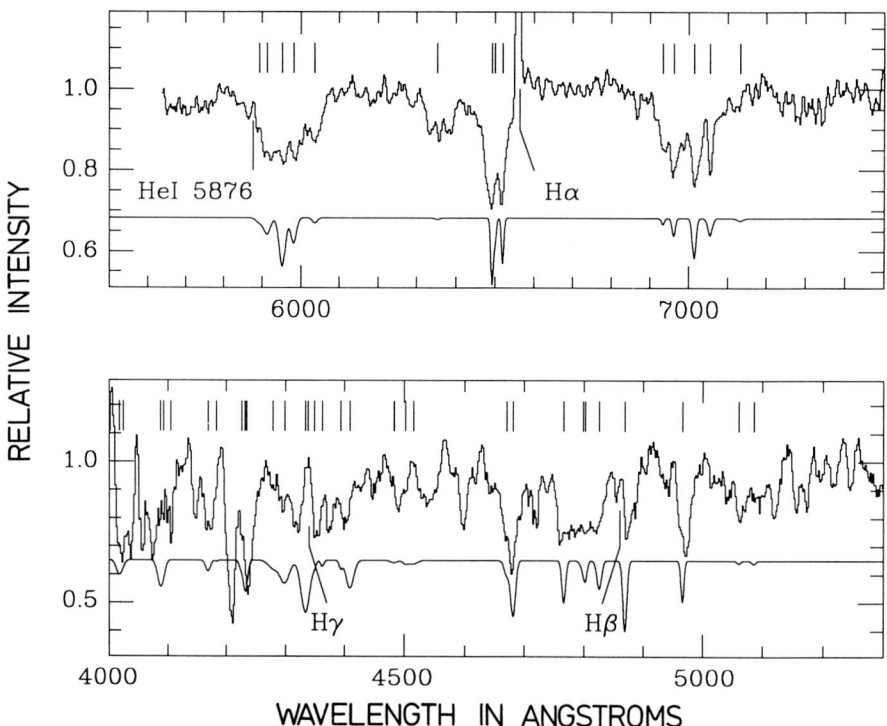

Fig. 7.12. Normalized spectrum of the AM Herculis system MR Serpentis. Predicted positions of individual Zeeman lines of H_α, H_β and H_γ are marked by vertical lines above the spectra for a magnetic field strength of 27.3 MG. Below the observed spectrum averaged absorption coefficients smeared assuming a Gaussian field distribution around the central value with width 0.3 MG are shown. (From Schwope et al. 1993)

7.2.7 LB 11146 (PG 0945+245)

This exotic stellar object also is a binary system, but consisting of *two* white dwarf stars: one a strongly polarized object with an inferred magnetic field strength among the largest yet found on a white dwarf (≥ 300 MG), the other a normal DA white dwarf with no detectable field (Liebert et al. 1993). Figure 7.13 shows the spectrum of the magnetic component (LB 11146b) together with calculated wavelengths for hydrogen. While certain of the features of LB 11146b are found to be due to hydrogen, the strong absorption feature around 5800 Å is not explicable by hydrogen and thus requires the presence of a second atmospheric component. The feature resembles the deep, red-shaded absorption features seen in GD 229 (whose spectrum is also shown in Fig. 7.13), for which helium in a strong magnetic field was proposed as an explanation. Further progress in the analysis of the spectra will require accurate data for wavelengths of neutral helium in magnetic fields of that strength.

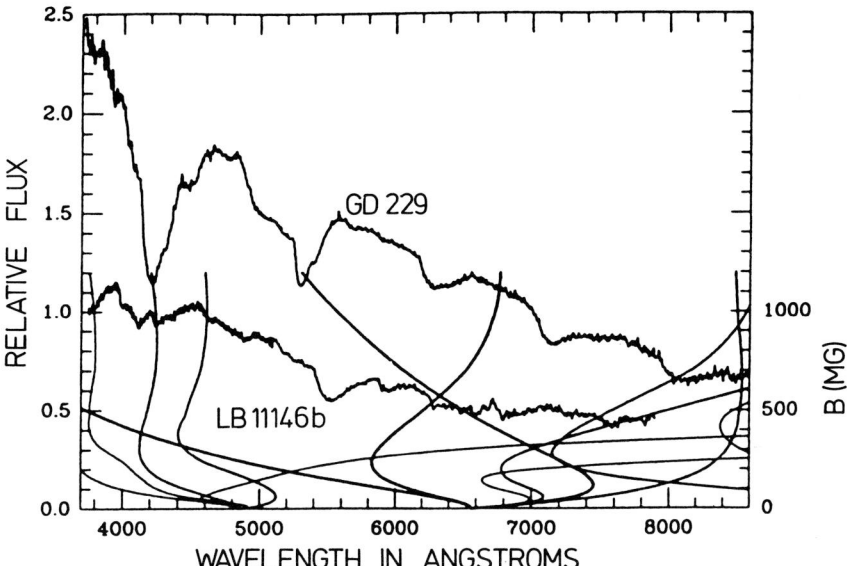

Fig. 7.13. Spectrum of the magnetic component LB 11146b of the double-degenerate binary system LB 11146 along with the magnetic-field dependence of the wavelengths of transitions in hydrogen in the range up to 1200 MG, compared with the highly polarized magnetic white dwarf GD 229. Predicted relative strengths of the Balmer transitions are indicated by line thickness. Some UV and optical features can be attributed to hydrogen in fields above 300 MG, while the broad feature around 5800 Å may be due to helium. Note the similarity of this feature and the broad features in GD 229, for which helium was proposed as an explanation. (From Liebert et al. 1993)

7.3 Table of Magnetic White Dwarfs

To conclude this chapter, in Table 7.2 we have listed the chemical composition, the effective temperature T_{eff}, the polar magnetic field strength, the rotation period and the brightness of all magnetic white dwarfs known until now.

Table 7.2. Parameters of magnetized white dwarfs

Object	chem. comp.	T_{eff} in K	B in MG	Period	V in magn.
LHS 1044	H	6 000	16.7		15.89
Feige 7	H, He	17 800	35	2.2h	14.5
PG 0136+251[3]	H	45 000	1.3?		15.8
MWD 0159−032[1]	H	26 000	6		17.1
KPD 0253+5052	H	18 000	18	4.1h	15.2
MWD 0307−428[1]	H	25 000	10		16.3
KUV 03292+0035	H	15 000	12		16.7
LHS 1734[4]	H	5 230	7.3		16.01
G 99−37	C_2,CH	6 200	10		14.5
G 99−47	H	5 600	25		14.1
WD 0637+447[5]	H	10 000	1.2		15.0 (ph)
GD 90	H	15 000	9		15.6
G 195−19		8 000	≈ 100	1.3d	13.8
PG 1015+01	H	10 000	120	1.65h	16
GD 116	H	16 000	65		16 (B)
PG 1031+234[2]	H	15 000	1000	3.4h	15.4
LP 790−29	C_2	8 600	200		15.6
ESO 439−162/163	C_2	6 300	≈ 100		18.77
LBQS 1136−0132	H		24		17.7 (B)
HS 1254+3430	H	15 000	18		17
PG 1312+098	H	15 000	≈ 10	5.43h	15
G 62−46[6]	H	6 040	7.4		17
PG 1533−057	H	17 000	31	≈ 1d	15.4
GD 356	H	7 500	14		15.1
PG 1658+441[3]	H	30 000	2.8		14.8
BPM 25114	H	8 600	36	2.8d	15.7
G 240−72		6 000	≈ 200		14.2
G 141−2	H	5 600	5		14.0
G 227−35[7]		7 000	130		16.0
Grw+70°8247[3]	H	16 000	320		13.1
GD 229	H	16 000	60?		14.8
KUV 2316+123[3]	H	10 400	32	17.9d	15.5
LB 11146b[8]	H, He	16 000	≥ 300		15.0
SBS 1349+5434[9]	H		760		16.4
MR Serpentis[10]	H	10 000	28		16

Literature: Koester, D. and Chanmugam, G. (1990): *Rep. Prog. Phys.* **53**, 837; Landstreet, J.D. (1992): *Astron. Astrophys. Rev.* **4**, 35; Chanmugam, G. (1992): *Ann. Rev. Astron. Astrophys.* **30**, 143; [1]Achilleos, N., Remillard, R.A., and Wickramasinghe, D.T. (1991): *Mon. Not. R. astr. Soc.* **253**, 522; [2]Östreicher, R., Seifert, W., Friedrich, S., Ruder, H., Schaich, M., Wolf, and D., Wunner, G. (1992): *Astron. Astrophys.* **257**, 353; [3]Friedrich, S. (1993): *PhD thesis*, Univ. Tübingen; [4]Bergeron, P., Ruiz, M-T., and Leggett, S.K. (1992): *Astrophys. J.* **400**, 315; [5]Schmidt, G.D., Stockman, H.S., and Smith, P.S. (1992): *Astrophys. J. Lett.* **398**, L57; [6]Bergeron, P., Ruiz, M-T., and Leggett, S.K. (1993): *Astrophys. J.* **407**, 733; [7]Cohen, M.H.,

Putney, A., and Goodrich, R.W. (1993): *Astrophys. J. Lett.* **405**, L67; [8]Liebert, J., Bergeron, P., Schmidt, G. D., Saffer, R. A. (1993): *Astrophys. J.* **418**, 426; [9]Liebert, J., Schmidt, G. D., Lesser, M., Stepanian, J. A., Lipovetsky, V. A., Chaffee, F. H., Foltz, C. B., Bergeron, P. (1994): *Astrophys. J.* **421**, 733; [10]Schwope, A. D., Beuermann, K., Jordan, S., and Thomas, H.-C. (1993): *Astron. Astrophys.* **278**, 487.

7.4 Future Work

The selected examples presented in the foregoing section have provided overwhelming evidence that unusual absorption features in low-resolution optical spectra of magnetized white dwarfs can be explained in terms of hydrogen lines in strong magnetic fields. However, recent observations of magnetic white dwarfs with high spectral resolution (of about 0.6 Å) (Friedrich et al. 1994) seem to reveal small blueshifts of the absorption edges with respect to the theoretically predicted positions of stationary wavelengths, e.g., of about 5–6 Å for the H_α and H_β transitions in the spectrum of the magnetic white dwarf Grw+70°8247 (Fig. 7.14). The origin of these shifts is not yet completely clear, but one possible explanation is offered by the presence of randomly fluctuating local electric fields acting on the absorbing hydrogen atoms. The fields are caused by free electrons and ions in the white dwarf atmosphere and can be on the order of up to 10^9 V/m. Obviously these plasma effects could produce shifts of the positions of stationary absorption edges.

Preliminary calculations for *parallel* electric and magnetic fields have not been able so far to explain shifts of that size. Since the electric field is random in its relative orientation with respect to the magnetic field, future calculations must take this fact into account and be performed for *arbitrary* angles between the electric and the magnetic field. However, in the presence of non-parallel components of the electric field neither parity nor the magnetic quantum number m are conserved, and thus symmetries that were essential to solving the problem of hydrogen atoms in pure magnetic fields are lost. As a consequence, the problem of calculating energies, wave functions and transition rates for hydrogen atoms in *combined* electric and magnetic fields with *arbitrary* orientation is at least one order of magnitude more complicated than the pure magnetic field situation.

Another outstanding problem is the calculation of the bound-free opacities of hydrogen atoms in white dwarf magnetic fields. These are prerequisite for a quantitative understanding of the *continuous* part of the spectra of magnetic degenerates, and their polarization spectra. Great efforts are being made to calculate these opacities, which includes determining the eigenstates of the hydrogen atom in magnetic fields in the positive-energy region. A very promising method for solving this problem is the complex-coordinate rotation method, by aid of which the many resonance states in the continuum which accumulate below the different Landua thresholds can be well described. Of course the opacity data must be determined on a dense grid of magnetic field strengths again to

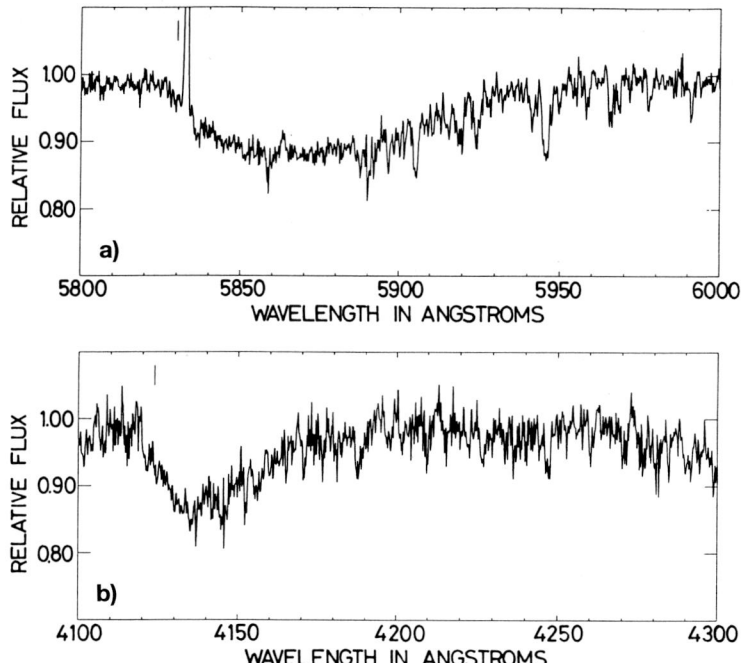

Fig. 7.14a,b. Balmer H_α (a) and Balmer H_β (b) absorption features in the spectrum of the magnetic white dwarf Grw+70°8247. The spectral resolution is about 0.6 Å . The positions of the stationary wavelengths are marked by vertical bars. The strong emission feature in (a) is due to a cosmic.

cope with the variation of the field strength across the observed hemisphere of a white dwarf star. First results for Balmer and Paschen bound-free opacities have recently become available (Merani et al. 1994) at seven field strengths in the range 117 to 235 MG, but obviously this is only a first step to building a complete data base of opacities of hydrogen in arbitrary magnetic fields of white dwarfs.

To summarize, accurate calculations both of the effects of random electric fields and of bound-free opacities are urgently needed, and intense efforts are being made to meet this challenge. It is only with these data that a conclusive and complete understanding of the spectra of magnetic white dwarfs will finally be achieved.

8 Relativistic Effects, Nuclear Mass Effects, and Landau-Excited States

8.1 Spin-Orbit Coupling

In nearly all papers on hydrogen atoms in strong magnetic fields, the problem is treated within the framework of non-relativistic quantum mechanics. The justification for this approach was furnished by Lindgren and Virtamo (1979), who, by numerically solving Dirac's equation, showed that relativistic effects on the Coulomb binding energies remain negligible even for magnetic field strengths of up to 4.7×10^9 T. From this it is obvious that the inclusion of relativistic corrections in Schrödinger's equation modifies the energy values only insignificantly in the total range of the magnetic field. The purpose of this section is to demonstrate that, in spite of its smallness, the spin-orbit interaction can produce drastic effects because it reduces the invariance group of the Hamiltonian (see also Wunner et al. 1985a).

Without spin-orbit coupling, the Coulomb problem in a uniform magnetic field is invariant under the parity transformation and independent rotations about the z-axis in ordinary space and in spin space, giving rise to the conserved quantum numbers parity Π, m and m_s. Since the parity operation is equivalent to a rotation of 180° about the z-axis, followed by a reflection with respect to the xy-plane, it suffices to label the states with the z-parity π instead of with the full parity Π. With spin-orbit coupling, parity (but no longer z-parity) is conserved, and the Hamiltonian is invariant only under simultaneous rotations about the z-axis in ordinary and spin space, whence it is only the projection of the *total* angular momentum, $m_j = m + m_s$, which is a good quantum number.

As is well known, because of spin-orbit coupling, the levels are arranged in multiplets with sharp total angular momentum (fine-structure splitting) for $B = 0$, and the m_j degeneracy of these states is removed for weak fields by the anomalous Zeeman effect. Also, the restructuring of states into simple products of orbital and spin functions as one proceeds to the Paschen-Back regime, where spin-orbit coupling becomes smaller than, and eventually negligible compared with, the magnetic energy, is a topic of standard quantum mechanics (Bethe and Salpeter 1957). However, when the magnetic field is increased further, the question arises as to how these states evolve from the Paschen-Back regime on to very large magnetic fields, and, ultimately, to the limiting case $B \to \infty$. The general means for establishing correspondences of this kind is provided by the non-crossing rule of quantum mechanics, according to which two levels (considered here as functions of the magnetic field) are forbidden to cross if they possess the same symmetry labels of the exact symmetries of the underlying Hamiltonian. Without spin-orbit coupling the problem was solved by Simola and Virtamo (1978). In the situation they describe, the correspondence diagrams are easily obtained by successively connecting, for given π, m and m_s, the states in the two limiting cases in order of increasing energy. With spin-orbit coupling, however, π, m and m_s are no longer good quantum numbers, and the non-crossing rule must be applied to the subspaces with fixed Π and m_j. Thus the correspondence diagrams of the physically real hydrogen atom must be different from those introduced by Simola and Virtamo (1978).

To be specific, let us consider the two subspaces $m_j = -\frac{1}{2}$ and $\Pi = \pm 1$. In Figs. 8.1 and 8.2 the energy values of the eight lowest states in each subspace are plotted as functions of the magnetic field strength for $10^{-4} \leq \beta \leq 10^3$, using our accurately computed energy values. On the left, the states are labeled by the field-free quantum numbers. On the right, the states are labeled by the $B \to \infty$ asymptotic quantum numbers. Ignoring the spin-orbit interaction, the computed B-dependence of the energy levels leads, of course, to the correspondence scheme of Simola and Virtamo (1978). Since the subspaces considered in Figs. 8.1 and 8.2 are adapted to the symmetries of the Hamiltonian with spin-orbit coupling, the crossings of spin-up states with spin-down states marked by circles must, in reality, be anticrossings.

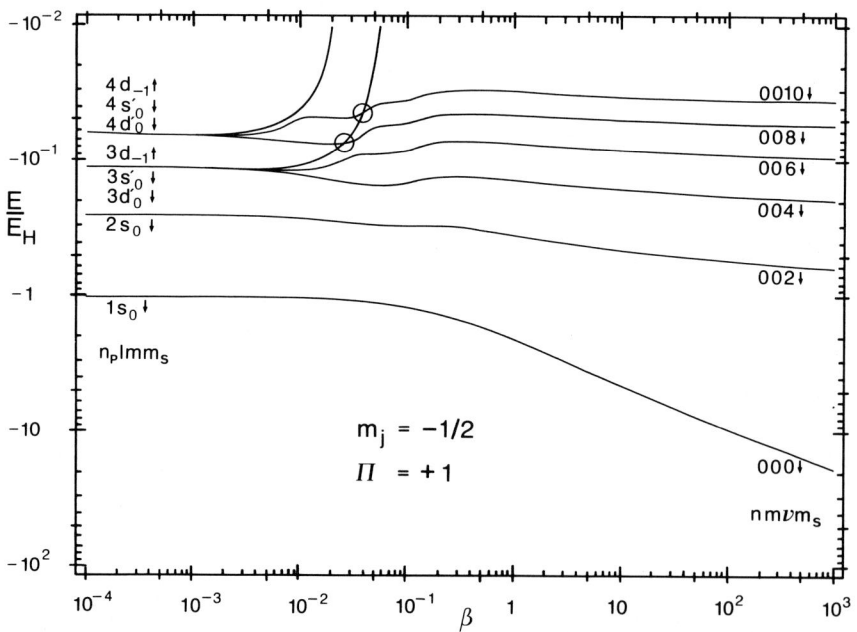

Fig. 8.1. Energy values (in units of the Rydberg energy E_H) of the lowest states in the $m_j = -\frac{1}{2}$, positive-parity subspace of the hydrogen atom as functions of the magnetic field strength. The states are characterized by their Paschen-Back quantum numbers on the low-field side and by the adiabatic approximation quantum numbers on the high-field side. Avoided crossings due to spin-orbit coupling are marked by circles.

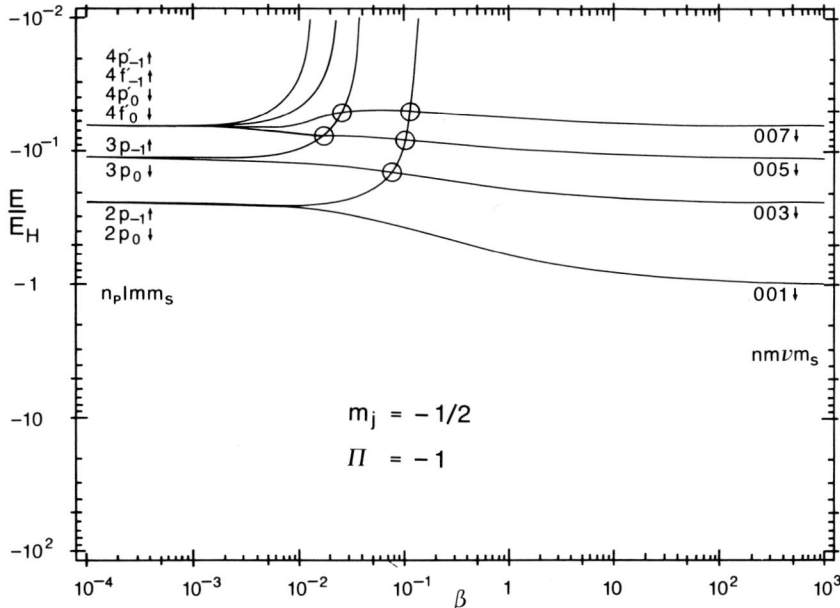

Fig. 8.2. Energy values (in units of the Rydberg energy E_H) of the lowest states in the $m_j = -\frac{1}{2}$, negative-parity subspace of the hydrogen atom as functions of the magnetic field strength. The states are characterized by their Paschen-Back quantum numbers on the low-field side and by the adiabatic approximation quantum numbers on the high-field side. Avoided crossings due to spin-orbit coupling are marked by circles.

An enlargement of the situation inside the circle which lies at the lowest energy in the $\Pi = -1$, $m_j = -\frac{1}{2}$ subspace is shown in Fig. 8.3. Energies and field strengths refer to the point E_0, β_0, where the $2p_{-1} \uparrow$ and $3p_0 \downarrow$ levels intersect if spin-orbit coupling is neglected, namely $E_0 = -0.1461477 E_H$, $\beta_0 = 0.0790531$.

The magnetic field dependence of the energy levels with spin-orbit coupling in the neighbourhood of E_0, β_0 was computed in the usual way (see, e.g., Messiah 1973) by diagonalizing the spin-orbit Hamiltonian

$$H_{ls} = \frac{\alpha}{r^3}[\boldsymbol{r} \times (\boldsymbol{p} - e\boldsymbol{A})] \cdot \boldsymbol{s} \tag{8.1}$$

in the basis of the $2p_{-1} \uparrow$, $3p_0 \downarrow$ states determined numerically without spin-orbit coupling. We emphasize the form (8.1) of the spin-orbit Hamiltonian in the presence of a magnetic field, which follows from a consistent expansion of Dirac's equation.

The diagonal matrix elements of H_{ls} produce slight shifts in the energy levels, while the coupling matrix element $\langle H_{ls} \rangle_{12}$ determines the minimum distance of the levels in the avoided crossing. Computed values for $3p_0 \downarrow$ and $2p_{-1} \uparrow$, as well as for two other crossings ($4f'_0 \downarrow$ and $3p_{-1} \uparrow$, $4d'_0 \downarrow$ and $3d_{-1} \uparrow$), are given in Table 8.1.

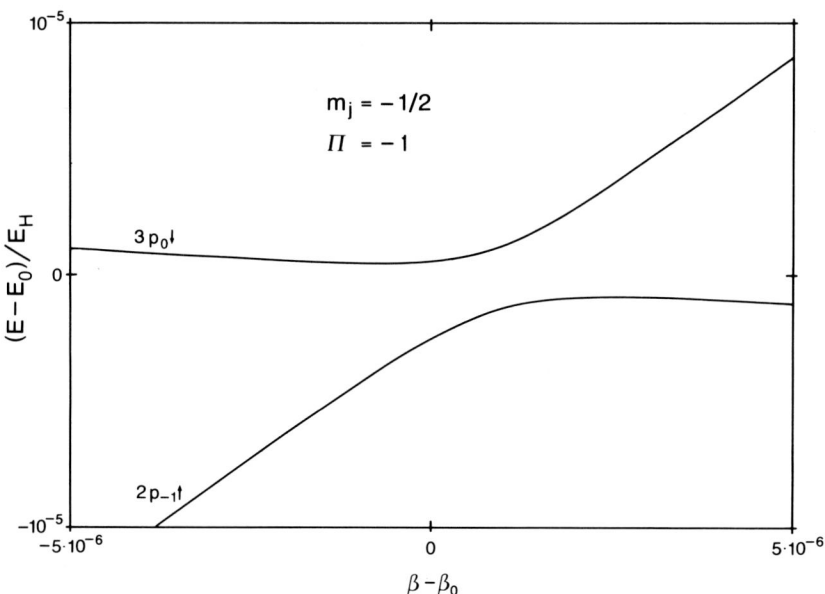

Fig. 8.3. Behaviour of the avoided crossing at the lowest energy in the $m_j = -\frac{1}{2}$, negative-parity subspace as a function of the magnetic field strength. If the spin-orbit coupling could be switched off, the two levels would intersect exactly at $\beta_0 = 0.0790531$, $E_0/E_H = -0.1461477$.

Table 8.1. Points of intersection E_0, β_0 of the energies of state 1 and state 2 if computed without spin-orbit coupling, and the matrix elements (in units of the Rydberg energy E_H) of the spin-orbit Hamiltonian in the basis of these states around E_0, β_0. The states are labeled by their Paschen-Back quantum numbers.

State 1	State 2	E_0	β_0	$\langle H_{ls}\rangle_{11}$	$\langle H_{ls}\rangle_{22}$	$\langle H_{ls}\rangle_{12}$
$2p_{-1}\uparrow$	$3p_0\downarrow$	-0.1461477	0.0790531	-1.38×10^{-6}	6.16×10^{-8}	1.15×10^{-6}
$3d_{-1}\uparrow$	$4d'_0\downarrow$	-0.0727262	0.0271432	-5.49×10^{-8}	3.03×10^{-8}	9.78×10^{-9}
$3p_{-1}\uparrow$	$4f'_0\downarrow$	-0.0759268	0.018256	-4.33×10^{-7}	1.20×10^{-8}	3.56×10^{-7}

The order of magnitude of the matrix elements ($\sim 10^{-6}$ of the binding energy) is caused essentially by the α^2 dependence of this relativistic effect. In spite of this smallness, the effect on the structure of the wave functions is tremendous: within a relative variation of the magnetic field of a few 10^{-6}, the characters of the wave functions abruptly change from one into the other. In particular, if we follow the energy routes, it is no longer the state $3p_0\downarrow$ that corresponds to the state $003\downarrow$ in the infinite field limit, but rather the state $2p_{-1}\uparrow$! Tracing further the path of $3p_0\downarrow$, we recognize from Fig. 8.2 that it soon anticrosses with $3p_{-1}\uparrow$ (which in turn has already gone through an avoided crossing with $4f'_0\downarrow$), whereby it is bent down again and converges, for $B\to\infty$, towards the state $005\downarrow$. Following the energy routes of the other levels in Figs. 8.1 and 8.2 we easily establish the correspondence scheme shown for the six lowest-lying states in the $m_j = -\frac{1}{2}, \Pi = \pm 1$ subspaces in Table 8.2.

Table 8.2. Correspondence between the Paschen-Back regime and the regime of the adiabatic approximation $(B \to \infty)$ for the six lowest states in the $m_j = -\frac{1}{2}$, $\Pi = \pm 1$ subspaces. Note the intrusion of spin-up states (indicated by daggers), which is brought about by spin-orbit coupling.

$m_j = -\frac{1}{2}$, $\Pi = -1$		$m_j = -\frac{1}{2}$, $\Pi = +1$	
(n_p, l, m, m_s)	(n, m, ν, m_s)	(n_p, l, m, m_s)	(n, m, ν, m_s)
$2p_0 \downarrow$	$001 \downarrow$	$1s_0 \downarrow$	$000 \downarrow$
$\dagger 2p_{-1} \uparrow$	$003 \downarrow$	$2s_0 \downarrow$	$002 \downarrow$
$3p_0 \downarrow$	$005 \downarrow$	$3d'_0 \downarrow$	$004 \downarrow$
$\dagger 3p_{-1} \uparrow$	$007 \downarrow$	$3s'_0 \downarrow$	$006 \downarrow$
$4f'_0 \downarrow$	$009 \downarrow$	$\dagger 3d_{-1} \uparrow$	$008 \downarrow$
$4p'_0 \downarrow$	$0011 \downarrow$	$4d'_0 \downarrow$	$0010 \downarrow$

These simple examples clearly demonstrate how spin-orbit coupling, though small in quantitative terms, completely upsets the low-field high-field correspondence discussed in Sect. 3.1.3.

Finally let us briefly comment on the modifications that arise when we look at the physically *real* hydrogen atom even more closely. If we consider the hyperfine interaction with the magnetic moment of the nucleus, the non-crossing rule must be applied to the subspaces with sharp projections m_f of the total angular momentum operator $\boldsymbol{f} = \boldsymbol{j} + \boldsymbol{i}$ ($\boldsymbol{i}$ = proton spin) and good parity. Moreover, if we consistently treat the hydrogen atom in a magnetic field as a two-body problem (see Chap. 2), the orbital angular momentum must be redefined in the usual way in terms of the relative orbital angular momentum of proton and electron. In the general case of a non-vanishing transverse momentum $\hbar \boldsymbol{K}_\perp$ of the atom, however, the two charged particles experience a motional electric field (proportional to $\boldsymbol{K}_\perp \times \boldsymbol{B}$) in their rest frame, which ultimately destroys the rotational symmetry of the Hamiltonian around the direction of the magnetic field. Thus, in this general situation, the only symmetry label that remains is parity.

8.2 Effects of the Finite Proton Mass and of Motion Perpendicular to the Magnetic Field

In Chap. 2 the foundations of the treatment of the two-body problem in the presence of a uniform magnetic field has been described. The purpose of this section is to discuss quantitatively the effects of the finite proton mass and to present results for the dependence of the lowest energy levels on the generalized momentum $\boldsymbol{K}_\perp$ of the system perpendicular to the magnetic field.

Starting point is the spectrum of the hydrogen atom with $\boldsymbol{K}_\perp = 0$ but finite proton mass. Using our results for infinite proton mass together with the scaling laws of Sect. 2.3 the energy level scheme is depicted in Fig. 8.4 for $B = 4.7 \times 10^8$ T ($\beta = 10^3$). The strong influence of the finiteness of the proton

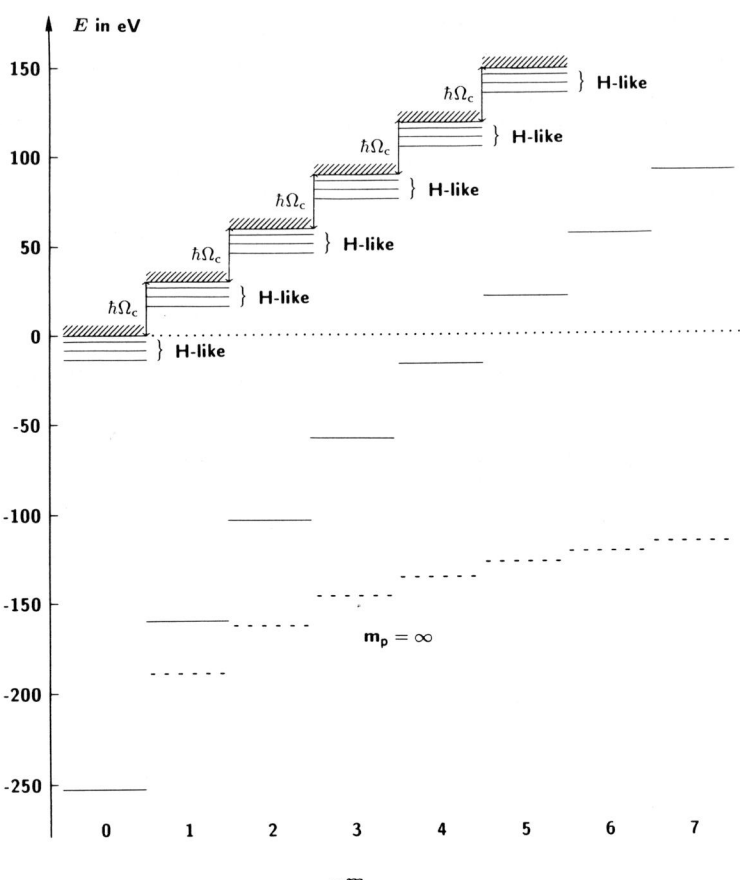

Fig. 8.4. Energy level scheme of the hydrogen atom for a magnetic field strength of $B = 4.7 \times 10^8$ T ($\beta = 10^3$). Bound states as well as continuum bands are shown for several values of the quantum number m (*solid lines*). The lower, "broken line" levels are the energy levels of the tightly bound states corresponding to an infinite proton mass. Taking into account the finite proton mass clearly results in a substantial modification of the energy spectrum. The bound and continuum states belonging to a given m are all displaced upward by the same amount $\hbar\Omega_c(-m) \simeq 29.6\,\text{eV}\,|m|$.

mass is clearly seen; one essential effect is that the tightly bound states are more and more raised with increasing $|m|$. For this value of the magnetic field strength the tightly bound states with $|m| \geq 5$ lie in the continuum of the subspaces with smaller $|m|$ and thus become autoionizing states.

Since these effects are physically important only for very strong magnetic fields ($\beta \gg 1$), for the numerical treatment of the eigenvalue equation (2.12) we use an expansion of $\Phi(r)$ in terms of Landau states:

$$\Phi(\mathbf{r}) = \sum_{n=0}^{\infty} \sum_{m=-\infty}^{n} c_{nm} g_{nm}(z) \phi_{nm}^{\text{Lan}}(\varrho, \varphi) \quad . \tag{8.2}$$

Whether it is better to use undisplaced Landau states as in the ansatz (8.2) or displaced Landau states as discussed in Sect. 2.2.2 depends on the mass ratio m_-/m_+. It can be shown (Herold et al. 1981) that for the mass ratio m_e/m_p the expansion (8.2) with undisplaced Landau states possesses a larger range of admissible $\boldsymbol{K}_\perp$ values. For $\beta \gg 1$ the Landau level distance is much larger than the Coulomb binding energies. Thus, the coupling between different n values will be very small and, consequently, it is a very good approximation to take only one value of n in (8.2). Additionally we restrict ourselves to the physically most important Landau ground state $n = 0$.

Forming the expectation value of the Hamiltonian of (2.12) with the wave function (8.2) and varying with respect to the longitudinal functions $g_{0m}(z)$ and the coefficients c_{0m} we obtain a system of coupled eigenvalue equations for the functions $g_{0m}(z)$

$$-\frac{\hbar^2}{2\mu}\frac{\partial^2}{\partial z^2}g_{0m}(z) + \sum_{m'}\frac{c_{0m'}}{c_{0m}}V_{\text{eff}}^{mm'}(z,\boldsymbol{K}_\perp)g_{0m'}(z) = \varepsilon_m g_{0m}(z) \tag{8.3a}$$

and an algebraic eigenvalue problem for the coefficients c_{0m}

$$\int g_{0m}(z)\left(-\frac{\hbar^2}{2\mu}\frac{\partial^2}{\partial z^2}\right)g_{0m}(z)dz\, c_{0m}$$
$$+\sum_{m'}\int g_{0m}(z)V_{\text{eff}}^{mm'}(z,\boldsymbol{K}_\perp)g_{0m'}(z)dz\, c_{0m'} = E\, c_{0m} \quad . \tag{8.3b}$$

The quantities ε_m and the total energy E – which is measured with respect to the ionization threshold – are Lagrange multipliers which originate from the normalization of the functions g_{0m} and of the total wave function $\Phi(\boldsymbol{r})$, respectively. The effective potentials $V_{\text{eff}}^{mm'}(z,\boldsymbol{K}_\perp)$ are defined by

$$V_{\text{eff}}^{mm'}(z,\boldsymbol{K}_\perp) = \frac{\hbar^2}{2M}K_\perp^2\delta_{mm'} + \frac{\hbar eB}{M}K_\perp\langle\phi_{0m}^{\text{Lan}}|x|\phi_{0m'}^{\text{Lan}}\rangle \tag{8.4}$$
$$+\frac{\hbar eB}{m_p}(-m)\delta_{mm'} - \frac{e^2}{4\pi\varepsilon_0}\langle\phi_{0m}^{\text{Lan}}|\frac{1}{r}|\phi_{0m}^{\text{Lan}}\rangle\delta_{mm'} \quad .$$

Without loss of generality we have assumed that $\boldsymbol{K}_\perp$ points in y direction, $\boldsymbol{K}_\perp = K_\perp \boldsymbol{e}_y$. The constant energy contribution $\hbar^2 K_z^2/2M$ of the centre-of-mass motion parallel to the magnetic field has been omitted.

The diagonal elements of the effective potentials (8.4) contain the well-known Coulomb matrix elements of (3.7) specialized to $n = n' = 0$, whereas the nondiagonal elements are z independent and read

$$V_{\text{eff}}^{mm'}(\boldsymbol{K}_\perp) = \frac{\hbar eB}{M}\frac{1}{2}K_\perp a_{\text{L}}\left(\sqrt{m+1}\,\delta_{m',m+1} + \sqrt{m}\,\delta_{m',m-1}\right) \quad . \tag{8.5}$$

The system (8.3) of coupled equations was solved by a Hartree-Fock-like method analogous to that described in Sect. 3.2.1 taking into account up to 30 m values. As results, in Fig. 8.5 the energies of the ground state, the first excited state and the first hydrogen-like state are shown in their dependence on the transverse generalized momentum as solid curves.

For small values of $K_\perp$ the energies exhibit a quadratic dependence on $K_\perp$ of the form:

$$E(K_\perp) = E(K_\perp = 0) + \frac{\hbar^2}{2M_{\text{eff}}} K_\perp^2 \quad . \tag{8.6}$$

Employing a standard perturbation treatment this behaviour can easily be verified. The dashed curves in Fig. 8.5 represent this quadratic approximation obtained by a fit to the numerical values yielding the strongly state-dependent effective masses M_{eff} for the motion of the hydrogen atom perpendicular to the magnetic field. In Table 8.3 for some more states the numerical values of M_{eff} are given. States with an odd node number ν do not depend very strongly on $K_\perp$ and, therefore, the effective mass of those states is much larger than that of states with even node number.

Table 8.3. Numerical values of M_{eff}/M for some states $(0, m, \nu)$ of the hydrogen atom

m	$\nu = 0$	$\nu = 1$	$\nu = 2$	$\nu = 3$	$\nu = 4$
0	1.46	273.2	61.4	1084	291.7
−1	3.79				
−2	9.50				
−3	24.5				
−4	85.6				

A consequence of these anisotropic and state-dependent masses is the fact that Doppler broadening leads to asymmetric atomic line shapes which, additionally, depend on the angle between the photon and the magnetic field direction.

8.3 Landau-Excited States

For very strong magnetic fields where the cyclotron energy is much larger than the Coulomb energies less effort has been made so far to study the physical properties of hydrogen atoms in which the electron is excited transverse to the magnetic field. It is the purpose of this section to discuss quantitatively the lifetimes, and thus the widths, of these Landau-excited states with respect to electromagnetic transitions. In doing so, we will also take into account the level shifts produced by the anomalous magnetic moment of the electron and the finite mass of the proton.

In treating the Schrödinger equation in such strong magnetic fields with field strengths $B \geq 10^6$ T the adiabatic approximation for the wave function, described in Sect. 3.1.2, is very good. Thus, the states are characterized by the quantum numbers n, m, ν. Including the spin contribution $\hbar \omega_c s_z$ with $s_z = \pm \frac{1}{2}$

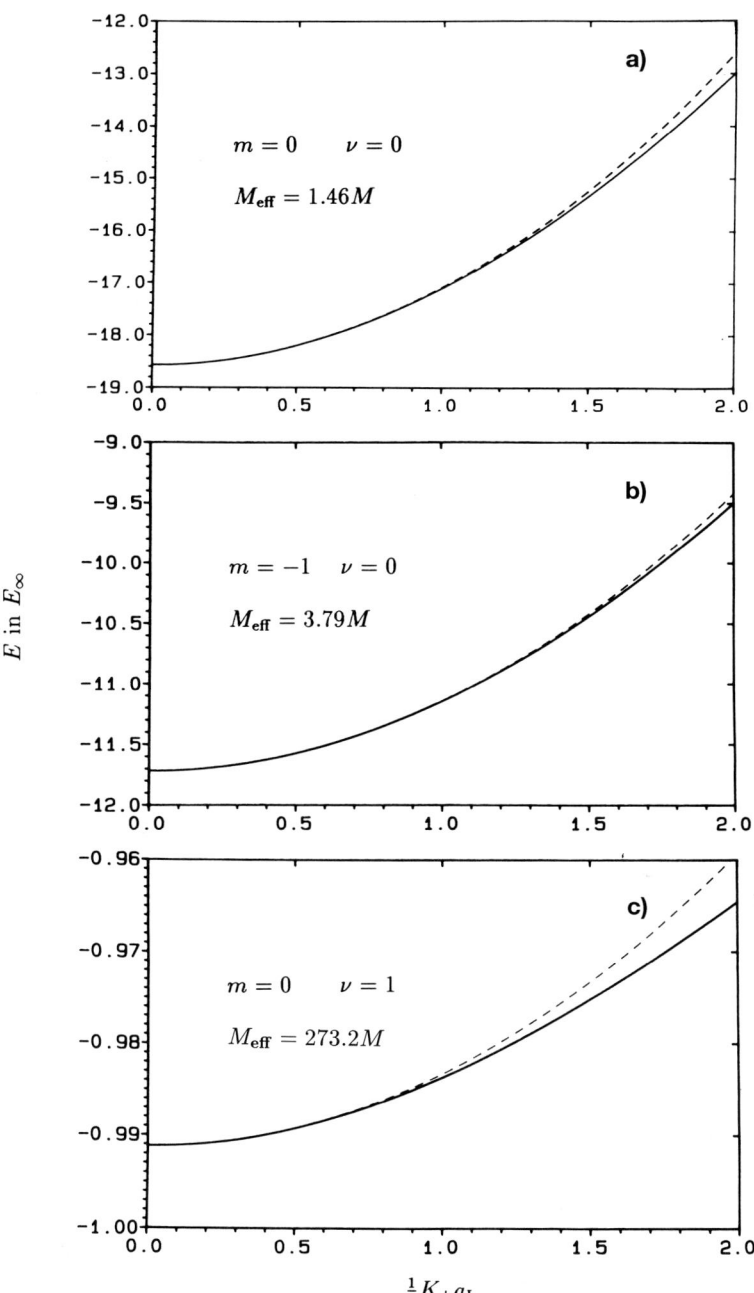

Fig. 8.5a–c. Energies of three low-lying states of the hydrogen atom as functions of $\frac{1}{2}K_\perp a_\mathrm{L}$ for $B = 4.7 \times 10^8$ T ($\beta = 10^3$): (**a**) ground state $(n, m, \nu) = (0, 0, 0)$, (**b**) first excited state $(n, m, \nu) = (0, -1, 0)$, (**c**) first hydrogen-like state $(n, m, \nu) = (0, 0, 1)$. The solid curves are the numerically calculated energies, the dashed curves represent the quadratic approximation (8.6) with fitted values of M_eff.

for spin-up and spin-down states, respectively, the total energy of a state is given by

$$E_{nm\nu s_z} = (n + s_z + \frac{1}{2})\hbar\omega_c + E_{mn\nu} = N\hbar\omega_c + E_{mn\nu} \quad , \tag{8.7}$$

where $E_{nm\nu}$ are the Coulomb bound state energies relative to the corresponding Landau threshold and $N = 0, 1, 2, \ldots$ is the general Landau quantum number. A schematic view of the level scheme in a very strong magnetic field is shown in Fig. 8.6.

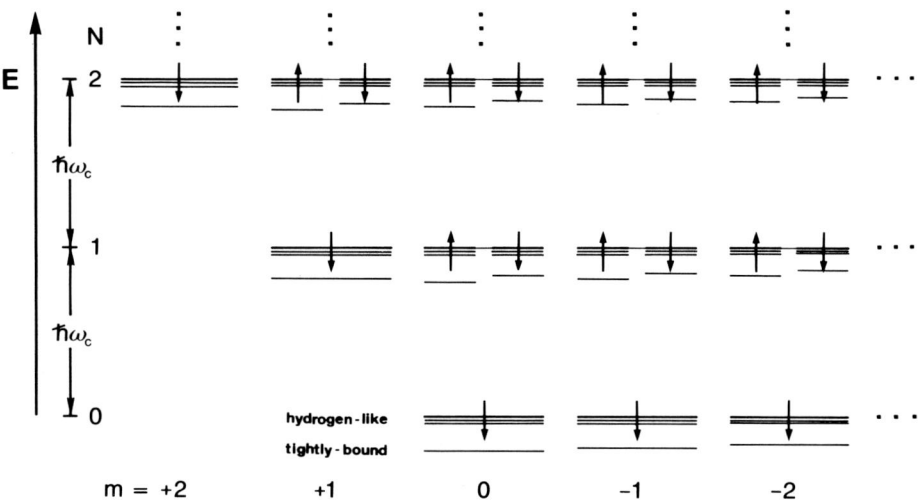

Fig. 8.6. Schematic representation (not to scale) of the energy eigenvalue spectrum (8.7) of an electron moving in the Coulomb potential of a fixed proton and in a very strong magnetic field. The arrows indicate the orientation of the spin with respect to the magnetic field ($s_z = \pm 1$). For every value of N, m, and s_z, there exists a complete set of Coulomb states: one *tightly bound* state with a binding energy relative to the Landau level of several Rydberg units, and infinitely many *hydrogen-like* states with energies less than E_∞. Note that for reasons of drawing accuracy the Coulomb energies have been strongly amplified compared with the Landau level distance $\hbar\omega_c = 4\beta E_\infty$. The maximum energy of a tightly bound state ($N = 0$, $m = 0$) is $7.5 E_\infty \approx 3.8 \times 10^{-2} \hbar\omega_c$ for $\beta = 50$, and $18.6 E_\infty \approx 4.7 \times 10^{-3} \hbar\omega_c$ for $\beta = 10^3$, for example.

In this regime, evidently the global structure of the level scheme is dominated by the equidistant Landau levels, separated by the cyclotron energy $\hbar\omega_c = 4\beta E_\infty$, while to each Landau level there is attached a substructure of (in principle infinitely many) discrete Coulomb states.

To calculate the lifetimes and widths of these Coulomb states the dipole strengths for $\Delta m = 0, +1, -1$ transitions (6.22) have to be considered. Transitions without change of the Landau energy ($n' = n$) are allowed in all three cases. Here the largest rates will occur for $n = 0 \to n' = 0$ transition since, as n is increased, the effective potentials V_{eff}^{nnm} of (3.7) become shallower, giving rise to lower transition energies and, thus, rates. Electromagnetic $n = 0 \to n' = 0$ transitions have been discussed in detail in Sect. 6.4. The $n' = n > 0$ transition rates are physically not interesting since the widths of $n > 0$ states are dominated by electromagnetic transitions to lower Landau level states. As is evident from (6.22), inter-n transitions are allowed only for $\Delta m = +1$ ($n \to n+1$, i.e. absorption) and $\Delta m = -1$ ($n \to n-1$, i.e. emission). In either case the size of the dipole strength is determined by the overlap of the longitudinal wave functions in the initial and final state. Since on the other hand the effective potentials $V_{\text{eff}}^{nnm}(z)$ and $V_{\text{eff}}^{n\pm 1\,n\pm 1\,m\pm 1}(z)$ differ from each other essentially only in the neighbourhood of the origin, the longitudinal wave functions exhibit some kind of quasi-orthogonality, viz. $|\langle g_{n\pm 1\,m\pm 1\,\nu'}|g_{nm\nu}\rangle|^2 \approx \delta_{\nu'\nu}$, as is verified by numerical computations (see Table 8.4).

Table 8.4. Overlaps $|\langle g_{n'm'\nu'}|g_{nm\nu}\rangle|^2$ of the longitudinal wave functions for $n = 0$, $m = 0$, $\nu = 0, \ldots, 5$ and $n' = 1$, $m' = 1$, $\nu' = 0, \ldots, 5$, for three values of the magnetic field strength ($\beta = 10, 10^2, 10^3$: upper, middle, and lower number of every $\nu' \to \nu$ entry, respectively). Evidently, as β increases the overlaps are better and better approximated by $\delta_{\nu'\nu}$.

	positive z-parity $\pi = +1$		
	$\nu = 0$	$\nu = 2$	$\nu = 4$
$\nu' = 0$	0.97524 0.98442 0.99033	6.72×10^{-3} 2.14×10^{-3} 6.86×10^{-4}	1.14×10^{-3} 3.68×10^{-4} 1.14×10^{-4}
$\nu' = 2$	4.56×10^{-3} 1.55×10^{-3} 5.26×10^{-4}	0.97670 0.99060 0.99605	7.95×10^{-3} 2.89×10^{-3} 1.14×10^{-3}
$\nu' = 4$	9.13×10^{-4} 2.89×10^{-4} 9.22×10^{-5}	5.64×10^{-3} 2.29×10^{-3} 9.87×10^{-4}	0.97670 0.99079 0.99612

	negative z-parity $\pi = -1$		
	$\nu = 1$	$\nu = 3$	$\nu = 5$
$\nu' = 1$	0.99405 0.99931 0.99997	2.34×10^{-3} 2.21×10^{-4} 1.00×10^{-5}	4.60×10^{-3} 4.55×10^{-5} 2.14×10^{-6}
$\nu' = 3$	1.98×10^{-3} 2.08×10^{-4} 1.01×10^{-5}	0.99268 0.99924 0.99996	2.44×10^{-3} 2.30×10^{-4} 1.08×10^{-5}
$\nu' = 5$	4.32×10^{-4} 4.40×10^{-5} 2.10×10^{-6}	2.04×10^{-3} 2.18×10^{-4} 1.06×10^{-5}	0.99260 0.99924 0.99996

Consequently we have for the dipole strengths of the emission process

$$|p^{(-1)}_{n-1\,m-1\,\nu',nm\nu}|^2 \approx \frac{n}{2\beta}\delta_{\nu'\nu} \tag{8.8}$$

and thus for the oscillator strengths

$$f^{(-1)}_{n-1\,m-1\,\nu',nm\nu} \approx -2n\delta_{\nu'\nu} \tag{8.9}$$

where use has been made of the fact that the Landau level distance is large compared with the Coulomb differences, in particular of states having the same ν, $\omega \approx \omega_c$. Comparing (8.9) with the sum rule (6.26c) it is found that the transition $nm\nu \to n-1\,m-1\,\nu$ almost completely exhausts the $\Delta m = -1$ sum rule of oscillator strengths. Finally, for the total transition probability we obtain

$$W^{(-1)}_{n-1\,m-1\,\nu',nm\nu} = \frac{32}{3}w_0\beta^2 n \tag{8.10}$$

leading to a width

$$\Gamma^{(-1)}_{n-1\,m-1\,\nu',nm\nu} = \frac{4}{3}\alpha m_e c^2 \left(\frac{B}{B_{\text{crit}}}\right)^2 \approx 5.64 \times 10^{-5}\text{eV}\,\beta^2 n \quad . \tag{8.11}$$

In fact, (8.11) is identical in form with the well-known non-relativistic result derived for the rates of cyclotron transitions of free electrons (cf. Daugherty and Ventura 1977, 1978; Herold et al. 1982). The widths following from (8.11) for the magnetic field strengths under consideration (e.g. $\beta = 50$: $\Gamma_{\text{cycl}} = 0.141$ eV; $\beta = 10^3$: $\Gamma_{\text{cycl}} = 56.4$ eV) turn out to be considerably larger than those of $\Delta n = \Delta m = 0$ transitions. Thus we have the result that, irrespective of whether the electron is bound to an ion or is moving freely along the magnetic field lines, it is the cyclotron process which predominantly determines the lifetime of a Landau-excited electron in a very strong magnetic field. In addition we find that for neutron star magnetic fields ($\beta \approx 10^3$) the cyclotron widths are almost comparable to or even larger than the corresponding Coulomb binding energies. Therefore, it no longer makes sense to consider these states as discrete states, with the exception of a few tightly bound states with sufficiently small values of m.

A further effect, which renders the situation even more complex, is caused by the anomalous magnetic moment of the electron, whereby the spin flip energy becomes different from the Landau level distance $\hbar\omega_c$. This gives rise to a spin-dependent splitting of the Landau levels for $N > 0$ and $m < N$ to lowest order in B/B_{crit} ("magnetic Lamb shift")

$$\Delta E_1 = \frac{1}{2}(g_e - 2)\hbar\omega_c \approx \frac{\alpha}{2\pi}\hbar\omega_c \approx 6.3 \times 10^{-2}\beta\,\text{eV} \quad . \tag{8.12}$$

It should be noted that the expression (8.12), which is linear in B, is a fairly good approximation for the energy splitting only up to about $\beta = 500$. The correct treatment of the B dependence up to first order in α has been given by

Constantinescu (1972), who has numerically evaluated only the $N = 0$ energy shift, however. A more complete calculation of $N > 0$ energy shifts up to $B/B_{\rm crit} = 10$ was performed by Geprägs et al. (1994).

Additionally, the finite mass of the proton yields a further energy shift ΔE_2 between states with fixed n but different m by multiples of the proton cyclotron energy $\hbar \Omega_{\rm c}$ (cf. Wunner et al. 1980, Herold et al. 1981)

$$\Delta E_2 = -\hbar \Omega_{\rm c}(\Delta m) \approx -2.96 \times 10^{-2} \beta (\Delta m)\,{\rm eV} \quad . \tag{8.13}$$

The whole situation is shown in Fig. 8.7 for the first excited Landau level and $\beta = 10^3$.

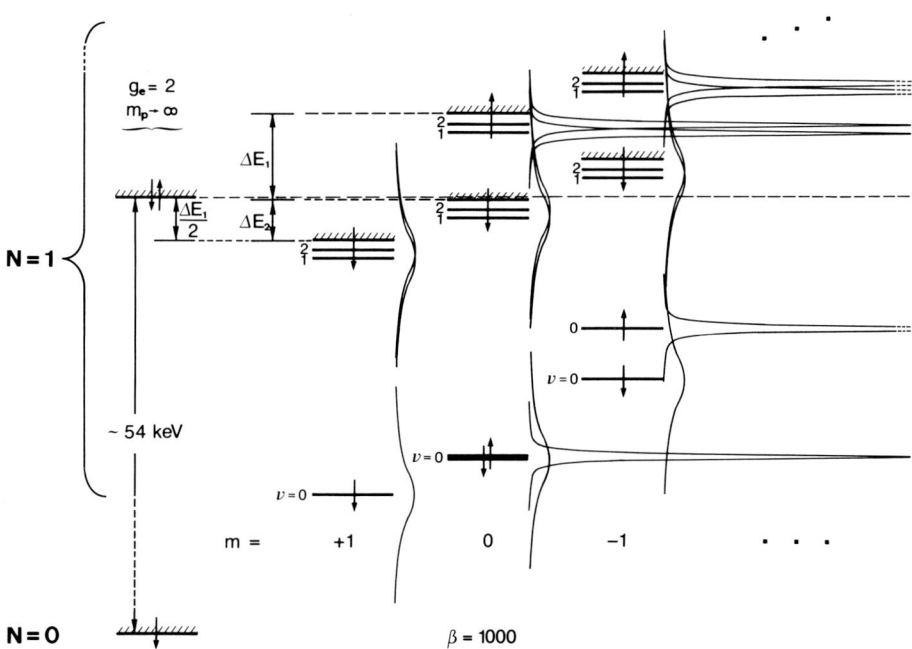

Fig. 8.7. Positions and widths of Landau-excited states ($N = 1$) of the hydrogen atom for $\beta = 10^3$. The splittings of the hatched continua for fixed m but different spin are caused by the anomalous magnetic moment of the electron (cf. (8.12)), the shifts of the continua with different m by the finite proton mass (cf. (8.13)). The decay widths of the states are essentially due to cyclotron transitions to $N = 0$. Note that the narrowness of $N = 1$ spin-up states originates in the reduction of spin-flip transitions. From the Lorentzians drawn it can be seen that only tightly bound states remain distinct states while hydrogen-like states melt into one another.

The narrow widths of the $N = 1$ spin-up states are caused by the fact that for $N = 0$ only spin-down states are present and, therefore, electric dipole transitions are not allowed. The non-relativistic spin-flip transition rate of these states turns out to be

$$\Gamma^{\rm spin-flip}_{N=1 \to N=0} = \frac{2}{3} \alpha m_{\rm e} c^2 (B/B_{\rm crit})^3 \approx 3.0 \times 10^{-9} \beta^3\,{\rm eV} \tag{8.14}$$

(cf. Herold et al. 1982). Obviously, this situation is a speciality of $N = 1$, since for $N \geq 2$, spin-up states, too, can perform dipole transitions.

The relation (8.11) for the natural line width of Landau-excited atomic states was derived from adiabatic approximation wave functions, which are adequate for strong fields. For lower field strengths, however, a multi-component form of the wave function – including several Landau states with z-dependent expansion functions – is more appropriate to the problem. Then Landau-excited states also contain small admixtures of longitudinal continuum wave functions belonging to lower Landau levels, and consequently possess a width with respect to autoionization (cf. Simola and Virtamo 1978). It is evident from our discussion that autoionization effects will become important in the same degree as the adiabatic approximation loses its validity (i.e. $\beta \leq 10\text{–}50$, depending on the respective state) and, by the rapid shrinking of the cyclotron width with decreasing β there will be a switch in the dominance of the widths of Landau-excited states from cyclotron to autoionization effects. As a matter of fact, autoionization widths have been calculated by Friedrich and Chu (1983) for the interval $0.05 \leq \beta \leq 0.5$, and by Greene (1983) for $\beta = 1$.

9 Helium-Like Atoms in Magnetic Fields of Arbitrary Strengths

As shown in the previous chapters our knowledge of hydrogen-like systems in magnetic fields of arbitrary strength is fairly good and complete although at some points further investigations are necessary. By contrast the knowledge of atomic data for helium-like atoms in strong magnetic fields is much more fragmentary. In those atoms the nonseparability of the three-body problem appears as an additional complicating fact to the competing symmetries of the Coulomb and Lorentz forces. As already discussed in our historical review in Sect. 1.2 only the lowest triplet and singlet states have been treated in some detail so far.

Besides the fundamental aspects of this problem there also exists the need for accurate atomic data for helium-like systems in strong magnetic fields from astrophysical observations. This has been stressed recently both in the context of multiple absorption features in gamma-ray burst spectra by Fenimore et al. (1988) and of features in the spectra of magnetic white dwarfs such as GD229 by Schmidt et al. (1990). Furthermore atomic data of magnetized helium are necessary for detailed calculations of the spectra of isolated neutron stars with effective surface temperatures of the order of 10^6 K (Ventura 1989).

In this chapter and in Appendix 4 we will present the results of Hartree-Fock calculations for the two-electron problem in magnetic fields ranging from $\beta_Z = 10^{-4}$ up to $\beta_Z = 10^3$ with $\beta_Z = \beta/Z^2$ for the ground state and excited states of He and the helium-like ions H$^-$, O^{6+}, Si^{12+} and Fe^{24+}. As well as in the hydrogen problem the correspondence between the field-free states and the states in the high-field limit is an important point and will be discussed at first.

9.1 Correspondence Diagrams

9.1.1 Short Review of the One-Electron Problem

In the field-free case, the conserved quantum numbers of the one-electron system (neglecting spin-orbit coupling and other relativistic effects) are the principal quantum number n_p, the angular momentum l, the magnetic quantum number m, the spin quantum number spin $s = 1/2$ which will be omitted in the following, and the quantum number m_s of the z-component of the spin. The eigenstates are labelled by n_p, l, m, m_s. In the limit of very strong magnetic fields the problem is separable in cylindrical coordinates and the states can be labelled by the Landau quantum number n, the magnetic quantum number m, the number of nodes ν of the longitudinal wave function, $s = 1/2$ and m_s, hence the states are designated by n, m, ν, m_s.

The correspondence between the field-free states and the states in the high-field limit and their quantum numbers using the non-crossing rule of quantum mechanics is discussed in Sect. 3.1.3 and, taking into account the spin and effects of the spin-orbit coupling in Sect. 8.1. According to that rule, two energy levels with the same symmetry quantum numbers of a Hamiltonian that depend continuously on a parameter cannot cross when this parameter is varied. Since the energy scheme of the hydrogen atom serves as starting point, in Fig. 9.1 the positions of low-lying levels are shown for five different magnetic field strengths on a linear energy scale. For each value of m one obtains one state of even parity and vanishing number of nodes for the longitudinal wave function. The energies of these tightly bound states decrease logarithmically with the magnetic field strength. The other hydrogen-like states have absolute energies less than the Rydberg energy E_∞.

9.1.2 The Two-Electron System in a Magnetic Field

The nonrelativistic Hamiltonian for two electrons in the Coulomb potential of a nucleus with infinite mass and charge Z and in a uniform magnetic field reads in Z-scaled units E_Z, a_Z, and B_Z

$$H = \sum_{i=1}^{2}\left(-\Delta_i - \frac{2}{r_i} + \beta_Z^2 \varrho_i^2\right) + 2\beta_Z(L_z + 2S_z) + \frac{1}{Z}\frac{2}{|\mathbf{r}_1 - \mathbf{r}_2|} \qquad (9.1)$$

with the z-components $L_z = l_{1z} + l_{2z}$ and $S_z = s_{1z} + s_{2z}$ of the total angular momentum and the total spin, respectively.

We first consider the behaviour of the energy E as a function of the magnetic field strength parameter β_Z and of the quantity $1/Z$ which characterizes the strength of the electron-electron interaction, i.e. $E = E(\beta_Z, 1/Z)$. The conserved quantum numbers are given by the symmetries of the system, and are therefore different in the low-field and high-field limits. Thus, the first step in the investigation of the continuous energy dependence is accomplished by finding the asymptotic correspondence between the quantum numbers of the low-field ($\beta_Z \to 0$) and the high-field ($\beta_Z \to \infty$) limits. This correspondence can be found by applying the non-crossing rule to the two continuous parameters β_Z and $1/Z$.

Without a magnetic field the total angular momentum L, its z-component $M = m_1 + m_2$, the total spin S, its z-component $M_s = m_{s1} + m_{s2}$, and the parity Π (alternatively the z-parity π) are good quantum numbers for the non-relativistic Hamiltonian (9.1). Due to the rotational invariance of the problem the energies are degenerate with respect to M and M_s. In the spherical Hartree-Fock (independent particle) picture we have the relations $\Pi = (-1)^{l_1+l_2}$, $\pi = (-1)^{l_1+m_1+l_2+m_2}$ and $M = m_1 + m_2$. The spherical Hartree-Fock states are in general designated by $n_{p_1}l_1m_1n_{p_2}l_2m_2$.

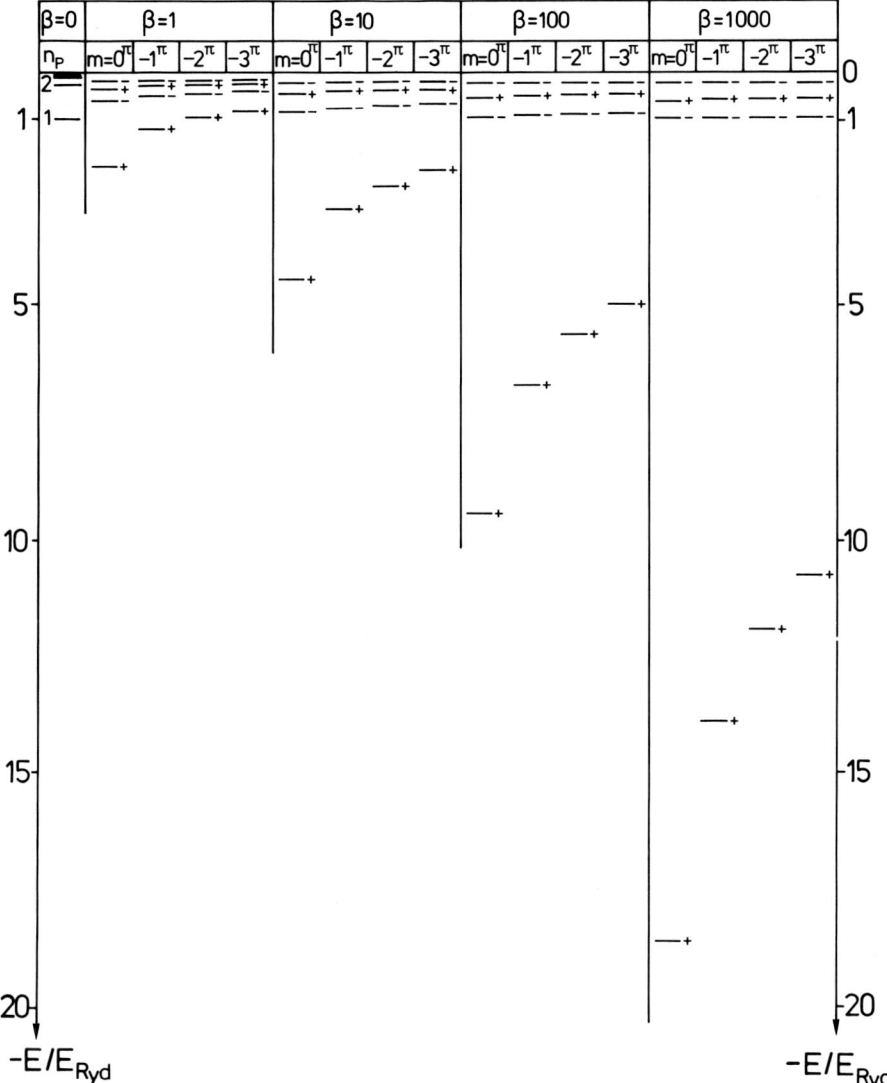

Fig. 9.1. Energy spectrum of the hydrogen atom for five different magnetic field strengths

For a non-vanishing magnetic field the rotational invariance is broken and the only conserved quantum numbers are M, S, M_s, Π or π. Furthermore the M- and M_s-degeneracy of the energy levels are removed. The cylindrical Hartree-Fock states are designated by $n_1 m_1 \nu_1 n_2 m_2 \nu_2$.

In the following we restrict our calculations to bound states where in the strong-field limit the two electrons are in the lowest Landau level ($n_1 = n_2 = 0$, $M_s = -1$), which means that we consider only triplet states, since singlet states are shifted, due to the magnetic energy of spin-up states, to the first Landau level. Therefore, we omit the Landau quantum numbers and quantum numbers for the z-components of the spins. In addition, only those states are included, which cannot autoionize. This condition can be satisfied if either both electrons are tightly bound (double tightly bound, DTB) or one electron is tightly bound and the other one is in a hydrogen-like state (single tightly bound, STB).

9.1.3 The Correspondence for $1/Z = 0$

Before considering the correspondence in general, we discuss the case of vanishing electronic interaction. For $1/Z = 0$ the projections of angular momenta m_i and the z-parities π_i are both conserved and the two-electron states can be labelled by using the single-particle quantum numbers. The correspondence can be simply accomplished by applying the non-crossing rule, discussed in Sect. 3.1.3 and by Simola and Virtamo (1978) for the one-electron system separately to both electrons. Using the Z-scaling law and the well-known magnetic field dependence of the one-electron states one obtains the energy behaviour shown in Fig. 9.2 for the DTB-states with $M = -1$, $M = -2$, $M = -3$ and the lowest STB-state with $M = -3$. All states possess positive parity. The states $1s_0 5f_{-3}/0\ 0\ -3\ 2$ and $2p_{-1}3d_{-2}/-1\ 0\ -2\ 0$ have the same magnetic quantum number $M = m_1 + m_2$ and cross at $\beta_Z \sim 2.0$. However, this is not in contradiction to the non-crossing rule, as the two states can be distinguished by their single-particle quantum numbers m_i. The state $2p_{-1}3d_{-2}$ is an intruder, which lies, for weak magnetic fields in the continuum of the $1s_0$-state, depicted by the dashed line in Fig. 9.2. With growing magnetic field strength this state intrudes into the region of bound states and crosses all $0\ 0\ -3\ \nu$ states (ν even) until it comes to lie in the vicinity of the state $0\ 0\ -3\ 0$.

For every M-value for which there exists more than one DTB-state there are additional intruders, e.g. for

$$
\begin{aligned}
&M = -4: \quad -1\ 0\ -3\ 0 \\
&M = -5: \quad -1\ 0\ -4\ 0 \quad \text{and} \quad -2\ 0\ -3\ 0 \\
&M = -6: \quad -1\ 0\ -5\ 0 \quad \text{and} \quad -2\ 0\ -4\ 0 \\
&M = -7: \quad -1\ 0\ -6\ 0, \quad -2\ 0\ -5\ 0 \quad \text{and} \quad -3\ 0\ -4\ 0
\end{aligned}
\tag{9.2}
$$

having properties similar to those described above.

Next we compare the states of the type $1s_0 n_{p2} l_2 m_2$ to those of the type $1s_0 n'_{p2} l'_2 m_2$ with $n'_{p2} = n_{p2} + 1$, $l'_2 = l_2 + 2$ and $M = m_1 + m_2 = -l_2$. If one computes these states by a one-configuration ansatz, the energy of $1s_0 n_{p2} l_2 m_2$ will be lower for weak magnetic fields than that of $1s_0 n'_{p2} l'_2 m_2$. With growing magnetic field strength the two states would cross each other, hence the state with the larger angular momentum would have the lower energy. Because the two states have identical conserved single-particle quantum numbers this level-crossing is forbidden. By the diamagnetic potential of the Hamiltonian level repulsion occurs and both states never cross, but exchange their angular momentum character at the point of smallest energy difference (anticrossing). The states $1s_0 n_{p2} l_2 m_2$ and $1s_0 n_{p2} l'_2 m_2$ with $l'_2 = l_2 + 2$ and $m_2 = -l_2$ are degenerate for $\beta_Z = 0$ and the correct states are obtained by degenerate perturbation

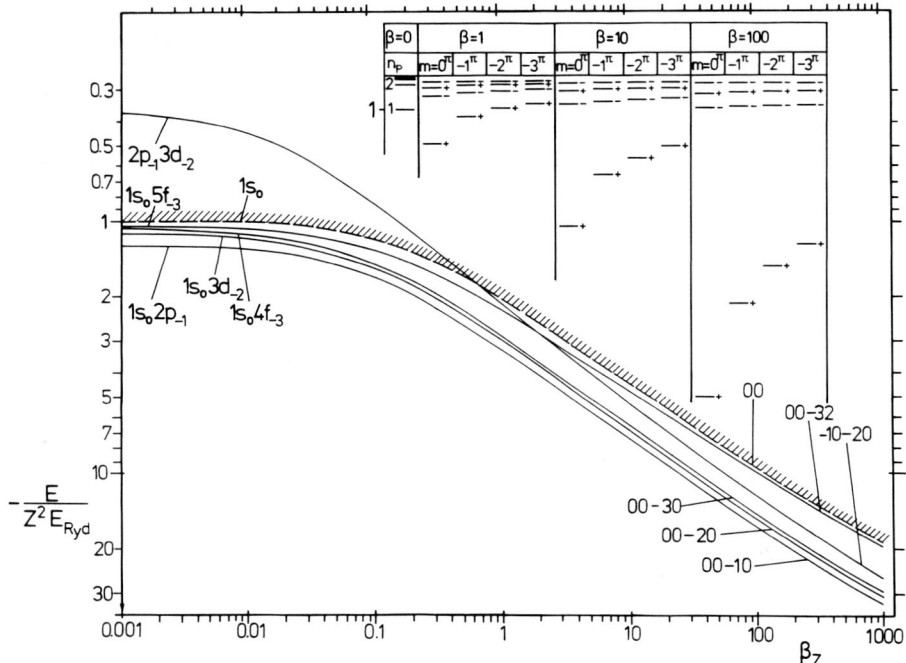

Fig. 9.2. Energies of some low-lying states of the two-electron problem for vanishing electron-electron interaction ($Z \to \infty$) as functions of the magnetic field strength. The states are labelled by their field-free and high-field quantum numbers.

theory. The coefficients of the correct linear combinations for low-lying states with $n_p \leq 6$ can be found in Table 4.1. For $\beta_Z \neq 0$ the state $1s_0 n_{p2} l'_2 m_2$ has a lower energy than $1s_0 n_{p2} l_2 m_2$.

9.1.4 The Correspondence for the General Case

If the electrostatic repulsion between the electrons is turned on, i.e. if one leaves the ($1/Z = 0$) plane, the single-particle quantum numbers are not conserved, and the only remaining good quantum numbers are M and π. In contrast with the ($1/Z = 0$) case the crossings between the intruders and the states $00m_2\nu_2$ (ν_2 even) are not allowed. In addition new anticrossings will appear, because the electrostatic electron-electron interaction will destroy the degeneracy of $1s_0 n_p lm$ and $1s_0 n_p l'm$ for $\beta_Z = 0$ and their splitting is opposite to the magnetic splitting.

Taking the anticrossings into account the following correspondence rule can be deduced for arbitrary values of $Z < \infty$ and M:

$$1s_0 n_p l M \to m_1 \; 0 \; m_2 \; \nu_2 \quad , \tag{9.3}$$

where ν_2 and $m_{1,2}$ are obtained as follows. Let $\lambda_{n_p l M}$ be the number of states of the two-electron problem with z-parity $\pi = (-1)^{l+M}$, magnetic quantum number M, and with energies below the energy of the state $1s_0 n_p l M$, including that state. By counting the states of given π and M in the order of increasing n_p and l we easily obtain

for $\pi = +1$:

$$\lambda_{n_p l M} = \sum_{n=N_{\min}}^{n_p-1} \left[\frac{n+M+1}{2}\right] + \frac{l+M}{2} + 1 \tag{9.4}$$

with $N_{\min} = 1 - M$

for $\pi = -1$:

$$\lambda_{n_p l M} = \sum_{n=N_{\min}}^{n_p-1} \left[\frac{n+M}{2}\right] + \frac{l+M+1}{2} \tag{9.5}$$

with $N_{\min} = 2 - M$. Here $[x]$ is the largest integer less than or equal to x.

According to the non-crossing rule the states for $\beta_Z = 0$ and $\beta_Z \to \infty$ of a given symmetry class labelled by π and M correspond to each other in the order of growing energy, hence we have to consider the ordering of the strong-fields states. The states of even parity belong to two groups: the DTB-states of the type $m_1 \; 0 \; M - m_1 \; 0$, and the states of the type $0 \; 0 \; M \; \nu_2$ with ν_2 even ($\neq 0$). The number N_D of DTB states for given magnetic quantum number M is $N_D = [\frac{1}{2}(-M+1)]$. The states of odd parity are all of the type $0 \; 0 \; M \; \nu_2$ with ν_2 odd. We therefore have as final result

for $\pi = +1$:

if $\lambda_{n_p l M} \le N_D$ then $m_1 = 1 - \lambda_{n_p l M}$, $m_2 = M + \lambda_{n_p l M} - 1$, and $\nu_2 = 0$ else $m_1 = 0$, $m_2 = M$, and $\nu_2 = 2(\lambda_{n_p l M} - N_D)$;

for $\pi = -1$:

$m_1 = 0$, $m_2 = M$, and $\nu_2 = 2\lambda_{n_p l M} - 1$.

The correspondence between the field-free and strong-field states is given for $M = 0$ to $M = -5$ in Table 9.1 and the qualitative behaviour of the energies is shown for $M = 0, -1, -2,$ and -3 in Figs. 9.3–9.6. The avoided crossings have been marked by circles or boxes. The states in the five regions of the two-dimensional parameter space of β_Z and Z are classified using the conserved quantum numbers besides M or using the single-particle quantum numbers of that zero-field or $(Z = \infty)$-state from which the respective state could be generated by adiabatically turning on the respective parameter or (for $\beta_Z \gg 1$ and $1/Z > 0$) by counting the states with increasing energy.

Table 9.1. Correspondence between field-free and strong-field states for the two-electron problem

M	\multicolumn{5}{c}{$\beta_Z = 0$}	\multicolumn{4}{c}{$\beta_Z \to \infty$}
	n_{p1} l_1 m_1	n_{p2} l_2 m_2	^{2S+1}L	n_1 m_1 ν_1	n_2 m_2 ν_2	π

M	n_{p1}	l_1	m_1	n_{p2}	l_2	m_2	^{2S+1}L	n_1	m_1	ν_1	n_2	m_2	ν_2	π
0	1	s	0	2	s	0	^3S	0	0	0	0	0	2	+1
				2	p	0	^3P				0	0	1	−1
				3	s	0	^3S				0	0	4	+1
				3	p	0	^3P				0	0	3	−1
				3	d	0	^3D				0	0	6	+1
−1	1	s	0	2	p	−1	^3P	0	0	0	0	−1	0	+1
				3	p	−1	^3P				0	−1	2	+1
				3	d	−1	^3D				0	−1	1	−1
				4	p	−1	^3P				0	−1	4	+1
				4	d	−1	^3D				0	−1	3	−1
				4	f	−1	^3F				0	−1	6	+1
−2	1	s	0	3	d	−2	^3D	0	0	0	0	−2	0	+1
				4	d	−2	^3D				0	−2	2	+1
				4	f	−2	^3F				0	−2	1	−1
				5	d	−2	^3D				0	−2	4	+1
				5	f	−2	^3F				0	−2	3	−1
				5	g	−2	^3G				0	−2	6	+1
−3	1	s	0	4	f	−3	^3F	0	0	0	0	−3	0	+1
				5	f	−3	^3F	0	−1	0	0	−2	0	+1
				5	g	−3	^3G	0	0	0	0	−3	1	−1
				6	f	−3	^3F				0	−3	2	+1
				6	g	−3	^3G				0	−3	3	−1
				6	h	−3	^3H				0	−3	4	+1
−4	1	s	0	5	g	−4	^3G	0	0	0	0	−4	0	+1
				6	g	−4	^3G	0	−1	0	0	−3	0	+1
				6	h	−4	^3H	0	0	0	0	−4	1	−1
				7	g	−4	^3G				0	−4	2	+1
				7	h	−4	^3H				0	−4	3	−1
				7	i	−4	^3I				0	−4	4	+1
−5	1	s	0	6	h	−5	^3H	0	0	0	0	−5	0	+1
				7	h	−5	^3H	0	−1	0	0	−4	0	+1
				7	i	−5	^3I	0	0	0	0	−5	1	−1
				8	h	−5	^3H	0	−2	0	0	−3	0	+1
				8	i	−5	^3I	0	0	0	0	−5	3	−1
				8	j	−5	^3J	0	0	0	0	−5	2	+1

9.2 Method of Solution

To determine energies and wave functions for the low-lying states of the two-electron problem in a magnetic field for field strengths in the two regimes $\beta_Z \lesssim 1$ and $\beta_Z \gtrsim 1$ we used the multi-configuration-Hartree-Fock method (MCHF) (Froese-Fischer 1977, 1978) and adapted the single-particle orbits to the respective regimes. A short description of our approach will be given below.

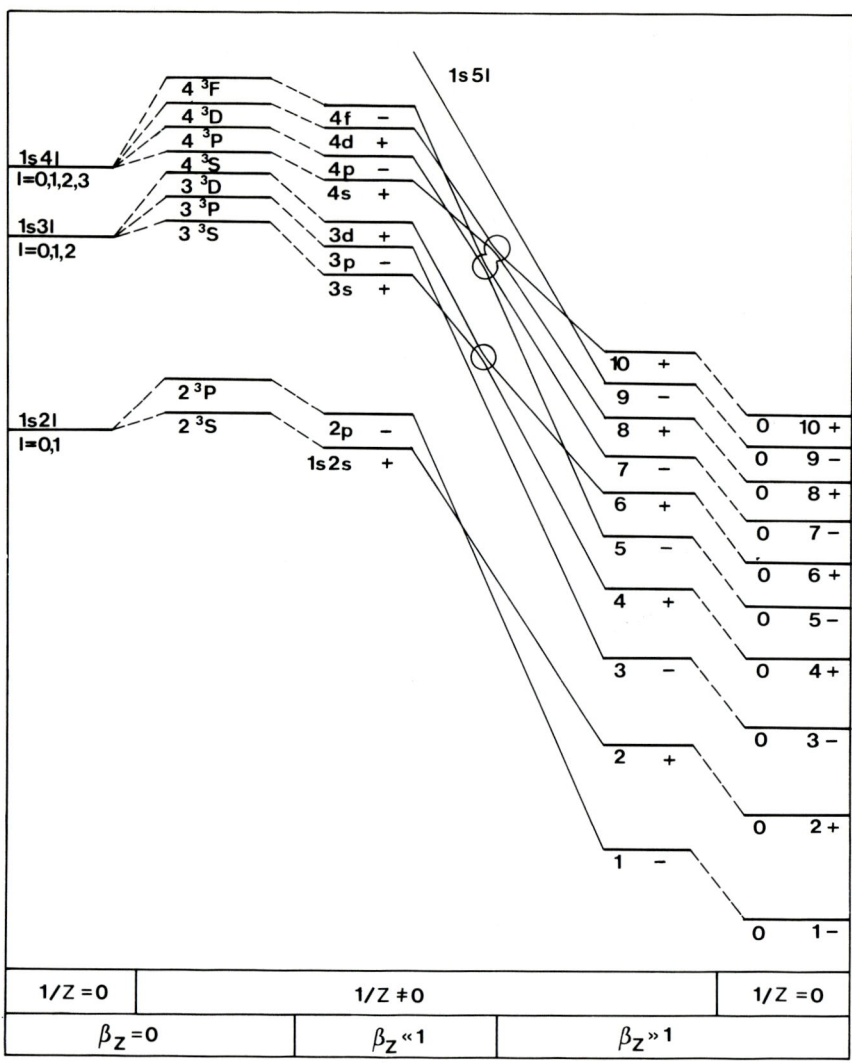

Fig. 9.3. Qualitative energy behaviour of the two-electron problem in dependence on the electron-electron interaction and the magnetic field strength for $M = 0$

9.2.1 The Hartree-Fock Method for Low to Intermediate Field Strengths ($\beta_Z \lesssim 1$)

In this regime we start from the Hamiltonian (9.1) in spherical coordinates and the single-particle wave functions are taken as

$$u_{n_p l m m_s}(r, \vartheta, \varphi, s) = \frac{1}{r} f_{n_p l m}(r) Y_{lm}(\vartheta, \varphi) \chi_{\frac{1}{2} m_s}(s) \quad . \tag{9.6}$$

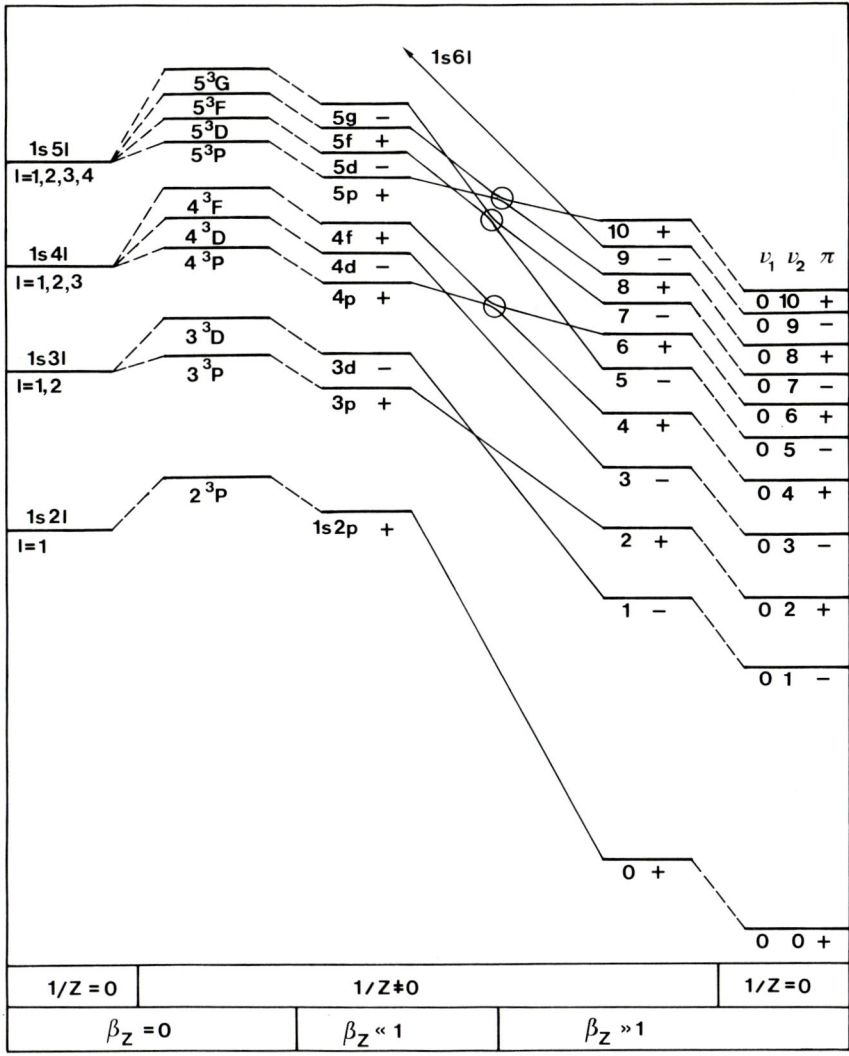

Fig. 9.4. Qualitative energy behaviour of the two-electron problem in dependence on the electron-electron interaction and the magnetic field strength for $M = -1$

To solve the MCHF equations resulting from this ansatz we used the MCHF77 code, which we extended to take into account the effects of the diamagnetic terms in the Hamiltonian. The paramagnetic terms lead only to a constant energy shift and, therefore, are easy to implement. The diamagnetic terms change the asymptotic character of the Hartree-Fock equations and introduce additional couplings between different configurations.

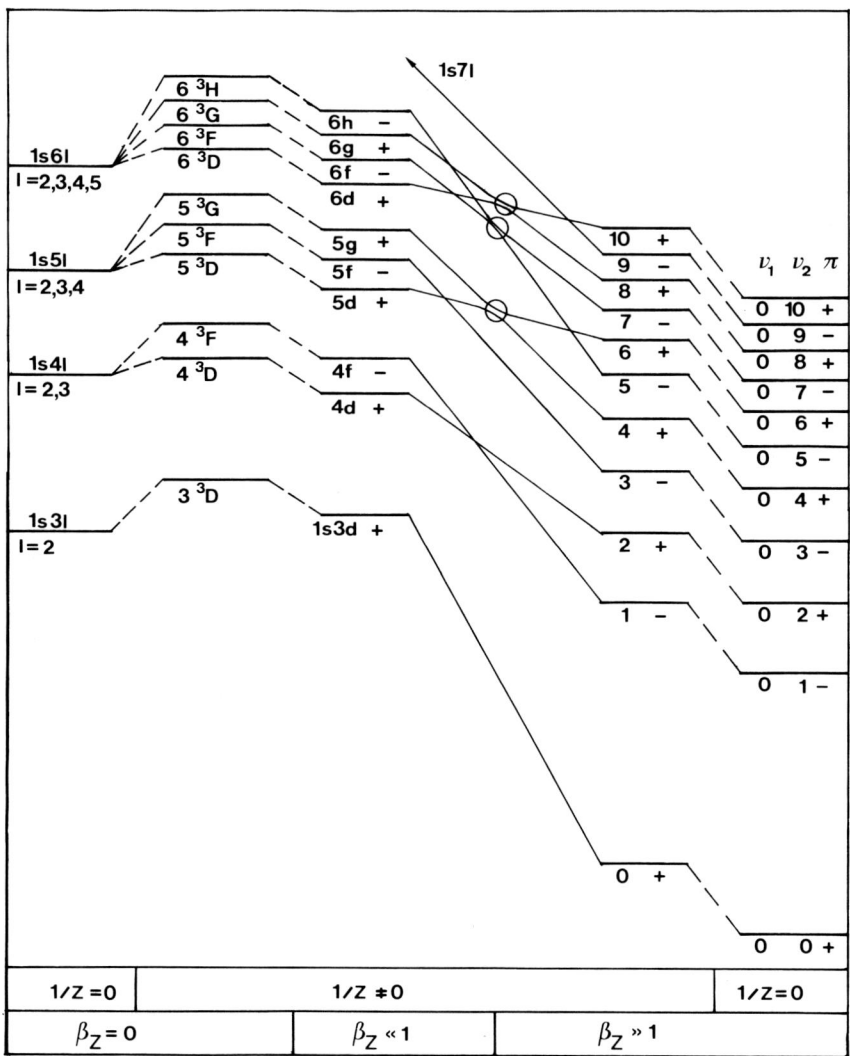

Fig. 9.5. Qualitative energy behaviour of the two-electron problem in dependence on the electron-electron interaction and the magnetic field strength for $M = -2$

9.2.2 The Hartree-Fock Method for High Field Strengths ($\beta_Z \gtrsim 1$)

For very strong magnetic field strengths ($\beta_Z \gtrsim 1$) it is more appropriate to express the Hamiltonian in terms of cylindrical coordinates:

$$H = \sum_{i=1}^{2} \left[-\nabla^2(\varrho_i, \varphi_i) - \frac{\partial^2}{\partial z_i^2} - \frac{2}{\sqrt{\varrho_i^2 + z_i^2}} + \beta_Z^2 \varrho_i^2 + 2\beta_Z(l_{z_i} + 2s_{z_i}) \right] + \frac{2}{Z} \frac{1}{|\boldsymbol{r}_1 - \boldsymbol{r}_2|} \quad . \quad (9.7)$$

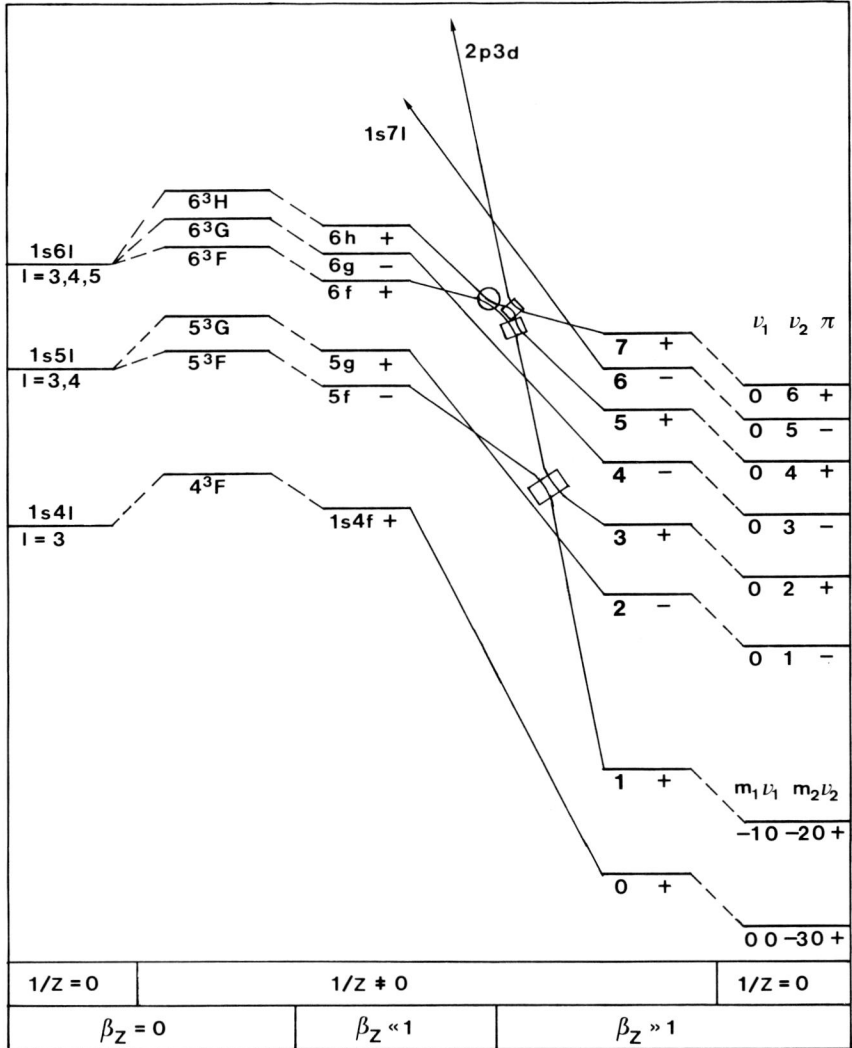

Fig. 9.6. Qualitative energy behaviour of the two-electron problem in dependence on the electron-electron interaction and the magnetic field strength for $M = -3$

In this regime the single-particle wave functions are chosen as simple products of a function depending on z and a Landau orbital, which describes the motion of an electron under the sole influence of a magnetic field (this product form constitutes the adiabatic approximation):

$$u_{m\nu m_s}(\varrho, z, \varphi, s) = g_{m\nu}(z)\phi^{\text{Lan}}_{0m}(\varrho, \varphi)\chi_{\frac{1}{2}m_s}(s) \quad . \tag{9.8}$$

As mentioned we restrict our considerations to the lowest Landau level, $n = 0$. Therefore we also have the restrictions $m \leq 0$ and $m_s = -1/2$.

To solve the resulting Hartree-Fock equations an extended version of the IMHF (intense magnetic field Hartree-Fock) code (Pröschel et al. 1982) was used. This code is also based on the MCHF77 code, but due to the change from spherical to cylindrical coordinates, the original code had to be modified considerably. In contrast to the spherical calculations only single configurations were used, i.e. no higher Landau channels were added and no M-mixing between different possibilities of obtaining M by adding m_1 and m_2 were included.

9.3 Dipole Strengths, Oscillator Strengths, and Transition Probabilities

In this section we present definitions and formulae referring to the quantities related to electromagnetic dipole transitions. In addition, we give the selection rules resulting from the Hartree-Fock method for both the spherical and the cylindrical ansatz. We use Z-scaled atomic units. The dipole matrix element for the $\Delta M = q$ dipole transition between two states f and i is defined by

$$p_{fi}^{(q)} = \left\langle f \left| \sum_{j=1}^{2} \frac{r_j^{(q)}}{a_Z} \right| i \right\rangle , \tag{9.9}$$

where $r_j^{(q)} = \sqrt{(4\pi/3)}\, r_j Y_{1q}(\vartheta_j, \varphi_j)$ stands for the spherical component of the position vector r_j, and a_Z is the Bohr radius for nuclear charge Z. The dipole strength $d_{fi}^{(q)}$, oscillator strength $f_{fi}^{(q)}$ and transition probability $w_{fi}^{(q)}$ for a $\Delta M = q$ dipole transition between two states f and i are defined by (cf. (6.2)–(6.4))

$$d_{fi}^{(q)} = |p_{fi}^{(q)}|^2 , \tag{9.10}$$

$$f_{fi}^{(q)} = \left(\frac{\hbar\omega}{E_{Z\infty}}\right) d_{fi}^{(q)} , \tag{9.11}$$

$$w_{fi}^{(q)} = \frac{1}{3} w_Z \left(\frac{\hbar\omega}{E_{Z\infty}}\right)^3 d_{fi}^{(q)} , \tag{9.12}$$

where the Z-dependent unit $w_Z = Z^4 w_0$ with $w_0 = \alpha^5 m_e c^2/2\hbar \approx 8.03 \times 10^9\ \mathrm{s}^{-1}$ for the transition probability has been introduced.

Using the fact that the antisymmetrizing operator

$$A = \frac{1}{2}(1 - P_{12}) , \tag{9.13}$$

where P_{12} exchanges the coordinates, is idempotent and commutes with the sum of the dipole operators, the dipole matrix elements between two Slater determinants can be written in the form

$$p_{fi}^{(q)} = \left\langle \Psi_1'(r_1)\Psi_2'(r_2) \left| \sum_{j=1}^{2} r_j^{(q)}(1 - P_{12}) \right| \Psi_1(r_1)\Psi_2(r_2) \right\rangle . \tag{9.14}$$

9.3.1 Selection Rules for the Spherical Ansatz

If we assume the initial and final states to be single Slater determinants

$$\Psi_i = \sqrt{2}A \frac{f_{1s_0}(r_1)}{r_1} \chi_{\frac{1}{2}-\frac{1}{2}}(s_1) \frac{f_{nlm}(r_2)}{r_2} Y_{lm}(\vartheta_2, \varphi_2) \chi_{\frac{1}{2}-\frac{1}{2}}(s_2) \quad , \tag{9.15a}$$

$$\Psi_f = \sqrt{2}A \frac{f'_{1s_0}(r_1)}{r_1} \chi_{\frac{1}{2}-\frac{1}{2}}(s_1) \frac{f'_{n'l'm'}(r_2)}{r_2} Y_{l'm'}(\vartheta_2, \varphi_2) \chi_{\frac{1}{2}-\frac{1}{2}}(s_2) \tag{9.15b}$$

we obtain for the dipole matrix elements

$$\begin{aligned}
p_{fi}^{(q)} &= \langle f'_{1s_0}|f_{1s_0}\rangle \langle f'_{n'l'm'}|r|f_{nlm}\rangle \langle Y_{l'm'}|\frac{r^{(q)}}{r}|Y_{lm}\rangle \\
&- \delta_{l'0}\delta_{m'0} \langle f'_{1s_0}|r|f_{nlm}\rangle \langle f'_{n'l'm'}|f_{1s_0}\rangle \langle Y_{00}|\frac{r^{(q)}}{r}|Y_{lm}\rangle \\
&- \delta_{l0}\delta_{m0} \langle f'_{1s_0}|f_{nlm}\rangle \langle f'_{n'l'm'}|r|f_{1s_0}\rangle \langle Y_{l'm'}|\frac{r^{(q)}}{r}|Y_{00}\rangle \quad ,
\end{aligned} \tag{9.16}$$

where the angular integrals are given in terms of Clebsch-Gordan coefficients:

$$\langle Y_{l'm'}|\frac{r^{(q)}}{r}|Y_{lm}\rangle = \sqrt{\frac{2l+1}{2l'+1}}(l,0;1,0|l',0)(1,m;1,q|l'm') \quad . \tag{9.17}$$

As is well known, these coefficients vanish unless $l + l' + 1$ is even, $|l - l'| = 1$ and $\Delta m = m' - m = q$. The results for the case of configuration mixing can be easily deduced from the above.

9.3.2 Selection Rules for the Cylindrical Ansatz

The initial and final states are given by the single Slater determinants

$$\Psi_i = \tag{9.18a}$$
$$\sqrt{2}A g_{m_1\nu_1}(z_1)\phi^{\text{Lan}}_{0m_1}(\varrho_1,\varphi_1)\chi_{\frac{1}{2}-\frac{1}{2}}(s_1) g_{m_2\nu_2}(z_2)\phi^{\text{Lan}}_{0m_2}(\varrho_1,\varphi_1)\chi_{\frac{1}{2}-\frac{1}{2}}(s_2) \quad ,$$

$$\Psi_f = \tag{9.18b}$$
$$\sqrt{2}A g'_{m'_1\nu'_1}(z_1)\phi^{\text{Lan}}_{0m'_1}(\varrho_1,\varphi_1)\chi_{\frac{1}{2}-\frac{1}{2}}(s_1) g'_{m'_2\nu'_2}(z_2)\phi^{\text{Lan}}_{0m'_2}(\varrho_2,\varphi_2)\chi_{\frac{1}{2}-\frac{1}{2}}(s_2) \quad ,$$

with $m_{1,2}, m'_{1,2} \leq 0$. Using (9.14) and expressing the dipole operators for $q \neq 0$ in terms of the ladder operators for the Landau functions, (cf. Canuto and Ventura 1977) we arrive at the following expressions for the dipole matrix elements:

For $\Delta M = 0$ transitions:

$$\begin{aligned}
p_{fi}^{(0)} &= \delta_{m_1,m'_1}\delta_{m_2,m'_2} \\
&\times \left(\langle g'_{m'_1\nu'_1}|z|g_{m_1\nu_1}\rangle \langle g'_{m'_2\nu'_2}|g_{m_2\nu_2}\rangle + \langle g'_{m'_1\nu'_1}|g_{m_1\nu_1}\rangle \langle g'_{m'_2\nu'_2}|z|g_{m_2\nu_2}\rangle \right) \\
&- \delta_{m_2,m'_1}\delta_{m_1,m'_2} \\
&\times \left(\langle g'_{m'_1\nu'_1}|z|g_{m_2\nu_2}\rangle \langle g'_{m'_2\nu'_2}|g_{m_1\nu_1}\rangle + \langle g'_{m'_1\nu'_1}|g_{m_2\nu_2}\rangle \langle g'_{m'_2\nu'_2}|z|g_{m_1\nu_1}\rangle \right) \quad ,
\end{aligned} \tag{9.19a}$$

for $\Delta M = 1$ transitions:
$$p_{fi}^{(1)} = -\langle g'_{m'_1\nu'_1}|g_{m_1\nu_1}\rangle\langle g'_{m'_2\nu'_2}|g_{m_2\nu_2}\rangle \qquad (9.19b)$$
$$\times \left(\delta_{m_2,m'_2}\delta_{m_1+1,m'_1}\sqrt{\frac{|m_1|}{2\beta_Z}} + \delta_{m_1,m'_1}\delta_{m_2+1,m'_2}\sqrt{\frac{|m_2|}{2\beta_Z}}\right)$$
$$+\langle g'_{m'_1\nu'_1}|g_{m_2\nu_2}\rangle\langle g'_{m'_2\nu'_2}|g_{m_1\nu_1}\rangle$$
$$\times \left(\delta_{m_1,m'_2}\delta_{m_2+1,m'_1}\sqrt{\frac{|m_2|}{2\beta_Z}} + \delta_{m_2,m'_1}\delta_{m_1+1,m'_2}\sqrt{\frac{|m_1|}{2\beta_Z}}\right),$$

for $\Delta M = -1$ transitions:
$$p_{fi}^{(-1)} = \langle g'_{m'_1\nu'_1}|g_{m_1\nu_1}\rangle\langle g'_{m'_2\nu'_2}|g_{m_2\nu_2}\rangle \qquad (9.19c)$$
$$\times \left(\delta_{m_2,m'_2}\delta_{m_1-1,m'_1}\sqrt{\frac{|m'_1|}{2\beta_Z}} + \delta_{m_1,m'_1}\delta_{m_2-1,m'_2}\sqrt{\frac{|m'_2|}{2\beta_Z}}\right)$$
$$-\langle g'_{m'_1\nu'_1}|g_{m_2\nu_2}\rangle\langle g'_{m'_2\nu'_2}|g_{m_1\nu_1}\rangle$$
$$\times \left(\delta_{m_1,m'_2}\delta_{m_2-1,m'_1}\sqrt{\frac{|m'_1|}{2\beta_Z}} + \delta_{m_2,m'_1}\delta_{m_1-1,m'_2}\sqrt{\frac{|m'_2|}{2\beta_Z}}\right).$$

Additional obvious selection rules result from the z-parity of the longitudinal wave functions involved.

9.4 Results for the Two-Electron Problem

9.4.1 Energy Levels Calculated with a Single-Configuration Ansatz

For states with $M = 0, -1, -2, -3$ of the two-electron systems Fe^{24+}, Si^{12+}, O^{6+}, He and H^- we present the results of SCHF (single-configuration-Hartree-Fock) calculations. In Tables A4.1–A4.5 of Appendix 4 the total energies of low-lying states of these two-electron systems are tabulated in the range $10^{-4} \leq \beta_Z \leq 10^3$. Above the underlined number, or above the first gap, in each column of the tables the results of the spherical calculations are given, below those of the cylindrical calculations. The states are designated by their one-electron quantum numbers as described in Chap. 3. A double underline indicates the crossing of two states. The energies above or below the double line belong to the the upper and lower state of the label, respectively. To assess the accuracy of the results we have included in brackets for some energy values for the ($M = -2$) states of Fe^{24+} and He and for the state $2p_{-1}/-1\,0$ of H^- the difference in per cent with respect to two-configuration calculations.

For illustrational purposes in Figs. 9.7–9.9 the continuous dependencies of energies of the DTB states with $M = -1, -2$, and -3, resulting from the spherical and cylindrical calculations, are depicted for $Z = 26, 2$ and 1 as functions of β_Z.

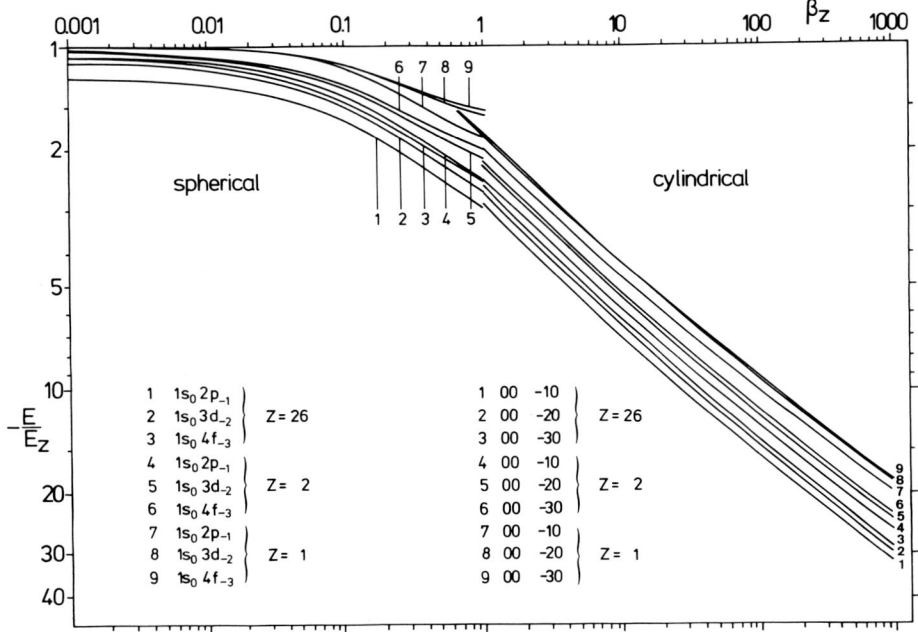

Fig. 9.7. Energies (in units of E_Z) of the three lowest double tightly bound (DTB)-states as functions of β_Z for $Z = 26$, 2, and 1

In Figs. 9.8 and 9.9 the ranges of weak ($10^{-3} \leq \beta_Z \leq 10^{-1}$) and intermediate ($2 \times 10^{-1} \leq \beta_Z \leq 2$) field strengths, respectively, have been magnified for purposes of better inspection. It is evident from Fig. 9.9 that the single-configuration spherical and cylindrical results do not match smoothly. The β_Z value for which the results of both methods intersect decreases with increasing $|m|$ and decreasing Z.

9.4.2 Influence of Configuration Mixing

The following states were calculated with more than one configuration:

$$M = 0: \quad 1s_0 2p_0 \;, \quad 1s_0 2s_0 \quad,$$
$$M = -1: \quad 1s_0 2p_{-1}, \quad 1s_0 3d_{-1} \;,$$
$$M = -2: \quad 1s_0 3d_{-2}, \quad 1s_0 4f_{-2} \;,$$
$$M = -3: \quad 1s_0 4f_{-3} \;.$$

All these states were calculated with two configurations, except $1s_0 2p_{-1}$ for which three configurations were used. The energies resulting from these calculations are given in Tables A4.6 (Fe^{24+}) and A4.7 (He). In the tables the last digit which is identical in the one- and the multi-configuration calculation is underlined. If no digit is underlined, the results differ already in the first digit. In Fig. 9.10 the ionization energies, relative to the hydrogen values of the lowest DTB states of He with $M = -1, -2$ and -3 resulting from one- to

138   9 Helium-Like Atoms in Magnetic Fields of Arbitrary Strengths

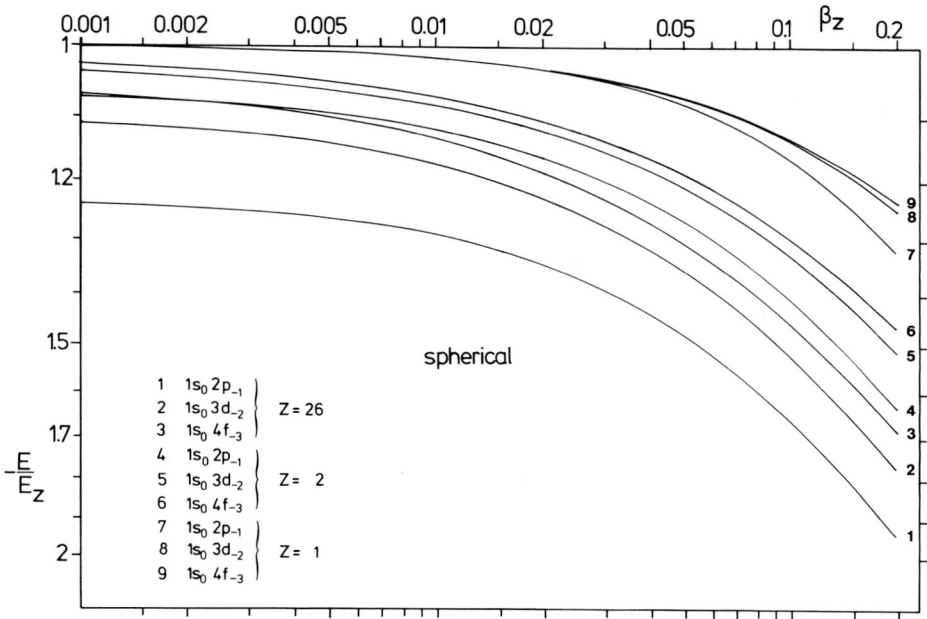

Fig. 9.8. Energies (in units of E_Z) of the three lowest DTB states as functions of β_Z in the range $10^{-3} \leq \beta_Z \leq 0.2$ for $Z = 26$, 2, and 1

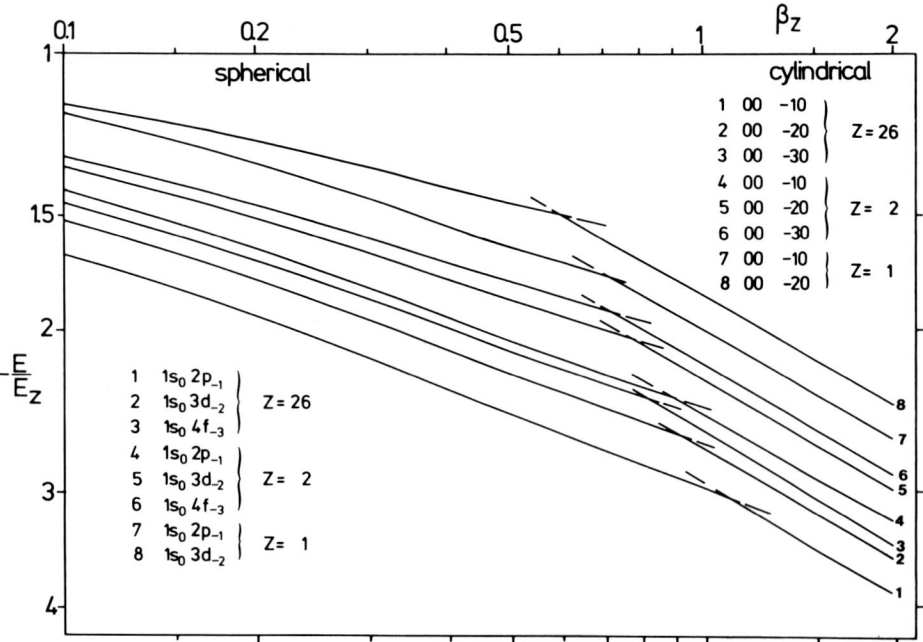

Fig. 9.9. Energies (in units of E_Z) of the three lowest DTB states as functions of β_Z in the range $0.1 \leq \beta_Z \leq 2$ for $Z = 26$, 2, and 1

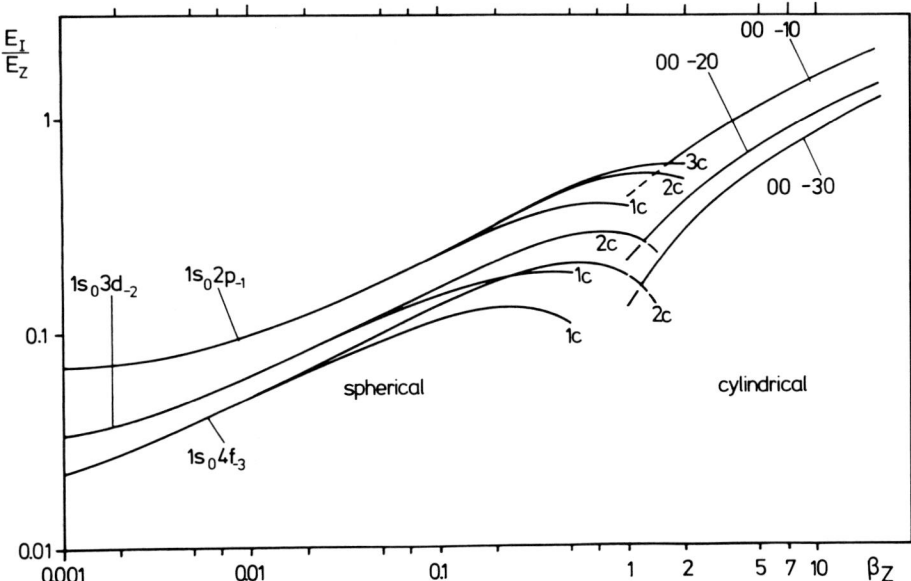

Fig. 9.10. Ionization energies (in units of E_Z) of the three lowest DTB states of He as functions of β_Z. The number of configurations is indicated by 1c, 2c, and 3c.

three-configuration calculations are plotted as functions of β_Z. One recognizes that the more one moves into a range where the magnetic field rearranges the structure of the wave functions the more important becomes the inclusion of more configurations of the spherical basis. The energy of the state $1s_0 2p_{-1}$ is lowered by mixing with the configuration $1s_0 4f_{-1}$ for $\beta_Z = 1$ and $Z = 2$ (26) by 6.08% (4.50%), and the additional mixing with $1s_0 6h_{-1}$ results in a further lowering of 1.35% (0.82%). For $\beta_Z = 0.5$ the mixing of $1s_0 4f_{-1}$ with $1s_0 2p_{-1}$, of $1s_0 5g_{-2}$ with $1s_0 3d_{-2}$ and of $1s_0 6h_{-3}$ with $1s_0 4f_{-3}$ results only in lowering the energies by 3.14%, 4.68% and 5.35%, respectively. Note that the results on the high-field side were obtained with one configuration.

In Fig. 9.11 the ionization energies, resulting from one- and two-configuration calculations, of the STB states $1s_0 2p_0$, $1s_0 2s_0$, $1s_0 3d_{-1}$, $1s_0 3p_{-1}$, $1s_0 4f_{-2}$ and $1s_0 4d_{-2}$ are shown for He as functions of β_Z. The ionization energies of single-configuration calculations have been determined with respect to single-configuration hydrogen atom results, the multi-configuration ionization energies with respect to the correct hydrogen values. For comparison the ionization energy of the DTB state $1s_0 2p_{-1}/0\ 0\ -1\ 0$ is rendered in Fig. 9.11 as dashed curve. Again the results on the high-field side were obtained with one configuration.

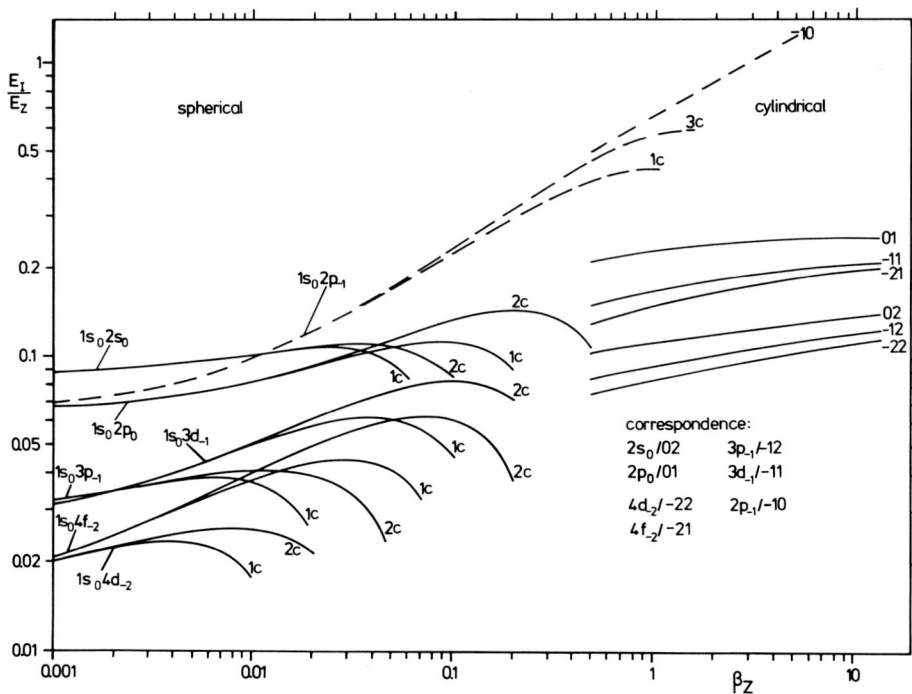

Fig. 9.11. Ionization energies (in units of E_Z) of some STB states of He as functions of β_Z. The number of configurations is indicated by 1c, 2c, and 3c. The dashed curves give the ionization energy of the DTB state $1s_0 2p_{-1}/0\ 0\ -1\ 0$.

9.4.3 Comparison of Energy Values Obtained by Different Methods

The comparison of energy values calculated by different authors is often handicapped by the technical reason that the results are presented on different magnetic field grids. To overcome this difficulty we interpolated our energy values using a polynomial fit of eighth order for $\beta_Z < 1$ and a double logarithmic polynomial fit of fifth order for $\beta_Z \geq 1$. For the region of very strong magnetic fields ($\beta_Z > 1$) Table 9.2 shows a comparison of our Hartree-Fock energies for He from Tables A4.4a,b with those calculated by Miller and Neuhauser (1991), who also used a Hartree-Fock approach in the adiabatic (i.e. one-configuration) approximation. Their energies are smaller in absolute value by several percent, and therefore inferior. As these authors work in the framework of the same physical ansatz, the discrepancy can only be accounted for by inaccuracies of their numerical procedure.

Table 9.2. Comparison of the total Hartree-Fock energies (in units of $-E_Z$) for Helium (column a) with those by Miller and Neuhauser (1991) in column b for $\beta_Z = 12.5, 50$ and 250.

β_Z	B in 10^8 T	\multicolumn{4}{c}{state $m_1\nu_1\ m_2\nu_2$}			
		\multicolumn{2}{c}{0 0 −1 0}		\multicolumn{2}{c}{0 0 −1 1}	
		a	b	a	b
12.5	0.235	6.42	6.30 (−1.9%)	4.87	4.76 (−2.3%)
50	0.940	10.36	10.22 (−1.4%)	7.71	7.57 (−1.8%)
250	4.701	17.29	17.11 (−1.0%)	12.68	12.50 (−1.4%)
		\multicolumn{2}{c}{0 0 −1 2}	\multicolumn{2}{c}{0 0 0 1}
12.5	0.235	4.77	4.67 (−2.1%)	4.91	4.79 (−2.4%)
50	0.940	7.63	7.49 (−1.8%)	7.75	7.60 (−1.9%)
250	4.701	12.54	12.43 (−0.9%)	12.71	12.52 (−1.5%)
		\multicolumn{2}{c}{0 0 0 2}		
12.5	0.235	4.80	4.69 (−2.3%)		
50	0.940	7.65	7.51 (−1.8%)		
250	4.701	12.62	12.45 (−1.3%)		

Table 9.3. Comparison of the total Hartree-Fock energies (in units of $-E_Z$) for Helium (column a) with the energies by Mueller et al. (1975) in column b, Vincke and Baye (1989) in column c and Larsen (1979) in column d.

β_Z	B in 10^8 T	\multicolumn{4}{c}{state $m_1\nu_1\ m_2\nu_2$}			
		\multicolumn{4}{c}{0 0 −1 0}			
		a	b	c	d
0.025	4.701(-4)	1.1805			1.1817 (0.10%)
0.0625	1.175(-3)	1.3082			1.3093 (0.08%)
0.125	2.351(-3)	1.4797			1.4819 (0.15%)
0.25	4.701(-3)	1.7470			1.7532 (0.35%)
0.5318	0.01	2.170	1.97 (−9.2%)	2.176 (0.28%)	
2.659	0.05	3.615	3.67 (1.5%)		
5.318	0.1	4.697	4.74 (0.9%)	4.850 (3.2%)	
26.59	0.5	8.362	8.27 (−1.1%)		
53.18	1.0	10.571	10.3 (−2.6%)	10.676 (0.99%)	
265.9	5.0	17.618	16.8 (−4.6%)		
531.8	10.0	21.636	20.29 (−6.2%)		
		\multicolumn{4}{c}{0 0 −2 0}			
0.5318	0.01	1.975		2.001 (1.3 %)	
5.318	0.1	4.315		4.460 (3.4 %)	
53.18	1.0	9.781		9.885 (1.1 %)	

Table 9.3 shows a comparison of our energies for the states 0 0 −1 0 and 0 0 −2 0 of He in the low, intermediate and high-field range with those obtained by Mueller et al. (1975), Larsen (1979) and Vincke and Baye (1989), together with the differences in per cent. For magnetic fields lower or equal than 10^6 T, which corresponds to $\beta_Z = 0.5318$, the interpolation was applied to the multi-configuration results of Table A4.7 for $\beta_Z \leq 0.7$. Compared to the variational results of Mueller et al. (1975) our energies are better at higher field strengths and slightly worse at smaller field strengths, with the exception of the multi-configuration result at 10^6 T. The variational results of Larsen for the ground state of He are found to be better by less than half a per cent. The results of Vincke and Baye (1989), who included longitudinal and transverse mixing, are seen to be better by one to three per cent.

Finally, in Table 9.4 we compare our single-configuration energies for H$^-$ for the same states with the results of the same authors as in Table 9.3. Similar to that table, the variational results by Mueller et al. (1975) are worse at higher field strengths and better at lower field strengths. The results of Larsen are, in contrast to Table 9.3, especially for larger values of β, better by several per cent. The reason is that our values are, in contrast to Table 9.3, from a single-configuration calculation. The results of Vincke and Baye are seen to be better by two to seven per cent.

Table 9.4. Comparison of the total Hartree-Fock energies (in units of $-E_{\text{Ryd}}$) for the negative hydrogen ion (*column a*) with the energies by Mueller et al. (1975) in *column b*, Vincke and Baye (1989) in *column c* and Larsen (1979) in *column d*.

		state $m_1\nu_1\ m_2\nu_2$			
β	B in 10^8 T	0 0 −1 0			
		a	b	c	d
0.05	2.351(-4)	1.0799			1.0983 (1.7%)
0.1	4.701(-4)	1.1645			1.1952 (2.6%)
0.25	1.175(-3)	1.3832			1.4481 (4.7%)
0.5	2.351(-3)	1.6188			1.7600 (8.7%)
2.127	0.01	2.669	2.83 (6.0%)	2.834 (6.2%)	
10.64	0.05	4.813	5.0 (3.9%)		
21.27	0.1	6.123	6.2 (1.3%)	6.286 (2.7%)	
106.4	0.5	10.19	10.1 (−0.9%)		
212.7	1.0	12.83	12.3 (−4.1%)	13.150 (2.5%)	
		0 0 −2 0			
2.127	0.01	2.449		2.629 (7.3%)	
21.27	0.1	5.767		5.847 (1.4%)	
212.7	1.0	12.11		12.312 (1.7%)	

9.4.4 Wavelengths, Dipole Strengths, Oscillator Strengths, and Transition Probabilities

In this section we present the results of our Hartree-Fock calculations of electromagnetic dipole transitions for two-electron systems in magnetic fields of arbitrary strength. Tables A4.8a–d contain, for He, the wavelengths (in Å) and oscillator strengths of the transitions from $1s_03d_{m'}$, $1s_04d_{m'}$, $1s_05d_{m'}$, $1s_03s_{m'}$, $1s_04s_{m'}$ and $1s_05s_{m'}$ to $1s_02p_m$ with $m', m \leq 0$. Our calculated wavelengths are about 3% larger than those calculated by Kemic (1974b). The reason for this difference is that Kemic explicitly inserted the experimental zero-field Coulomb energy values into his calculations. Our energy differences calculated in the limit $B \to 0$ are larger than the accurate results (Froese-Fischer 1977, 1978; Pekeris 1958) and the more recent high-precision results (Krivec et al. 1991, Drake 1993), since they stem from few-configuration-Hartree-Fock calculations. If our energy differences are corrected for this zero-field discrepancy, wavelengths are obtained (given in Table A4.8a as third entry in every block) that agree with those of Kemic within the first four digits. The behaviour of the wavelengths of the transitions for He covered by the tables is shown in Fig. 9.12 as continuous functions of the magnetic field strength on a double logarithmic scale. For $10^{-5} \leq \beta_{Z=2} \lesssim 3 \times 10^{-4}$ the Zeeman split lines are still separated, for higher fields the wavelengths of the different Zeeman components begin to intersect. This reflects the onset of the breakdown of the spherical symmetry.

To check in how far the wavelengths calculated in the low-field regime with a spherical basis can be joined to those in the high-field regime calculated with a cylindrical basis we show in Fig. 9.13, in the range $10^{-2} \leq \beta_{Z=2} \leq 10$, the wavelengths of the transitions $2s_0 \to 2p_0$, $2s_0 \to 2p_{-1}$, $3d_0 \to 2p_{-1}$, $3d_{-1} \to 2p_0$, $3d_{-1} \to 2p_{-1}$, $3d_{-2} \to 2p_{-1}$ and of the corresponding transitions $0\,2 \to 0\,1$, $0\,2 \to -1\,0$, $0\,4 \to -1\,0$, $-1\,1 \to 0\,1$, $-1\,1 \to -1\,0$ and $-2\,0 \to -1\,0$. (For simplicity the $1s_0/0\,0$ state is omitted in the notation.) From the correspondence of the states the qualitative behaviour of the wavelengths can be guessed in the transitional region, quantitative applications, however, e. g. to white dwarfs, evidently require the inclusion of more configurations and the use of more refined numerical methods. This is particularly true for transitions involving the $2s_0$ state.

We now turn to electromagnetic transition probabilities. On the low-field side a comparison with previously published (Garstang and Kemic, 1974) oscillator strengths is possible, therefore in Fig. 9.14 our oscillator strengths of the transitions $3d_{m'} \to 2p_m$ are shown as functions of $\beta_{Z=2}$ together with the results of Garstang and Kemic, which are indicated by crosses. In view of the different methods used the agreement is satisfactory.

In the rest of this section we discuss the results for electromagnetic transitions in high magnetic fields obtained with cylindrical single-configuration calculations. The transitions to be considered $-1\,1 \to -1\,0$, $-2\,1 \to -2\,0$, $-3\,1 \to -3\,0$, $-2\,0 \to -1\,0$, $-3\,0 \to -2\,0$, $0\,2 \to -1\,0$, are drawn, in Fig. 9.15, into the level diagram of He for a representative value $\beta_Z = 100$ neglecting the electron-electron interaction.

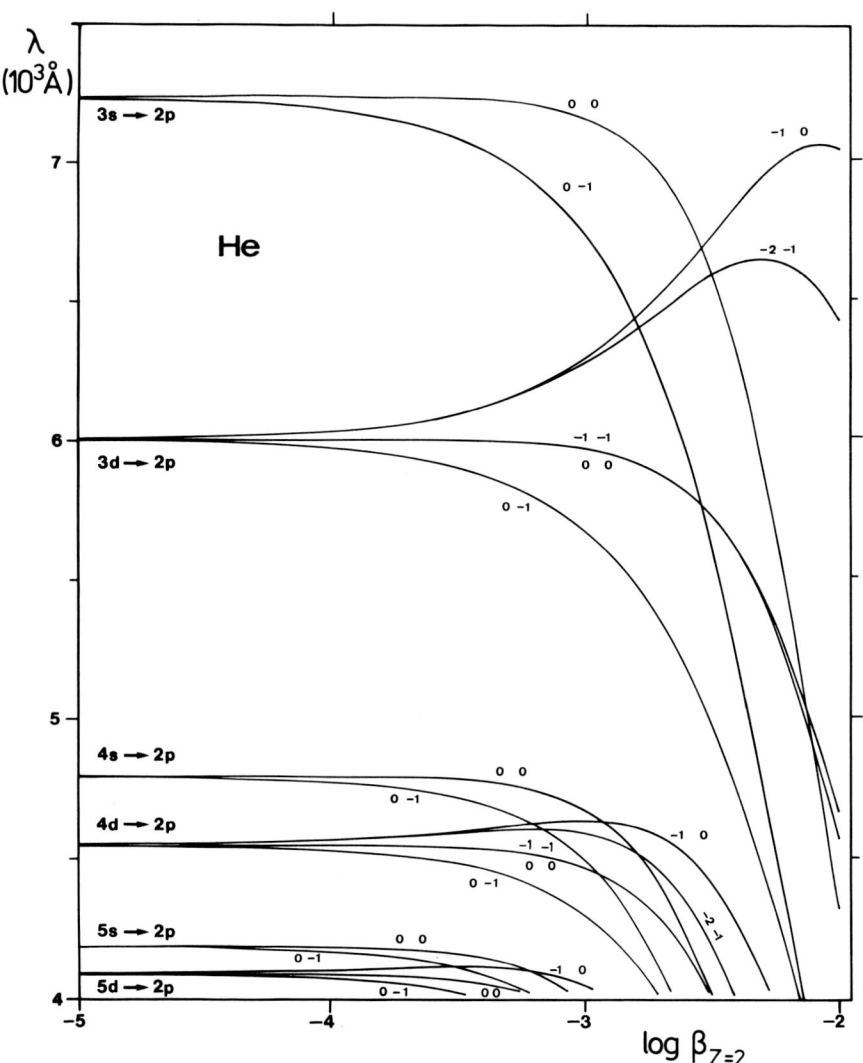

Fig. 9.12. Wavelengths of some transitions $\tau \leftrightarrow \tau'$ for He as functions of β_Z. The digits indicate the combinations (m, m').

In Tables A4.9a–f the values of dipole strengths, oscillator strengths, transition probabilities and transition energies are listed for $0.1 \leq \beta_Z \leq 10^3$ and for nuclear charges $Z = 2, 8, 14, 26, \infty$. In addition, in Table A4.10 these values are given for the transition $-1\,1 \to 0\,1$ for He only, and $\beta_{Z=2} \geq 10$. For all these transitions in Figs. 9.16 to 9.21 we have plotted the oscillator strengths as functions of β_Z for different Z values (a) and as functions of the nuclear charge Z for different values of β_Z (b), and the transition probabilities w as functions of β_Z for different Z (c).

9.4 Results for the Two-Electron Problem 145

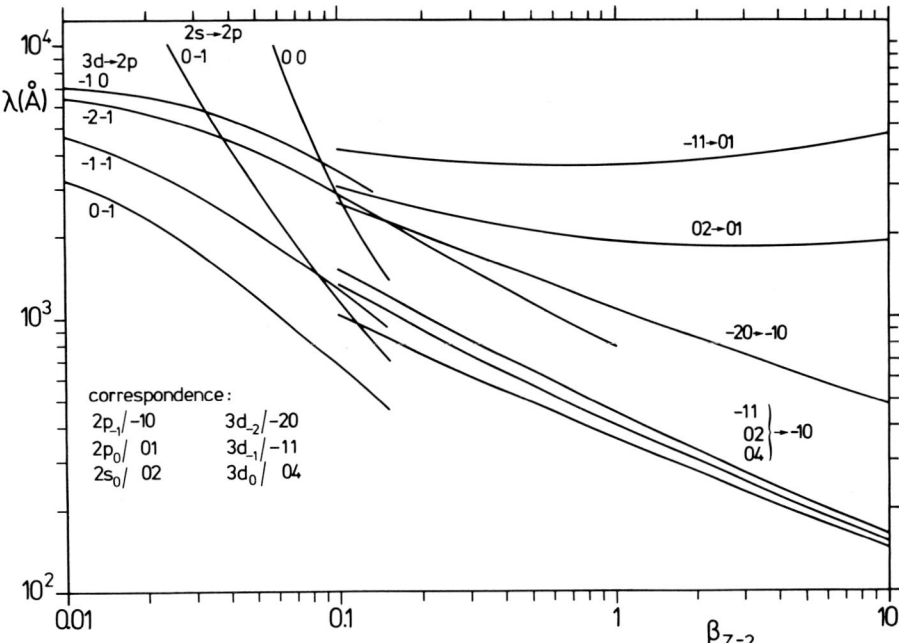

Fig. 9.13. Comparison of the wavelengths of some transitions obtained with the spherical and with the cylindrical basis for He as functions of β_Z

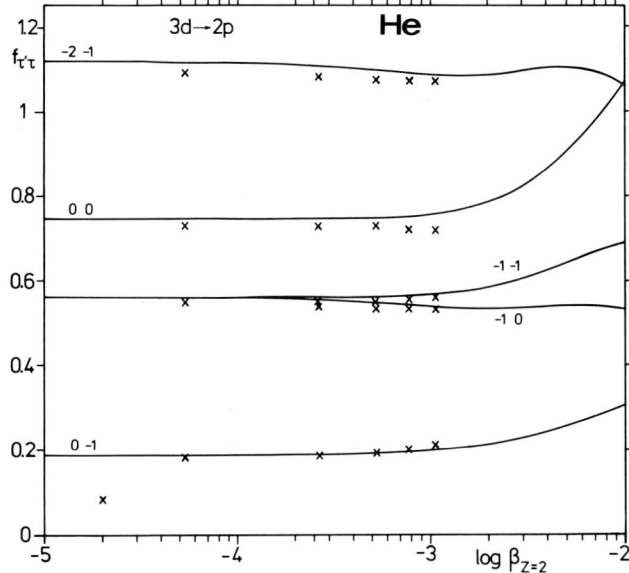

Fig. 9.14. Oscillator strengths of the transitions $3d_m \leftrightarrow 2p_{m'}$ as functions of β_Z. The digits indicate the combinations (m, m'). Crosses represent the results of Garstang and Kemic (1974).

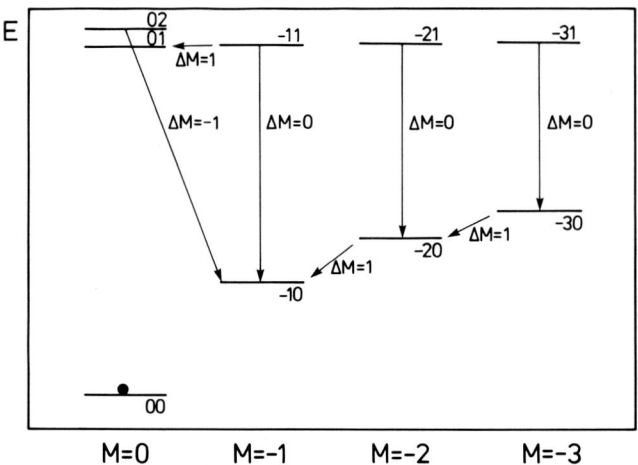

Fig. 9.15. Level diagram for the transitions to be considered for the case of vanishing interaction. The *black circle* in the lower left corner symbolizes the tightly bound electron in the 0 0 state.

The following global properties can be read off the tables and figures. In $\Delta M = 0$ transitions, the oscillator strengths are of order unity and decrease by an order of magnitude over the range $\beta_Z = 1$ to $\beta_Z = 10^3$. This can be explained as follows: The integrals $\langle g_{m1}|z|g_{m0}\rangle$, that determine the behaviour of the dipole matrix elements (the overlap integrals involved are always of order unity), decrease with growing β_Z due to the concentration of the function g_{m0} at the origin. For the oscillator strengths, defined as product of the transition energy and of the square of the dipole matrix elements, the decrease of the dipole matrix elements overcompensates the increase in the transition energy due to the strong lowering of the $\nu = 0$ states with growing β_Z, resulting in the observed moderate decrease. For all values of β_Z the oscillator strengths show a rapid saturation above $Z = 10$. Due to the increase in the transition energies the transition probabilities strongly increase with β_Z and exhibit a strong Z dependence for small values of Z. In $\Delta M = 1$ transitions the oscillator strengths are found to be smaller by about an order of magnitude, to decrease more strongly with β_Z and to show only a slight Z dependence, with rapid saturation. The oscillator strengths are smaller than for the $\Delta M = 0$ transitions, because they involve the transverse dimensions, which, especially for larger values of β_Z, are much smaller than the longitudinal dimensions. In the behaviour of the transition probabilities the effects of increasing transition energies and decreasing oscillator strengths partially compensate so that the transition probabilities go through a maximum and eventually decrease for large β_Z. In the $\Delta M = -1$ transitions the oscillator strength is smaller by yet one more order of magnitude, and exhibits an even stronger decrease with β_Z, otherwise the behaviour is similar to that of $\Delta M = +1$ transitions.

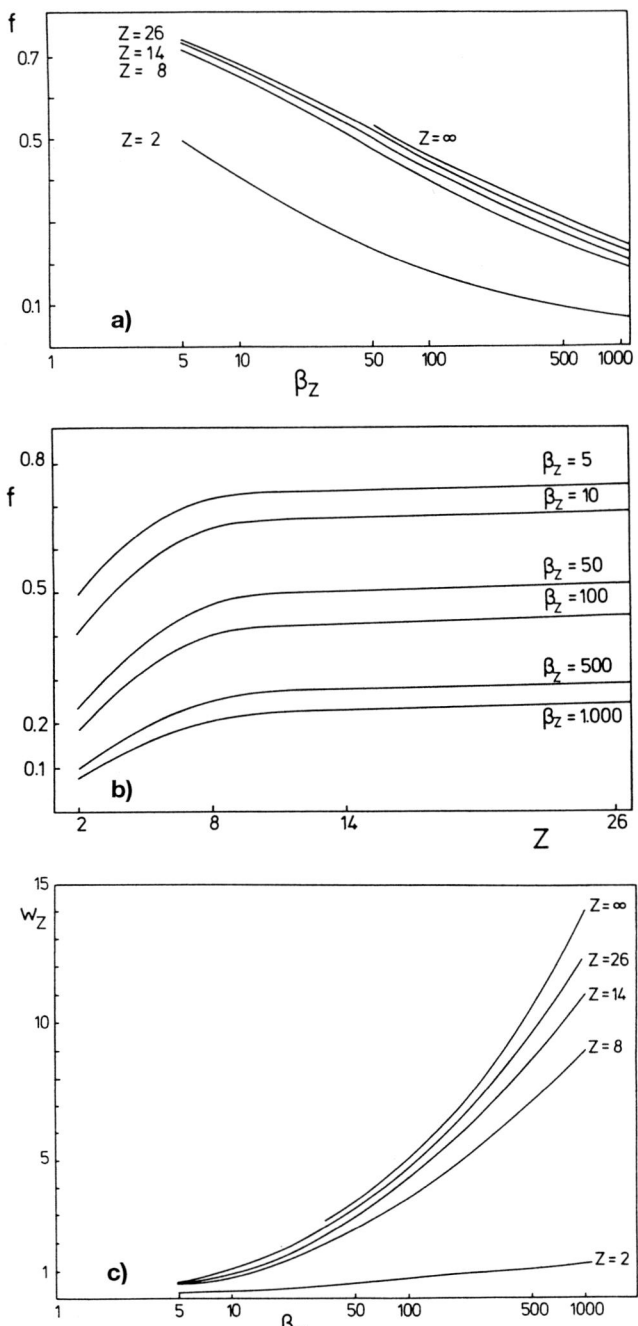

Fig. 9.16a–c. Oscillator strengths f (**a**, **b**) and transition probabilities w in units of w_Z (**c**) for the $\Delta M = 0$ transitions $0\ 0\ -1\ 1 \leftrightarrow 0\ 0\ -1\ 0$ as functions of β_Z (**a**, **c**) and as functions of the nuclear charge Z (**b**).

148   9 Helium-Like Atoms in Magnetic Fields of Arbitrary Strengths

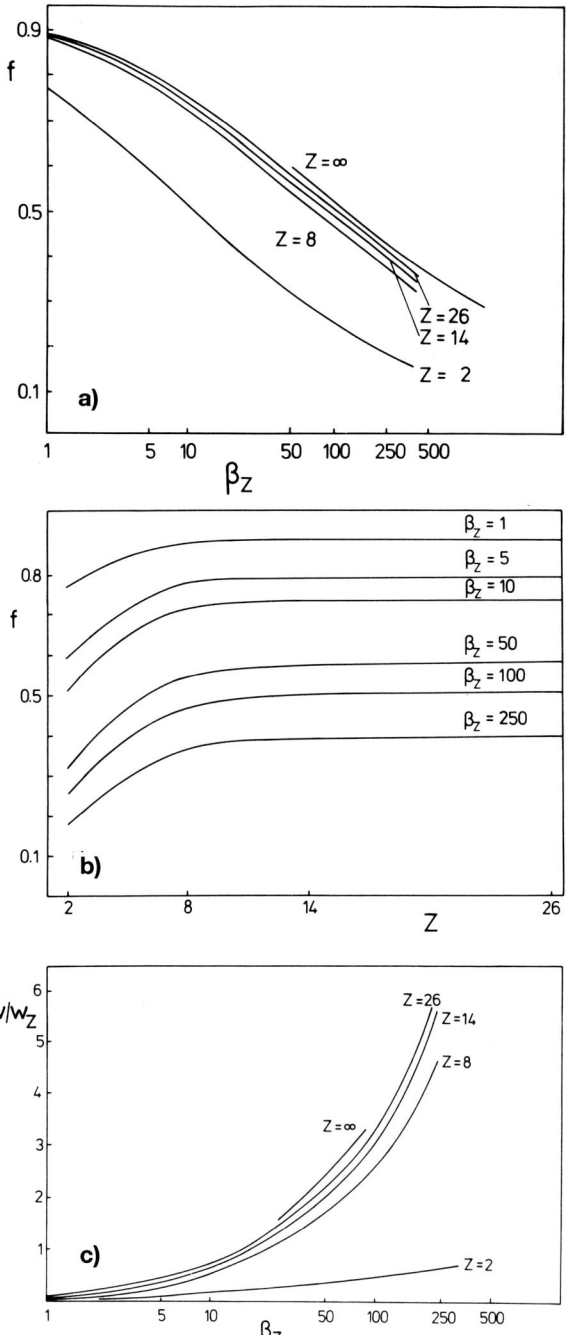

Fig. 9.17a–c. Oscillator strengths f (**a**, **b**) and transition probabilities w in units of w_Z (**c**) for the $\Delta M = 0$ transitions $0\ 0\ -2\ 1 \leftrightarrow 0\ 0\ -2\ 0$ as functions of β_Z (**a**, **c**) and as functions of the nuclear charge Z (**b**).

9.4 Results for the Two-Electron Problem

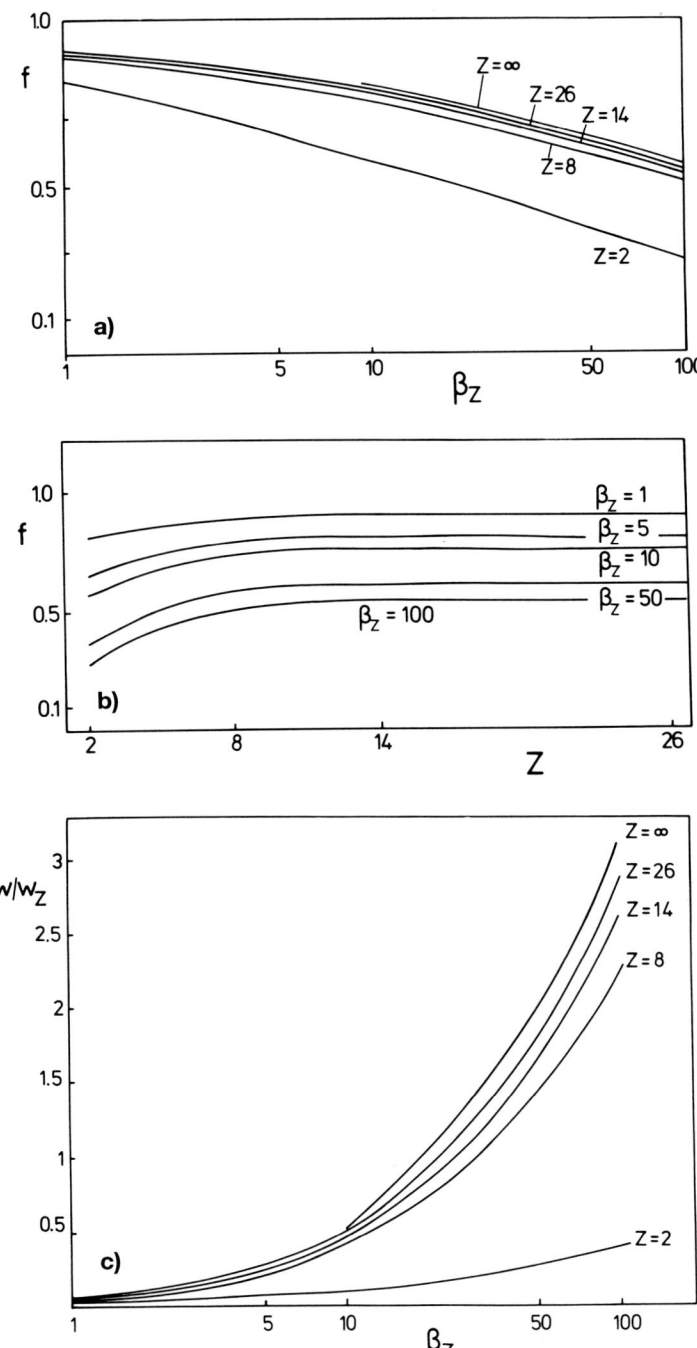

Fig. 9.18a–c. Oscillator strengths f (**a**, **b**) and transition probabilities w in units of w_Z (**c**) for the $\Delta M = 0$ transitions $0\ 0\ -3\ 1 \leftrightarrow 0\ 0\ -3\ 0$ as functions of β_Z (**a**, **c**) and as functions of the nuclear charge Z (**b**).

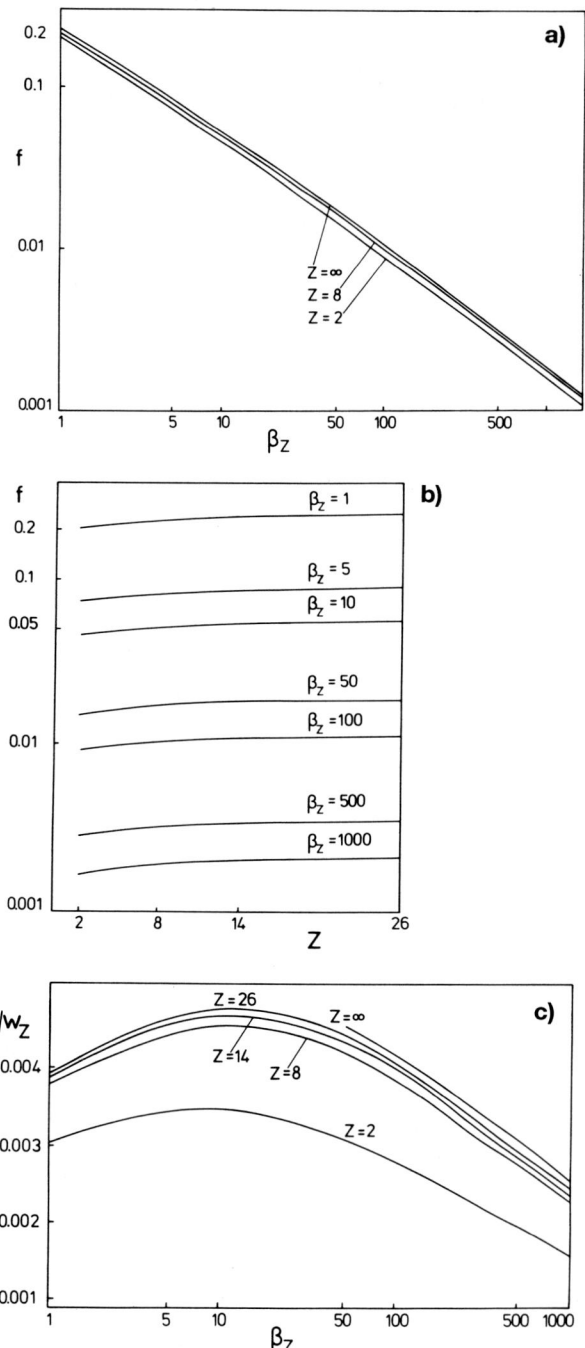

Fig. 9.19a–c. Oscillator strengths f (**a**, **b**) and transition probabilities w in units of w_Z (**c**) for the $\Delta M = 1$ transitions $0\ 0\ -2\ 0 \leftrightarrow 0\ 0\ -1\ 0$ as functions of β_Z (**a**, **c**) and as functions of the nuclear charge Z (**b**).

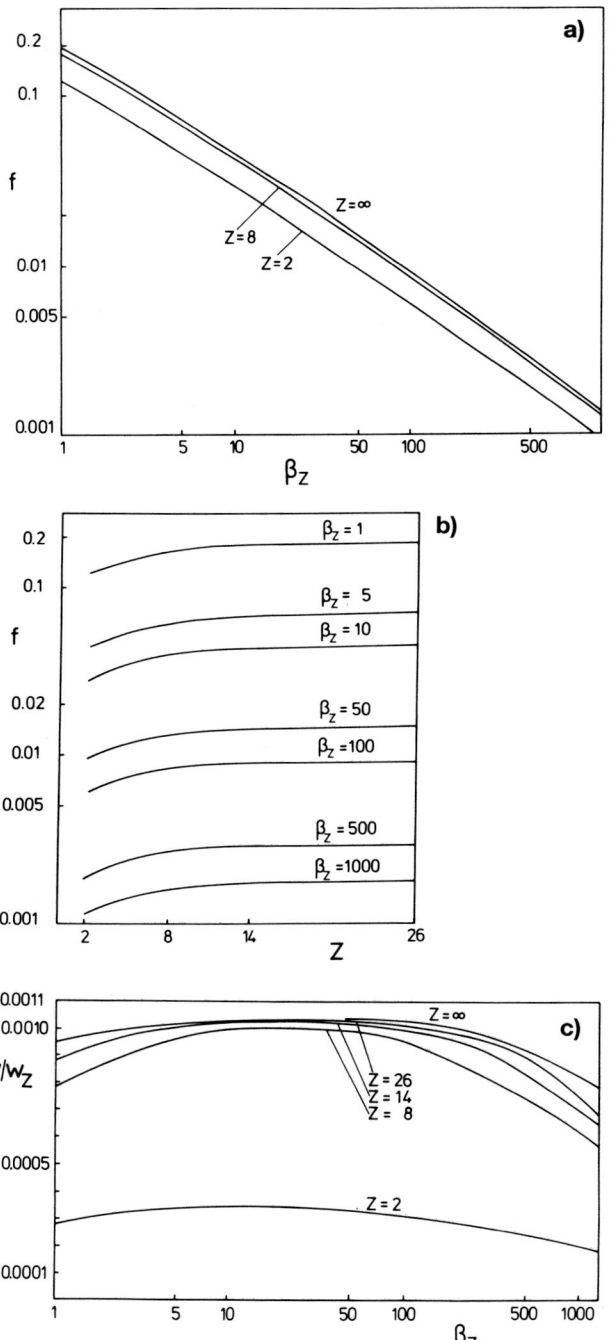

Fig. 9.20a–c. Oscillator strengths f (**a**, **b**) and transition probabilities w in units of w_Z (**c**) for the $\Delta M = 1$ transitions $0\ 0\ -3\ 0 \leftrightarrow 0\ 0\ -2\ 0$ as functions of β_Z (**a**, **c**) and as functions of the nuclear charge Z (**b**).

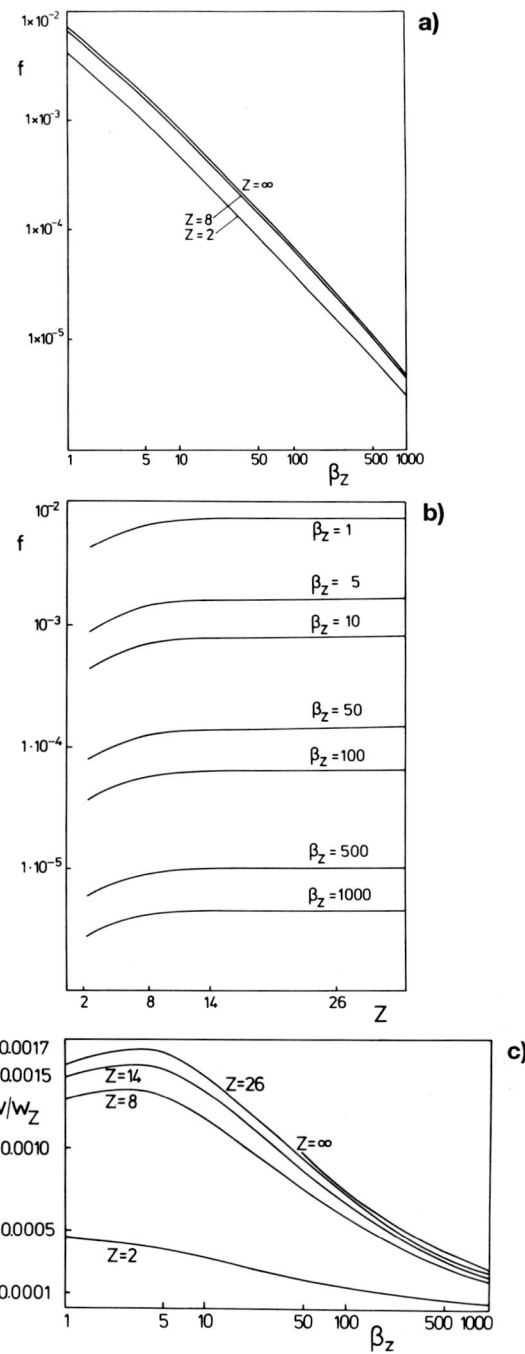

Fig. 9.21a–c. Oscillator strengths f (**a**, **b**) and transition probabilities w in units of w_Z (**c**) for the $\Delta M = -1$ transitions $0\;0\;0\;2 \leftrightarrow 0\;0\;-1\;0$ as functions of β_Z (**a**, **c**) and as functions of the nuclear charge Z (**b**).

10 Highly Excited States

In the previous chapters we have discussed in much detail the influence of strong magnetic fields on *low-lying* states. The crucial point in the whole discussion was that the atoms were allowed to be exposed to magnetic fields of such strengths as to make the effects of the magnetic field of the same order of magnitude as, or even larger than, the Coulomb binding forces acting in low-lying states in the atom. We had already noted in Sect. 3.1.1 that the reference magnetic field, which is obtained by setting electrons on Bohr orbits and requiring the equality of Lorentz and Coulomb forces, scales with the inverse cube of the principal quantum number n_p

$$B_{n_\mathrm{p}} = \frac{B_0}{2n_\mathrm{p}^3} \approx \frac{4.70 \times 10^5\,\mathrm{T}}{2n_\mathrm{p}^3} \approx 8.3 \left(\frac{30}{n_\mathrm{p}}\right)^3\,\mathrm{T} \;, \tag{10.1}$$

so that the strong-field situation, which is encountered for low-lying states only in the field strengths of white dwarfs or neutron stars, can be realized for highly excited states ($n_\mathrm{p} \geq 30$, e.g.) even in terrestrial laboratory field strengths of several Tesla. The discussion of the influence of strong laboratory magnetic fields on *highly excited states* of the hydrogen atom is the subject of this chapter.

Although uncovering the behaviour of highly excited states in strong magnetic fields is an interesting problem of physics in its own right, these "Rydberg" states have gained additional importance in recent years in connection with the fundamental question of whether or not there is *chaos* in quantum mechanics. It turns out that the hydrogen atom in a strong magnetic field is a prototype of a nonintegrable system with two degrees of freedom which classically undergoes a transition to chaos so that all the topics that are causing so much excitement in investigations of the relation between classical chaos and quantum mechanical behaviour in nonlinear systems can be studied in this real and simple system both theoretically and experimentally. Therefore a large part of this chapter will also be devoted to the discussion of the importance of highly excited states of the hydrogen atom in magnetic fields to the topical research area of "quantum chaos".

10.1 Results

10.1.1 Energy Levels

We begin by presenting results for the behaviour of the energies of highly excited states as a function of the magnetic field strength. As we have to discuss hundreds of levels in their magnetic field dependence in this part of the bound spectrum, we shall give the results in graphical rather than in tabular form. Numerical methods for solving the magnetized Coulomb problem in the range of high excitation have been discussed in detail in Chap. 3. Because of the large number of levels that one wants to describe the diagonalization of the Hamiltonian in a large complete basis set is the method of choice.

As a representative example Fig. 10.1 shows the behaviour of the energy levels of the hydrogen atom originating from multiplets with principal quantum numbers between 38 and 46 in the magnetic field range 2 to 7 T.

It is evident that the diamagnetic interaction causes a mixing of the different multiplets already below 2 T, from where on n_p is no longer a good quantum number (n_p-mixing regime). Nevertheless it is possible to trace the individual levels over a wide range of the field, an indication of the fact that in spite of the competition of the Coulomb and Lorentz forces in the atom the problem is still quasi-integrable. In other words, for magnetic fields not too strong there still exists an approximate third constant of motion other than the energy and the z-component of the angular momentum. This approximate constant of motion Λ, which is related to the Runge-Lenz vector $\boldsymbol{A}$, was discovered as late as 1982 by Solov'ev (1982a,b) and is given by

$$\Lambda(\boldsymbol{A}) \equiv 4\left(A_x^2 + A_y^2\right) - A_z^2 = 4\boldsymbol{A}^2 - 5A_z^2 \;, \tag{10.2}$$

where A_x, A_y and A_z are the three Cartesian components of the Runge-Lenz vector, which can be written with $\boldsymbol{l} = \boldsymbol{r} \times \boldsymbol{p}$ in the form

$$\boldsymbol{A} = \boldsymbol{p} \times \boldsymbol{l} - \frac{\boldsymbol{r}}{r} = \left(p^2 - \frac{1}{r}\right)\boldsymbol{r} - (\boldsymbol{r}\cdot\boldsymbol{p})\boldsymbol{p} \;. \tag{10.3}$$

Loosely speaking, Λ takes the place of the angular momentum as a constant of motion for not too strong magnetic fields. Since, in classical terms, $|\boldsymbol{A}| = \sqrt{1 + 2El^2}$ represents the eccentricity of a Kepler ellipse, less than unity for $E < 0$, the allowed values of Λ are found to lie in the range $-1 \leq \Lambda \leq 4$. The levels of a given n_p-manifold are ordered according to ascending values of Λ.

The level crossings in Fig. 10.1 are always avoided crossings, but it is only beyond the breakdown of quasi-integrability that the anticrossings become sufficiently large to be visible in this figure (see inset). It can also be recognized that in every diamagnetic multiplet the lowest state possesses the minimum rate of variation of its energy as a function of the field: it is the state whose wave function has maximum spatial probability distribution along the magnetic field axis; conversely, the state with the maximum variation of energy with the field is the one with maximum spatial probability distribution of the

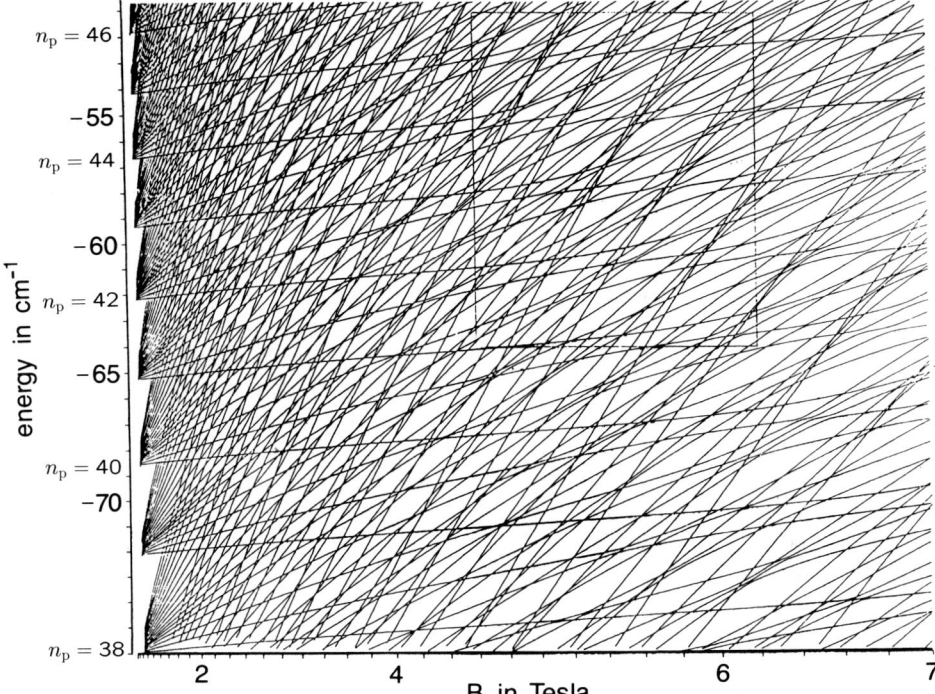

Fig. 10.1. Level scheme of the hydrogen atom in the energy range $-76\,\mathrm{cm}^{-1}$ to $-50\,\mathrm{cm}^{-1}$ (corresponding to principal quantum numbers $38 \leq n_\mathrm{p} \leq 46$ in the field-free case) as a function of the (square of the) magnetic field strength. It is recognized that the diamagnetic multiplets belonging to different principal quantum numbers are completely interwoven at high field strengths. (From Friedrich and Wintgen 1989)

electron in the plane perpendicular to the direction of the magnetic field. It is obvious that this distribution in configuration space will react most readily to the diamagnetic term in the Hamiltonian. The general impression conveyed by Fig. 10.1 is that in a strong magnetic field there is a complete interweaving of levels originating from different principal quantum numbers with the result that the density of states is high and the mean level spacing is small compared to the field-free case.

10.1.2 Transition Probabilities and Comparison with Experiments

With the eigenvectors, following from the diagonalization procedure, at hand it is a straightforward task to compute dipole strengths, oscillator strengths etc. (cf. Chap. 6) for electromagnetic transitions to highly excited states. In our own calculations, we increased the basis sizes until an overall accuracy of oscillator strengths was obtained to within two to three significant digits up to the field-free ionization threshold. The maximum basis sizes we used were $\sim 300\,000$, the number of matrix elements different from zero was about 10^8;

these calculations of course required the use of a powerful supercomputer, in our case the Cray 2 at the University of Stuttgart. We note that in the results presented below the mass scaling laws valid in a magnetic field (Sect. 2.3), which relate the field strength, energies, and oscillator strengths pertaining to the infinite-nuclear-mass Hamiltonian to those belonging to finite core mass, are taken into account.

As an example Figs. 10.2 and 10.3 show the spectra calculated for $B = 6$ T (10.2) and for $B = 5.96$ T (10.3) for the oscillator strengths of $\Delta m = 0$ Balmer transitions from $2p_0$ to Rydberg states with $m = 0$ and even parity in the energy range $-190\,\text{cm}^{-1}$ to $-20\,\text{cm}^{-1}$ (10.2) and $-80\,\text{cm}^{-1}$ to $0\,\text{cm}^{-1}$ (10.3). The figures also contain, for comparison, the results of corresponding experiments carried out by Welge's group at the University of Bielefeld (Holle et al. 1986, Main et al. 1986). In these experiments the Rydberg states were excited by resonant two-photon absorption: in a first step a (frequency-doubled) Lyman excitation was performed from the ground state to the Paschen-Back resolved $2p$ substates, from where a tunable dye-laser took the electron up to close below or above the ionization threshold. The experiments were conducted with deuterium to reduce possible motional electric fields by the higher atomic mass.

A total of 177 lines are compared in Fig. 10.2, which covers the energy range between $-190\,\text{cm}^{-1}$ and $-70\,\text{cm}^{-1}$. The theoretical results shown in the figures were obtained for a value of the magnetic field parameter $\beta = 1.275 \times 10^{-5}$ corresponding, for deuterium, to $B = 5.997$ T. We encourage the reader to go through the figure line by line. The agreement between the theoretical predictions and the experimental values for the positions and intensities of the lines evidently is excellent. Moreover, theory reveals where neighbouring lines were no longer resolved in the experiment. Slight differences in the relative intensities of a few lines are due to the fact that saturation effects occurred in the experiment in strong lines, and, also, to the uncertainty in the strength of the magnetic field. The range of energy considered in Fig. 10.2 extends from the onset of the overlap of neighbouring n_p manifolds deeply into the n_p-mixing regime. The overall structure of the spectrum still seems relatively ordered, with the clusters of lines belonging to the different n_p manifolds interpenetrating, at least at the beginning, without substantial mutual perturbation. This behaviour, which was already pointed out in the earlier calculations of Clark and Taylor (1982), and even allows labelling the states in terms of weak-field quantum numbers, is again closely related to the existence of the approximate constant of motion discussed above, and thus to the approximate integrability of the problem in this range of energy.

The comparison between experimental and theoretical spectra shown in Fig. 10.3 in the energy interval $[-80\,\text{cm}^{-1},\,0\,\text{cm}^{-1}]$ proceeds to regions where this approximate integrability is rapidly destroyed, as witnessed by the disappearance of very narrow avoided crossings of the energy levels when viewed as functions of the field. In this case the experimental data were taken at a magnetic field strength of 5.96 T, and the theoretical results were determined for $\beta = 1.2675 \times 10^{-5}$ ($B = 5.9619$ T). A total of 639 lines contributes to the theoretical spectrum in that energy range. As one moves up in energy in

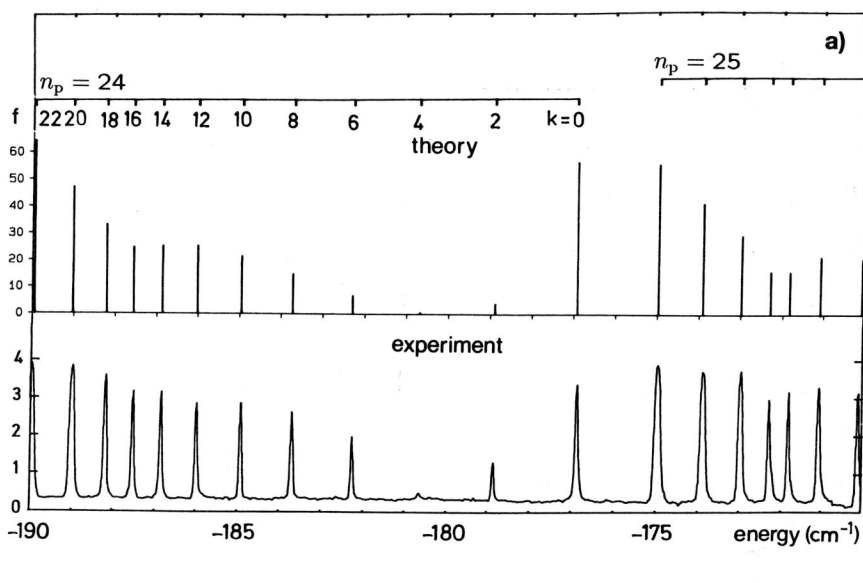
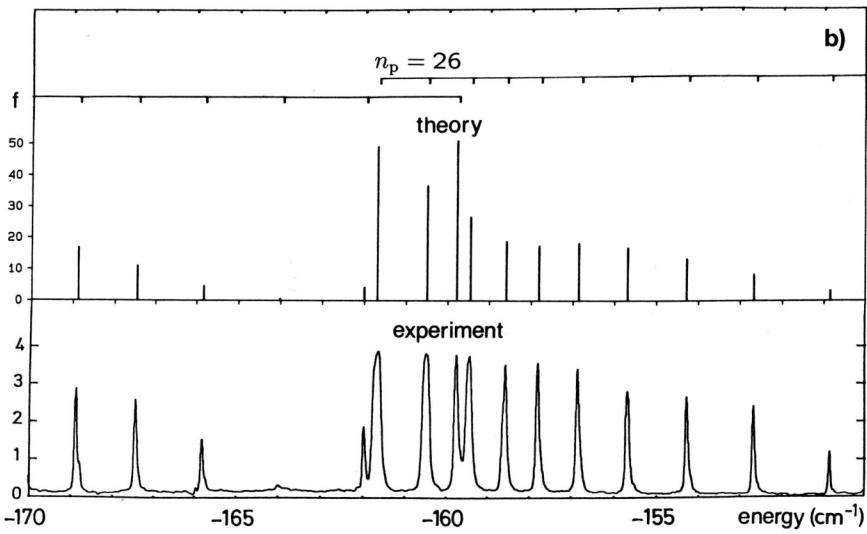

Fig. 10.2a–f. Deuterium Rydberg atoms in a magnetic field of 6.0 T: comparison between the theoretical oscillator strength spectrum and the experimental photoabsorption spectrum for $\Delta m = 0$ Balmer transitions to $m = 0$, even-parity Rydberg states, over the range of energy $-190\,\text{cm}^{-1}$ to $-70\,\text{cm}^{-1}$. Oscillator strengths are given in units of 10^{-6}, the experimental intensity scale is in arbitrary units. The six successive energy intervals cover the range from the onset of n_p-mixing (in **a** the $n_p = 24$ and $n_p = 25$ manifolds are still just separated from each other, while in **b** the uppermost state of the $n_p = 25$ manifold lies already inside the $n_p = 26$ manifold) to the regime where the approximate integrability of the problem is destroyed to an ever increasing extent (**f**). A total of 177 lines contribute to the spectrum shown. The index K labels the states within a given diamagnetic n_p-multiplet, even or odd values of K distinguish states with even or odd parity, respectively. (From Holle et al. 1987)

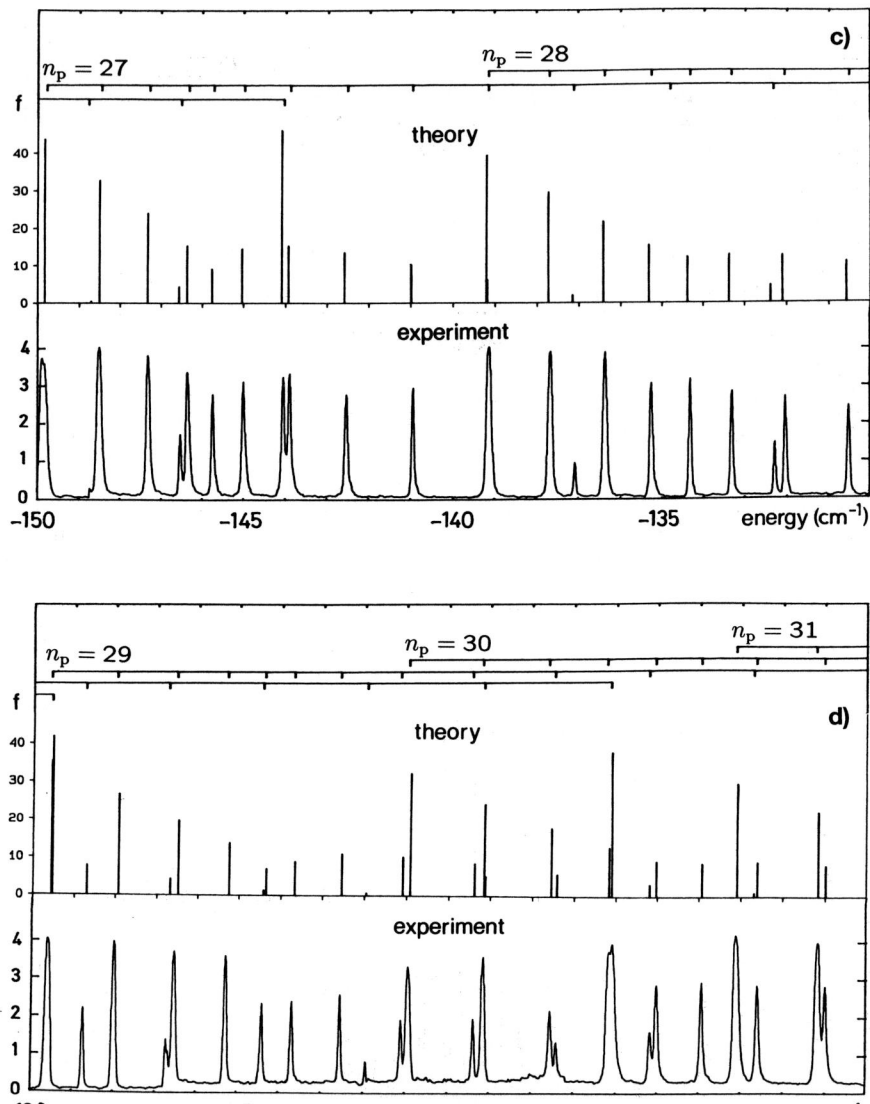

Fig. 10.2a–f. (Continued)

Fig. 10.3, the line density is clearly growing, and the structure of the spectrum becomes increasingly complicated. Nevertheless it is also here that we find practically complete agreement between theory and experiment, with the experiment being limited, as regards the number of detectable lines, by the finite resolution. Furthermore, it is recognized from Fig. 10.3 that the experiment still reproduces, in spite of the too low resolution, the essential features of the numerically computed spectrum even up to the field-free ionization threshold.

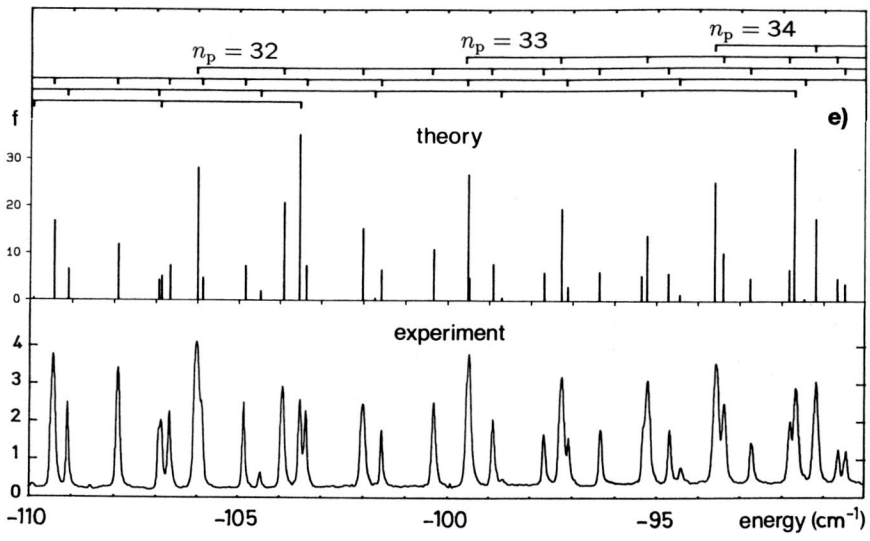

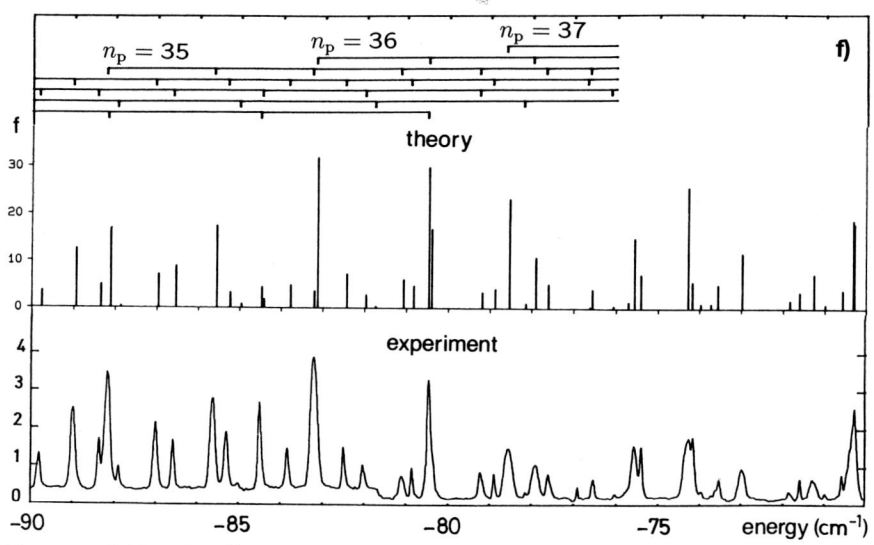

Fig. 10.2a–f. (Continued)

For the sake of completeness we also present the theoretical and experimental spectra for σ-transitions to the Rydberg states of deuterium in a field of 6 T (Fig. 10.4). Also for this type of transitions we find excellent agreement between the numerically computed and the measured energies and oscillator strengths. All in all more than a thousand lines are successfully compared in Figs. 10.2–4!

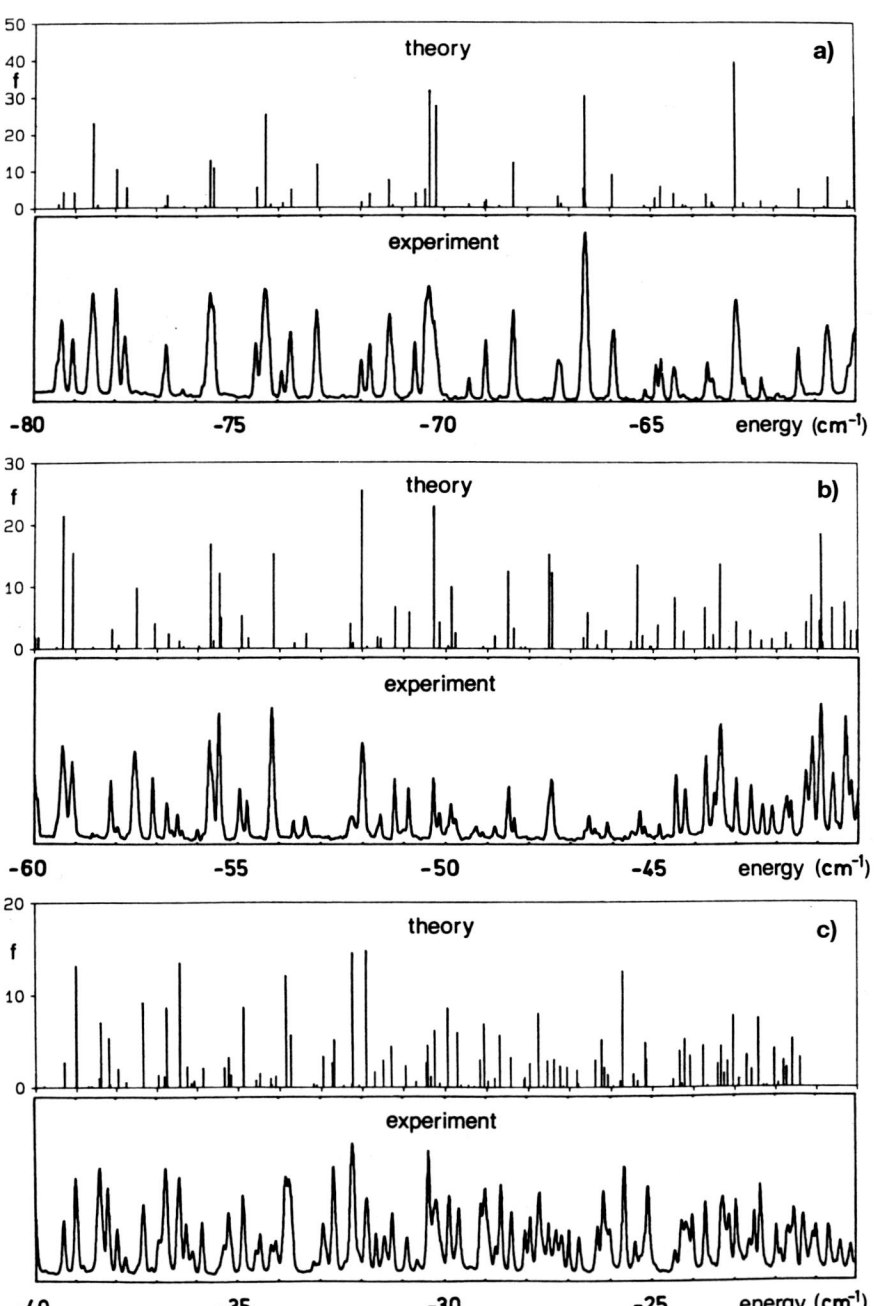

Fig. 10.3a–e. Same as Fig. 10.2 but for a magnetic field of 5.96 T and energies of the Rydberg states between $-80\,\mathrm{cm}^{-1}$ and $0\,\mathrm{cm}^{-1}$, the field-free ionization threshold. Note that at the end of the energy range (above ~ -25 cm^{-1}) the corresponding classical system becomes completely chaotic. 639 lines are shown in this spectrum. (From Holle et al. 1987 and Zeller 1990)

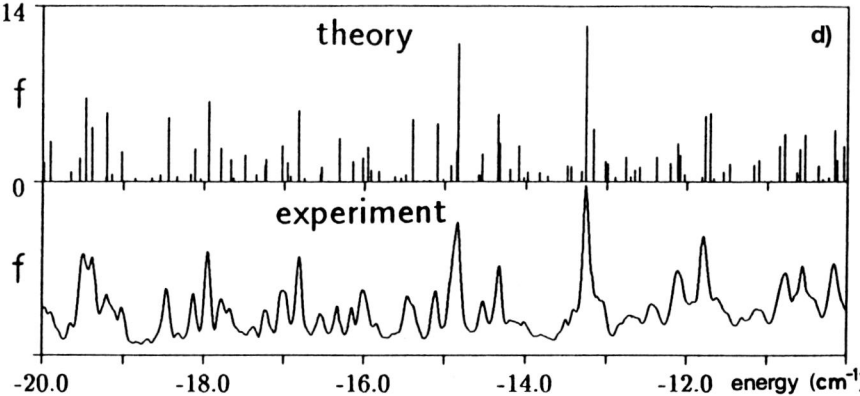

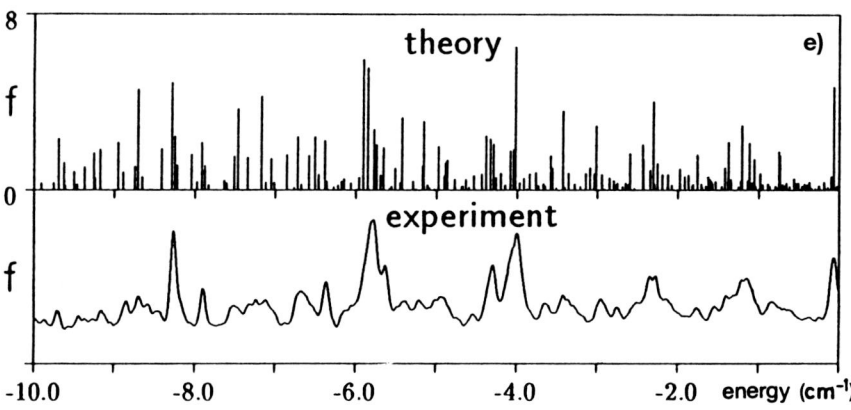

Fig. 10.3a–e. (Continued)

While these comparisons focussed on the bound-state part of the spectrum, progress in the numerical methods has made it possible to compute transition rates also for continuum states, and thus extend the comparison into the positive-energy region. Figure 10.5 shows, for 6 T and the energy range $-30\,\text{cm}^{-1}$ to $+30\,\text{cm}^{-1}$, the comparison between the experimental spectrum obtained by the MIT group (Iu et al. 1991) for odd-parity states of lithium (which are almost hydrogen-like because of the small quantum defects of odd parity states) and the spectrum computed by Delande et al. (1991). To calculate the photoionization cross section in the positive-energy region these authors adapted the complex-rotation method (cf. Reinhardt 1982), which consists of formally replacing the spatial coordinates r with $re^{i\theta}$ in the Hamiltonian and the wave functions. Continuum wave functions can then be described by square-integrable basis functions with complex arguments, and the diagonalization methods in large, complete basis sets that proved so efficient for bound states can directly be transferred to the positive-energy region. Again the agreement

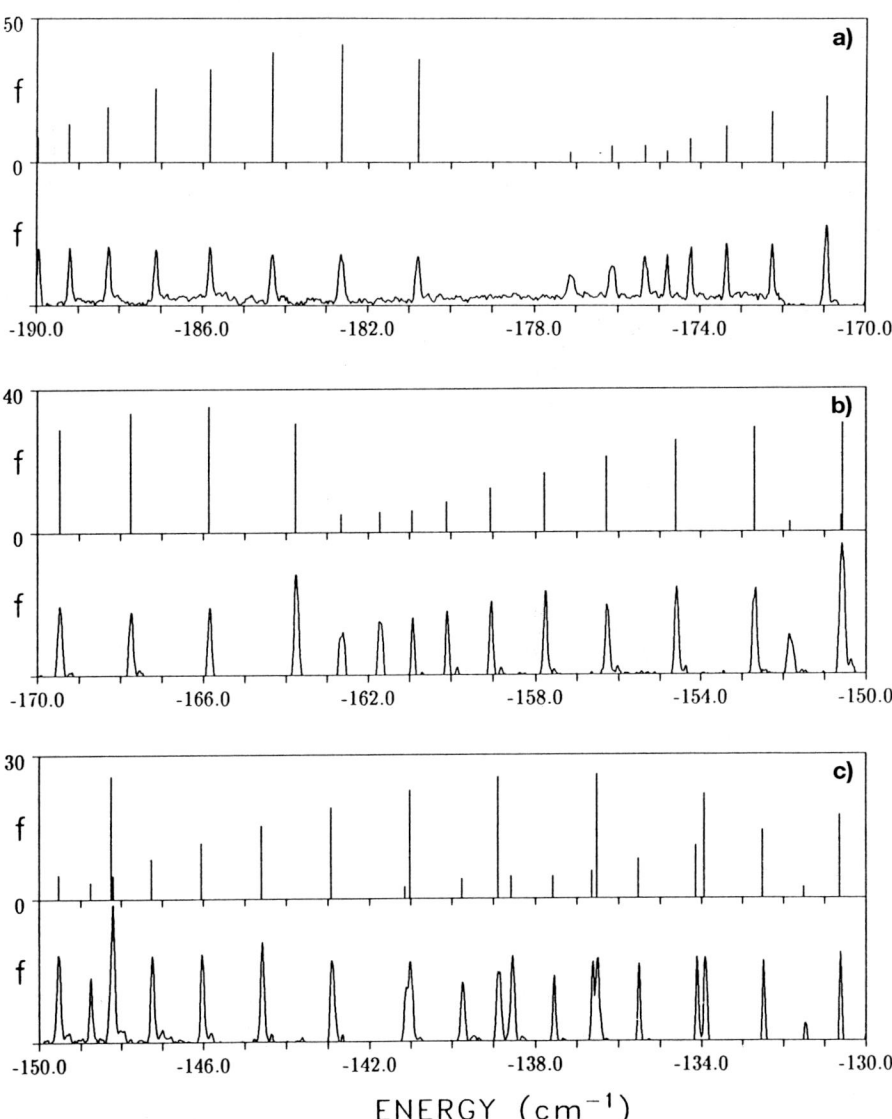

Fig. 10.4a–i. Theoretical and experimental spectrum of $\Delta m = -1$ transitions in deuterium from $2p_0$ to Rydberg states with $m = -1$ and even parity in the energy range $-190\,\text{cm}^{-1}$ to $-10\,\text{cm}^{-1}$ in a magnetic field of 5.96 T. Experimental data (unpublished) by courtesy of K.-H. Welge. (From Zeller 1990)

between theory and experiment is excellent. Figure 10.5 again testifies to the very complicated, disorderly looking structure and high level density of the level scheme of hydrogen atoms in strong magnetic fields.

It is of course extremely gratifying to have such an excellent agreement between the results of theory and experiment – after all it proves that quantum

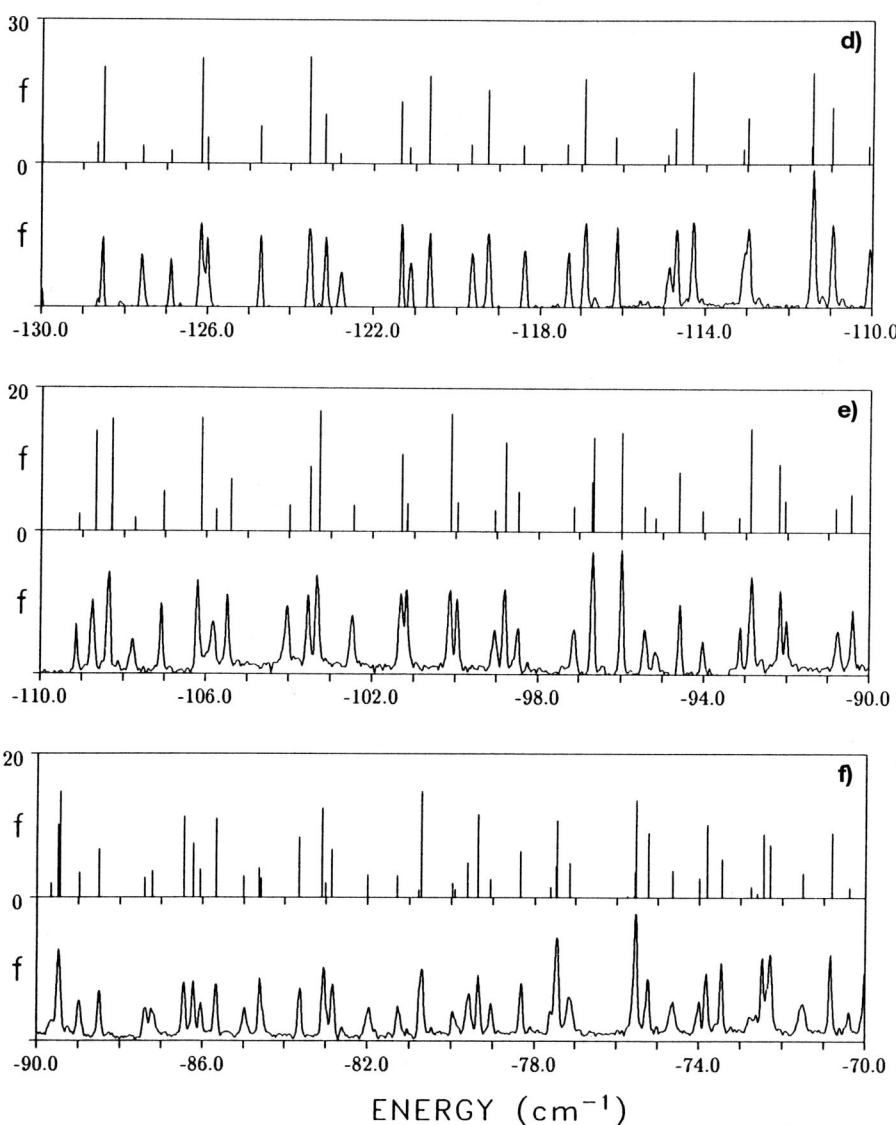

Fig. 10.4a–i. (Continued)

theory "works" even in this highly nonintegrable regime. The more profound importance, however, of these theoretical and experimental investigations of the Rydberg states of the hydrogen atom in strong laboratory magnetic fields lies in the fact that one has the rare opportunity to look at a real quantum system in a range of parameters where the corresponding classical system exhibits *chaotic* behaviour. For this reason we will outline in the following sections the importance of the hydrogen atom in strong magnetic fields to studies of

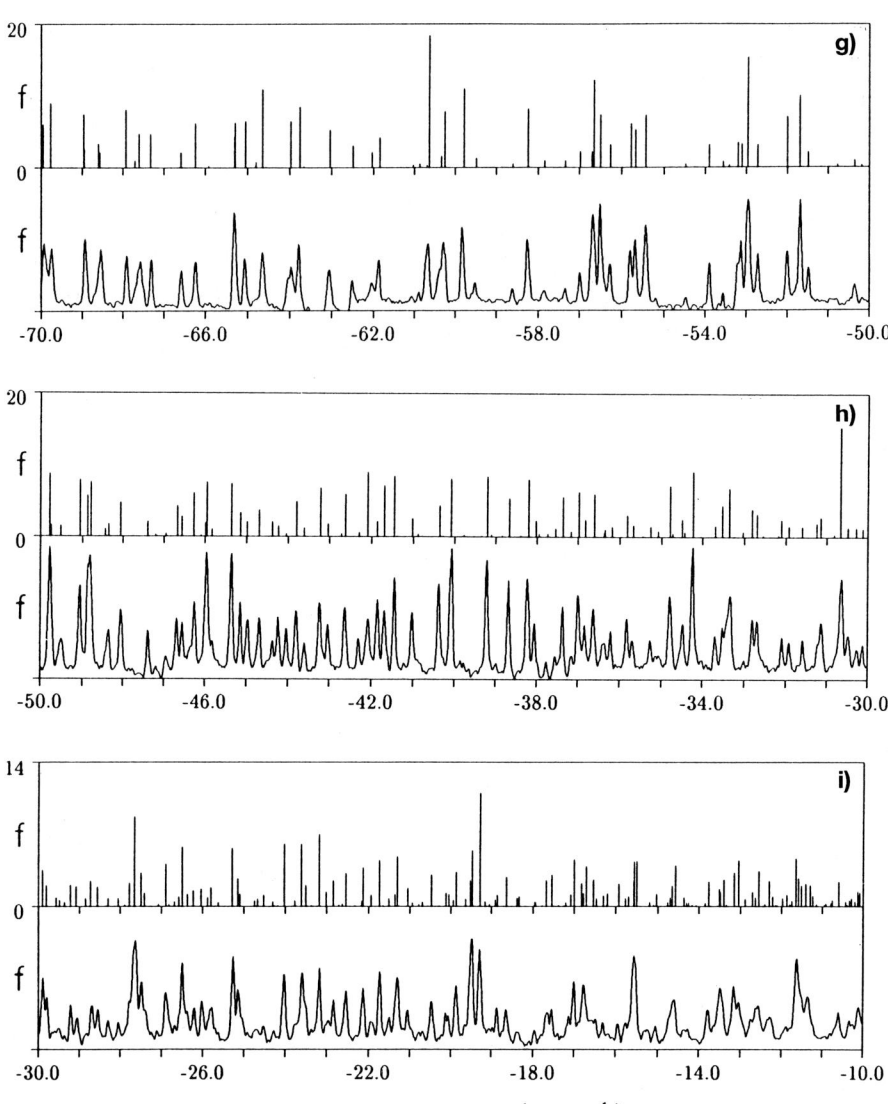

Fig. 10.4a–i. (Continued)

the problem of *quantum chaos*. This topic has been the object of intensive investigations in recent years, and we must restrict ourselves to sketching the essential lines of approach and results. For more detailed accounts we refer the reader, e.g., to the review papers by Friedrich and Wintgen (1989) and Hasegawa et al. (1989) and the books by Gutzwiller (1990), Gay (1990) and Haake (1991). To put these studies into perspective we shall discuss them in the broader context of the question "Is there chaos in quantum mechanics?".

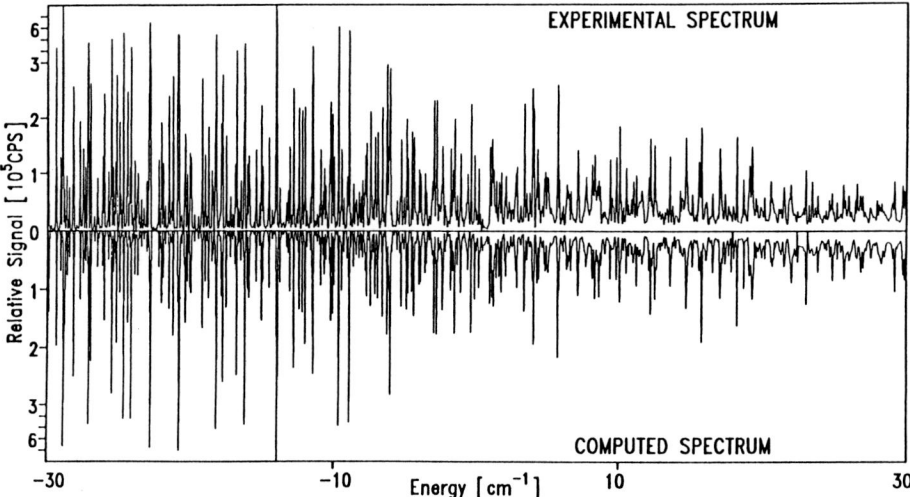

Fig. 10.5. Comparison of the experimental (Iu et al. 1991) and the theoretical (Delande et al. 1991) spectrum for dipole transitions from the 3s state of lithium to odd-parity Rydberg states in a magnetic field of 6 T in the energy range $-30\,\text{cm}^{-1}$ to $+30\,\text{cm}^{-1}$

10.2 Is There Chaos in Quantum Mechanics?

Can we tell from the properties of a quantum system in the limit of large quantum numbers that the corresponding classical system behaves chaotically? Recent theoretical and experimental studies of hydrogen atoms – the "showpiece" of quantum mechanics – in intensive microwave fields and in strong magnetic fields have produced results that could help answer this question.

10.2.1 Introduction

Studies on the subject "chaos" have become very popular in recent years. One of the main reasons is that chaos is a universal phenomenon that turns up in the problems of the different branches of natural science – biology, chemistry, electrotechnology and physics – and therefore is able to bridge the gap between these branches. As a technical term, chaos describes a behaviour of deterministic systems that are *extremely sensitive to the choice of initial conditions*. *Deterministic* means that the temporal evolution of the system is described by differential equations, in a way that the past and the future are uniquely determined by the initial conditions.

The cause of the extreme sensitivity of the solutions to the choice of initial conditions is a characteristic, local instability of the motion, namely the *exponential* divergence of initially closely neighbouring trajectories. One measure of the "speed" of divergence is the (positive) Liapunov exponent of the trajectory (Fig. 10.6a). The "chaotic" or statistical aspect can be seen from the fact that, even with only infinitesimally different initial conditions, because of

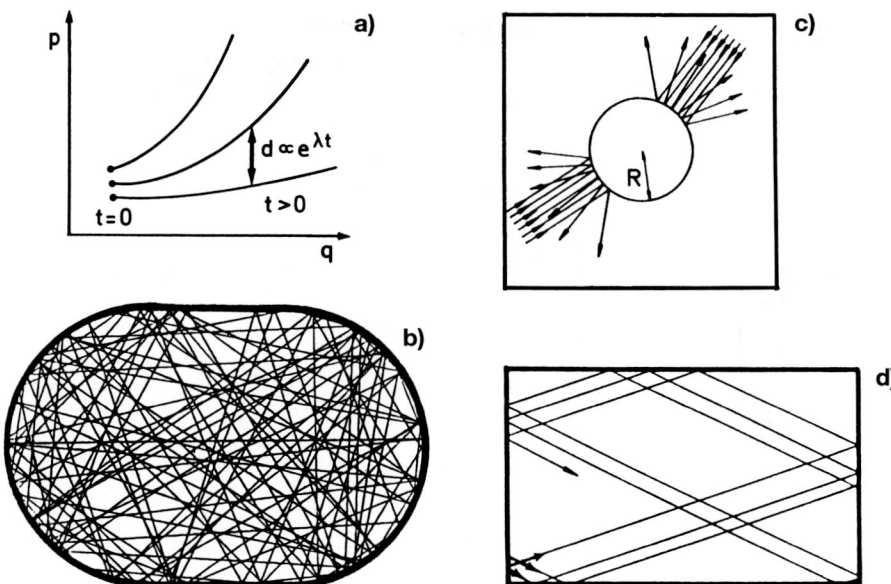

Fig. 10.6a–d. Chaos and order in classical model systems: (a) Classical chaos: orbits with only slightly different initial conditions diverge exponentially with time. The Liapunov exponent λ of a trajectory in phase space characterizes the average "speed" with which trajectories in the neighbourhood of the given trajectory exponentially separate. The size of the Liapunov exponent determines the degree of instability of an orbit. The mathematically strict definition and calculation of λ is realized via the stability matrix of the trajectory. (In regular systems, the divergence is linear, and the calculation of λ always gives the value zero.) (b) The stadium billiard as an example of a chaotic system with two degrees of freedom. The reflection on the semicircles causes a divergence of neighbouring groups of trajectories. (c) In the "Sinai billiard", the reflection on the disc produces defocusing, and with it divergence, of the trajectories. (d) The rectangular billiard as a standard example of a system with regular orbits only.

the immediate onset of exponential divergence of the solutions, the behaviour of the systems can be totally different over long periods of time, so that long-term predictions are impossible. The equations of motion that lead to chaos are necessarily nonlinear, and the systems can be either time-dependent, driven systems or conservative, nonintegrable Hamiltonian systems (*nonintegrable* means that no canonical transformation can be found so that the new Hamiltonian depends only on the generalized momenta, i.e., on action variables). Figure 10.6 shows the stadium billiard and the Sinai billiard as examples of simple chaotic systems in comparison to the "regular" rectangular billiard.

Chaos is a classical concept. We know, however, that as soon as we study phenomena on the molecular, atomic or nuclear level, classical mechanics must be replaced by quantum mechanics. Because, according to Bohr's correspondence principle, classical physics is contained within the quantum theory in the limit of large quantum numbers, the following question arises: in what way is classical chaos reflected in the characteristics of corresponding quantum systems? To date, no satisfactory definition of "quantum chaos" has been found (cf. Haake 1991), and in this chapter we will not attempt to find one. One of the main difficulties in finding such a definition lies in the fact that the Schrödinger equation is linear in the wave function that determines the temporal evolution of all observables, and therefore the nonlinearity of the classical systems can come into being only in the limiting process $\hbar \to 0$. The current research standpoint is pragmatic: one is looking for new semiclassical, but nonclassical, phenomena characteristic of quantum systems whose classical counterparts exhibit chaos. Berry (1987, 1989) coined the term "quantum chaology" for this field of study ("chaology" being a theological term borrowed from the 19th century: it describes the study of the condition of the universe at a time in which it was still "chaotic" or "without form and void").

It is obvious that the (numerical) solution to the Schrödinger equation of nonintegrable systems in the range of large quantum numbers is a very complicated problem that can be dealt with only in a few systems. As we have already pointed out, in addition to the chaotic model systems (billiards, nonlinear oscillators), in atomic physics "real" physical quantum systems exist in which we can look for symptoms of chaos, theoretically as well as experimentally, namely the hydrogen atom in a strong microwave field as an example of a *driven* chaotic system, and in a uniform magnetic field as an example of a nonintegrable *conservative* Hamiltonian system.

Using these examples, we will explain several of the theoretically, as well as experimentally, ascertained phenomena that have turned up as candidates for typical behaviour of quantum systems with classically chaotic counterparts.

10.2.2 Microwave Ionization of Rydberg States of the Hydrogen Atom

Bayfield and Koch (1974) conducted an experiment in which a hydrogen atomic beam that had been prepared for principal quantum numbers $n_p \simeq 66$ was sent through a microwave cavity, and the ionization of the atoms was measured as a function of microwave intensity. With the low frequency used, ≈ 10 GHz (i.e., $\lambda \approx 3$ cm), ≈ 100 microwave photons were necessary for ionization. The dimensions of the cavity were chosen so that the atoms were exposed to several hundred microwave oscillations. The experiments showed a sharp increase in the ionization signal at a certain threshold value F of the microwave field strength.

Classical calculations of electron orbits (Leopold and Percival 1979) provided a surprising explanation for the results: below F, the electrons moved under the combined influence of the Coulomb attraction of the nucleus and the external radiation field with periodic absorption of energy in captured orbits around the atomic nucleus; above F, however, chaotic orbits appeared that quickly removed the electron from the nucleus and could lead to premature ionization of the electron. The energy absorption in these orbits occurred more or less statistically and in a diffusion-like manner. For the range of excited states with principal quantum numbers from $n_p = 32$ to $n_p = 90$, Fig. 10.7 shows a comparison of the experimental threshold field strengths for the onset

168   10 Highly Excited States

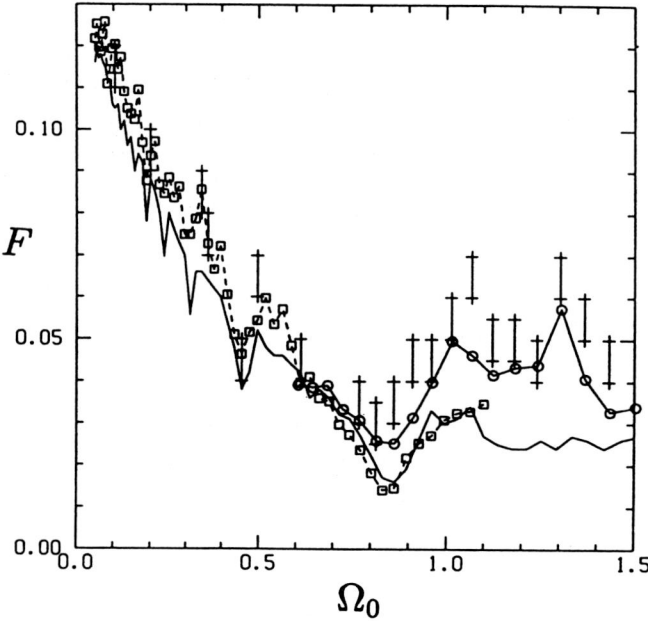

Fig. 10.7. "Stochastic" ionization of highly excited hydrogen atoms with principal quantum numbers $n_p = 32\text{--}90$ by intensive microwaves with frequency 9.92 GHz ($\square$) (van Leeuwen et al. 1985) and 36 GHz ($\bigcirc$) (Galvez et al. 1988). The graph shows the field strength F (in units of the Coulomb field strength, which would be felt by an electron in a Kepler orbit with the principal quantum number n_r) at which 10 percent of the atoms were ionized after passing the microwave cavity, as a function of the ratio Ω_0 of the microwave frequency to the Kepler orbital frequency. Each experimental data point indicates a different initial quantum state. For $\Omega_0 > 1$ the experimental results are in very good agreement with the classical predictions (*solid curve*) and the (less accurate) quantum calculations (*crosses with large error bars*). For $\Omega_0 < 1$ the experimental thresholds continue to show good detailed agreement with the quantum calculations but at fields much larger than the classical thresholds. (From Jensen et al. 1991)

of microwave ionization with the numerically calculated classical and quantum thresholds of chaotic ionization in a one-dimensional model of the experiment (Jensen et al. 1991).

It is seen from Fig. 10.7 that the classical predictions for the onset of chaotic ionization are in remarkably good agreement with the quantum ionization thresholds (both experimental and theoretical) as long as the microwave frequency is less than the Kepler orbital frequency. These results clearly indicate that, although quantum mechanics cannot satisfy the strict definitions of classical chaos, the quantum dynamics may nevertheless mimic the chaotic classical ionization at least on the limited timescales of these experiments. Do these experiments illustrate "quantum chaos"? No, because chaos is unpredictability that persists, and in these experiments the atomic electrons cross only a short distance of the microwave field and therefore diffuse for only a short time (≈ 500 microwave periods).

10.2 Is There Chaos in Quantum Mechanics?

The surprise came with quantum mechanical calculations for microwave frequencies greater than the Kepler orbital frequency (Blümel and Smilanski 1984, 1987; Casati et al. 1986). These revealed that, for short times, the highly excited electron indeed absorbs the energy as in the classical case (in a diffusion-like manner), but that after a certain "break-time" the energy is absorbed more slowly, i.e., the classical chaotic diffusion is ultimately suppressed by quantum mechanics. Casati et al (1988) showed that the quantum interference effects responsible for this suppression of chaotic classical ionization are related to the mechanism of Anderson localization in solid-state physics. The results shown in Fig. 10.7 illustrate that for high principal quantum numbers the quantum system indeed tends to be more stable against ionization than the classical system.

An analogous quantum suppression of classical chaotic diffusion was found earlier in a model system, the so-called "kicked rotator" (Casati et al. 1979), in which a particle on a ring is periodically kicked, with a strength that is dependent on the angular position of the particle, i.e., the potential is given by

$$V(\phi,t) = k F(\phi) \sum_n \delta(t - n T_0) \quad . \tag{10.4}$$

For strong kicks (large k), the angular momentum of the classical rotator shows a diffusion-like behaviour, and the energy grows linearly with time. In the quantum mechanical treatment within the range of large quantum numbers, on the other hand, the energy increases for only a certain length of time and then performs complicated, quasiperiodic oscillations.

These results illustrate a general phenomenon: the quantum suppression of classical chaos. Such a suppression is obviously expected. Classical chaos can be considered as the appearance of increasingly complicated structures in increasingly finer ranges of the phase space: smooth curves, which depict groups of orbits, develop increasingly intricate structures like cream in coffee. But quantum mechanics puts an end to the refinements, because, in the phase-space cells smaller than Planck's constant, every structure is obliterated.

Although no chaotic temporal evolution of quantum systems is found in these examples, a new phenomenon does crystallize in the semiclassical range: the quantum suppression of classical chaotic diffusion.

10.2.3 Statistical Analysis of Energy-Level Sequences

We here leave temporal evolution and focus our attention on the *energy eigenvalues of systems* whose classical counterparts behave chaotically. This is an important difference: energies are associated with stationary states, and a stationary state, in quantum mechanics, is something that goes on forever. Here a connection to (persisting) classical chaos might be hoped for.

The remarkable result is that the distribution of energy eigenvalues in the semiclassical range points to *universality*. In order to produce comparable conditions, however, the spectra must be transformed in a way that the average level spacing becomes the same. This is accomplished as follows: the average level density $\rho(E)$ is determined, the average number of levels up to energy E is calculated from it,

$$\bar{N}(E) = \int_{-\infty}^{E} \rho(E)\, dE \quad , \tag{10.5}$$

and then one considers the spectrum $x_j = \bar{N}(E_j)$, which, by construction, has a constant average level spacing that is chosen as the energy unit. The cumulative level density $n(x)$ (number of levels up to x) then is a staircase function that fluctuates around a straight line with slope 1.

Universality is found in the *statistics* of the level sequences. One such statistic, a short range one, is the probability distribution $P(S)$ of neighbouring energy levels, i.e., the distribution of $S_j = x_{j+1} - x_j$. Figure 10.8a shows the distribution $P(S)$ (Bohigas et al. 1984a,b) calculated from several hundred levels of the stadium billiard and, above it, the distribution for the Sinai billiard.

The distributions are obviously the same. The solid curve that so exactly approximates the data is the Wigner distribution

$$P_W(S) = \frac{\pi}{2} S e^{-\pi/4 S^2} \quad . \tag{10.6}$$

The probability of finding two levels in the same location, therefore, tends to zero linearly with S. Another statistic, a long-range one, is the spectral rigidity $\Delta_3(L)$. By this one means the mean-square deviation of the staircase function $n(x)$ from the straight line best approximating it,

$$\Delta_3(L) = \langle \frac{1}{L} \min_{A,B} \int_{-\frac{L}{2}}^{+\frac{L}{2}} [n(x) - Ax - B]^2\, dx \rangle \quad . \tag{10.7}$$

Figure 10.8b shows the rigidity for both these chaotic systems; they are equal and suggest universality.

The solid curve in Fig. 10.8b, as well as the Wigner distribution (10.6), is borrowed from the eigenvalue statistics of infinitely large, real symmetric matrices, whose elements are random numbers (GOE: Gaussian Orthogonal Ensemble). The theory of random matrices was developed in the 1960s to simulate by means of a model the complicated many-body Hamiltonians of atomic nuclei. It is amazing that the same results can also describe exactly the quantum mechanical energy level sequences of classically chaotic systems with few degrees of freedom.

For systems whose classical motion is not chaotic, the corresponding level statistics are much different. Figure 10.9 shows the results for the rectangular billiard (Berry 1987). Here, the statistical results are equivalent to those found in random *numbers*, i.e., of the Poisson type: the distribution of the neighbouring energy levels reaches its maximum at a level distance of zero, and $\Delta_3(L)$ grows linearly.

A comparison of Figs. 10.8 and 10.9 suggests that level repulsion and, as a result of this, a maximum probable level distance different from zero is characteristic of quantum systems if the corresponding classical system behaves chaotically; level accumulation, on the other hand, proves to be typical of quantum

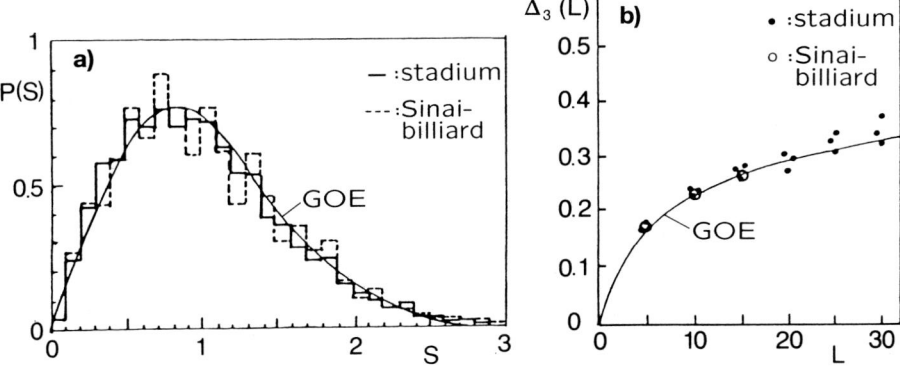

Fig. 10.8a,b. (a) Histograms $P(S)$ of the distribution of level distances of the quantum energies of the stadium and Sinai billiard (Bohigas et al. 1984a,b). The solid curve is the result for the distribution of nearest-neighbour spacings of symmetric random matrices (Gaussian Orthogonal Ensemble, *GOE*) and is approximated very well by the Wigner distribution (10.6). (b) Spectral rigidity $\Delta_3(L)$ for the eigenvalue sequences of the stadium and Sinai billiard in comparison to the spectral rigidity of the eigenvalue sequences of the real symmetric random matrices (asymptotically $\Delta_3(L) \to (1/\pi^2) \ln L + $ const. for $L \to \infty$).

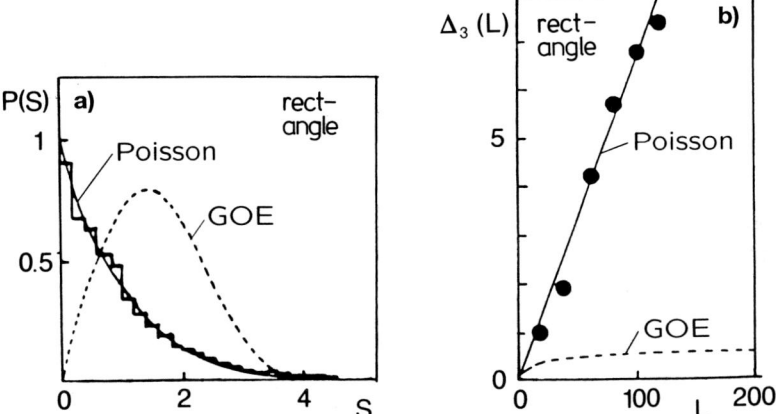

Fig. 10.9. Nearest-neighbour distributions $P(S)$ (histogram) and spectral rigidity $\Delta_3(L)$ (*black dots*) for the eigenvalue of the rectangular billiard (Berry 1987). The solid curves depict the corresponding statistics for Poisson-distributed eigenvalues ($P(S) = e^{-S}$ and $\Delta_3(L) = L/15$), the dotted curves show those for real symmetric random matrices.

systems with classically regular behaviour. The concepts developed by aid of these model systems have been excellently confirmed through recent theoretical and experimental studies of the hydrogen atom in magnetic fields. We will review these results in the following sections.

10.2.4 Order and Chaos in the Hydrogen Atom in a Magnetic Field

In order to find chaos in a hydrogen atom, we must, of course, go beyond the usual perturbation-theoretical Zeeman range to magnetic fields in which the effects of the magnetic fields are comparable to those of the Coulomb field. An estimate of the necessary sizes of the fields in dependence on the principal quantum number was given in (10.1). At such field strengths, the two different symmetries of the problem – the spherical symmetry of the Coulomb field and the cylindrical symmetry of the magnetic field – prevent even an *approximate* separation of the variables (only the z-component of the angular momentum is conserved), and we have to deal with the prototype of a nonintegrable system with two degrees of freedom.

The appearance of chaotic trajectories in the hydrogen atom in a magnetic field can be illustrated with the method of Poincaré surfaces of section. The equations of motion that result from the diamagnetic Hamiltonian (3.1) are solved numerically, plane sections are laid in the three-dimensional subspace of four-dimensional phase space defined by the given value of the conserved energy (energy shell), and the points of intersection of the orbits with the plane are marked in that plane. In Fig. 10.10, Poincaré surfaces of section calculated for a magnetic field of 6 T show that, with decreasing binding energy of the electron, the regions filled with regular orbits gradually break up, until no more regular trajectories are recognizable.

To give a quantitative measure of the increasing collapse of regularity, in Fig. 10.11 the area fraction of ranges filled with regular orbits is shown as a function of energy for a magnetic field of 6 T. Figure 10.11 is actually universal and not limited to this magnetic field strength. This is the result of a remarkable scaling property of the Hamiltonian (3.1): through the substitutions of the spatial, e.g. cylindrical, coordinates, $(\rho, z) \to \beta^{2/3}(\rho, z)$, $(p_\rho, p_z) \to \beta^{-1/3}(p_\rho, p_z)$ it assumes a form that is no longer explicitly dependent on the magnetic field, but only on the *scaled* energy $\varepsilon = E/(2\beta)^{2/3}$. The intuitive meaning of this transformation is that the *relative* strengths of the Coulomb attraction and the Lorentz forces are held constant at a given scaled energy. This means that the classical dynamics is "frozen" at given ε, apart from the similarity transformation given above. The upper abscissa in Fig. 10.11 gives the scale for ε. It can be read off Fig. 10.11 that in the diamagnetic Kepler problem the breakdown of regularity begins around $\varepsilon = -1$, and the last islands of stability visible in the Poincaré surfaces of section disappear around $\varepsilon = -0.25$. The latter turns out to be the case when the periodic orbit perpendicular to the direction of the magnetic field finally becomes unstable (the exact value of ε being -0.2545372). The strong dip in the behaviour of the function shown in Fig. 10.11 is related to the merging of an unstable periodic orbit with the stable periodic orbit perpendicular to the field, which appears (Schweizer et al. 1993) at $\varepsilon = -0.632372$, when the winding number of the trajectory assumes a rational value (viz. 1/2). The merging is reflected in a strong reduction of the size of the regular island around the orbit perpendicular to the field shortly below this value of the scaled energy.

10.2 Is There Chaos in Quantum Mechanics?

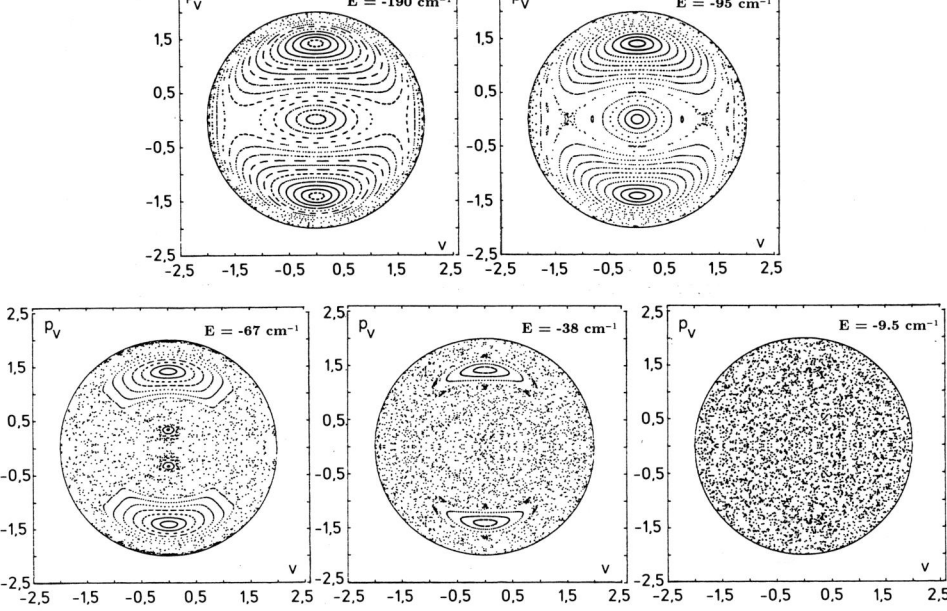

Fig. 10.10. Poincaré surfaces of section of classical orbits in the hydrogen atom in a magnetic field of 6 T and at different energies. To avoid the singularity at the origin caused by the Coulomb potential, the equations of motion were solved in semiparabolic coordinates u, v ($u^2 = r + z$, $v^2 = r - z$) and the v-p_v-plane ($u = 0$) was chosen as the section. The collapse of regularity with decreasing binding energy is clearly visible. As the ground state energy in spectroscopic units is 109737 cm^{-1} the energy values lie in the range of highly excited states.

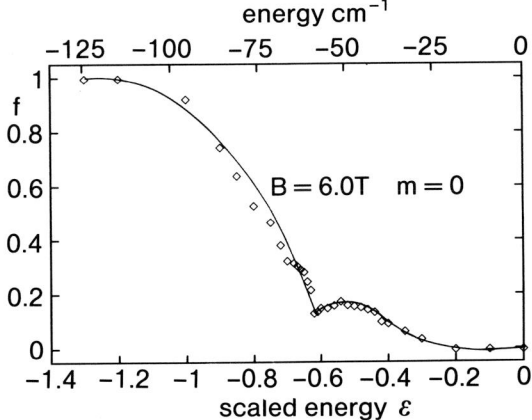

Fig. 10.11. Area fraction of regular regions in the Poincaré surfaces of section as a function of energy for a magnetic field of 6 T (*lower abscissa*) and on the scaled energy $\varepsilon = (E/R_\infty)/(B/2.35 \times 10^5 \text{ T})^{2/3}$. The dip in the otherwise more or less smooth behaviour of the function is caused by the confluence of an unstable periodic orbit with the stable periodic orbit perpendicular to the field at a value of $\varepsilon = -0.632372$. (From Schweizer et al. 1993)

10.2.5 Level Statistics for the Hydrogen Atom in Magnetic Fields

In Figs. 10.2–4 we had shown examples of theoretical and experimental quantum spectra of hydrogenic atoms in strong laboratory fields. The figures conveyed a *qualitative* impression of the growing complexity of the quantum spectrum of even the simplest systems, when we push forward to field strengths in which classical movement becomes more and more chaotic.

Inspired by corresponding results in model systems, several authors subjected the energy spectra to a statistical analysis in order to reach a *quantitative* measure of this complexity (Delande and Gay 1986, Wintgen and Friedrich 1986c, Wunner et al. 1986b). Figure 10.12 shows histograms of the distribution of neighbouring energy levels of the hydrogen atom in magnetic fields as a function of the scaled energy (Wintgen and Friedrich 1987). Here, too, the rearrangement from a Poisson-like to a Wigner-like distribution when the classical system changes from regularity to chaos was confirmed. This rearrangement was thus demonstrated for the first time in a *real* physical system. The Δ_3-statistics of the energy levels (Wunner et al. 1986b) also demonstrated the expected behaviour as based on the model systems (see Fig. 10.13).

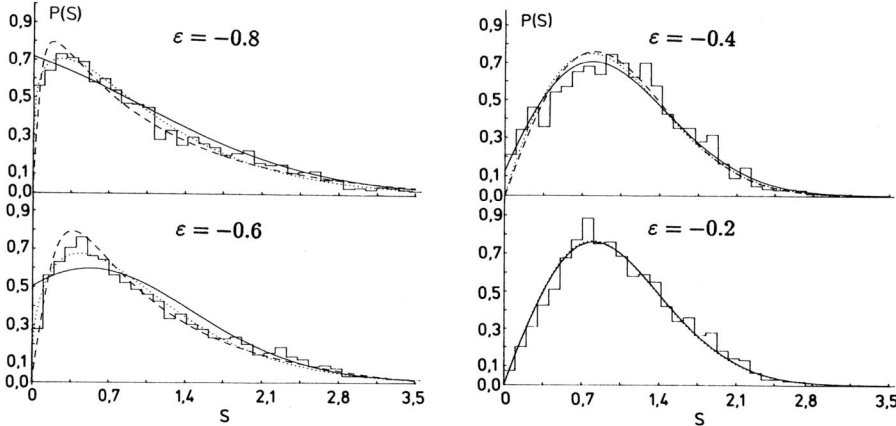

Fig. 10.12. Histograms $P(S)$ of the level distances of quantum energies in the hydrogen atom in magnetic fields at different values of the scaled energy (Wintgen and Friedrich 1987). The smooth curves are the results of fits to the histograms using various formulae derived in literature that interpolate between the Poisson and the Wigner distributions in the transition range to chaos.

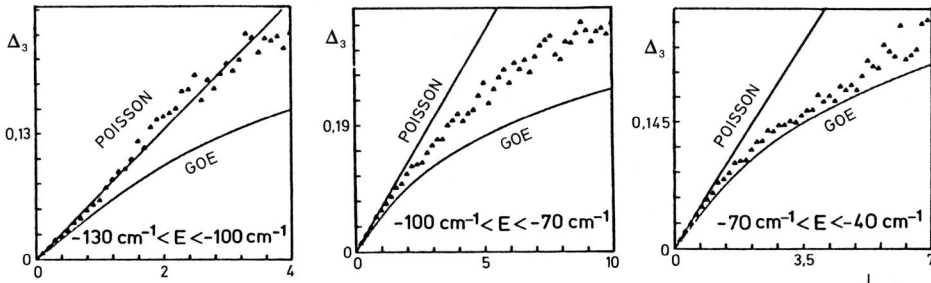

Fig. 10.13. Spectral rigidity $\Delta_3(L)$ for energy level sequences of the hydrogen atom in a magnetic field of 6 T in three different energy intervals. The swing to the GOE-distribution (cf. Fig. 10.8b) as soon as the classical motion becomes chaotic is also visible here.

10.2.6 Resonances in Chaos – the Role of Periodic Orbits

The Fourier transformation of spectra, such as those of Figs. 10.2–4, yields a surprising result: one finds distinct peaks, an indication that different "wavelengths" ΔE with periodicities not visible to the human eye are (apparently) present in irregular spectra. More surprising, however, is the fact that these modulations of orbital frequencies can be attributed to *classically periodic*, mostly *unstable* orbits, unstable meaning here that the Liapunov exponent is different from zero. These orbits are embedded in the sea of chaotic orbits.

Figure 10.14 shows, as experimental results, the Fourier-transformed spectra of Balmer transitions to Rydberg states taken at a constant scaled energy, in comparison to the positions of peaks as they follow from the calculation of relevant classical periodic orbits (Holle et al. 1988). One finds an almost perfect correlation between the positions of the observed structures and the results of the classical calculations.

How is this correlation to be understood? In the attempt to generalize the semiclassical quantization conditions valid only for stable periodic orbits, Gutzwiller (1980) and others, using the Feynman path integral formalism, were able to derive an asymptotic formula that allows expressing the spectral density of quantum levels in terms of classical, closed orbits. The level density can accordingly be divided into an averaged part, and an oscillatory part to which *all* closed, classical orbits contribute:

$$\rho(E) = \bar{\rho}(E) + \sum_j \sum_k a_{jk}(E) \cos[k(S_j(E) - \alpha_j)/\hbar] \quad . \tag{10.8}$$

The sum over j runs over all closed orbits, the sum over k over all repetitions, S_j is the action and α_j a corresponding phase shift (Maslov index) of the primitive orbit j. The amplitudes a_{jk} are dependent on, among others, the Liapunov exponents. Between a quantum level and a closed orbit, there is therefore generally no longer a one-to-one correspondence; instead, each closed orbit describes a *periodic accumulation* of levels on a scale ΔE, which is determined by the condition $\Delta E/\hbar \, d(kS_j)/dE = 2\pi$. Because the derivative of the action with

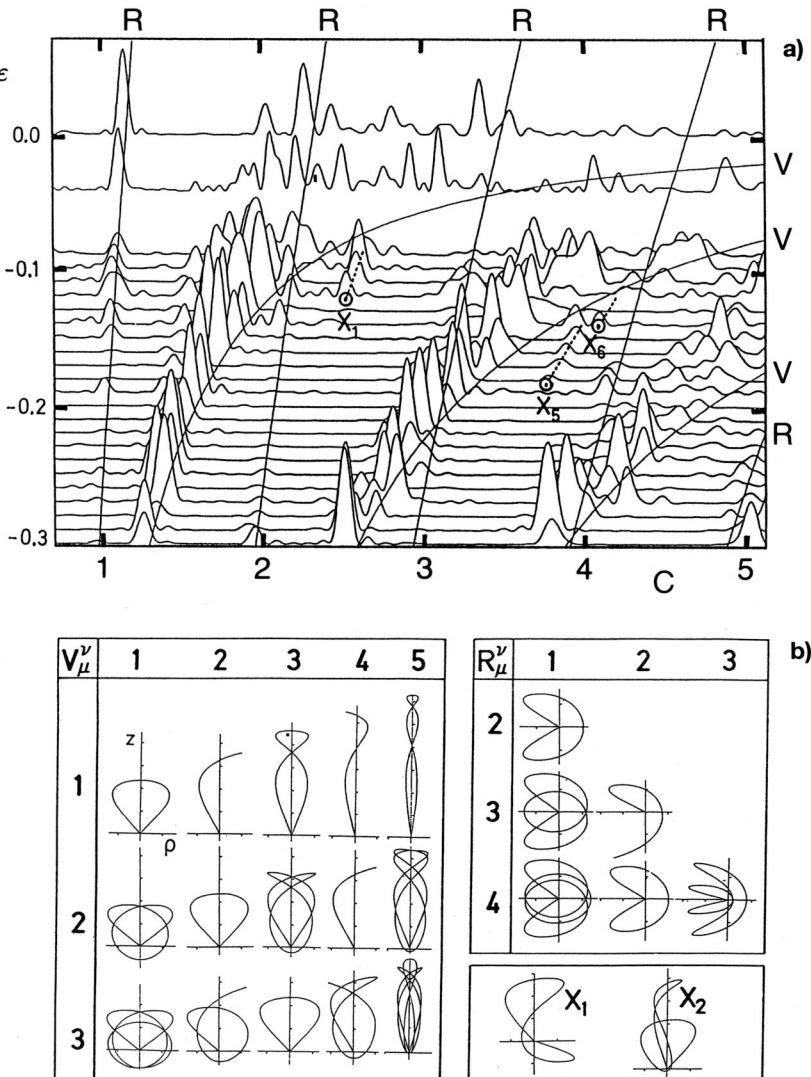

Fig. 10.14a,b. Resonances in the quantum spectrum of the "chaotic" hydrogen atom ($m = 0$, even parity) (Holle et al. 1988): **(a)** Fourier-transformed photoabsorption spectra at different scaled energies as functions of the scaled action C ($C = n(2\beta)^{-2/3}$). Each resonance can be attributed to a closed classical orbit (designated, for historical reasons, "rotational" (R), "vibrational" (V) or "exotic" (X) by the authors). The solid curves represent the expected location of the resonances according to classical calculations. Further "vibrational" curves fan out to the left from each particular base curve and coincide with the other observed resonances. The figure shows the quantum support of regularity in the chaotic range. **(b)** Several closed, mainly unstable ($\lambda \neq 0$), periodic orbits (in cylindrical coordinates ρ, z) of the hydrogen atom in the magnetic field that lead to the observed resonances.

respect to the energy gives the orbital periods T_j, one has $\Delta E = h/(kT_j)$, and the modulations are given by the orbital periods and their repetitions.

Although the periodic orbits in the Poincaré surfaces of section are no longer recognizable in the background of chaotic orbits (they are infinite in number, but are isolated and have measure zero), it is just these orbits that produce the *structure* of the quantum spectrum in the chaotic range. This is a further phenomenon that is typical of the behaviour of quantum systems with a chaotic counterpart: the quantum support of "regularity" in the chaotic range.

10.2.7 "Scarring" of Wave Functions

What do wave functions in the chaotic range look like? Because the orbit of the electron in this range comes arbitrarily close to every phase-space point in the course of time, it was very early on suspected that "chaotic" wave functions essentially are delocalized. But Heller (1984) discovered that, in the quantum mechanical treatment of the stadium problem, wave functions along classical, unstable closed orbits are "scarred" (cf. Fig. 10.15), i.e., show an increased probability of presence: the smaller the "discrepancy per period" λT of the classical trajectory, the more pronounced is this behaviour.

Figure 10.15 shows examples of wave functions of the hydrogen atom in the chaotic range together with the projection of a selected orbit of the corresponding classical system. Dominant structures in the quantum mechanical system along the (unstable) classical orbit can be clearly recognized. Thus scars are also discovered in the wave functions of highly excited states of the hydrogen atom in strong magnetic fields, which is an indication that scarring of wave functions along unstable, closed orbits is another universal "chaotic" quantum phenomenon.

Even more fundamental insight into the effect of scarring by classical orbits is gained from the inspection of wave functions in *phase space*. This is suggested by the following reasoning. The path of a classical trajectory in phase space yields unique points in coordinate or momentum space but not vice versa since, e.g., to a point in coordinate space there generally belongs a bundle of trajectories separated by different momenta but with equal kinetic energy. Therefore phase space is the natural setting for studies of classical dynamics; and hence, in quest of quantal scarring, it is more sensible to investigate, instead of wave functions in coordinate or momentum space, quantum mechanical *phase space* distribution functions, such as the Wigner distribution function P_W (see e.g. Takahashi 1989a,b)

$$P_W(\boldsymbol{x},\boldsymbol{p}) = \frac{1}{(\hbar\pi)^f} \int_{-\infty}^{\infty} db_1 \ldots \int_{-\infty}^{\infty} db_f \psi^*(\boldsymbol{x}+\boldsymbol{b})\psi(\boldsymbol{x}-\boldsymbol{b}) \exp(2\frac{i}{\hbar}\boldsymbol{p}\cdot\boldsymbol{b}) \quad (10.9)$$

($\boldsymbol{b}$ is a coordinate-like shift vector, f denotes the number of degrees of freedom of the system under consideration). It must be emphasized, however, that the Wigner distribution function cannot be regarded as the probability of finding a particle at point $\boldsymbol{x}$ with momentum $\boldsymbol{p}$, as this would contradict the Heisenberg uncertainty principle, but integrating over momentum space variables yields the

10 Highly Excited States

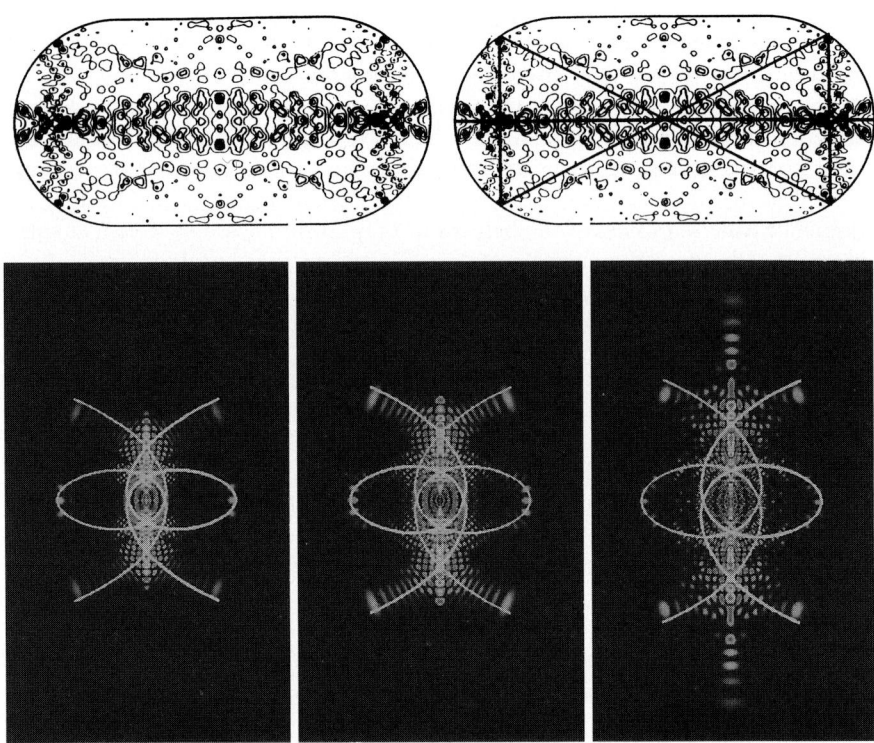

Fig. 10.15. "Scarring" of wave functions along unstable classical periodic orbits in the chaotic range. *Top*: Contours of a wave function of the stadium billiard and the classical orbits that contribute most to it (Heller 1984). *Bottom*: Quantum mechanical spatial probability distributions of the electron in $m = 0$ Rydberg states for B = 4.7 T together with the projections of selected orbits of the corresponding classical system (in cylindrical coordinates, the magnetic field points upward): 429th excited state ($\varepsilon = -0.2985$), superimposed classical orbit Liapunov-stable (*left*), 434th excited state ($\varepsilon = -0.2897$), classical orbit Liapunov-stable (*middle*), 448th excited state ($\varepsilon = -0.2736$), classical orbit Liapunov-unstable (*right*)

probability of finding a particle at the point x in coordinate space, and vice versa.

For conservative systems with two degrees of freedom, such as the diamagnetic hydrogen atom, phase space is four-dimensional, and the energy shell is three-dimensional. Thus, by fixing one phase space coordinate we select a two-dimensional subspace, the Poincaré surface of section Γ

$$\Gamma(E) = \{(x_1, p_1) | H(\boldsymbol{x}, \boldsymbol{p}) = E, x_2 = 0 \wedge p_2 \geq 0\} \quad , \tag{10.10}$$

where $\boldsymbol{x}$ and $\boldsymbol{p}$ are the canonically conjugate coordinates, H is the Hamiltonian of the physical system and E is its energy. By following a trajectory as a function of time this surface of section will be crossed and a sequence of intersections will be created. By knowing one point of this sequence (i.e. the initial conditions) any other point can be calculated, as classical dynamics is deterministic. Hence this sequence carries information on the corresponding trajectory, and the global structure of the Hamiltonian system at fixed energy is mirrored by the structure of a Poincaré surface of section.

As the classical Poincaré structure of the diamagnetic Kepler system is most suitably uncovered in semiparabolic coordinates we transform the quantal wave functions to these coordinates and calculate the correlation matrix $\psi^*(\boldsymbol{x} + \boldsymbol{b})\psi(\boldsymbol{x} - \boldsymbol{b})$ over a semiparabolic space-grid. Note that the Wigner distribution function can be regarded as a Fourier transformation of the correlation matrix. As we want to compare Poincaré-like structures, the further calculation can be simplified by setting the non-Poincaré surface coordinate x_2 equal to zero. The final Wigner distribution function on the Poincaré surface of section is then computed via fast Fourier transformation of the discretized correlation matrix.

For semiclassical considerations the most important periodic trajectories are those with small action. For the diamagnetic Kepler problem at sufficiently low scaled energies these are the Liapunov-stable straight line orbits parallel and perpendicular to the magnetic field, and the Liapunov-unstable almost circular, so-called, orbit (Schweizer et al. 1988). These are simple periodic orbits with one fixed point on the Poincaré surface of section. A weak variation of their initial conditions gives rise to elliptic structures for the Liapunov-stable and to a hyperbolic structure for the unstable one. In Fig. 10.16 we compare, for $\varepsilon = -0.6$, the classical structures with single Wigner distribution functions of diamagnetic Rydberg states. At this scaled energy the phase space of the classical system is still partially regular (cf. Fig. 10.11), and similar structures can be found in the classical and the quantum system.

Classical Poincaré surfaces of section are created by superposition of structures due to different periodic and non-periodic orbits. To quantum mechanically mimicry this classical behaviour, in our calculations we denormalize the Wigner distribution functions in such a way that the maximum values of the different distribution functions are equal. We do this since we want to visualize different structures via Poincaré-like surfaces, and large-scale structures of individual distribution functions would become too low-valued compared to peak structures of others. Figure 10.17 shows the results for two different scaled energies. In both cases the quantal Poincaré surface of section was produced by superposition of five successive denormalized Wigner distribution functions. The length of this sequence was chosen in such a way that including an additional Wigner distribution function would have doubled one of the former structures. The classical and the quantal picture coincide only in the sense that in the regular case similar structures can be recognized, whereas in the irregular case no dominant structures are detectable.

In conclusion the calculation of Wigner distribution functions provides an excellent mathematical tool for quantum chaological studies. With their help, it can be documented that enhanced probabilities close to classical trajectories are an important element of structure for the wave functions. Additionally the validity of semiclassical quantization by orbits can be tested by calculating quantal Wigner distribution functions and analyzing their structure with respect to the importance of classical trajectories (Schweizer et al. 1993).

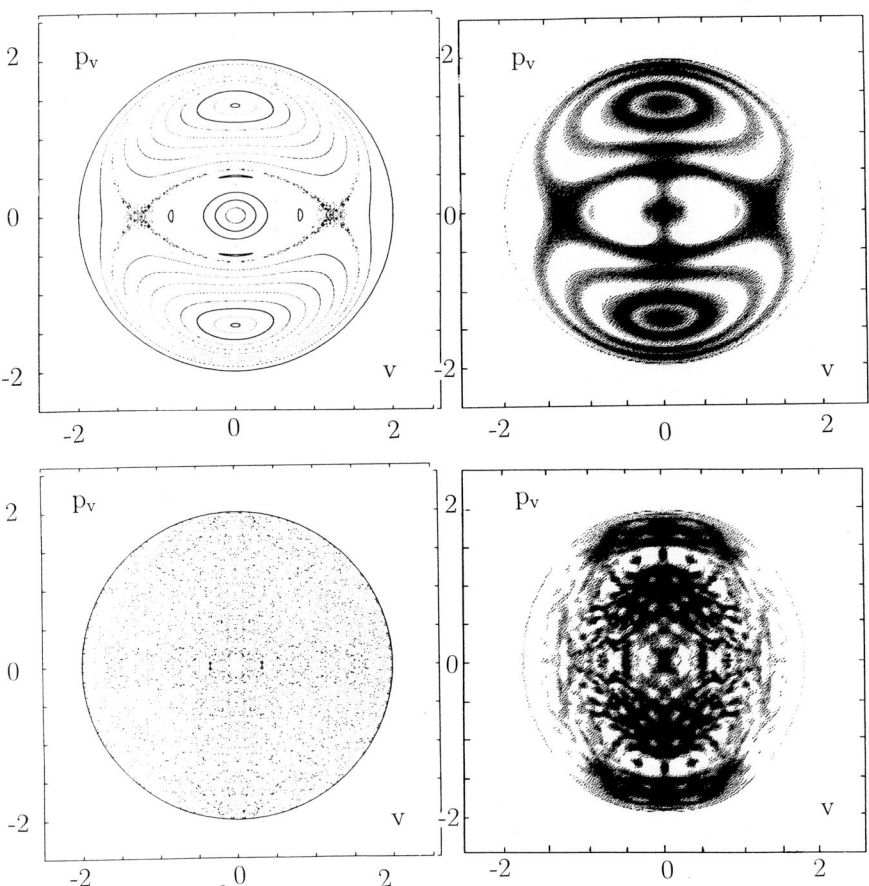

Fig. 10.16. Classical (*left*) and quantal (*right*) Poincaré surfaces of section in semiparabolic coordinates at scaled energy $\varepsilon = -1.0$ (*upper part*) and $\varepsilon = -0.2$ (*lower part*). The quantal surfaces of section were calculated by superimposing the grey-coded Wigner distribution functions of the $m = 0$, positive-parity eigenstates with numbers 131–135 (for $\varepsilon = -1.0$) and 243–247 (for $\varepsilon = -0.2$). Note the strong similarity between the classical and quantum mechanical pictures both in the regular and in the irregular regime.

As examples of the characteristic behaviour of quantum systems in the classical chaotic range, we have here reported on the quantum suppression of chaotic diffusion in driven systems, on level statistics, on the scarring of wave functions, and on the appearance of resonances in the quantum spectrum caused by unstable periodic orbits. This is of course not the complete list. Further characteristics that are still being analyzed are, e.g., the parameter sensitivity of spectra, the distribution of wave function values, and the statistics of transition probabilities in the chaotic range.

10.2 Is There Chaos in Quantum Mechanics?

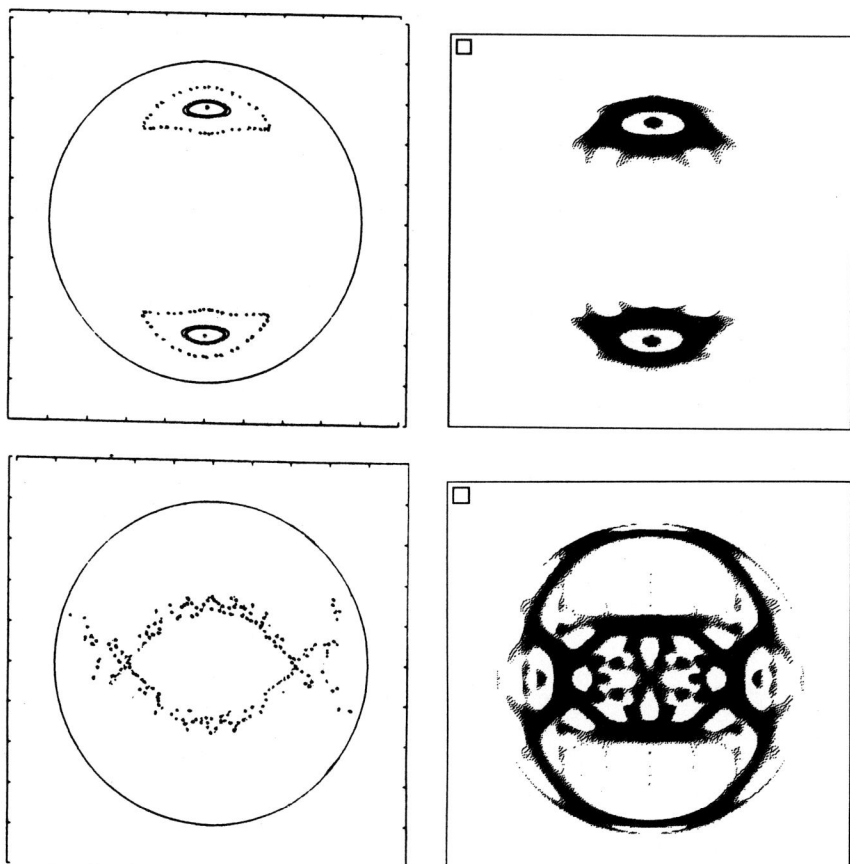

Fig. 10.17. Classical (*left*) and quantal (*right*) Poincaré surfaces of section in semiparabolic coordinates at $\varepsilon = -0.6$. On the left, from top to bottom, there are two classical Poincaré structures visible, corresponding to the straight-line orbit perpendicular to the magnetic field axis and the Liapunov-unstable almost circular periodic orbit. On the left the corresponding quantal structures of grey-coded Wigner distribution functions are shown. The agreement between both is impressive. Note that since the coordinates on the axes are canonically conjugate these must fulfill the uncertainty relation. The square on the left top corner of the quantal picture indicates the size of $\hbar$.

The question of "chaos in quantum mechanics" has certainly not been handled conclusively to date. But the results reported on here can give an impression of the studies being done around the world on nonintegrable quantum systems – in mathematical theory, with numerical calculations and in laboratory experiments. The particular fascination of these studies can certainly be attributed to the fact that, more than 70 years after the development of quantum mechanics, they still allow new and unexpected insights into the connection between the world of classical physics and that of quantum theory.

Outlook

In the past 70 years, since the late twenties of this century, much work has been done in this field of research, and a lot of problems have been solved, not least due to the powerful computers that have become available over the decades. As is evident from this book, substantial progress has been made since Garstang's comprehensive review from 1976 about "Atoms in high magnetic fields", which appeared in *Reports on Progress in Physics*. This fact is best demonstrated by comparing the last sentence in his report *"Finally we look forward to the day when someone will find an interpretation of the unidentified features in certain magnetic white dwarf spectra and further elucidate the nature of their magnetic fields"* with the results given in Chap. 7. It cannot be overlooked, however, that in spite of these successes much work still remains to be done. As is common in Science, the whole field is far from being completed, and we therefore want to conclude by pointing out a few open problems.

As already mentioned, the accuracy in calculating the properties of helium and helium-like systems in magnetic fields of arbitrary strength must be improved in order to keep pace with the progress of observations. This, however, is not a trivial task. Classical diagonalization methods seem to be exhausted, increasing computer facilities notwithstanding, and new numerical methods must be developed. One very promising way for obtaining far better results for helium-like systems is the use of finite element methods. But again, the extension of this approach to higher atoms is by no means obvious. Thus the calculation of the properties of these atoms in magnetic fields with spectroscopic accuracy *is* an unsolved problem. Progress may come from density functional methods which have been extended to electronic systems in strong magnetic fields by the formulation of the current-density functional theory. Within this framework a reliable treatment of binding energies of one-dimensional atomic chains also seems to be possible. Such investigations are necessary for a quantitative understanding of neutron star surfaces.

Another broad and important complex of problems of atomic physics in the presence of strong magnetic fields is the calculation of cross sections for bound-free and free-free transitions, which are prerequisite for modelling the atmospheres of magnetized white dwarfs and neutron stars. Here a lot of work has been done, but the results are not yet in a state for final reviewing.

Finally, to motivate young scientists, we emphasize that even the hydrogen atom itself still offers challenges, namely the accurate calculation of all properties in the presence of magnetic and *electric* fields of arbitrary strength and orientation, a situation in which all symmetries are broken. Calculations of this kind are not only a playground for theoreticians and computer freaks but are absolutely necessary for the quantitative understanding of the profiles of absorption lines in the spectra of magnetized white dwarfs, from which a wealth of information about the physics of these objects can be extracted.

Appendix 1: Energy Values

A1.1 Tables of the Energy Values

In Table A1.1 we have listed the relations between states, range of β, and the following tables and figures. In the column "range of β" (s) denotes the standard and (f) the finer mesh.

In the following tables, the energies (in units of $-E_\infty$) of the 38 lowest states of an electron in a static Coulomb potential and a magnetic field with strengths ranging from $\beta = 10^{-4}$ ($B \approx 47.01$ T) to $\beta = 10^3$ ($B \approx 4.701 \times 10^8$ T) (Tables A1.2a–h) and, in a finer mesh, from $\beta = 10^{-3}$ ($B \approx 4.701 \times 10^2$ T) to $\beta = 10^{-1}$ ($B \approx 4.701 \times 10^4$ T) (Tables A1.3a–e) are listed. (All tables of the book are available by "anonymous ftp" over Internet. On a network-ready machine you should execute the command *ftp atoms.tat.physik.uni-tuebingen.de* and login as *anonymous*, using your e-mail address as a password. Then execute the commands *cd pub/atoms* and *mget ** to download the relevant .tex files of the tables. Should you require values that cannot be extracted from the tables, you can obtain these, if numerically feasible, from the authors on request.) Except for possible uncertainties in the seventh significant figure, the numbers without slashes are accurate within the quoted digits. In cases where the last digits still change with the maximum number of expansion terms employed, the tables give both the corresponding value and, separated by a slash, the figures still changing if the two energy values last computed are linearly extrapolated to $1/n_c \to 0$. The accurate value lies between these two limits.

Parentheses behind the energy values indicate an expansion in terms of the spherical basis (3.2) and square brackets of the cylindrical one (3.3). The first number in parentheses denotes the number of spherical harmonics used in the expansion and the second number is the number of orthonormal functions $G_{nl}^\zeta(r)$ of (3.8) used for the expansion of each l-dependent radial function. The sign ∞ stands for the direct integration method for solving the system of coupled differential equations (3.4), which, in a numerical sense, yields exact radial functions. The product of the two numbers in the parentheses is the total number of expansion functions employed and thus, equivalent to the dimension of the matrix for which eigenvalues and eigenvectors have to be computed. The numbers in the square brackets are the numbers of Landau channels taken into account in the cylindrical expansion (3.3). The longitudinal functions $g_{nm}(z)$ were always determined by numerical integration of the coupled system (3.5). The number of expansion terms given are sufficient to guarantee the digits quoted. In those cases where the last digits still change with the maximum

Appendix 1: Energy Values

number of expansion terms employed, we list both the corresponding value and, separated by a slash, the figures still changing if the two last energy values computed are linearly extrapolated to 1/(number of expansion terms) → 0 in the energy versus 1/(number of expansion terms) plot (cf. Fig. 3.1). This latter value presents a lower bound, and the true value lies in-between. The continuous dependence of the energies of all these states on the magnetic field strength is shown in Fig. A1.1. The energies of states with positive m have been included by simply adding $4\beta m$ to the corresponding energies with $-m$.

Table A1.1. Relations between states, range of β, tables, and figures

m	π	states $n_p lm/nm\nu$			range of β	Table	Figure
0	+1	$1s_0/0\,0\,0$,	$2s_0/0\,0\,2$,	$3d'_0/0\,0\,4$	$10^{-4} - 10^3$, (s)	A1.2a	A1.2a
0	+1	$3s'_0/0\,0\,6$,	$4d'_0/0\,0\,8$,	$4s'_0/0\,0\,10$	$10^{-4} - 10^3$, (s)	A1.2a	A1.2a
0	+1	$5g'_0/0\,0\,12$,	$5d'_0/0\,0\,14$,	$5s'_0/0\,0\,16$	$10^{-4} - 10^3$, (s)	A1.2a	A1.2a
0	−1	$2p_0/0\,0\,1$,	$3p_0/0\,0\,3$,	$4f'_0/0\,0\,5$	$10^{-4} - 10^3$, (s)	A1.2b	A1.2b
0	−1	$4p'_0/0\,0\,7$,	$5f'_0/0\,0\,9$,	$5p'_0/0\,0\,11$	$10^{-4} - 10^3$, (s)	A1.2b	A1.2b
−1	+1	$2p_{-1}/0\,-1\,0$, $3p_{-1}/0\,-1\,2$, $4f'_{-1}/0\,-1\,4$			$10^{-4} - 10^3$, (s)	A1.2c	A1.2c
−1	+1	$4p'_{-1}/0\,-1\,6$, $5f'_{-1}/0\,-1\,8$, $5p'_{-1}/0\,-1\,10$			$10^{-4} - 10^3$, (s)	A1.2c	A1.2c
−1	−1	$3d_{-1}/0\,-1\,1$, $4d_{-1}/0\,-1\,3$, $5g'_{-1}/0\,-1\,5$			$10^{-4} - 10^3$, (s)	A1.2d	A1.2d
−1	−1	$5d'_{-1}/0\,-1\,7$			$10^{-4} - 10^3$, (s)	A1.2d	A1.2d
−2	+1	$3d_{-2}/0\,-2\,0$, $4d_{-2}/0\,-2\,2$, $5g'_{-2}/0\,-2\,4$			$10^{-4} - 10^3$, (s)	A1.2e	A1.2e
−2	+1	$5d'_{-2}/0\,-2\,6$, $6g'_{-2}/0\,-2\,8$			$10^{-4} - 10^3$, (s)	A1.2e	A1.2e
−2	−1	$4f_{-2}/0\,-2\,1$			$10^{-4} - 10^3$, (s)	A1.2e	A1.2f
−2	−1	$5f_{-2}/0\,-2\,3$, $6h'_{-2}/0\,-2\,5$, $6f'_{-2}/0\,-2\,7$			$10^{-4} - 10^3$, (s)	A1.2f	A1.2f
−3	+1	$4f_{-3}/0\,-3\,0$, $5f_{-3}/0\,-3\,2$			$10^{-4} - 10^3$, (s)	A1.2g	A1.2g
−3	−1	$5g_{-3}/0\,-3\,1$			$10^{-4} - 10^3$, (s)	A1.2g	A1.2h
−4	+1	$5g_{-4}/0\,-4\,0$			$10^{-4} - 10^3$, (s)	A1.2h	A1.2i
0	+1	$2s_0/0\,0\,2$,	$3d'_0/0\,0\,4$,	$3s'_0/0\,0\,6$	$10^{-3} - 10^{-1}$, (f)	A1.3a	A1.2a
0	+1	$4d'_0/0\,0\,8$,	$4s'_0/0\,0\,10$,	$5g'_0/0\,0\,12$	$10^{-3} - 10^{-1}$, (f)	A1.3a	A1.2a
0	+1	$5d'_0/0\,0\,14$,	$5s'_0/0\,0\,16$		$10^{-3} - 10^{-1}$, (f)	A1.3a	A1.2a
0	−1	$2p_0/0\,0\,1$,	$3p_0/0\,0\,3$,	$4f'_0/0\,0\,5$	$10^{-3} - 10^{-1}$, (f)	A1.3b	A1.2b
0	−1	$4p'_0/0\,0\,7$,	$5f'_0/0\,0\,9$,	$5p'_0/0\,0\,11$	$10^{-3} - 10^{-1}$, (f)	A1.3b	A1.2b
−1	+1	$2p_{-1}/0\,-1\,0$, $3p_{-1}/0\,-1\,2$, $4f'_{-1}/0\,-1\,4$			$10^{-3} - 10^{-1}$, (f)	A1.3c	A1.2c
−1	+1	$4p'_{-1}/0\,-1\,6$, $5f'_{-1}/0\,-1\,8$, $5p'_{-1}/0\,-1\,10$			$10^{-3} - 10^{-1}$, (f)	A1.3c	A1.2c
−1	−1	$3d_{-1}/0\,-1\,1$, $4d_{-1}/0\,-1\,3$, $5g'_{-1}/0\,-1\,5$			$10^{-3} - 10^{-1}$, (f)	A1.3d	A1.2d
−1	−1	$5d'_{-1}/0\,-1\,7$			$10^{-3} - 10^{-1}$, (f)	A1.3d	A1.2d
−2	+1	$3d_{-2}/0\,-2\,0$, $4d_{-2}/0\,-2\,2$, $5g'_{-2}/0\,-2\,4$			$10^{-3} - 10^{-1}$, (f)	A1.3e	A1.2e
−2	+1	$5d'_{-2}/0\,-2\,6$			$10^{-3} - 10^{-1}$, (f)	A1.3e	A1.2e

A1.1 Tables of the Energy Values

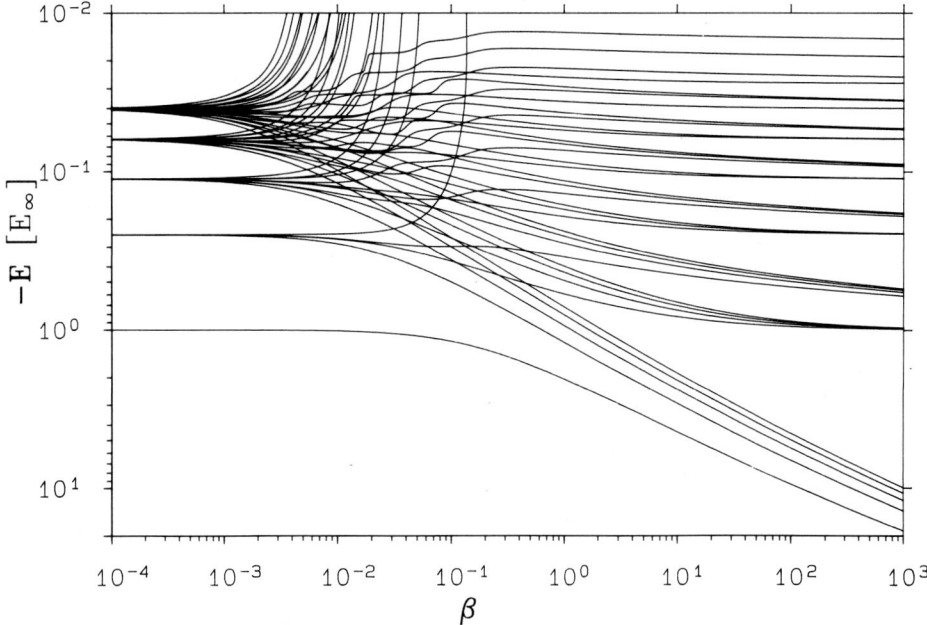

Fig. A1.1. Energy values (in units of E_∞) of all states of an electron in a static Coulomb potential with principal quantum numbers $n_p \leq 5$ as continuous functions of the magnetic field parameter $\beta = B/(4.701 \times 10^5 \,\text{T})$. The Rydberg series corrected by the splitting of the linear and quadratic Zeeman effect is well recognized for small values of β. At β-values around 10^{-2} a complete rearrangement of the energy levels occurs. States with positive m are shifted to higher Landau levels, and for $\beta \geq 10$ the level structure of the strong-field regime is assumed, where states with $\nu > 0$ converge towards the Rydberg series again (*hydrogen-like states*), while the z-nodeless ground states in every $m \leq 0$ band are strongly lowered (*tightly bound states*) and diverge logarithmically in the limit $\beta \to \infty$.

Table A1.2a. Energies (in units of $-E_\infty$) of the lowest $m = 0$, z-parity $\pi = +1$ states of an electron in a static Coulomb potential and a magnetic field with strength parameter β

	$m = 0 \quad \pi = +1$		
β	$1s_0/0\ 0\ 0$	$2s_0/0\ 0\ 2$	$3d'_0/0\ 0\ 4$
1×10^{-4}	1.000200 $(1,\infty)$	0.2501997 $(1,\infty)$	0.1113107 $(2,\infty)$
1.5×10^{-4}	1.000300 $(1,\infty)$	0.2502994 $(1,\infty)$	0.1114102 $(2,\infty)$
2×10^{-4}	1.000400 $(1,\infty)$	0.2503989 $(1,\infty)$	0.1115095 $(2,\infty)$
3×10^{-4}	1.000600 $(1,\infty)$	0.2505975 $(1,\infty)$	0.1117074 $(2,\infty)$
5×10^{-4}	1.000999 $(1,\infty)$	0.2509930 $(1,\infty)$	0.1121008 $(3,\infty)$
7×10^{-4}	1.001399 $(1,\infty)$	0.2513863 $(1,\infty)$	0.1124908 $(3,\infty)$
1×10^{-3}	1.001998 $(1,\infty)$	0.2519720 $(2,\infty)$	0.1130698 $(3,\infty)$
1.5×10^{-3}	1.002996 $(1,\infty)$	0.2529370 $(2,\infty)$	0.1140182 $(3,\infty)$
2×10^{-3}	1.003992 $(1,\infty)$	0.2538881 $(3,\infty)$	0.1149461 $(3,\infty)$
3×10^{-3}	1.005982 $(1,\infty)$	0.2557484 $(2,\infty)$	0.1167413 $(4,\infty)$
5×10^{-3}	1.009950 $(2,\infty)$	0.2593031 $(2,\infty)$	0.1200958 $(4,\infty)$
7×10^{-3}	1.013902 $(1,\infty)$	0.2626399 $(2,\infty)$	0.1231530 $(4,\infty)$
1×10^{-2}	1.019800 $(2,\infty)$	0.2672484 $(3,\infty)$	0.1272339 $(5,\infty)$
1.5×10^{-2}	1.029550 $(3,\infty)$	0.2739309 $(3,\infty)$	0.1328897 $(6,\infty)$
2×10^{-2}	1.039201 $(2,\infty)$	0.2794796 $(4,\infty)$	0.1373908 $(7,\infty)$
3×10^{-2}	1.058207 $(2,\infty)$	0.2877269 $(5,\infty)$	0.1438266 $(9,\infty)$
5×10^{-2}	1.095053 $(3,\infty)$	0.2961783 $(7,\infty)$	0.1498760 $(12,\infty)$
7×10^{-2}	1.130396 $(3,\infty)$	0.2985628 $(8,\infty)$	0.1501991 $(14,\infty)$
1×10^{-1}	1.180763 $(3,\infty)$	0.2979734 $(10,\infty)$	0.1451139 $(17,\infty)$
1.5×10^{-1}	1.258373 $(4,\infty)$	0.2967346 $(12,\infty)$	0.1353428 $(35,35)$
2×10^{-1}	1.329211 $(4,\infty)$	0.2983327 $(17,\infty)$	0.1309745 $(39,39)$
3×10^{-1}	1.454925 $(7,\infty)$	0.3055311 $(18,\infty)$	0.1297957 $(70,70)$
5×10^{-1}	1.662338 $(7,\infty)$	0.3209379 $(23,\infty)$	0.13202/63 [12]
7×10^{-1}	1.832332 $(9,\infty)$	0.3334492 $(25,25)$	0.13499/537 [12]
1	2.044428 $(11,\infty)$	0.3478880 $(24,\infty)$	0.13841/65 [12]
1.5	2.329066 $(11,\infty)$	0.365131/87 $(23,\infty)$	0.14245/60 [12]
2	2.561596 $(12,\infty)$	0.37762/85 $(23,\infty)$	0.14533/45 [12]
3	2.936492 $(14,\infty)$	0.39530/63 [12]	0.149356/431 [12]
5	3.495594 $(19,\infty)$	0.41777/98 [12]	0.154286/333 [12]
7	3.922425 $(21,\infty)$	0.43237/53 [12]	0.157425/60 [12]
1×10^1	4.430797 $(20,\infty)$	0.44761/73 [12]	0.160647/74 [12]
1.5×10^1	5.084843 $(24,\infty)$	0.464570/660 [12]	0.164173/92 [12]
2×10^1	5.602058/60 $(24,\infty)$	0.476352/425 [12]	0.166585/600 [12]
3×10^1	6.4103/32 [12]	0.492573/627 [12]	0.169856/67 [12]
5×10^1	7.5781/805 [12]	0.512339/77 [12]	0.1737679/754 [12]
7×10^1	8.4443/64 [12]	0.524938/68 [12]	0.1762202/60 [12]
1×10^2	9.4531/50 [12]	0.537921/45 [12]	0.1787145/90 [12]
1.5×10^2	10.7206/22 [12]	0.552213/31 [12]	0.1814221/55 [12]
2×10^2	11.7023/39 [12]	0.562050/65 [12]	0.1832631/59 [12]
3×10^2	13.2090/104 [12]	0.575488/99 [12]	0.1857486/507 [12]
5×10^2	15.3241/53 [12]	0.5917099/180 [12]	0.1887047/62 [12]
7×10^2	16.8569/80 [11]	0.6019690/754 [12]	0.1905495/506 [12]
1×10^3	18.60896/986 [12]	0.6124793/844 [12]	0.1923843 [1]

Table A1.2a. (Continued)

	$m=0 \quad \pi=+1$		
β	$3s'_0/0\ 0\ 6$	$4d'_0/0\ 0\ 8$	$4s'_0/0\ 0\ 10$
1×10^{-4}	0.1113095 $(3,\infty)$	0.06269851 $(3,\infty)$	0.06269478 $(2,\infty)$
1.5×10^{-4}	0.1114076 $(2,\infty)$	0.06279664 $(3,\infty)$	0.06278825 $(3,\infty)$
2×10^{-4}	0.1115048 $(2,\infty)$	0.06289403 $(3,\infty)$	0.06287910 $(3,\infty)$
3×10^{-4}	0.1116970 $(2,\infty)$	0.06308656 $(3,\infty)$	0.06305300 $(3,\infty)$
5×10^{-4}	0.1120720 $(3,\infty)$	0.06346272 $(4,\infty)$	0.06336963 $(3,\infty)$
7×10^{-4}	0.1124344 $(3,\infty)$	0.06382705 $(4,\infty)$	0.06364498 $(3,\infty)$
1×10^{-3}	0.1129547 $(3,\infty)$	0.06435166 $(4,\infty)$	0.06398170 $(3,\infty)$
1.5×10^{-3}	0.1137600 $(3,\infty)$	0.06516911 $(4,\infty)$	0.06434530 $(4,\infty)$
2×10^{-3}	0.1144886 $(3,\infty)$	0.06591861 $(4,\infty)$	0.06447397 $(4,\infty)$
3×10^{-3}	0.1157213 $(3,\infty)$	0.06723194 $(5,\infty)$	0.06409349 $(5,\infty)$
5×10^{-3}	0.1173387 $(4,\infty)$	0.06925742 $(7,\infty)$	0.06127474 $(6,\infty)$
7×10^{-3}	0.1179415 $(4,\infty)$	0.07072114 $(7,\infty)$	0.05656138 $(8,\infty)$
1×10^{-2}	0.1172685 $(5,\infty)$	0.07228304 $(9,\infty)$	0.04861447 $(13,\infty)$
1.5×10^{-2}	0.1130664 $(6,\infty)$	0.07388960 $(11,\infty)$	0.04742543 $(16,\infty)$
2×10^{-2}	0.1065606 $(20,\infty)$	0.07449760 $(14,\infty)$	0.04763838 $(20,\infty)$
3×10^{-2}	0.09343459 $(18,\infty)$	0.07107623 $(17,\infty)$	0.04723598 $(24,\infty)$
5×10^{-2}	0.08713337 $(19,\infty)$	0.05614951 $(29,29)$	0.03888676 $(37,37)$
7×10^{-2}	0.08622458 $(37,37)$	0.05399226 $(39,39)$	0.03637441 $(51,51)$
1×10^{-1}	0.08303951 $(35,35)$	0.05211949 $(59,59)$	0.03519757 $(61,61)$
1.5×10^{-1}	0.07568627 $(45,45)$	0.04764542 $(77,77)$	0.03249431 $(75,75)$
2×10^{-1}	0.07207175 $(80,80)$	0.04521073 $(80,80)$	0.0308884 $(80,80)$
3×10^{-1}	0.07069836 $(72,72)$	0.04401/33 [12]	0.03011/29 [12]
5×10^{-1}	0.07128/53 [12]	0.04447/59 [12]	0.030345/415 [12]
7×10^{-1}	0.07238/53 [12]	0.044983/5057 [12]	0.030630/72 [12]
1	0.073658/753 [12]	0.045597/643 [12]	0.030971/97 [12]
1.5	0.075172/231 [12]	0.046322/50 [12]	0.031373/89 [12]
2	0.076249/91 [12]	0.046836/56 [12]	0.031657/69 [12]
3	0.077741/69 [12]	0.047546/60 [12]	0.0320496/570 [12]
5	0.079554/72 [12]	0.0484050/133 [12]	0.0325222/67 [12]
7	0.080699/712 [12]	0.0489446/507 [12]	0.0328182/215 [12]
1×10^{1}	0.0818668/763 [12]	0.0494926/71 [12]	0.0331180/204 [12]
1.5×10^{1}	0.0831347/415 [12]	0.0500854/86 [12]	0.0334415/32 [12]
2×10^{1}	0.0839962/40016 [12]	0.0504867/92 [12]	0.0336599/613 [12]
3×10^{1}	0.0851577/616 [12]	0.0510257/75 [12]	0.03395267/365 [12]
5×10^{1}	0.0865357/84 [12]	0.0516624/36 [12]	0.03429748/813 [12]
7×10^{1}	0.0873934/54 [12]	0.05205712/805 [12]	0.03449260 [1]
1×10^{2}	0.0882609/25 [12]	0.05245513/85 [12]	0.03471117 [1]
1.5×10^{2}	0.0891972/84 [12]	0.05286352 [1]	0.03494498 [1]
2×10^{2}	0.08983051/146 [12]	0.05315587 [1]	0.03510200 [1]
3×10^{2}	0.09068136/206 [12]	0.05354676 [1]	0.03531158 [1]
5×10^{2}	0.09166797 [1]	0.05400622 [1]	0.03555739 [1]
7×10^{2}	0.09229609 [1]	0.05429009 [1]	0.03570897 [1]
1×10^{3}	0.09292960 [1]	0.05457573 [1]	0.03586129 [1]

Table A1.2a. (Continued)

	$m=0 \quad \pi=+1$		
β	$5g'_0/0\ 0\ 12$	$5d'_0/0\ 0\ 14$	$5s'_0/0\ 0\ 16$
1×10^{-4}	0.04019773 $(23,\infty)$	0.04019450 $(23,\infty)$	0.04018677 $(23,\infty)$
1.5×10^{-4}	0.04029490 $(23,\infty)$	0.04028763 $(23,\infty)$	0.04027025 $(23,\infty)$
2×10^{-4}	0.04039094 $(23,\infty)$	0.04037802 $(23,\infty)$	0.04034713 $(23,\infty)$
3×10^{-4}	0.04057962 $(22,\infty)$	0.04055061 $(22,\infty)$	0.04048123 $(22,\infty)$
5×10^{-4}	0.04094350 $(22,\infty)$	0.04086334 $(22,\infty)$	0.04067166 $(22,\infty)$
7×10^{-4}	0.04128959 $(20,\infty)$	0.04113368 $(20,\infty)$	0.04076094 $(20,\infty)$
1×10^{-3}	0.04177604 $(25,25)$	0.04146285 $(25,25)$	0.04071412 $(25,25)$
1.5×10^{-3}	0.04250325 $(25,25)$	0.04182340 $(25,25)$	0.04019641 $(25,25)$
2×10^{-3}	0.04313314 $(24,24)$	0.04197747 $(24,24)$	0.03920361 $(24,24)$
3×10^{-3}	0.04413822 $(24,24)$	0.04179624 $(24,24)$	0.03611944 $(24,24)$
5×10^{-3}	0.04538022 $(27,27)$	0.04016530 $(27,27)$	0.03115953 $(27,27)$
7×10^{-3}	0.04595995 $(25,25)$	0.03763604 $(30,30)$	0.03098342 $(30,30)$
1×10^{-2}	0.04520191 $(25,25)$	0.03397457 $(28,28)$	0.02905155 $(28,28)$
1.5×10^{-2}	0.03311284 $(30,30)$	0.02983707 $(30,30)$	0.02403859 $(33,33)$
2×10^{-2}	0.03269779 $(30,30)$	0.02373009 $(40,40)$	0.01798546 $(49,49)$
3×10^{-2}	0.03250226 $(40,40)$	0.02354430 $(50,50)$	0.01778495 $(58,58)$
5×10^{-2}	0.02820337 $(58,58)$	0.02116420 $(75,75)$	0.01635199 $(80,80)$
7×10^{-2}	0.02596886 $(80,80)$	0.019395 $(80,80)$	0.015003/7 [96]
1×10^{-1}	0.025188 $(80,80)$	0.018851/7 [96]	0.014618/21 [96]
1.5×10^{-1}	0.023456/95 [96]	0.017708/34 [96]	0.013826/44 [96]
2×10^{-1}	0.022385/407 [96]	0.016962/77 [96]	0.013290/301 [96]
3×10^{-1}	0.021962/9 [96]	0.016665/70 [96]	0.013076/9 [96]
5×10^{-1}	0.022016/59 [12]	0.016697/726 [12]	0.0131095/106 [96]
7×10^{-1}	0.022190/216 [12]	0.016811/28 [12]	0.013174/86 [12]
1	0.022398/414 [12]	0.016948/58 [12]	0.0132689/761 [12]
1.5	0.0226439/535 [12]	0.0171087/150 [12]	0.0133801/45 [12]
2	0.0228176/246 [12]	0.0172224/70 [12]	0.0134586/618 [12]
3	0.0230566/612 [12]	0.0173788/818 [12]	0.0135664/85 [12]
5	0.0233439/67 [12]	0.0175663/81 [12]	0.0136956/68 [12]
7	0.0235235/55 [12]	0.0176833/47 [12]	0.01377601/92 [12]
1×10^1	0.0237051/66 [12]	0.01780151/248 [12]	0.01385718/84 [12]
1.5×10^1	0.0239006/16 [12]	0.01792855/923 [12]	0.01394434/80 [12]
2×10^1	0.02403235/318 [12]	0.01801409/63 [12]	0.01400291/328 [12]
3×10^1	0.02420872/931 [12]	0.01812840/78 [12]	0.01408121/47 [12]
5×10^1	0.02441599/639 [12]	0.01826258/84 [12]	0.01416691 [1]
7×10^1	0.02454394/425 [12]	0.01833829 [1]	0.01422474 [1]
1×10^2	0.02466412 [1]	0.01842298 [1]	0.01428252 [1]
1.5×10^2	0.02480405 [1]	0.01851325 [1]	0.01434412 [1]
2×10^2	0.02489790 [1]	0.01857374 [1]	0.01438543 [1]
3×10^2	0.02502302 [1]	0.01865432 [1]	0.01444034 [1]
5×10^2	0.02516955 [1]	0.01874859 [1]	0.01450451 [1]
7×10^2	0.02525977 [1]	0.01880659 [1]	0.01454396 [1]
1×10^3	0.02535036 [1]	0.01886477 [1]	0.01458352 [1]

A1.1 Tables of the Energy Values

Table A1.2b. Energies (in units of $-E_\infty$) of the lowest $m = 0$, z-parity $\pi = -1$ states of an electron in a static Coulomb potential and a magnetic field with strength parameter β

	$m=0 \quad \pi=-1$		
β	$2p_0$/0 0 1	$3p_0$/0 0 3	$4f'_0$/0 0 5
1×10^{-4}	0.2501999 $(1,\infty)$	0.1113104 $(1,\infty)$	0.06269893 $(3,\infty)$
1.5×10^{-4}	0.2502997 $(1,\infty)$	0.1114095 $(1,\infty)$	0.06279760 $(3,\infty)$
2×10^{-4}	0.2503995 $(1,\infty)$	0.1115082 $(2,\infty)$	0.06289573 $(3,\infty)$
3×10^{-4}	0.2505989 $(1,\infty)$	0.1117046 $(1,\infty)$	0.06309039 $(3,\infty)$
5×10^{-4}	0.2509970 $(1,\infty)$	0.1120931 $(2,\infty)$	0.06347332 $(4,\infty)$
7×10^{-4}	0.2513941 $(1,\infty)$	0.1124759 $(2,\infty)$	0.06384775 $(4,\infty)$
1×10^{-3}	0.2519880 $(1,\infty)$	0.1130392 $(2,\infty)$	0.06439356 $(4,\infty)$
1.5×10^{-3}	0.2529730 $(2,\infty)$	0.1139497 $(2,\infty)$	0.06526151 $(4,\infty)$
2×10^{-3}	0.2539520 $(3,\infty)$	0.1148249 $(2,\infty)$	0.06607842 $(4,\infty)$
3×10^{-3}	0.2558921 $(3,\infty)$	0.1164720 $(3,\infty)$	0.06756588 $(5,\infty)$
5×10^{-3}	0.2597008 $(2,\infty)$	0.1193757 $(3,\infty)$	0.07001254 $(6,\infty)$
7×10^{-3}	0.2634152 $(2,\infty)$	0.1218140 $(4,\infty)$	0.07186766 $(7,\infty)$
1×10^{-2}	0.2688129 $(3,\infty)$	0.1247571 $(5,\infty)$	0.07380055 $(9,\infty)$
1.5×10^{-2}	0.2773627 $(3,\infty)$	0.1283037 $(6,\infty)$	0.07548642 $(11,\infty)$
2×10^{-2}	0.2853874 $(4,\infty)$	0.1308128 $(7,\infty)$	0.07607329 $(18,\infty)$
3×10^{-2}	0.3000325 $(4,\infty)$	0.1344065 $(10,\infty)$	0.07613841 $(22,\infty)$
5×10^{-2}	0.3248202 $(6,\infty)$	0.1397834 $(12,\infty)$	0.07647957 $(20,\infty)$
7×10^{-2}	0.3452365 $(7,\infty)$	0.1442643 $(14,\infty)$	0.07750589 $(50,50)$
1×10^{-1}	0.3703681 $(8,\infty)$	0.1498509 $(16,\infty)$	0.07918961 $(60,60)$
1.5×10^{-1}	0.4030083 $(10,\infty)$	0.1570134 $(18,\infty)$	0.081535/97 [12]
2×10^{-1}	0.4285310 $(12,\infty)$	0.1624457 $(21,\infty)$	0.083385/421 [12]
3×10^{-1}	0.4673569 $(13,\infty)$	0.1703659 $(22,\infty)$	0.086074/92 [12]
5×10^{-1}	0.5200132 $(16,\infty)$	0.180441/63 [12]	0.0894514/93 [12]
7×10^{-1}	0.5562670 $(22,\infty)$	0.186979/93 [12]	0.0916065/113 [12]
1	0.5954219 $(23,\infty)$	0.1937062/144 [12]	0.0937956/84 [12]
1.5	0.6400802 $(24,\infty)$	0.2010007/54 [12]	0.0961373/89 [12]
2	0.6713901/4 $(24,\infty)$	0.2059002/34 [12]	0.0976912/31 [12]
3	0.714320/31 [12]	0.2123613/32 [12]	0.09972083/145 [12]
5	0.7652975/3036 [12]	0.2196909/18 [12]	0.1019939/42 [12]
7	0.7963735/76 [12]	0.2239975/81 [12]	0.1033160/2 [12]
1×10^1	0.8267545/72 [12]	0.2281021/5 [12]	0.1045671/2 [12]
1.5×10^1	0.8577958/74 [12]	0.2321977/80 [12]	0.1058072 [12]
2×10^1	0.8774672/83 [12]	0.2347462/4 [12]	0.1065747 [12]
3×10^1	0.9018598/604 [12]	0.2378603 [12]	0.1075085 [12]
5×10^1	0.9272354/8 [12]	0.2410508 [12]	0.1084608 [12]
7×10^1	0.9409211/3 [12]	0.2427530 [12]	0.1089671 [12]
1×10^2	0.9530640/1 [12]	0.2442537 [12]	0.1094125 [11]
1.5×10^2	0.9642544 [12]	0.2456296 [12]	0.1098201 [10]
2×10^2	0.9707261 [12]	0.2464227 [12]	0.1100547 [11]
3×10^2	0.9780941 [12]	0.2473237 [12]	0.1103210 [9]
5×10^2	0.9849900 [12]	0.2481656 [12]	0.1105694 [1]
7×10^2	0.9883800 [12]	0.2485793 [9]	0.1106916 [1]
1×10^3	0.9911896 [12]	0.2489221 [10]	0.1107929 [1]

Table A1.2b. (Continued)

	$m=0 \quad \pi=-1$		
β	$4p'_0/0\ 0\ 7$	$5f'_0/0\ 0\ 9$	$5p'_0/0\ 0\ 11$
1×10^{-4}	0.06269691 (2,∞)	0.04019726 (30,30)	0.04019124 (30,30)
1.5×10^{-4}	0.06279305 (3,∞)	0.04029384 (30,30)	0.04028030 (30,30)
2×10^{-4}	0.06288764 (3,∞)	0.04038905 (30,30)	0.04036500 (30,30)
3×10^{-4}	0.06307219 (3,∞)	0.04057540 (30,30)	0.04052137 (30,30)
5×10^{-4}	0.06342285 (3,∞)	0.04093191 (30,30)	0.04078259 (30,30)
7×10^{-4}	0.06374907 (3,∞)	0.04126727 (30,30)	0.04097678 (30,30)
1×10^{-3}	0.06419317 (4,∞)	0.04173218 (30,30)	0.04114829 (30,30)
1.5×10^{-3}	0.06481602 (3,∞)	0.04241287 (30,30)	0.04114336 (30,30)
2×10^{-3}	0.06529908 (4,∞)	0.04298996 (30,30)	0.04082708 (30,30)
3×10^{-3}	0.06588500 (5,∞)	0.04389787 (30,30)	0.03949718 (30,30)
5×10^{-3}	0.06583575 (5,∞)	0.04506068 (30,30)	0.03544809 (30,30)
7×10^{-3}	0.06467752 (8,∞)	0.04568721 (31,31)	0.03272679 (31,31)
1×10^{-2}	0.06188510 (11,∞)	0.04593216 (30,30)	0.03231503 (30,30)
1.5×10^{-2}	0.05682263 (13,∞)	0.04477764 (30,30)	0.03183342 (30,30)
2×10^{-2}	0.05309329 (16,∞)	0.04203484 (30,30)	0.03088286 (30,30)
3×10^{-2}	0.04943232 (40,40)	0.03573297 (40,40)	0.02748068 (40,40)
5×10^{-2}	0.04777979 (42,42)	0.03253585 (48,48)	0.02352624 (57,57)
7×10^{-2}	0.04790683 (50,50)	0.03240971 (60,60)	0.02333971 (67,67)
1×10^{-1}	0.04854600 (60,60)	0.03269674 (70,70)	0.02348586 (84,84)
1.5×10^{-1}	0.049565/95 [12]	0.033224/41 [12]	0.023793/804 [12]
2×10^{-1}	0.050392/410 [12]	0.0336634/730 [12]	0.0240538/97 [12]
3×10^{-1}	0.0516011/96 [12]	0.0343067/114 [12]	0.0244364/92 [12]
5×10^{-1}	0.0531137/74 [12]	0.0351106/25 [12]	0.0249137/49 [12]
7×10^{-1}	0.0540730/52 [12]	0.0356184/96 [12]	0.02521448/518 [12]
1	0.0550418/31 [12]	0.03612954/3023 [12]	0.02551650/91 [12]
1.5	0.05607162/234 [12]	0.03667081/119 [12]	0.02583550/72 [12]
2	0.05675158/205 [12]	0.03702699/725 [12]	0.02604493/508 [12]
3	0.05763435/63 [12]	0.03748798/812 [12]	0.02631541/9 [12]
5	0.05861742/56 [12]	0.03799943/50 [12]	0.02661472/6 [12]
7	0.05918635/43 [12]	0.03829449/54 [12]	0.02678703/6 [12]
1×10^{1}	0.05972282/7 [12]	0.03857211/4 [12]	0.02694891/2 [12]
1.5×10^{1}	0.06025272/5 [12]	0.03884575/6 [12]	0.02710823 [12]
2×10^{1}	0.06057983/5 [12]	0.03901437/8 [12]	0.02720629 [12]
3×10^{1}	0.06097688/9 [12]	0.03921876 [12]	0.02732504 [12]
5×10^{1}	0.06138077 [12]	0.03942636 [12]	0.02744552 [12]
7×10^{1}	0.06159510 [12]	0.03953543 [1]	0.02750878 [1]
1×10^{2}	0.06178343 [12]	0.03963240 [1]	0.02756498 [1]
1.5×10^{2}	0.06195561 [12]	0.03972091 [1]	0.02761626 [1]
2×10^{2}	0.06205419 [1]	0.03977178 [1]	0.02764571 [1]
3×10^{2}	0.06216674 [1]	0.03982943 [1]	0.02767909 [1]
5×10^{2}	0.06227172 [1]	0.03988318 [1]	0.02771020 [1]
7×10^{2}	0.06232324 [1]	0.03990955 [1]	0.02772546 [1]
1×10^{3}	0.06236591 [1]	0.03993139 [1]	0.02773810 [1]

Table A1.2c. Energies (in units of $-E_\infty$) of the lowest $m = -1$, z-parity $\pi = +1$ states of an electron in a static Coulomb potential and a magnetic field with strength parameter β

	$m=-1$ $\pi=+1$					
β	$2p_{-1}/0\ -1\ 0$		$3p_{-1}/0\ -1\ 2$		$4f'_{-1}/0\ -1\ 4$	
1×10^{-4}	0.2503998	$(1,\infty)$	0.1115097	$(1,\infty)$	0.06289828	$(2,\infty)$
1.5×10^{-4}	0.2505995	$(1,\infty)$	0.1117079	$(2,\infty)$	0.06309613	$(3,\infty)$
2×10^{-4}	0.2507990	$(1,\infty)$	0.1119054	$(2,\infty)$	0.06329312	$(3,\infty)$
3×10^{-4}	0.2511978	$(1,\infty)$	0.1122982	$(1,\infty)$	0.06368452	$(3,\infty)$
5×10^{-4}	0.2519940	$(1,\infty)$	0.1130751	$(2,\infty)$	0.06445703	$(4,\infty)$
7×10^{-4}	0.2527882	$(1,\infty)$	0.1138406	$(2,\infty)$	0.06521587	$(4,\infty)$
1×10^{-3}	0.2539760	$(2,\infty)$	0.1149673	$(2,\infty)$	0.06632869	$(4,\infty)$
1.5×10^{-3}	0.2559460	$(2,\infty)$	0.1167883	$(2,\infty)$	0.06811656	$(4,\infty)$
2×10^{-3}	0.2579041	$(2,\infty)$	0.1185387	$(2,\infty)$	0.06982313	$(4,\infty)$
3×10^{-3}	0.2617843	$(3,\infty)$	0.1218330	$(3,\infty)$	0.07300534	$(5,\infty)$
5×10^{-3}	0.2694023	$(2,\infty)$	0.1276402	$(3,\infty)$	0.07855087	$(6,\infty)$
7×10^{-3}	0.2768327	$(2,\infty)$	0.1325105	$(4,\infty)$	0.08319589	$(7,\infty)$
1×10^{-2}	0.2876352	$(3,\infty)$	0.1383502	$(4,\infty)$	0.08886648	$(9,\infty)$
1.5×10^{-2}	0.3047682	$(3,\infty)$	0.1451588	$(6,\infty)$	0.09577849	$(9,\infty)$
2×10^{-2}	0.3208951	$(3,\infty)$	0.1495447	$(7,\infty)$	0.1003289	$(10,\infty)$
3×10^{-2}	0.3505288	$(4,\infty)$	0.1547466	$(9,\infty)$	0.1040563	$(12,\infty)$
5×10^{-2}	0.4016913	$(5,\infty)$	0.1623424	$(12,\infty)$	0.09805689	$(23,23)$
7×10^{-2}	0.4452239	$(6,\infty)$	0.1703942	$(15,\infty)$	0.09239537	$(27,27)$
1×10^{-1}	0.5010782	$(7,\infty)$	0.1816837	$(17,\infty)$	0.09312162	$(50,50)$
1.5×10^{-1}	0.5781850	$(8,\infty)$	0.1969821	$(19,\infty)$	0.09719/36	[12]
2×10^{-1}	0.6427096	$(9,\infty)$	0.2089556	$(21,\infty)$	0.10081/91	[12]
3×10^{-1}	0.7492475	$(11,\infty)$	0.2269580	$(23,23)$	0.106211/65	[12]
5×10^{-1}	0.9131941	$(13,\infty)$	0.250922/1013	[12]	0.113156/83	[12]
7×10^{-1}	1.042015	$(13,\infty)$	0.267276/340	[12]	0.117701/19	[12]
1	1.199226	$(15,\infty)$	0.284886/932	[12]	0.122440/53	[12]
1.5	1.407093	$(17,\infty)$	0.305078/110	[12]	0.1276917/7003	[12]
2	1.575651	$(17,\infty)$	0.319423/48	[12]	0.1313145/211	[12]
3	1.846584	$(19,\infty)$	0.339552/71	[12]	0.1362605/52	[12]
5	2.250845	$(22,22)$	0.364598/611	[12]	0.1422109/39	[12]
7	2.56045/65	[12]	0.3808198/296	[12]	0.1459555/78	[12]
1×10^{1}	2.93095/113	[12]	0.3977233/309	[12]	0.1497736/54	[12]
1.5×10^{1}	3.41052/67	[12]	0.4165303/60	[12]	0.1539277/89	[12]
2×10^{1}	3.79211/25	[12]	0.4295900/46	[12]	0.1567573/83	[12]
3×10^{1}	4.39389/402	[12]	0.4475749/84	[12]	0.1605843/50	[12]
5×10^{1}	5.26948/59	[12]	0.4695043/68	[12]	0.1651464/70	[12]
7×10^{1}	5.924947/5044	[12]	0.4834936/56	[12]	0.1679996/80000	[12]
1×10^{2}	6.694255/342	[12]	0.4979219/35	[12]	0.1708975/8	[12]
1.5×10^{2}	7.669181/259	[12]	0.5138189/201	[12]	0.1740390/2	[12]
2×10^{2}	8.430228/99	[12]	0.5247707/17	[12]	0.1761728/30	[12]
3×10^{2}	9.607359/422	[12]	0.5397448/56	[12]	0.1790510/1	[12]
5×10^{2}	11.27681/7	[12]	0.5578434/9	[12]	0.1824706	[12]
7×10^{2}	12.49785/90	[12]	0.5693007/11	[12]	0.1846026	[12]
1×10^{3}	13.90394/8	[12]	0.5810478/81	[12]	0.1867629	[12]

Table A1.2c. (Continued)

	$m=-1 \quad \pi=+1$		
β	$4p'_{-1}/0 -1\ 6$	$5f'_{-1}/0 -1\ 8$	$5p'_{-1}/0 -1\ 10$
1×10^{-4}	0.06289500 $(2,\infty)$	0.04039487 $(30,30)$	0.04038713 $(30,30)$
1.5×10^{-4}	0.06308875 $(2,\infty)$	0.04058847 $(30,30)$	0.04057106 $(30,30)$
2×10^{-4}	0.06328001 $(2,\infty)$	0.04077951 $(30,30)$	0.04074857 $(30,30)$
3×10^{-4}	0.06365504 $(3,\infty)$	0.04115396 $(30,30)$	0.04108446 $(30,30)$
5×10^{-4}	0.06437527 $(3,\infty)$	0.04187261 $(30,30)$	0.04168058 $(30,30)$
7×10^{-4}	0.06505601 $(3,\infty)$	0.04255176 $(30,30)$	0.04217827 $(30,30)$
1×10^{-3}	0.06600411 $(4,\infty)$	0.04349936 $(30,30)$	0.04274891 $(30,30)$
1.5×10^{-3}	0.06739511 $(4,\infty)$	0.04490366 $(30,30)$	0.04327185 $(30,30)$
2×10^{-3}	0.06856110 $(3,\infty)$	0.04611619 $(30,30)$	0.04333186 $(30,30)$
3×10^{-3}	0.07028153 $(4,\infty)$	0.04808911 $(30,30)$	0.04237864 $(30,30)$
5×10^{-3}	0.07174309 $(5,\infty)$	0.05086873 $(30,30)$	0.03813726 $(30,30)$
7×10^{-3}	0.07135482 $(8,\infty)$	0.05274327 $(30,30)$	0.03727151 $(30,30)$
1×10^{-2}	0.06897776 $(11,\infty)$	0.05427046 $(30,30)$	0.03790900 $(30,30)$
1.5×10^{-2}	0.06578793 $(14,\infty)$	0.05203992 $(30,30)$	0.03833929 $(30,30)$
2×10^{-2}	0.06628064 $(16,\infty)$	0.04689582 $(31,31)$	0.03596993 $(30,30)$
3×10^{-2}	0.06885616 $(22,\infty)$	0.04551478 $(40,40)$	0.03179807 $(40,40)$
5×10^{-2}	0.06896141 $(45,45)$	0.04650423 $(45,45)$	0.03218946 $(50,50)$
7×10^{-2}	0.05981763 $(40,40)$	0.04276147 $(60,60)$	0.03110816 $(61,61)$
1×10^{-1}	0.05587986 $(60,60)$	0.03703858 $(80,80)$	0.02627028 $(81,81)$
1.5×10^{-1}	0.057104/95 [12]	0.037381/432 [12]	0.026312/44 [12]
2×10^{-1}	0.058577/625 [12]	0.038109/35 [12]	0.026723/38 [12]
3×10^{-1}	0.060814/38 [12]	0.039237/49 [12]	0.0273692/766 [12]
5×10^{-1}	0.063665/77 [12]	0.0406709/69 [12]	0.0281895/930 [12]
7×10^{-1}	0.0655045/122 [12]	0.0415890/929 [12]	0.0287123/45 [12]
1	0.0673983/4035 [12]	0.0425280/306 [12]	0.0292446/60 [12]
1.5	0.0694688/723 [12]	0.0435471/88 [12]	0.02981955/2052 [12]
2	0.0708806/32 [12]	0.0442375/88 [12]	0.03020747/819 [12]
3	0.0727869/87 [12]	0.04516416/504 [12]	0.03072606/56 [12]
5	0.0750495/507 [12]	0.04625566/621 [12]	0.03133392/423 [12]
7	0.07645675/762 [12]	0.04693004/45 [12]	0.03170786/809 [12]
1×10^{1}	0.07787876/941 [12]	0.04760806/37 [12]	0.03208259/76 [12]
1.5×10^{1}	0.07941153/200 [12]	0.04833503/25 [12]	0.03248298/311 [12]
2×10^{1}	0.08044721/58 [12]	0.04882399/417 [12]	0.03275149/59 [12]
3×10^{1}	0.08183723/50 [12]	0.04947740/53 [12]	0.03310929/36 [12]
5×10^{1}	0.08347847/65 [12]	0.05024473/81 [12]	0.03352799/804 [12]
7×10^{1}	0.08449624/39 [12]	0.05071830/6 [12]	0.03378561/4 [12]
1×10^{2}	0.08552319/30 [12]	0.05119437/43 [12]	0.03404397/400 [12]
1.5×10^{2}	0.08662887/94 [12]	0.05170498/501 [12]	0.03431647 [1]
2×10^{2}	0.08737536/42 [12]	0.05204855/8 [12]	0.03450279 [1]
3×10^{2}	0.08837652/7 [12]	0.05250338 [1]	0.03475119 [1]
5×10^{2}	0.08955053 [1]	0.05304436 [1]	0.03504210 [1]
7×10^{2}	0.09028359 [1]	0.05337818 [1]	0.03522122 [1]
1×10^{3}	0.09102246 [1]	0.05371375 [1]	0.03540096 [1]

Table A1.2d. Energies (in units of $-E_\infty$) of the lowest $m = -1$, z-parity $\pi = -1$ states of an electron in a static Coulomb potential and a magnetic field with strength parameter β

	$m = -1 \quad \pi = -1$		
β	$3d_{-1}/0\ -1\ 1$	$4d_{-1}/0\ -1\ 3$	$5g'_{-1}/0\ -1\ 5$
1×10^{-4}	0.1115104 $(1,\infty)$	0.06289712 $(2,\infty)$	0.04039645 $(20,\infty)$
1.5×10^{-4}	0.1117095 $(1,\infty)$	0.06309352 $(2,\infty)$	0.04059201 $(25,25)$
2×10^{-4}	0.1119082 $(1,\infty)$	0.06328848 $(2,\infty)$	0.04078581 $(25,25)$
3×10^{-4}	0.1123046 $(1,\infty)$	0.06367410 $(2,\infty)$	0.04116808 $(25,25)$
5×10^{-4}	0.1130931 $(2,\infty)$	0.06442814 $(2,\infty)$	0.04191152 $(25,25)$
7×10^{-4}	0.1138759 $(2,\infty)$	0.06515940 $(2,\infty)$	0.04262710 $(25,25)$
1×10^{-3}	0.1150392 $(3,\infty)$	0.06621414 $(2,\infty)$	0.04364933 $(25,25)$
1.5×10^{-3}	0.1169495 $(2,\infty)$	0.06786259 $(3,\infty)$	0.04522220 $(25,25)$
2×10^{-3}	0.1188244 $(3,\infty)$	0.06938046 $(3,\infty)$	0.04664226 $(25,25)$
3×10^{-3}	0.1224693 $(2,\infty)$	0.07205993 $(4,\infty)$	0.04907694 $(25,25)$
5×10^{-3}	0.1293563 $(3,\infty)$	0.07626436 $(5,\infty)$	0.05264977 $(25,25)$
7×10^{-3}	0.1357450 $(3,\infty)$	0.07939045 $(6,\infty)$	0.05494537 $(25,25)$
1×10^{-2}	0.1445071 $(4,\infty)$	0.08291265 $(8,\infty)$	0.05676235 $(25,25)$
1.5×10^{-2}	0.1573240 $(5,\infty)$	0.08733824 $(11,\infty)$	0.05746923 $(25,25)$
2×10^{-2}	0.1684264 $(6,\infty)$	0.09099912 $(11,\infty)$	0.05757337 $(25,25)$
3×10^{-2}	0.1870550 $(7,\infty)$	0.09715607 $(14,\infty)$	0.05882242 $(30,30)$
5×10^{-2}	0.2156242 $(8,\infty)$	0.1064913 $(16,\infty)$	0.06216807 $(42,42)$
7×10^{-2}	0.2375816 $(10,\infty)$	0.1133887 $(18,\infty)$	0.064872/97 [12]
1×10^{-1}	0.2635700 $(12,\infty)$	0.1211637 $(21,\infty)$	0.067938/50 [12]
1.5×10^{-1}	0.2964437 $(14,\infty)$	0.1304018 $(22,\infty)$	0.0715141/98 [12]
2×10^{-1}	0.3218464 $(15,\infty)$	0.1371163/244 [12]	0.0740577/612 [12]
3×10^{-1}	0.3604109 $(17,\infty)$	0.1466881/925 [12]	0.0775984/6002 [12]
5×10^{-1}	0.4131347 $(19,\infty)$	0.1587188/210 [12]	0.08191633/717 [12]
7×10^{-1}	0.4499882/935 [12]	0.1665115/29 [12]	0.08464021/73 [12]
1	0.4904806/42 [12]	0.1745723/31 [12]	0.08740211/42 [12]
1.5	0.5377152/76 [12]	0.1833992/7 [12]	0.09036613/32 [12]
2	0.5716053/70 [12]	0.1893973/7 [12]	0.09234658/71 [12]
3	0.6192621/32 [12]	0.1974174/7 [12]	0.09495473/80 [12]
5	0.6779122/9 [12]	0.2067080/1 [12]	0.09792253/7 [12]
7	0.7149455/60 [12]	0.2122872 [12]	0.09967884/7 [12]
1×10^{1}	0.7522395/9 [12]	0.2177073 [12]	0.1013675 [12]
1.5×10^{1}	0.7916242/4 [12]	0.2232369 [12]	0.1030730 [12]
2×10^{1}	0.8173526/7 [12]	0.2267512 [12]	0.1041483 [11]
3×10^{1}	0.8502184 [12]	0.2311378 [12]	0.1054815 [11]
5×10^{1}	0.8857423 [12]	0.2357618 [11]	0.1068763 [10]
7×10^{1}	0.9055753 [12]	0.2382954 [11]	0.1076361 [11]
1×10^{2}	0.9236433 [12]	0.2405763 [11]	0.1083177 [8]
1.5×10^{2}	0.9407524 [12]	0.2427145 [11]	0.1089544 [7]
2×10^{2}	0.9508809 [11]	0.2439711 [12]	0.1093278 [11]
3×10^{2}	0.9626562 [12]	0.2454245 [9]	0.1097588 [7]
5×10^{2}	0.9739559 [12]	0.2468126 [7]	0.1101694 [1]
7×10^{2}	0.9796294 [10]	0.2475075 [7]	0.1103750 [1]
1×10^{3}	0.9844024 [9]	0.2480914 [5]	0.1105475 [1]

Table A1.2d. (Continued)

β	$5d'_{-1}/0\ -1\ 7$	
\multicolumn{3}{c}{$m=-1\quad \pi=-1$}		
1×10^{-4}	0.04039155	(25,25)
1.5×10^{-4}	0.04058100	(25,25)
2×10^{-4}	0.04076624	(25,25)
3×10^{-4}	0.04112415	(25,25)
5×10^{-4}	0.04179027	(25,25)
7×10^{-4}	0.04239165	(25,25)
1×10^{-3}	0.04317793	(25,25)
1.5×10^{-3}	0.04420646	(25,25)
2×10^{-3}	0.04493160	(25,25)
3×10^{-3}	0.04569294	(25,25)
5×10^{-3}	0.04568510	(25,25)
7×10^{-3}	0.04514560	(25,25)
1×10^{-2}	0.04474022	(25,25)
1.5×10^{-2}	0.04392754	(25,25)
2×10^{-2}	0.04174100	(25,25)
3×10^{-2}	0.03931285	(36,36)
5×10^{-2}	0.04034289	(42,42)
7×10^{-2}	0.041607/21	[12]
1×10^{-1}	0.0430801/65	[12]
1.5×10^{-1}	0.0447951/80	[12]
2×10^{-1}	0.0460044/61	[12]
3×10^{-1}	0.04767024/112	[12]
5×10^{-1}	0.04967341/81	[12]
7×10^{-1}	0.05092144/68	[12]
1	0.05217499/514	[12]
1.5	0.05350743/52	[12]
2	0.05439055/61	[12]
3	0.05554510/3	[12]
5	0.05684749/50	[12]
7	0.05761270/1	[12]
1×10^{1}	0.05834463	[12]
1.5×10^{1}	0.05908022	[12]
2×10^{1}	0.05954211	[12]
3×10^{1}	0.06011280	[12]
5×10^{1}	0.06070758	[12]
7×10^{1}	0.06103064	[11]
1×10^{2}	0.06131984	[11]
1.5×10^{2}	0.06158903	[1]
2×10^{2}	0.06174750	[11]
3×10^{2}	0.06192943	[1]
5×10^{2}	0.06210299	[1]
7×10^{2}	0.06218971	[1]
1×10^{3}	0.06226248	[1]

A1.1 Tables of the Energy Values 197

Table A1.2e. Energies (in units of $-E_\infty$) of the lowest $m=-2$ states of an electron in a static Coulomb potential and a magnetic field with strength parameter β

	$m=-2 \quad \pi=+1$					
β	$3d_{-2}/0\ -2\ 0$		$4d_{-2}/0\ -2\ 2$		$5g'_{-2}/0\ -2\ 4$	
1×10^{-4}	0.1117100	$(1,\infty)$	0.06309568	$(2,\infty)$	0.04059530	$(20,20)$
1.5×10^{-4}	0.1120087	$(1,\infty)$	0.06339028	$(2,\infty)$	0.04088942	$(20,20)$
2×10^{-4}	0.1123068	$(1,\infty)$	0.06368273	$(2,\infty)$	0.04118119	$(20,20)$
3×10^{-4}	0.1129014	$(1,\infty)$	0.06426115	$(2,\infty)$	0.04175771	$(20,20)$
5×10^{-4}	0.1140841	$(1,\infty)$	0.06539221	$(2,\infty)$	0.04288284	$(20,20)$
7×10^{-4}	0.1152582	$(2,\infty)$	0.06648914	$(2,\infty)$	0.04397125	$(20,20)$
1×10^{-3}	0.1170033	$(2,\infty)$	0.06807135	$(2,\infty)$	0.04553685	$(20,20)$
1.5×10^{-3}	0.1198689	$(3,\infty)$	0.07054451	$(3,\infty)$	0.04797659	$(20,20)$
2×10^{-3}	0.1226814	$(2,\infty)$	0.07282233	$(3,\infty)$	0.05022155	$(20,20)$
3×10^{-3}	0.1281506	$(2,\infty)$	0.07684462	$(4,\infty)$	0.05420840	$(20,20)$
5×10^{-3}	0.1384944	$(3,\infty)$	0.08314204	$(5,\infty)$	0.06063510	$(20,20)$
7×10^{-3}	0.1481133	$(3,\infty)$	0.08774656	$(6,\infty)$	0.06558319	$(20,20)$
1×10^{-2}	0.1613717	$(4,\infty)$	0.09267376	$(7,\infty)$	0.07103225	$(25,25)$
1.5×10^{-2}	0.1809811	$(4,\infty)$	0.09831937	$(9,\infty)$	0.07609616	$(25,25)$
2×10^{-2}	0.1982491	$(5,\infty)$	0.1029996	$(10,\infty)$	0.07718960	$(30,30)$
3×10^{-2}	0.2279625	$(6,\infty)$	0.1118481	$(12,\infty)$	0.07308603	$(35,35)$
5×10^{-2}	0.2756790	$(8,\infty)$	0.1267353	$(15,\infty)$	0.07247905	$(46,46)$
7×10^{-2}	0.3143615	$(8,\infty)$	0.1382654	$(17,\infty)$	0.076396/457	[12]
1×10^{-1}	0.3626412	$(9,\infty)$	0.1516684	$(19,\infty)$	0.081280/308	[12]
1.5×10^{-1}	0.4279525	$(10,\infty)$	0.1681594	$(21,\infty)$	0.087144/57	[12]
2×10^{-1}	0.4819653	$(13,\infty)$	0.1805607/14	$(22,\infty)$	0.0913913/4001	[12]
3×10^{-1}	0.5705063	$(14,\infty)$	0.198876/91	[12]	0.0974108/60	[12]
5×10^{-1}	0.7060961	$(17,\infty)$	0.2230932/1024	[12]	0.1049430/59	[12]
7×10^{-1}	0.8124766	$(17,\infty)$	0.2395678/747	[12]	0.1098225/46	[12]
1	0.9423439	$(19,\infty)$	0.2573472/523	[12]	0.1148922/37	[12]
1.5	1.114303	$(20,20)$	0.2778026/63	[12]	0.1205001/12	[12]
2	1.254019	$(20,20)$	0.2923869/98	[12]	0.1243665/73	[12]
3	1.479163	$(21,21)$	0.3129273/95	[12]	0.1296447/53	[12]
5	1.816423/47	[12]	0.3386011/27	[12]	0.1359965/9	[12]
7	2.075819/40	[12]	0.3552934/46	[12]	0.1399951/4	[12]
1×10^1	2.387260/79	[12]	0.3727360/70	[12]	0.1440734/7	[12]
1.5×10^1	2.792052/69	[12]	0.3921955/62	[12]	0.1485119/21	[12]
2×10^1	3.115394/409	[12]	0.4057380/6	[12]	0.1515361/3	[12]
3×10^1	3.627363/76	[12]	0.4244241/5	[12]	0.1556272	[12]
5×10^1	4.376331/42	[12]	0.4472607/10	[12]	0.1605059	[12]
7×10^1	4.939837/48	[12]	0.4618565/8	[12]	0.1635580	[12]
1×10^2	5.603995/4005	[12]	0.4769319/21	[12]	0.1666587	[12]
1.5×10^2	6.449578/86	[12]	0.4935662/4	[12]	0.1700211	[12]
2×10^2	7.112423/31	[12]	0.5050404/6	[12]	0.1723056	[12]
3×10^2	8.141986/93	[12]	0.5207473	[12]	0.1753878	[12]
5×10^2	9.610219/25	[12]	0.5397592	[12]	0.1790510	[12]
7×10^2	10.68942	[12]	0.5518098	[12]	0.1813356	[12]
1×10^3	11.93718	[12]	0.5641765	[12]	0.1836506	[12]

Table A1.2e. (Continued)

β	$m=-2\quad \pi=+1$				$m=-2\quad \pi=-1$	
	$5d'_{-2}/0\ -2\ 6$		$6g'_{-2}/0\ -2\ 8$		$4f_{-2}/0\ -2\ 1$	
1×10^{-4}	0.04058821	(20,20)	0.02836569	(20,20)	0.06309760	$(1,\infty)$
1.5×10^{-4}	0.04087348	(20,20)	0.02865100	(20,20)	0.06339460	$(2,\infty)$
2×10^{-4}	0.04115288	(20,20)	0.02892952	(20,20)	0.06369040	$(2,\infty)$
3×10^{-4}	0.04169413	(20,20)	0.02946956	(20,20)	0.06427841	$(2,\infty)$
5×10^{-4}	0.04270729	(20,20)	0.03048038	(20,20)	0.06544008	$(3,\infty)$
7×10^{-4}	0.04363019	(20,20)	0.03140371	(20,20)	0.06658272	$(4,\infty)$
1×10^{-3}	0.04485307	(20,20)	0.03264060	(20,20)	0.06826133	$(2,\infty)$
1.5×10^{-3}	0.04649747	(20,20)	0.03437273	(20,20)	0.07096657	$(3,\infty)$
2×10^{-3}	0.04771481	(20,20)	0.03578994	(20,20)	0.07356017	$(3,\infty)$
3×10^{-3}	0.04915093	(20,20)	0.03799744	(20,20)	0.07843472	$(3,\infty)$
5×10^{-3}	0.04957300	(20,20)	0.04088441	(20,20)	0.08710544	$(4,\infty)$
7×10^{-3}	0.04907996	(25,25)	0.04181078	(25,25)	0.09463770	$(6,\infty)$
1×10^{-2}	0.05024426	(25,25)	0.03923162	(25,25)	0.1043856	$(5,\infty)$
1.5×10^{-2}	0.05365790	(25,25)	0.03736572	(25,25)	0.1178355	$(7,\infty)$
2×10^{-2}	0.05638095	(25,25)	0.03849359	(40,40)	0.1289963	$(7,\infty)$
3×10^{-2}	0.05837407	(40,40)	0.04050527	(60,60)	0.1471341	$(9,\infty)$
5×10^{-2}	0.04705283	(50,50)	0.0347801	(80,80)	0.1742216	$(11,\infty)$
7×10^{-2}	0.047775/816	[12]	0.032500/30	[12]	0.1947770	$(12,\infty)$
1×10^{-1}	0.049913/27	[12]	0.0335646/733	[12]	0.2190099	$(15,\infty)$
1.5×10^{-1}	0.0525417/82	[12]	0.0349472/508	[12]	0.2496743	$(17,\infty)$
2×10^{-1}	0.0544226/67	[12]	0.0359327/49	[12]	0.2734510	$(18,\infty)$
3×10^{-1}	0.0570420/43	[12]	0.0372929/42	[12]	0.3097628	$(19,\infty)$
5×10^{-1}	0.0602433/45	[12]	0.03893458/524	[12]	0.3599433	(21,21)
7×10^{-1}	0.06227444/532	[12]	0.03996454/99	[12]	0.3954483/90	[12]
1	0.06435170/232	[12]	0.04100889/920	[12]	0.4349087/93	[12]
1.5	0.06661271/313	[12]	0.04213567/88	[12]	0.4815891/5	[12]
2	0.06815039/72	[12]	0.04289625/41	[12]	0.5155466/9	[12]
3	0.07022328/51	[12]	0.04391444/55	[12]	0.5640115/7	[12]
5	0.07267971/86	[12]	0.04511076/83	[12]	0.6249143/4	[12]
7	0.07420587/99	[12]	0.04584856/62	[12]	0.6641646	[12]
1×10^{1}	0.07574702/10	[12]	0.04658944/8	[12]	0.7043852	[12]
1.5×10^{1}	0.07740712/9	[12]	0.04738287/90	[12]	0.7477027	[12]
2×10^{1}	0.07852828/33	[12]	0.04791604/6	[12]	0.7765219	[12]
3×10^{1}	0.08003237/41	[12]	0.04862794/5	[12]	0.8140130	[12]
5×10^{1}	0.08180745/7	[12]	0.04946315/6	[12]	0.8555170	[11]
7×10^{1}	0.08290782/4	[12]	0.04997822	[12]	0.8792049	[12]
1×10^{2}	0.08401784/6	[12]	0.05049574	[12]	0.9011594	[11]
1.5×10^{2}	0.08521269/70	[12]	0.05105049	[12]	0.9223281	[10]
2×10^{2}	0.08601924	[12]	0.05142360	[12]	0.9350591	[10]
3×10^{2}	0.08710079	[12]	0.05191970	[1]	0.9500760	[9]
5×10^{2}	0.08837628	[12]	0.05250594	[1]	0.9647413	[8]
7×10^{2}	0.08916324	[1]	0.05286760	[1]	0.9722164	[8]
1×10^{3}	0.08996028	[1]	0.05323104	[1]	0.9785735	[6]

Table A1.2f. Energies (in units of $-E_\infty$) of the lowest $m = -2$, z-parity $\pi = -1$ states of an electron in a static Coulomb potential and a magnetic field with strength parameter β

β	\multicolumn{2}{c	}{$m = -2 \quad \pi = -1$}				
	\multicolumn{2}{c	}{$5f_{-2}/0\ -2\ 3$}	\multicolumn{2}{c	}{$6h'_{-2}/0\ -2\ 5$}	\multicolumn{2}{c	}{$6f'_{-2}/0\ -2\ 7$}
1×10^{-4}	0.04059250	(20,20)	0.02836934	(20,20)	0.02835959	(20,20)
1.5×10^{-4}	0.04088314	(25,25)	0.02865881	(25,25)	0.02863690	(25,25)
2×10^{-4}	0.04117003	(25,25)	0.02894408	(25,25)	0.02890520	(25,25)
3×10^{-4}	0.04173266	(35,35)	0.02950212	(35,35)	0.02941515	(35,35)
5×10^{-4}	0.04281372	(35,35)	0.03056890	(35,35)	0.03033167	(35,35)
7×10^{-4}	0.04383713	(38,38)	0.03157198	(40,40)	0.03111881	(40,40)
1×10^{-3}	0.04526866	(38,38)	0.03296393	(40,40)	0.03208397	(40,40)
1.5×10^{-3}	0.04740035	(38,38)	0.03501289	(40,40)	0.03322685	(40,40)
2×10^{-3}	0.04925438	(40,40)	0.03677006	(40,40)	0.03394791	(40,40)
3×10^{-3}	0.05230904	(40,40)	0.03958487	(40,40)	0.03468901	(40,40)
5×10^{-3}	0.05675986	(40,40)	0.04319485	(40,40)	0.03558753	(40,40)
7×10^{-3}	0.06008522	(40,40)	0.04498287	(40,40)	0.03667988	(40,40)
1×10^{-2}	0.06416955	(40,40)	0.04579351	(40,40)	0.03767593	(40,40)
1.5×10^{-2}	0.06977200	(40,40)	0.04645264	(40,40)	0,03588273	(40,40)
2×10^{-2}	0.07439268	(40,40)	0.047939/8013	[12]	0.03372713	(40,40)
3×10^{-2}	0.08170596	(40,40)	0.051054/76	[12]	0.034592/608	[12]
5×10^{-2}	0.09201792	(41,41)	0.0556338/400	[12]	0.0368846/83	[12]
7×10^{-2}	0.0993441/500	[12]	0.0588238/69	[12]	0.0385085/102	[12]
1×10^{-1}	0.1074765/97	[12]	0.0622685/701	[12]	0.04024313/99	[12]
1.5×10^{-1}	0.1170688/705	[12]	0.06620171/250	[12]	0.04219280/322	[12]
2×10^{-1}	0.1240340/51	[12]	0.06897411/61	[12]	0.04354695/721	[12]
3×10^{-1}	0.1339807/13	[12]	0.07282093/120	[12]	0.04539932/45	[12]
5×10^{-1}	0.1465589/92	[12]	0.07751390/403	[12]	0.04761975/82	[12]
7×10^{-1}	0.1547684/7	[12]	0.08048379/87	[12]	0.04900399/403	[12]
1	0.1633230/1	[12]	0.08350753/9	[12]	0.05039765/8	[12]
1.5	0.1727748	[12]	0.08677110/3	[12]	0.05188499/500	[12]
2	0.1792545	[12]	0.08896503/5	[12]	0.05287544	[12]
3	0.1880007	[12]	0.09187438/40	[12]	0.05417769	[12]
5	0.1982692	[12]	0.09521933	[12]	0.05565978	[12]
7	0.2045188	[12]	0.09722030	[12]	0.05653890	[12]
1×10^{1}	0.2106611	[12]	0.09916272	[12]	0.05738712	[12]
1.5×10^{1}	0.2170114	[12]	0.1011470	[11]	0.05824847	[12]
2×10^{1}	0.2210990	[12]	0.1024119	[12]	0.05879488	[11]
3×10^{1}	0.2262679	[12]	0.1039982	[9]	0.05947725	[10]
5×10^{1}	0.2318130	[11]	0.1056840	[8]	0.06019905	[10]
7×10^{1}	0.2349024	[11]	0.1066165	[7]	0.06059680	[10]
1×10^{2}	0.2377213	[8]	0.1074633	[6]	0.06095708	[9]
1.5×10^{2}	0.2404021	[8]	0.1082651	[6]	0.06129745	[7]
2×10^{2}	0.2419983	[7]	0.1087409	[8]	0.06149909	[7]
3×10^{2}	0.2438667	[7]	0.1092965	[6]	0.06173404	[1]
5×10^{2}	0.2456782	[5]	0.1098338	[6]	0.06196119	[1]
7×10^{2}	0.2465972	[5]	0.1101057	[1]	0.06207609	[1]
1×10^{3}	0.2473769	[4]	0.1103363	[1]	0.06217340	[1]

Table A1.2g. Energies (in units of $-E_\infty$) of the lowest $m = -3$, z-parity $\pi = \pm 1$ states of an electron in a static Coulomb potential and a magnetic field with strength parameter β

	$m = -3 \quad \pi = +1$		$m = -3 \quad \pi = -1$
β	$4f_{-3}/0\ -3\ 0$	$5f_{-3}/0\ -3\ 2$	$5g_{-3}/0\ -3\ 1$
1×10^{-4}	0.06329680 $(1,\infty)$	0.04079000 $(2,\infty)$	0.04079400 $(2,\infty)$
1.5×10^{-4}	0.06369280 $(2,\infty)$	0.04117751 $(2,\infty)$	0.04118651 $(3,\infty)$
2×10^{-4}	0.06408720 $(2,\infty)$	0.04156005 $(2,\infty)$	0.04157602 $(2,\infty)$
3×10^{-4}	0.06487122 $(2,\infty)$	0.04231023 $(2,\infty)$	0.04234610 $(2,\infty)$
5×10^{-4}	0.06642013 $(2,\infty)$	0.04375172 $(2,\infty)$	0.04385075 $(3,\infty)$
7×10^{-4}	0.06794371 $(2,\infty)$	0.04511649 $(3,\infty)$	0.04530884 $(2,\infty)$
1×10^{-3}	0.07018210 $(2,\infty)$	0.04702602 $(3,\infty)$	0.04741145 $(3,\infty)$
1.5×10^{-3}	0.07379034 $(2,\infty)$	0.04987130 $(4,\infty)$	0.05070415 $(4,\infty)$
2×10^{-3}	0.07725159 $(3,\infty)$	0.05234762 $(4,\infty)$	0.05375750 $(4,\infty)$
3×10^{-3}	0.08376660 $(3,\infty)$	0.05642111 $(5,\infty)$	0.05926136 $(5,\infty)$
5×10^{-3}	0.09541529 $(4,\infty)$	0.06223079 $(6,\infty)$	0.06849160 $(5,\infty)$
7×10^{-3}	0.1056341 $(4,\infty)$	0.06633317 $(9,\infty)$	0.07613206 $(6,\infty)$
1×10^{-2}	0.1190464 $(6,\infty)$	0.07124914 $(10,\infty)$	0.08571207 $(8,\infty)$
1.5×10^{-2}	0.1379718 $(6,\infty)$	0.07853026 $(12,\infty)$	0.09860592 $(9,\infty)$
2×10^{-2}	0.1540911 $(7,\infty)$	0.08499403 $(13,\infty)$	0.1091478 $(10,\infty)$
3×10^{-2}	0.1811574 $(7,\infty)$	0.09565628 $(16,\infty)$	0.1261270 $(10,\infty)$
5×10^{-2}	0.2237204 $(9,\infty)$	0.1112881 $(18,\infty)$	0.1513434 $(12,\infty)$
7×10^{-2}	0.2578335 $(10,\infty)$	0.1227951 $(19,\infty)$	0.1704607 $(16,\infty)$
1×10^{-1}	0.3001767 $(11,\infty)$	0.1359704 $(22,22)$	0.1930321 $(17,\infty)$
1.5×10^{-1}	0.3572597 $(13,\infty)$	0.1520866/914 [12]	0.2216907 $(18,\infty)$
2×10^{-1}	0.4043935 $(15,\infty)$	0.1642002/37 [12]	0.2440063 $(20,\infty)$
3×10^{-1}	0.4816205 $(17,\infty)$	0.1821291/315 [12]	0.2782685/8 [12]
5×10^{-1}	0.5999380 $(19,\infty)$	0.2059326/41 [12]	0.3260134/6 [12]
7×10^{-1}	0.6928731 $(19,\infty)$	0.2222063/75 [12]	0.3600931/2 [12]
1	0.8064806 $(21,21)$	0.2398448/57 [12]	0.3982760 [12]
1.5	0.9571828 $(23,23)$	0.2602339/45 [12]	0.4438838 [12]
2	1.079851/6 [12]	0.2748307/12 [12]	0.4773755 [12]
3	1.277935/9 [12]	0.2954678/82 [12]	0.5256619 [12]
5	1.575537/41 [12]	0.3213801/4 [12]	0.5872126 [12]
7	1.805078/82 [12]	0.3382897/9 [12]	0.6274404 [12]
1×10^{1}	2.081357/60 [12]	0.3560056/8 [12]	0.6691614 [12]
1.5×10^{1}	2.441458/61 [12]	0.3758193/4 [12]	0.7147136 [11]
2×10^{1}	2.729853/6 [12]	0.3896357/8 [12]	0.7454104 [11]
3×10^{1}	3.187728/31 [12]	0.4087326 [12]	0.7858645 [11]
5×10^{1}	3.860012/4 [12]	0.4321176 [12]	0.8314254 [12]
7×10^{1}	4.367535/6 [12]	0.4470880 [12]	0.8578488 [9]
1×10^{2}	4.967402/4 [12]	0.4625689 [12]	0.8826523 [9]
1.5×10^{2}	5.733520/1 [12]	0.4796711 [12]	0.9068934 [9]
2×10^{2}	6.335773/4 [12]	0.4914803 [12]	0.9216476 [11]
3×10^{2}	7.273893/4 [12]	0.5076612 [12]	0.9392450 [7]
5×10^{2}	8.616740/1 [12]	0.5272704 [12]	0.9566662 [7]
7×10^{2}	9.607129 [12]	0.5397122 [12]	0.9656517 [6]
1×10^{3}	10.75536 [12]	0.5524904 [12]	0.9733591 [6]

Table A1.2h. Energies (in units of $-E_\infty$) of the lowest $m = -4$, z-parity $\pi = +1$ states of an electron in a static Coulomb potential and a magnetic field with strength parameter β

β	$m = -4 \quad \pi = +1$	
	$5g_{-4}/0\ -4\ 0$	
1×10^{-4}	0.04099250	(25,25)
1.5×10^{-4}	0.04148313	(20,20)
2×10^{-4}	0.04197003	(20,20)
3×10^{-4}	0.04293264	(20,20)
5×10^{-4}	0.04481358	(20,20)
7×10^{-4}	0.04663656	(20,20)
1×10^{-3}	0.04926633	(20,20)
1.5×10^{-3}	0.05338904	(20,20)
2×10^{-3}	0.05772057	(20,20)
3×10^{-3}	0.06416051	(20,20)
5×10^{-3}	0.07594078	(20,20)
7×10^{-3}	0.08586279	(20,20)
1×10^{-2}	0.09855312	(20,20)
1.5×10^{-2}	0.1161034	(20,20)
2×10^{-2}	0.1308699	(30,30)
3×10^{-2}	0.155463/89	[12]
5×10^{-2}	0.193906/19	[12]
7×10^{-2}	0.2246228/318	[12]
1×10^{-1}	0.2627031/95	[12]
1.5×10^{-1}	0.3140146/91	[12]
2×10^{-1}	0.3563865/902	[12]
3×10^{-1}	0.4258455/83	[12]
5×10^{-1}	0.5323754/75	[12]
7×10^{-1}	0.6161573/90	[12]
1	0.7187017/32	[12]
1.5	0.8549335/48	[12]
2	0.9659872/83	[12]
3	1.145599	[12]
5	1.416052	[12]
7	1.625099	[12]
1×10^{1}	1.877174	[12]
1.5×10^{1}	2.206420	[12]
2×10^{1}	2.470615	[12]
3×10^{1}	2.890917	[12]
5×10^{1}	3.509712	[12]
7×10^{1}	3.978035	[12]
1×10^{2}	4.532740	[12]
1.5×10^{2}	5.242840	[12]
2×10^{2}	5.802243	[12]
3×10^{2}	6.675490	[12]
5×10^{2}	7.929019	[12]
7×10^{2}	8.855907	[12]
1×10^{3}	9.932766	[12]

Table A1.3a. Energies (in units of $-E_\infty$) of the lowest $m=0$, z-parity $\pi=+1$ states of an electron in a static Coulomb potential and a magnetic field with strength parameter β

	$m=0 \quad \pi=+1$		
β	$2s_0/0\ 0\ 2$	$3d'_0/0\ 0\ 4$	$3s'_0/0\ 0\ 6$
1×10^{-3}	0.2519720 (2,∞)	0.1130698 (3,∞)	0.1129547 (3,∞)
1.5×10^{-3}	0.2529370 (2,∞)	0.1140182 (3,∞)	0.1137600 (3,∞)
2×10^{-3}	0.2538881 (3,∞)	0.1149461 (3,∞)	0.1144886 (3,∞)
2.5×10^{-3}	0.2548252 (26,26)	0.1158538 (26,26)	0.1151419 (26,26)
3×10^{-3}	0.2557484 (2,∞)	0.1167413 (4,∞)	0.1157213 (3,∞)
3.5×10^{-3}	0.2566577 (27,27)	0.1176090 (27,27)	0.1162287 (27,27)
4×10^{-3}	0.2575533 (27,27)	0.1184571 (27,27)	0.1166659 (27,27)
4.5×10^{-3}	0.2584351 (27,27)	0.1192860 (27,27)	0.1170352 (27,27)
5×10^{-3}	0.2593031 (2,∞)	0.1200958 (4,∞)	0.1173387 (4,∞)
5.5×10^{-3}	0.2601576 (35,35)	0.1208871 (35,35)	0.1175787 (35,35)
6×10^{-3}	0.2609985 (35,35)	0.1216601 (35,35)	0.1177576 (35,35)
6.5×10^{-3}	0.2618259 (35,35)	0.1224153 (35,35)	0.1178777 (35,35)
7×10^{-3}	0.2626399 (2,∞)	0.1231530 (4,∞)	0.1179415 (4,∞)
8×10^{-3}	0.2642282 (35,35)	0.1245775 (35,35)	0.1179092 (35,35)
9×10^{-3}	0.2657640 (35,35)	0.1259367 (35,35)	0.1176790 (35,35)
1×10^{-2}	0.2672484 (3,∞)	0.1272339 (5,∞)	0.1172685 (5,∞)
1.05×10^{-2}	0.2679715 (47,47)	0.1278601 (47,47)	0.1170008 (47,47)
1.1×10^{-2}	0.2686821 (35,35)	0.1284719 (35,35)	0.1166943 (35,35)
1.15×10^{-2}	0.2693802 (47,47)	0.1290697 (47,47)	0.1163509 (47,47)
1.2×10^{-2}	0.2700661 (35,35)	0.1296538 (35,35)	0.1159724 (35,35)
1.25×10^{-2}	0.2707399 (47,47)	0.1302246 (47,47)	0.1155607 (47,47)
1.3×10^{-2}	0.2714016 (35,35)	0.1307823 (35,35)	0.1151176 (35,35)
1.35×10^{-2}	0.2720514 (47,47)	0.1313274 (47,47)	0.1146449 (47,47)
1.4×10^{-2}	0.2726895 (35,35)	0.1318601 (35,35)	0.1141444 (35,35)
1.45×10^{-2}	0.2733159 (47,47)	0.1323808 (47,47)	0.1136176 (47,47)
1.5×10^{-2}	0.2739309 (3,∞)	0.1328897 (6,∞)	0.1130664 (6,∞)
1.55×10^{-2}	0.2745346 (47,47)	0.1333871 (47,47)	0.1124922 (47,47)
1.6×10^{-2}	0.2751270 (47,47)	0.1338734 (47,47)	0.1118968 (47,47)
1.65×10^{-2}	0.2757083 (47,47)	0.1343487 (47,47)	0.1112817 (47,47)
1.7×10^{-2}	0.2762787 (47,47)	0.1348133 (47,47)	0.1106486 (47,47)
1.75×10^{-2}	0.2768384 (47,47)	0.1352676 (47,47)	0.1099990 (47,47)
1.8×10^{-2}	0.2773873 (47,47)	0.1357116 (47,47)	0.1093345 (47,47)
1.81×10^{-2}	0.2774959 (47,47)	0.1357992 (47,47)	0.1092000 (47,47)
1.82×10^{-2}	0.2776040 (47,47)	0.1358864 (47,47)	0.1090650 (47,47)
1.83×10^{-2}	0.2777117 (47,47)	0.1359733 (47,47)	0.1089294 (47,47)
1.84×10^{-2}	0.2778190 (47,47)	0.1360597 (47,47)	0.1087934 (47,47)
1.85×10^{-2}	0.2779258 (47,47)	0.1361458 (47,47)	0.1086569 (47,47)
1.86×10^{-2}	0.2780323 (47,47)	0.1362314 (47,47)	0.1085199 (47,47)

Table A1.3a. (Continued)

	$m=0 \quad \pi=+1$		
β	$2s_0$/0 0 2	$3d_0'$/0 0 4	$3s_0'$/0 0 6
1.87×10^{-2}	0.2781383 (47,47)	0.1363167 (47,47)	0.1083824 (47,47)
1.88×10^{-2}	0.2782439 (47,47)	0.1364015 (47,47)	0.1082446 (47,47)
1.89×10^{-2}	0.2783491 (47,47)	0.1364861 (47,47)	0.1081062 (47,47)
1.90×10^{-2}	0.2784539 (47,47)	0.1365702 (47,47)	0.1079675 (47,47)
1.91×10^{-2}	0.2785583 (47,47)	0.1366539 (47,47)	0.1078284 (47,47)
1.92×10^{-2}	0.2786623 (47,47)	0.1367373 (47,47)	0.1076889 (47,47)
1.93×10^{-2}	0.2787659 (47,47)	0.1368203 (47,47)	0.1075490 (47,47)
1.94×10^{-2}	0.2788691 (47,47)	0.1369029 (47,47)	0.1074088 (47,47)
1.95×10^{-2}	0.2789719 (47,47)	0.1369851 (47,47)	0.1072682 (47,47)
1.96×10^{-2}	0.2790742 (47,47)	0.1370670 (47,47)	0.1071273 (47,47)
1.97×10^{-2}	0.2791762 (47,47)	0.1371485 (47,47)	0.1069861 (47,47)
1.98×10^{-2}	0.2792777 (47,47)	0.1372296 (47,47)	0.1068445 (47,47)
1.99×10^{-2}	0.2793789 (47,47)	0.1373104 (47,47)	0.1067027 (47,47)
2×10^{-2}	0.2794796 (4,∞)	0.1373908 (7,∞)	0.1065606 (20,∞)
2.2×10^{-2}	0.2814128 (47,47)	0.1389246 (47,47)	0.1036807 (47,47)
2.4×10^{-2}	0.2831957 (47,47)	0.1403253 (47,47)	0.1008082 (47,47)
2.6×10^{-2}	0.2848368 (47,47)	0.1416032 (47,47)	0.09806473 (47,47)
2.8×10^{-2}	0.2863446 (47,47)	0.1427676 (47,47)	0.09557263 (47,47)
2.9×10^{-2}	0.2870509 (30,30)	0.1433098 (30,30)	0.09445448 (30,30)
3.0×10^{-2}	0.2877269 (5,∞)	0.1438266 (9,∞)	0.09343459 (18,∞)
3.2×10^{-2}	0.2889916 (60,60)	0.1447873 (60,60)	0.09170608 (60,60)
3.25×10^{-2}	0.2892902 (47,47)	0.1450129 (47,47)	0.09133849 (47,47)
3.4×10^{-2}	0.2901459 (60,60)	0.1456561 (60,60)	0.09038085 (60,60)
3.5×10^{-2}	0.2906838 (47,47)	0.1460579 (47,47)	0.08985337 (47,47)
3.6×10^{-2}	0.2911968 (60,60)	0.1464387 (60,60)	0.08940382 (60,60)
3.75×10^{-2}	0.2919211 (47,47)	0.1469721 (47,47)	0.08885437 (47,47)
3.8×10^{-2}	0.2921510 (60,60)	0.1471401 (60,60)	0.08869988 (60,60)
4×10^{-2}	0.2930148 (60,60)	0.1477650 (60,60)	0.08819737 (60,60)
4.25×10^{-2}	0.2939767 (60,60)	0.1484448 (60,60)	0.08776554 (60,60)
4.5×10^{-2}	0.2948177 (60,60)	0.1490189 (60,60)	0.08747719 (60,60)
4.75×10^{-2}	0.2955484 (60,60)	0.1494939 (60,60)	0.08727825 (60,60)
5×10^{-2}	0.2961783 (7,∞)	0.1498760 (12,∞)	0.08713337 (19,∞)
5.5×10^{-2}	0.2971717 (60,60)	0.1503837 (60,60)	0.08692015 (60,60)
6×10^{-2}	0.2978635 (60,60)	0.1505826 (60,60)	0.08672800 (60,60)
7×10^{-2}	0.2985628 (8,∞)	0.1501991 (14,∞)	0.08622458 (37,37)
8×10^{-2}	0.2986462 (60,60)	0.1489952 (60,60)	0.08544935 (60,60)
9×10^{-2}	0.2983857 (60,60)	0.1472217 (60,60)	0.08437534 (60,60)
1×10^{-1}	0.2979734 (10,∞)	0.1451139 (17,∞)	0.08303951 (35,35)

Table A1.3a. (Continued)

	$m=0 \quad \pi=+1$		
β	$4d'_0/0\ 0\ 8$	$4s'_0/0\ 0\ 10$	$5g'_0/0\ 0\ 12$
1×10^{-3}	0.06435166 $(4,\infty)$	0.06398170 $(3,\infty)$	0.04177604 $(25,25)$
1.5×10^{-3}	0.06516911 $(4,\infty)$	0.06434530 $(4,\infty)$	0.04250325 $(25,25)$
2×10^{-3}	0.06591861 $(4,\infty)$	0.06447397 $(4,\infty)$	0.04313314 $(24,24)$
2.5×10^{-3}	0.06660456 $(26,26)$	0.06438411 $(26,26)$	0.04367496 $(26,26)$
3×10^{-3}	0.06723194 $(5,\infty)$	0.06409349 $(5,\infty)$	0.04413822 $(24,24)$
3.5×10^{-3}	0.06780591 $(27,27)$	0.06362008 $(27,27)$	0.04453216 $(27,27)$
4×10^{-3}	0.06833158 $(27,27)$	0.06298146 $(27,27)$	0.04486540 $(27,27)$
4.5×10^{-3}	0.06881389 $(27,27)$	0.06219440 $(27,27)$	0.04514577 $(27,27)$
5×10^{-3}	0.06925742 $(7,\infty)$	0.06127474 $(6,\infty)$	0.04538022 $(27,27)$
5.5×10^{-3}	0.06966635 $(35,35)$	0.06023744 $(35,35)$	0.04557469 $(35,35)$
6×10^{-3}	0.07004445 $(35,35)$	0.05909683 $(35,35)$	0.04573404 $(35,35)$
6.5×10^{-3}	0.07039507 $(35,35)$	0.05786679 $(35,35)$	0.04586182 $(35,35)$
7×10^{-3}	0.07072114 $(7,\infty)$	0.05656138 $(8,\infty)$	0.04595995 $(25,25)$
8×10^{-3}	0.07130954 $(35,35)$	0.05378762 $(35,35)$	0.04606137 $(35,35)$
9×10^{-3}	0.07182616 $(35,35)$	0.05095970 $(35,35)$	0.04595782 $(35,35)$
1×10^{-2}	0.07228304 $(9,\infty)$	0.04861447 $(13,\infty)$	0.04520191 $(25,25)$
1.05×10^{-2}	0.07249186 $(47,47)$	0.04795629 $(47,47)$	0.04427096 $(47,47)$
1.1×10^{-2}	0.07268873 $(35,35)$	0.04762597 $(35,35)$	0.04299892 $(35,35)$
1.15×10^{-2}	0.07287431 $(47,47)$	0.04747177 $(47,47)$	0.04154679 $(47,47)$
1.2×10^{-2}	0.07304912 $(35,35)$	0.04740107 $(35,35)$	0.04001708 $(35,35)$
1.25×10^{-2}	0.07321360 $(47,47)$	0.04737194 $(47,47)$	0.03846421 $(47,47)$
1.3×10^{-2}	0.07336807 $(35,35)$	0.04736520 $(35,35)$	0.03692800 $(35,35)$
1.35×10^{-2}	0.07351278 $(47,47)$	0.04737135 $(47,47)$	0.03546574 $(47,47)$
1.4×10^{-2}	0.07364789 $(35,35)$	0.04738525 $(35,35)$	0.03422084 $(35,35)$
1.45×10^{-2}	0.07377348 $(47,47)$	0.04740390 $(47,47)$	0.03345022 $(47,47)$
1.5×10^{-2}	0.07388960 $(11,\infty)$	0.04742543 $(16,\infty)$	0.03311284 $(30,30)$
1.55×10^{-2}	0.07399622 $(47,47)$	0.04744859 $(47,47)$	0.03295770 $(47,47)$
1.6×10^{-2}	0.07409325 $(47,47)$	0.04747254 $(47,47)$	0.03287190 $(47,47)$
1.65×10^{-2}	0.07418057 $(47,47)$	0.04749665 $(47,47)$	0.03281799 $(47,47)$
1.7×10^{-2}	0.07425798 $(47,47)$	0.04752042 $(47,47)$	0.03278147 $(47,47)$
1.75×10^{-2}	0.07432524 $(47,47)$	0.04754348 $(47,47)$	0.03275558 $(47,47)$
1.8×10^{-2}	0.07438208 $(47,47)$	0.04756552 $(47,47)$	0.03273671 $(47,47)$
1.81×10^{-2}	0.07439217 $(47,47)$	0.04756978 $(47,47)$	0.03273358 $(47,47)$
1.82×10^{-2}	0.07440182 $(47,47)$	0.04757398 $(47,47)$	0.03273062 $(47,47)$
1.83×10^{-2}	0.07441103 $(47,47)$	0.04757813 $(47,47)$	0.03272783 $(47,47)$
1.84×10^{-2}	0.07441981 $(47,47)$	0.04758222 $(47,47)$	0.03272520 $(47,47)$
1.85×10^{-2}	0.07442815 $(47,47)$	0.04758625 $(47,47)$	0.03272271 $(47,47)$
1.86×10^{-2}	0.07443603 $(47,47)$	0.04759022 $(47,47)$	0.03272036 $(47,47)$

Table A1.3a. (Continued)

	$m=0 \quad \pi=+1$		
β	$4d'_0/0\ 0\ 8$	$4s'_0/0\ 0\ 10$	$5g'_0/0\ 0\ 12$
1.87×10^{-2}	0.07444347 (47,47)	0.04759413 (47,47)	0.03271814 (47,47)
1.88×10^{-2}	0.07445046 (47,47)	0.04759797 (47,47)	0.03271604 (47,47)
1.89×10^{-2}	0.07445699 (47,47)	0.04760174 (47,47)	0.03271404 (47,47)
1.90×10^{-2}	0.07446305 (47,47)	0.04760545 (47,47)	0.03271216 (47,47)
1.91×10^{-2}	0.07446866 (47,47)	0.04760908 (47,47)	0.03271037 (47,47)
1.92×10^{-2}	0.07447380 (47,47)	0.04761264 (47,47)	0.03270868 (47,47)
1.93×10^{-2}	0.07447846 (47,47)	0.04761614 (47,47)	0.03270707 (47,47)
1.94×10^{-2}	0.07448266 (47,47)	0.04761955 (47,47)	0.03270554 (47,47)
1.95×10^{-2}	0.07448637 (47,47)	0.04762289 (47,47)	0.03270409 (47,47)
1.96×10^{-2}	0.07448960 (47,47)	0.04762615 (47,47)	0.03270271 (47,47)
1.97×10^{-2}	0.07449234 (47,47)	0.04762933 (47,47)	0.03270139 (47,47)
1.98×10^{-2}	0.07449459 (47,47)	0.04763243 (47,47)	0.03270014 (47,47)
1.99×10^{-2}	0.07449634 (47,47)	0.04763544 (47,47)	0.03269894 (47,47)
2×10^{-2}	0.07449760 (14,∞)	0.04763838 (20,∞)	0.03269779 (30,30)
2.2×10^{-2}	0.07440986 (47,47)	0.04767707 (47,47)	0.03268106 (47,47)
2.4×10^{-2}	0.07407559 (47,47)	0.04766989 (47,47)	0.03266389 (47,47)
2.6×10^{-2}	0.07344065 (47,47)	0.04760447 (47,47)	0.03263422 (47,47)
2.8×10^{-2}	0.07245097 (47,47)	0.04746603 (47,47)	0.03258314 (47,47)
2.9×10^{-2}	0.07181183 (30,30)	0.04736380 (30,30)	0.03254697 (30,30)
3.0×10^{-2}	0.07107623 (17,∞)	0.04723598 (24,∞)	0.03250226 (40,40)
3.2×10^{-2}	0.06934041 (60,60)	0.04689102 (60,60)	0.03238232 (60,60)
3.25×10^{-2}	0.06885966 (47,47)	0.04678365 (47,47)	0.03234499 (47,47)
3.4×10^{-2}	0.06733821 (60,60)	0.04640361 (60,60)	0.03221226 (60,60)
3.5×10^{-2}	0.06628313 (47,47)	0.04609762 (47,47)	0.03210430 (47,47)
3.6×10^{-2}	0.06522048 (60,60)	0.04574615 (60,60)	0.03197853 (60,60)
3.75×10^{-2}	0.06365868 (47,47)	0.04513056 (47,47)	0.03175202 (47,47)
3.8×10^{-2}	0.06315643 (60,60)	0.04490203 (60,60)	0.03166539 (60,60)
4×10^{-2}	0.06129156 (60,60)	0.04388386 (60,60)	0.03125722 (60,60)
4.25×10^{-2}	0.05937067 (60,60)	0.04245906 (60,60)	0.03060106 (60,60)
4.5×10^{-2}	0.05792883 (60,60)	0.04104836 (60,60)	0.02980810 (60,60)
4.75×10^{-2}	0.05689063 (60,60)	0.03983227 (60,60)	0.02896898 (60,60)
5×10^{-2}	0.05614951 (29,29)	0.03888676 (37,37)	0.02820337 (58,58)
5.5×10^{-2}	0.05521506 (60,60)	0.03768549 (60,60)	0.02712041 (60,60)
6×10^{-2}	0.05466621 (60,60)	0.03703482 (60,60)	0.02652641 (60,60)
7×10^{-2}	0.05399226 (39,39)	0.03637441 (51,51)	0.02596886 (80,80)
8×10^{-2}	0.05344537 (60,60)	0.03597503 (60,60)	0.0256769 (60,60)
9×10^{-2}	0.05284042 (60,60)	0.03560723 (60,60)	0.025435 (60,60)
1×10^{-1}	0.05211949 (59,59)	0.03519757 (61,61)	0.025188 (80,80)

Table A1.3a. (Continued)

	$m = 0 \quad \pi = +1$	
β	$5d'_0/0\ 0\ 14$	$5s'_0/0\ 0\ 16$
1×10^{-3}	0.04146285 (25,25)	0.04071412 (25,25)
1.5×10^{-3}	0.04182340 (25,25)	0.04019641 (25,25)
2×10^{-3}	0.04197747 (24,24)	0.03920361 (24,24)
2.5×10^{-3}	0.04195835 (26,26)	0.03781933 (26,26)
3×10^{-3}	0.04179624 (24,24)	0.03611944 (24,24)
3.5×10^{-3}	0.04151682 (27,27)	0.03417384 (27,27)
4×10^{-3}	0.04114110 (27,27)	0.03207224 (27,27)
4.5×10^{-3}	0.04068604 (27,27)	0.03114750 (27,27)
5×10^{-3}	0.04016530 (27,27)	0.03115953 (27,27)
5.5×10^{-3}	0.03959018 (35,35)	0.03115992 (35,35)
6×10^{-3}	0.03897050 (35,35)	0.03113290 (35,35)
6.5×10^{-3}	0.03831573 (35,35)	0.03107541 (35,35)
7×10^{-3}	0.03763604 (30,30)	0.03098342 (30,30)
8×10^{-3}	0.03625472 (35,35)	0.03066356 (35,35)
9×10^{-3}	0.03497274 (35,35)	0.03006490 (35,35)
1×10^{-2}	0.03397457 (28,28)	0.02905155 (28,28)
1.05×10^{-2}	0.03361165 (47,47)	0.02838613 (47,47)
1.1×10^{-2}	0.03332870 (35,35)	0.02765075 (35,35)
1.15×10^{-2}	0.03310734 (47,47)	0.02689340 (47,47)
1.2×10^{-2}	0.03292784 (35,35)	0.02617177 (35,35)
1.25×10^{-2}	0.03277001 (47,47)	0.02554169 (47,47)
1.3×10^{-2}	0.03260791 (35,35)	0.02503797 (35,35)
1.35×10^{-2}	0.03239027 (47,47)	0.02466153 (47,47)
1.4×10^{-2}	0.03197655 (35,35)	0.02438859 (35,35)
1.45×10^{-2}	0.03111132 (47,47)	0.02418942 (47,47)
1.5×10^{-2}	0.02983707 (30,30)	0.02403859 (33,33)
1.55×10^{-2}	0.02840785 (47,47)	0.02391546 (47,47)
1.6×10^{-2}	0.02694670 (47,47)	0.02379756 (47,47)
1.65×10^{-2}	0.02553442 (47,47)	0.02363289 (47,47)
1.7×10^{-2}	0.02442756 (47,47)	0.02317367 (47,47)
1.75×10^{-2}	0.02401086 (47,47)	0.02203937 (47,47)
1.8×10^{-2}	0.02388277 (47,47)	0.02063958 (47,47)
1.81×10^{-2}	0.02386720 (47,47)	0.02035438 (47,47)
1.82×10^{-2}	0.02385342 (47,47)	0.02007011 (47,47)
1.83×10^{-2}	0.02384110 (47,47)	0.01978810 (47,47)
1.84×10^{-2}	0.02382998 (47,47)	0.01951025 (47,47)
1.85×10^{-2}	0.02381987 (47,47)	0.01923946 (47,47)
1.86×10^{-2}	0.02381061 (47,47)	0.01898061 (47,47)

Table A1.3a. (Continued)

β	$m=0 \quad \pi=+1$	
	$5d'_0/0\ 0\ 14$	$5s'_0/0\ 0\ 16$
1.87×10^{-2}	0.02380209 (47,47)	0.01874177 (47,47)
1.88×10^{-2}	0.02379421 (47,47)	0.01853503 (47,47)
1.89×10^{-2}	0.02378688 (47,47)	0.01837275 (47,47)
1.90×10^{-2}	0.02378005 (47,47)	0.01825793 (47,47)
1.91×10^{-2}	0.02377366 (47,47)	0.01818116 (47,47)
1.92×10^{-2}	0.02376767 (47,47)	0.01812950 (47,47)
1.93×10^{-2}	0.02376202 (47,47)	0.01809326 (47,47)
1.94×10^{-2}	0.02375670 (47,47)	0.01806656 (47,47)
1.95×10^{-2}	0.02375167 (47,47)	0.01804597 (47,47)
1.96×10^{-2}	0.02374690 (47,47)	0.01802947 (47,47)
1.97×10^{-2}	0.02374238 (47,47)	0.01801583 (47,47)
1.98×10^{-2}	0.02373808 (47,47)	0.01800427 (47,47)
1.99×10^{-2}	0.02373399 (47,47)	0.01799426 (47,47)
2×10^{-2}	0.02373009 (40,40)	0.01798546 (49,49)
2.2×10^{-2}	0.02367739 (47,47)	0.01789981 (47,47)
2.4×10^{-2}	0.02364662 (47,47)	0.01786290 (47,47)
2.6×10^{-2}	0.02361967 (47,47)	0.01783741 (47,47)
2.8×10^{-2}	0.02358763 (47,47)	0.01781322 (47,47)
2.9×10^{-2}	0.02356727 (30,30)	0.01777342 (30,30)
3.0×10^{-2}	0.02354430 (50,50)	0.01778495 (58,58)
3.2×10^{-2}	0.02348422 (60,60)	0.01774858 (60,60)
3.25×10^{-2}	0.02346591 (47,47)	0.01773776 (47,47)
3.4×10^{-2}	0.02340156 (60,60)	0.01770042 (60,60)
3.5×10^{-2}	0.02334969 (47,47)	0.01767058 (47,47)
3.6×10^{-2}	0.02328953 (60,60)	0.01763643 (60,60)
3.75×10^{-2}	0.02318136 (47,47)	0.01757498 (47,47)
3.8×10^{-2}	0.02313993 (60,60)	0.01755185 (60,60)
4×10^{-2}	0.02294313 (60,60)	0.01744097 (60,60)
4.25×10^{-2}	0.02261609 (60,60)	0.01725547 (60,60)
4.5×10^{-2}	0.02219050 (60,60)	0.01700769 (60,60)
4.75×10^{-2}	0.02168617 (60,60)	0.01669798 (60,60)
5×10^{-2}	0.02116420 (75,75)	0.01635199 (80,80)
5.5×10^{-2}	0.02032533 (60,60)	0.01573354 (60,60)
6×10^{-2}	0.01983834 (60,60)	0.01534/5 (60,60)
7×10^{-2}	0.019395 (80,80)	0.015003/7 [96]
8×10^{-2}	0.01917 (60,60)	0.01484771/3 [96]
9×10^{-2}	0.01898 (60,60)	0.0147324/34 [96]
1×10^{-1}	0.018851/7 [96]	0.014618/21 [96]

Table A1.3b. Energies (in units of $-E_\infty$) of the lowest $m=0$, z-parity $\pi=-1$ states of an electron in a static Coulomb potential and a magnetic field with strength parameter β

	$m=0 \quad \pi=-1$		
β	$2p_0/0\ 0\ 1$	$3p_0/0\ 0\ 3$	$4f'_0/0\ 0\ 5$
1×10^{-3}	0.2519880 $(1,\infty)$	0.1130392 $(2,\infty)$	0.06439356 $(4,\infty)$
1.5×10^{-3}	0.2529730 $(2,\infty)$	0.1139497 $(2,\infty)$	0.06526151 $(4,\infty)$
2×10^{-3}	0.2539520 $(3,\infty)$	0.1148249 $(2,\infty)$	0.06607842 $(4,\infty)$
3×10^{-3}	0.2558921 $(3,\infty)$	0.1164720 $(3,\infty)$	0.06756588 $(5,\infty)$
5×10^{-3}	0.2597008 $(2,\infty)$	0.1193757 $(3,\infty)$	0.07001254 $(6,\infty)$
5.5×10^{-3}	0.2606382 $(35,35)$	0.1200262 $(35,35)$	0.07052666 $(35,35)$
6×10^{-3}	0.2615697 $(40,40)$	0.1206486 $(40,40)$	0.07100594 $(40,40)$
6.5×10^{-3}	0.2624954 $(40,40)$	0.1212442 $(40,40)$	0.07145232 $(40,40)$
7×10^{-3}	0.2634152 $(2,\infty)$	0.1218140 $(4,\infty)$	0.07186766 $(7,\infty)$
7.5×10^{-3}	0.2643292 $(30,30)$	0.1223594 $(30,30)$	0.07225373 $(30,30)$
8×10^{-3}	0.2652374 $(40,40)$	0.1228813 $(40,40)$	0.07261220 $(40,40)$
8.5×10^{-3}	0.2661398 $(35,35)$	0.1233809 $(35,35)$	0.07294466 $(35,35)$
1×10^{-2}	0.2688129 $(3,\infty)$	0.1247571 $(5,\infty)$	0.07380055 $(9,\infty)$
1.25×10^{-2}	0.2731559 $(30,30)$	0.1267015 $(30,30)$	0.07483214 $(30,30)$
1.3×10^{-2}	0.2740080 $(30,30)$	0.1270462 $(30,30)$	0.07498935 $(30,30)$
1.35×10^{-2}	0.2748547 $(30,30)$	0.1273780 $(30,30)$	0.07513250 $(30,30)$
1.4×10^{-2}	0.2756960 $(30,30)$	0.1276977 $(30,30)$	0.07526249 $(30,30)$
1.45×10^{-2}	0.2765320 $(30,30)$	0.1280061 $(30,30)$	0.07538020 $(30,30)$
1.5×10^{-2}	0.2773627 $(3,\infty)$	0.1283037 $(6,\infty)$	0.07548642 $(11,\infty)$
1.6×10^{-2}	0.2790083 $(40,40)$	0.1288692 $(40,40)$	0.07566755 $(40,40)$
1.7×10^{-2}	0.2806332 $(40,40)$	0.1293989 $(40,40)$	0.07581171 $(40,40)$
1.8×10^{-2}	0.2822379 $(40,40)$	0.1298970 $(40,40)$	0.07592425 $(40,40)$
1.9×10^{-2}	0.2838224 $(40,40)$	0.1303672 $(40,40)$	0.07601000 $(40,40)$
$2.\times10^{-2}$	0.2853874 $(4,\infty)$	0.1308128 $(7,\infty)$	0.07607329 $(18,\infty)$
2.1×10^{-2}	0.2869331 $(40,40)$	0.1312366 $(40,40)$	0.07611804 $(40,40)$
2.2×10^{-2}	0.2884598 $(40,40)$	0.1316412 $(40,40)$	0.07614772 $(40,40)$
2.3×10^{-2}	0.2899679 $(40,40)$	0.1320288 $(40,40)$	0.07616542 $(40,40)$
2.4×10^{-2}	0.2914577 $(40,40)$	0.1324015 $(40,40)$	0.07617371 $(40,40)$
2.5×10^{-2}	0.2929297 $(40,40)$	0.1327610 $(40,40)$	0.07617526 $(40,40)$
2.6×10^{-2}	0.2943840 $(40,40)$	0.1331088 $(40,40)$	0.07617176 $(40,40)$
2.7×10^{-2}	0.2958212 $(40,40)$	0.1334461 $(40,40)$	0.07616501 $(40,40)$
2.8×10^{-2}	0.2972414 $(40,40)$	0.1337742 $(40,40)$	0.07615644 $(40,40)$
2.9×10^{-2}	0.2986451 $(40,40)$	0.1340940 $(40,40)$	0.07614726 $(40,40)$
3×10^{-2}	0.3000325 $(4,\infty)$	0.1344065 $(10,\infty)$	0.07613841 $(22,\infty)$
3.2×10^{-2}	0.3027599 $(47,47)$	0.1350123 $(47,47)$	0.07612467 $(47,47)$
3.4×10^{-2}	0.3054262 $(47,47)$	0.1355963 $(47,47)$	0.07611956 $(47,47)$
3.6×10^{-2}	0.3080336 $(47,47)$	0.1361623 $(47,47)$	0.07612546 $(47,47)$
3.8×10^{-2}	0.3105845 $(47,47)$	0.1367131 $(47,47)$	0.07614338 $(47,47)$
4×10^{-2}	0.3130812 $(47,47)$	0.1372508 $(47,47)$	0.07617343 $(47,47)$
5×10^{-2}	0.3248202 $(6,\infty)$	0.1397834 $(12,\infty)$	0.07647957 $(20,\infty)$
6.3×10^{-2}	0.3384937 $(47,47)$	0.1427739 $(47,47)$	0.07711604 $(47,47)$
7×10^{-2}	0.3452365 $(7,\infty)$	0.1442643 $(14,\infty)$	0.07750589 $(50,50)$
8×10^{-2}	0.3542336 $(47,47)$	0.1462624 $(47,47)$	0.07807700 $(47,47)$
1×10^{-1}	0.3703681 $(8,\infty)$	0.1498509 $(16,\infty)$	0.07918961 $(60,60)$

Table A1.3b. (Continued)

β	$4p'_0/0\ 0\ 7$	$5f'_0/0\ 0\ 9$	$5p'_0/0\ 0\ 11$
	$m=0\quad \pi=-1$		
1×10^{-3}	0.06419317 (4,∞)	0.04173218 (30,30)	0.04114829 (30,30)
1.5×10^{-3}	0.06481602 (3,∞)	0.04241287 (30,30)	0.04114336 (30,30)
2×10^{-3}	0.06529908 (4,∞)	0.04298996 (30,30)	0.04082708 (30,30)
3×10^{-3}	0.06588500 (5,∞)	0.04389787 (30,30)	0.03949718 (30,30)
5×10^{-3}	0.06583575 (5,∞)	0.04506068 (30,30)	0.03544809 (30,30)
5.5×10^{-3}	0.06562845 (35,35)	0.04525842 (35,35)	0.03444516 (35,35)
6×10^{-3}	0.06536119 (40,40)	0.04542725 (40,40)	0.03361252 (40,40)
6.5×10^{-3}	0.06504180 (40,40)	0.04556958 (40,40)	0.03304557 (40,40)
7×10^{-3}	0.06467752 (8,∞)	0.04568721 (31,31)	0.03272679 (31,31)
7.5×10^{-3}	0.06427497 (30,30)	0.04578152 (30,30)	0.03255913 (30,30)
8×10^{-3}	0.06384024 (40,40)	0.04585354 (40,40)	0.03246723 (40,40)
8.5×10^{-3}	0.06337894 (35,35)	0.04590406 (35,35)	0.03241185 (35,35)
1×10^{-2}	0.06188510 (11,∞)	0.04593216 (30,30)	0.03231503 (30,30)
1.25×10^{-2}	0.05927135 (30,30)	0.04558679 (30,30)	0.03213390 (30,30)
1.3×10^{-2}	0.05875920 (30,30)	0.04546095 (30,30)	0.03208407 (30,30)
1.35×10^{-2}	0.05825645 (30,30)	0.04531676 (30,30)	0.03202908 (30,30)
1.4×10^{-2}	0.05776505 (30,30)	0.04515452 (30,30)	0.03196894 (30,30)
1.45×10^{-2}	0.05728665 (30,30)	0.04497466 (30,30)	0.03190368 (30,30)
1.5×10^{-2}	0.05682263 (13,∞)	0.04477764 (30,30)	0.03183342 (30,30)
1.6×10^{-2}	0.05594182 (40,40)	0.04433454 (40,40)	0.03167834 (40,40)
1.7×10^{-2}	0.05512824 (40,40)	0.04383097 (40,40)	0.03150482 (40,40)
1.8×10^{-2}	0.05438359 (40,40)	0.04327414 (40,40)	0.03131394 (40,40)
1.9×10^{-2}	0.05370647 (40,40)	0.04267240 (40,40)	0.03110651 (40,40)
$2.\times 10^{-2}$	0.05309329 (16,∞)	0.04203484 (30,30)	0.03088286 (30,30)
2.1×10^{-2}	0.05253917 (40,40)	0.04137084 (40,40)	0.03064270 (40,40)
2.2×10^{-2}	0.05203877 (40,40)	0.04068980 (40,40)	0.03038503 (40,40)
2.3×10^{-2}	0.05158682 (40,40)	0.04000110 (40,40)	0.03010814 (40,40)
2.4×10^{-2}	0.05117843 (40,40)	0.03931407 (40,40)	0.02980968 (40,40)
2.5×10^{-2}	0.05080924 (40,40)	0.03863813 (40,40)	0.02948705 (40,40)
2.6×10^{-2}	0.05047541 (40,40)	0.03798271 (40,40)	0.02913784 (40,40)
2.7×10^{-2}	0.05017359 (40,40)	0.03735694 (40,40)	0.02876058 (40,40)
2.8×10^{-2}	0.04990084 (40,40)	0.03676919 (40,40)	0.02835578 (40,40)
2.9×10^{-2}	0.04965453 (40,40)	0.03622631 (40,40)	0.02792681 (40,40)
3×10^{-2}	0.04943232 (40,40)	0.03573297 (40,40)	0.02748068 (40,40)
3.2×10^{-2}	0.04905187 (47,47)	0.03490088 (47,47)	0.02658175 (47,47)
3.4×10^{-2}	0.04874460 (47,47)	0.03426217 (47,47)	0.02576221 (47,47)
3.6×10^{-2}	0.04849799 (47,47)	0.03378328 (47,47)	0.02509977 (47,47)
3.8×10^{-2}	0.04830156 (47,47)	0.03342643 (47,47)	0.02460828 (47,47)
4×10^{-2}	0.04814655 (47,47)	0.03315958 (47,47)	0.02425734 (47,47)
5×10^{-2}	0.04777979 (42,42)	0.03253585 (48,48)	0.02352624 (57,57)
6.3×10^{-2}	0.04780160 (47,47)	0.03238897 (47,47)	0.02334478 (47,47)
7×10^{-2}	0.04790683 (50,50)	0.03240971 (60,60)	0.02333971 (67,67)
8×10^{-2}	0.04810286 (47,47)	0.03248565 (47,47)	0.02336780 (47,47)
1×10^{-1}	0.04854600 (60,60)	0.03269674 (70,70)	0.02348586 (84,84)

Table A1.3c. Energies (in units of $-E_\infty$) of the lowest $m = -1$, z-parity $\pi = +1$ states of an electron in a static Coulomb potential and a magnetic field with strength parameter β

β	$2p_{-1}/0\ -1\ 0$	$3p_{-1}/0\ -1\ 2$	$4f'_{-1}/0\ -1\ 4$
1×10^{-3}	0.2539760 $(2,\infty)$	0.1149673 $(2,\infty)$	0.06632869 $(4,\infty)$
1.5×10^{-3}	0.2559460 $(2,\infty)$	0.1167883 $(2,\infty)$	0.06811656 $(4,\infty)$
2×10^{-3}	0.2579041 $(2,\infty)$	0.1185387 $(2,\infty)$	0.06982313 $(4,\infty)$
2.5×10^{-3}	0.2598501 $(30,30)$	0.1202198 $(30,30)$	0.07145153 $(30,30)$
3×10^{-3}	0.2617843 $(3,\infty)$	0.1218330 $(3,\infty)$	0.07300534 $(5,\infty)$
4×10^{-3}	0.2656169 $(30,30)$	0.1248619 $(30,30)$	0.07590444 $(30,30)$
5×10^{-3}	0.2694023 $(2,\infty)$	0.1276402 $(3,\infty)$	0.07855087 $(6,\infty)$
6×10^{-3}	0.2731407 $(30,30)$	0.1301841 $(30,30)$	0.08097302 $(30,30)$
6.5×10^{-3}	0.2749925 $(30,30)$	0.1313734 $(30,30)$	0.08210795 $(30,30)$
7×10^{-3}	0.2768327 $(2,\infty)$	0.1325105 $(4,\infty)$	0.08319589 $(7,\infty)$
7.5×10^{-3}	0.2786614 $(30,30)$	0.1335974 $(30,30)$	0.08423942 $(30,30)$
8×10^{-3}	0.2804787 $(30,30)$	0.1346362 $(30,30)$	0.08524088 $(30,30)$
8.5×10^{-3}	0.2822846 $(30,30)$	0.1356289 $(30,30)$	0.08620242 $(30,30)$
9×10^{-3}	0.2840793 $(30,30)$	0.1365776 $(30,30)$	0.08712602 $(30,30)$
1×10^{-2}	0.2876352 $(3,\infty)$	0.1383502 $(4,\infty)$	0.08886648 $(9,\infty)$
1.1×10^{-2}	0.2911470 $(30,30)$	0.1399687 $(30,30)$	0.09047494 $(30,30)$
1.2×10^{-2}	0.2946155 $(30,30)$	0.1414466 $(30,30)$	0.09196206 $(30,30)$
1.3×10^{-2}	0.2980413 $(30,30)$	0.1427966 $(30,30)$	0.09333684 $(30,30)$
1.4×10^{-2}	0.3014253 $(30,30)$	0.1440304 $(30,30)$	0.09460683 $(30,30)$
1.5×10^{-2}	0.3047682 $(3,\infty)$	0.1451588 $(6,\infty)$	0.09577849 $(9,\infty)$
1.6×10^{-2}	0.3080709 $(30,30)$	0.1461917 $(30,30)$	0.09685729 $(30,30)$
1.7×10^{-2}	0.3113341 $(30,30)$	0.1471386 $(30,30)$	0.09784794 $(30,30)$
1.8×10^{-2}	0.3145586 $(30,30)$	0.1480078 $(30,30)$	0.09875448 $(30,30)$
1.9×10^{-2}	0.3177454 $(30,30)$	0.1488074 $(30,30)$	0.09958043 $(30,30)$
2×10^{-2}	0.3208951 $(3,\infty)$	0.1495447 $(7,\infty)$	0.1003289 $(10,\infty)$
2.1×10^{-2}	0.3240086 $(47,47)$	0.1502263 $(47,47)$	0.1010025 $(47,47)$
2.2×10^{-2}	0.3270867 $(47,47)$	0.1508587 $(47,47)$	0.1016038 $(47,47)$
2.3×10^{-2}	0.3301302 $(47,47)$	0.1514475 $(47,47)$	0.1021351 $(47,47)$
2.4×10^{-2}	0.3331398 $(47,47)$	0.1519981 $(47,47)$	0.1025983 $(47,47)$
2.5×10^{-2}	0.3361164 $(47,47)$	0.1525155 $(47,47)$	0.1029954 $(47,47)$
2.6×10^{-2}	0.3390606 $(47,47)$	0.1530040 $(47,47)$	0.1033284 $(47,47)$
2.7×10^{-2}	0.3419733 $(47,47)$	0.1534678 $(47,47)$	0.1035991 $(47,47)$
2.8×10^{-2}	0.3448551 $(47,47)$	0.1539106 $(47,47)$	0.1038093 $(47,47)$
2.9×10^{-2}	0.3477067 $(47,47)$	0.1543359 $(47,47)$	0.1039611 $(47,47)$
3×10^{-2}	0.3505288 $(4,\infty)$	0.1547466 $(9,\infty)$	0.1040563 $(12,\infty)$
3.2×10^{-2}	0.3560874 $(47,47)$	0.1555350 $(47,47)$	0.1040854 $(47,47)$
3.4×10^{-2}	0.3615358 $(47,47)$	0.1562941 $(47,47)$	0.1039141 $(47,47)$
3.6×10^{-2}	0.3668789 $(47,47)$	0.1570377 $(47,47)$	0.1035620 $(47,47)$
3.8×10^{-2}	0.3721215 $(47,47)$	0.1577760 $(47,47)$	0.1030505 $(47,47)$
4×10^{-2}	0.3772678 $(47,47)$	0.1585160 $(47,47)$	0.1024030 $(47,47)$
4.2×10^{-2}	0.3823220 $(47,47)$	0.1592624 $(47,47)$	0.1016449 $(47,47)$
4.4×10^{-2}	0.3872878 $(47,47)$	0.1600177 $(47,47)$	0.1008032 $(47,47)$

Table A1.3c. (Continued)

β	$m=-1 \quad \pi=+1$		
	$2p_{-1}/0\ -1\ 0$	$3p_{-1}/0\ -1\ 2$	$4f'_{-1}/0\ -1\ 4$
4.6×10^{-2}	0.3921691 (47,47)	0.1607828 (47,47)	0.09990583 (47,47)
4.8×10^{-2}	0.3969692 (47,47)	0.1615580 (47,47)	0.09898116 (47,47)
5×10^{-2}	0.4016913 (5,∞)	0.1623424 (12,∞)	0.09805689 (23,23)
5.2×10^{-2}	0.4063387 (47,47)	0.1631349 (47,47)	0.09715884 (47,47)
5.4×10^{-2}	0.4109140 (47,47)	0.1639340 (47,47)	0.09630964 (47,47)
5.6×10^{-2}	0.4154201 (47,47)	0.1647383 (47,47)	0.09552746 (47,47)
5.8×10^{-2}	0.4198595 (47,47)	0.1655462 (47,47)	0.09482521 (47,47)
$6.\times10^{-2}$	0.4242347 (47,47)	0.1663563 (47,47)	0.09421021 (47,47)
6.3×10^{-2}	0.4306821 (47,47)	0.1675723 (47,47)	0.09345490 (47,47)
7×10^{-2}	0.4452239 (6,∞)	0.1703942 (15,∞)	0.09239537 (27,27)
7.5×10^{-2}	0.4552155 (47,47)	0.1723797 (47,47)	0.09210106 (47,47)
8×10^{-2}	0.4649066 (47,47)	0.1743293 (47,47)	0.09206261 (47,47)
8.5×10^{-2}	0.4743196 (47,47)	0.1762371 (47,47)	0.09219497 (47,47)
9×10^{-2}	0.4834744 (47,47)	0.1780998 (47,47)	0.09243951 (47,47)
1×10^{-1}	0.5010782 (7,∞)	0.1816837 (17,∞)	0.09312162 (50,50)

Table A1.3c. (Continued)

β	$m=-1 \quad \pi=+1$		
	$4p'_{-1}/0\ -1\ 6$	$5f'_{-1}/0\ -1\ 8$	$5p'_{-1}/0\ -1\ 10$
1×10^{-3}	0.06600411 (4,∞)	0.04349936 (30,30)	0.04274891 (30,30)
1.5×10^{-3}	0.06739511 (4,∞)	0.04490366 (30,30)	0.04327185 (30,30)
2×10^{-3}	0.06856110 (3,∞)	0.04611619 (30,30)	0.04333186 (30,30)
2.5×10^{-3}	0.06951763 (30,30)	0.04716848 (30,30)	0.04300980 (30,30)
3×10^{-3}	0.07028153 (4,∞)	0.04808911 (30,30)	0.04237864 (30,30)
4×10^{-3}	0.07129918 (30,30)	0.04962703 (30,30)	0.04044776 (30,30)
5×10^{-3}	0.07174309 (5,∞)	0.05086873 (30,30)	0.03813726 (30,30)
6×10^{-3}	0.07172802 (30,30)	0.05189422 (30,30)	0.03715482 (30,30)
6.5×10^{-3}	0.07158033 (30,30)	0.05233947 (30,30)	0.03718537 (30,30)
7×10^{-3}	0.07135482 (8,∞)	0.05274327 (30,30)	0.03727151 (30,30)
7.5×10^{-3}	0.07106259 (30,30)	0.05310652 (30,30)	0.03737392 (30,30)
8×10^{-3}	0.07071447 (30,30)	0.05342894 (30,30)	0.03748190 (30,30)
8.5×10^{-3}	0.07032113 (30,30)	0.05370925 (30,30)	0.03759140 (30,30)
9×10^{-3}	0.06989329 (30,30)	0.05394517 (30,30)	0.03770021 (30,30)
1×10^{-2}	0.06897776 (11,∞)	0.05427046 (30,30)	0.03790900 (30,30)
1.1×10^{-2}	0.06805699 (30,30)	0.05437105 (30,30)	0.03809560 (30,30)
1.2×10^{-2}	0.06721938 (30,30)	0.05420837 (30,30)	0.03824734 (30,30)
1.3×10^{-2}	0.06653991 (30,30)	0.05375497 (30,30)	0.03835028 (30,30)

Table A1.3c. (Continued)

β	$4p'_{-1}/0\ -1\ 6$	$5f'_{-1}/0\ -1\ 8$	$5p'_{-1}/0\ -1\ 10$
	$m = -1 \quad \pi = +1$		
1.4×10^{-2}	0.06606192 (30,30)	0.05301572 (30,30)	0.03838785 (30,30)
1.5×10^{-2}	0.06578793 (14,∞)	0.05203992 (30,30)	0.03833929 (30,30)
1.6×10^{-2}	0.06568898 (30,30)	0.05091425 (30,30)	0.03817872 (30,30)
1.7×10^{-2}	0.06572383 (30,30)	0.04974402 (30,30)	0.03787650 (30,30)
1.8×10^{-2}	0.06585340 (30,30)	0.04863365 (30,30)	0.03740637 (30,30)
1.9×10^{-2}	0.06604661 (30,30)	0.04766809 (30,30)	0.03676089 (30,30)
2×10^{-2}	0.06628064 (16,∞)	0.04689582 (31,31)	0.03596993 (30,30)
2.1×10^{-2}	0.06653934 (47,47)	0.04632303 (47,47)	0.03510686 (47,47)
2.2×10^{-2}	0.06681141 (47,47)	0.04592473 (47,47)	0.03426947 (47,47)
2.3×10^{-2}	0.06708889 (47,47)	0.04566372 (47,47)	0.03354333 (47,47)
2.4×10^{-2}	0.06736613 (47,47)	0.04550428 (47,47)	0.03297166 (47,47)
2.5×10^{-2}	0.06763906 (47,47)	0.04541775 (47,47)	0.03255281 (47,47)
2.6×10^{-2}	0.06790470 (47,47)	0.04538289 (47,47)	0.03226000 (47,47)
2.7×10^{-2}	0.06816079 (47,47)	0.04538451 (47,47)	0.03206153 (47,47)
2.8×10^{-2}	0.06840561 (47,47)	0.04541181 (47,47)	0.03193064 (47,47)
2.9×10^{-2}	0.06863777 (47,47)	0.04545708 (47,47)	0.03184748 (47,47)
3×10^{-2}	0.06885616 (22,∞)	0.04551478 (40,40)	0.03179807 (40,40)
3.2×10^{-2}	0.06924774 (47,47)	0.04565225 (47,47)	0.03176454 (47,47)
3.4×10^{-2}	0.06957348 (47,47)	0.04580252 (47,47)	0.03178206 (47,47)
3.6×10^{-2}	0.06982701 (47,47)	0.04595279 (47,47)	0.03182600 (47,47)
3.8×10^{-2}	0.07000174 (47,47)	0.04609492 (47,47)	0.03188281 (47,47)
4×10^{-2}	0.07009049 (47,47)	0.04622315 (47,47)	0.03194450 (47,47)
4.2×10^{-2}	0.07008542 (47,47)	0.04633294 (47,47)	0.03200597 (47,47)
4.4×10^{-2}	0.06997813 (47,47)	0.04642022 (47,47)	0.03206370 (47,47)
4.6×10^{-2}	0.06976005 (47,47)	0.04648093 (47,47)	0.03211503 (47,47)
4.8×10^{-2}	0.06942324 (47,47)	0.04651063 (47,47)	0.03215767 (47,47)
5×10^{-2}	0.06896141 (45,45)	0.04650423 (45,45)	0.03218946 (50,50)
5.2×10^{-2}	0.06837149 (47,47)	0.04645570 (47,47)	0.03220815 (47,47)
5.4×10^{-2}	0.06765521 (47,47)	0.04635784 (47,47)	0.03221118 (47,47)
5.6×10^{-2}	0.06682083 (47,47)	0.04620213 (47,47)	0.03219551 (47,47)
5.8×10^{-2}	0.06588422 (47,47)	0.04597872 (47,47)	0.03215742 (47,47)
$6.\times10^{-2}$	0.06486933 (47,47)	0.04567683 (47,47)	0.03209227 (47,47)
6.3×10^{-2}	0.06327014 (47,47)	0.04505402 (47,47)	0.03193064 (47,47)
7×10^{-2}	0.05981763 (40,40)	0.04276147 (60,60)	0.03110816 (61,61)
7.5×10^{-2}	0.05807605 (47,47)	0.04076079 (47,47)	0.02999284 (47,47)
8×10^{-2}	0.05698955 (47,47)	0.03911474 (47,47)	0.02860555 (47,47)
8.5×10^{-2}	0.05637397 (47,47)	0.03807620 (47,47)	0.02747164 (47,47)
9×10^{-2}	0.05605335 (47,47)	0.03749973 (47,47)	0.02679685 (47,47)
1×10^{-1}	0.05587986 (60,60)	0.03703858 (80,80)	0.02627028 (81,81)

Table A1.3d. Energies (in units of $-E_\infty$) of the lowest $m = -1$, z-parity $\pi = -1$ states of an electron in a static Coulomb potential and a magnetic field with strength parameter β

	$m = -1 \quad \pi = -1$		
β	$3d_{-1}/0\ -1\ 1$	$4d_{-1}/0\ -1\ 3$	$5g'_{-1}/0\ -1\ 5$
1×10^{-3}	0.1150392 (3,∞)	0.06621414 (2,∞)	0.04364933 (25,25)
1.5×10^{-3}	0.1169495 (2,∞)	0.06786259 (3,∞)	0.04522220 (25,25)
2×10^{-3}	0.1188244 (3,∞)	0.06938046 (3,∞)	0.04664226 (25,25)
2.5×10^{-3}	0.1206641 (25,25)	0.07077639 (25,25)	0.04792283 (25,25)
3×10^{-3}	0.1224693 (2,∞)	0.07205993 (4,∞)	0.04907694 (25,25)
3.5×10^{-3}	0.1242404 (25,25)	0.07324096 (25,25)	0.05011660 (25,25)
4×10^{-3}	0.1259782 (25,25)	0.07432923 (25,25)	0.05105254 (25,25)
4.5×10^{-3}	0.1276832 (25,25)	0.07533412 (25,25)	0.05189418 (25,25)
5×10^{-3}	0.1293563 (3,∞)	0.07626436 (5,∞)	0.05264977 (25,25)
5.5×10^{-3}	0.1309982 (25,25)	0.07712808 (25,25)	0.05332658 (25,25)
6×10^{-3}	0.1326097 (25,25)	0.07793264 (25,25)	0.05393105 (25,25)
6.5×10^{-3}	0.1341917 (25,25)	0.07868475 (25,25)	0.05446891 (25,25)
7×10^{-3}	0.1357450 (3,∞)	0.07939045 (6,∞)	0.05494537 (25,25)
7.5×10^{-3}	0.1372704 (25,25)	0.08005510 (25,25)	0.05536522 (25,25)
8×10^{-3}	0.1387686 (25,25)	0.08068350 (25,25)	0.05573289 (25,25)
9×10^{-3}	0.1416872 (25,25)	0.08184811 (25,25)	0.05632816 (25,25)
1×10^{-2}	0.1445071 (4,∞)	0.08291265 (8,∞)	0.05676235 (25,25)
1.1×10^{-2}	0.1472343 (25,25)	0.08389897 (25,25)	0.05706439 (25,25)
1.2×10^{-2}	0.1498744 (25,25)	0.08482370 (25,25)	0.05726158 (25,25)
1.3×10^{-2}	0.1524328 (25,25)	0.08569934 (25,25)	0.05737951 (25,25)
1.4×10^{-2}	0.1549145 (25,25)	0.08653522 (25,25)	0.05744176 (25,25)
1.5×10^{-2}	0.1573240 (5,∞)	0.08733824 (11,∞)	0.05746923 (25,25)
1.6×10^{-2}	0.1596655 (25,25)	0.08811349 (25,25)	0.05747956 (25,25)
1.7×10^{-2}	0.1619431 (25,25)	0.08886475 (25,25)	0.05748669 (25,25)
1.8×10^{-2}	0.1641602 (25,25)	0.08959480 (25,25)	0.05750079 (25,25)
1.9×10^{-2}	0.1663203 (25,25)	0.09030572 (25,25)	0.05752848 (25,25)
2×10^{-2}	0.1684264 (6,∞)	0.09099912 (11,∞)	0.05757337 (25,25)
2.2×10^{-2}	0.1724879 (25,25)	0.09233797 (25,25)	0.05771845 (25,25)
2.4×10^{-2}	0.1763652 (47,47)	0.09361856 (47,47)	0.05793081 (47,47)
2.6×10^{-2}	0.1800757 (47,47)	0.09484592 (47,47)	0.05819554 (47,47)
2.8×10^{-2}	0.1836346 (47,47)	0.09602399 (47,47)	0.05849707 (47,47)
3×10^{-2}	0.1870550 (7,∞)	0.09715607 (14,∞)	0.05882242 (30,30)
3.5×10^{-2}	0.1950726 (47,47)	0.09980367 (47,47)	0.05968229 (47,47)
4×10^{-2}	0.2024388 (47,47)	0.1022211 (47,47)	0.06054659 (47,47)
4.5×10^{-2}	0.2092618 (47,47)	0.1044413 (47,47)	0.06137930 (47,47)
5×10^{-2}	0.2156242 (8,∞)	0.1064913 (16,∞)	0.06216807 (42,42)
5.5×10^{-2}	0.2215906 (47,47)	0.1083939 (47,47)	0.06291036 (47,47)
6×10^{-2}	0.2272125 (47,47)	0.1101677 (47,47)	0.06360763 (47,47)
6.5×10^{-2}	0.2325314 (47,47)	0.1118282 (47,47)	0.06426287 (47,47)
7×10^{-2}	0.2375816 (10,∞)	0.1133887 (18,∞)	0.064872/97 [12]
7.5×10^{-2}	0.2423917 (47,47)	0.1148600 (47,47)	0.06546083 (47,47)
8×10^{-2}	0.2469858 (47,47)	0.1162516 (47,47)	0.06601002 (47,47)
9×10^{-2}	0.2556045 (47,47)	0.1188260 (47,47)	0.06702255 (47,47)
1×10^{-1}	0.2635700 (12,∞)	0.1211637 (21,∞)	0.067938/50 [12]

Table A1.3d. (Continued)

β	$5d'_{-1}/0\ -1\ 7$	
\multicolumn{3}{c}{$m=-1\quad \pi=-1$}		

β	$5d'_{-1}/0\ -1\ 7$	
1×10^{-3}	0.04317793	(25,25)
1.5×10^{-3}	0.04420646	(25,25)
2×10^{-3}	0.04493160	(25,25)
2.5×10^{-3}	0.04540986	(25,25)
3×10^{-3}	0.04569294	(25,25)
3.5×10^{-3}	0.04582602	(25,25)
4×10^{-3}	0.04584790	(25,25)
4.5×10^{-3}	0.04579164	(25,25)
5×10^{-3}	0.04568510	(25,25)
5.5×10^{-3}	0.04555133	(25,25)
6×10^{-3}	0.04540858	(25,25)
6.5×10^{-3}	0.04527038	(25,25)
7×10^{-3}	0.04514560	(25,25)
7.5×10^{-3}	0.04503886	(25,25)
8×10^{-3}	0.04495124	(25,25)
9×10^{-3}	0.04482549	(25,25)
1×10^{-2}	0.04474022	(25,25)
1.1×10^{-2}	0.04466122	(25,25)
1.2×10^{-2}	0.04455885	(25,25)
1.3×10^{-2}	0.04441100	(25,25)
1.4×10^{-2}	0.04420307	(25,25)
1.5×10^{-2}	0.04392754	(25,25)
1.6×10^{-2}	0.04358373	(25,25)
1.7×10^{-2}	0.04317773	(25,25)
1.8×10^{-2}	0.04272223	(25,25)
1.9×10^{-2}	0.04223573	(25,25)
$2.\times 10^{-2}$	0.04174100	(25,25)
2.2×10^{-2}	0.04082203	(25,25)
2.4×10^{-2}	0.04011481	(47,47)
2.6×10^{-2}	0.03966064	(47,47)
2.8×10^{-2}	0.03941495	(47,47)
3×10^{-2}	0.03931285	(36,36)
3.5×10^{-2}	0.03938941	(47,47)
4×10^{-2}	0.03966378	(47,47)
4.5×10^{-2}	0.03999759	(47,47)
5×10^{-2}	0.04034289	(42,42)
5.5×10^{-2}	0.04068173	(47,47)
6×10^{-2}	0.04100693	(47,47)
6.5×10^{-2}	0.04131584	(47,47)
7×10^{-2}	0.041607/21	[12]
7.5×10^{-2}	0.04188044	(47,47)
8×10^{-2}	0.04213383	(47,47)
9×10^{-2}	0.04257092	(47,47)
1×10^{-1}	0.0430801/65	[12]

Table A1.3e. Energies (in units of $-E_\infty$) of the lowest $m = -2$, z-parity $\pi = +1$ states of an electron in a static Coulomb potential and a magnetic field with strength parameter β

	$m = -2 \quad \pi = +1$		
β	$3d_{-2}/0\ -2\ 0$	$4d_{-2}/0\ -2\ 2$	$5g'_{-2}/0\ -2\ 4$
1×10^{-3}	0.1170033 (2,∞)	0.06807135 (2,∞)	0.04553685 (20,20)
1.5×10^{-3}	0.1198689 (3,∞)	0.07054451 (3,∞)	0.04797659 (20,20)
2×10^{-3}	0.1226814 (2,∞)	0.07282233 (3,∞)	0.05022155 (20,20)
2.5×10^{-3}	0.1254417 (30,30)	0.07491771 (30,30)	0.05229234 (30,30)
3×10^{-3}	0.1281506 (2,∞)	0.07684462 (4,∞)	0.05420840 (20,20)
3.5×10^{-3}	0.1308091 (30,30)	0.07861719 (30,30)	0.05598698 (30,30)
4×10^{-3}	0.1334184 (30,30)	0.08024914 (30,30)	0.05764290 (30,30)
4.5×10^{-3}	0.1359797 (30,30)	0.08175342 (30,30)	0.05918875 (30,30)
5×10^{-3}	0.1384944 (3,∞)	0.08314204 (5,∞)	0.06063510 (20,20)
6×10^{-3}	0.1433890 (30,30)	0.08561546 (30,30)	0.06326340 (30,30)
7×10^{-3}	0.1481133 (3,∞)	0.08774656 (6,∞)	0.06558319 (20,20)
8×10^{-3}	0.1526782 (30,30)	0.08959929 (30,30)	0.06763379 (30,30)
9×10^{-3}	0.1570943 (30,30)	0.09122684 (30,30)	0.06944335 (30,30)
1×10^{-2}	0.1613717 (4,∞)	0.09267376 (7,∞)	0.07103225 (25,25)
1.1×10^{-2}	0.1655198 (30,30)	0.09397753 (30,30)	0.07241552 (30,30)
1.2×10^{-2}	0.1695471 (30,30)	0.09516978 (30,30)	0.07360464 (30,30)
1.3×10^{-2}	0.1734616 (30,30)	0.09627696 (30,30)	0.07460894 (30,30)
1.4×10^{-2}	0.1772707 (30,30)	0.09732095 (30,30)	0.07543675 (30,30)
1.5×10^{-2}	0.1809811 (4,∞)	0.09831937 (9,∞)	0.07609616 (25,25)
1.6×10^{-2}	0.1845989 (30,30)	0.09928600 (30,30)	0.07659564 (30,30)
1.7×10^{-2}	0.1881297 (47,47)	0.1002312 (47,47)	0.07694437 (47,47)
1.8×10^{-2}	0.1915786 (47,47)	0.1011623 (47,47)	0.07715231 (47,47)
1.9×10^{-2}	0.1949502 (47,47)	0.1020841 (47,47)	0.07723025 (47,47)
2×10^{-2}	0.1982491 (5,∞)	0.1029996 (10,∞)	0.07718960 (30,30)
2.2×10^{-2}	0.2046435 (47,47)	0.1048168 (47,47)	0.07680064 (47,47)
2.4×10^{-2}	0.2107898 (47,47)	0.1066156 (47,47)	0.07608546 (47,47)
2.6×10^{-2}	0.2167119 (47,47)	0.1083912 (47,47)	0.07515026 (47,47)
2.8×10^{-2}	0.2224302 (47,47)	0.1101371 (47,47)	0.07410953 (47,47)
3×10^{-2}	0.2279625 (6,∞)	0.1118481 (12,∞)	0.07308603 (35,35)
3.5×10^{-2}	0.2410752 (47,47)	0.1159513 (47,47)	0.07130991 (47,47)
4×10^{-2}	0.2533092 (47,47)	0.1197926 (47,47)	0.07101388 (47,47)
4.5×10^{-2}	0.2648076 (47,47)	0.1233805 (47,47)	0.07158339 (47,47)
5×10^{-2}	0.2756790 (8,∞)	0.1267353 (15,∞)	0.07247905 (46,46)
6×10^{-2}	0.2958638 (47,47)	0.1328372 (47,47)	0.07448107 (47,47)
7×10^{-2}	0.3143615 (8,∞)	0.1382654 (17,∞)	0.076396/457 [12]
8×10^{-2}	0.3315081 (47,47)	0.1431514 (47,47)	0.07819009 (47,47)
9×10^{-2}	0.3475425 (47,47)	0.1475940 (47,47)	0.07980699 (47,47)
1×10^{-1}	0.3626412 (9,∞)	0.1516684 (19,∞)	0.081280/308 [12]

Table A1.3e. (Continued)

β	$5d'_{-2}/0\ -2\ 6$	
	$m=-2\quad \pi=+1$	
1×10^{-3}	0.04485307	(20,20)
1.5×10^{-3}	0.04649747	(20,20)
2×10^{-3}	0.04771481	(20,20)
2.5×10^{-3}	0.04857775	(30,30)
3×10^{-3}	0.04915093	(20,20)
3.5×10^{-3}	0.04949025	(30,30)
4×10^{-3}	0.04964460	(30,30)
4.5×10^{-3}	0.04965817	(30,30)
5×10^{-3}	0.04957300	(20,20)
6×10^{-3}	0.04927479	(30,30)
7×10^{-3}	0.04907996	(25,25)
8×10^{-3}	0.04920460	(30,30)
9×10^{-3}	0.04963512	(30,30)
1×10^{-2}	0.05024426	(25,25)
1.1×10^{-2}	0.05092998	(30,30)
1.2×10^{-2}	0.05163601	(30,30)
1.3×10^{-2}	0.05233414	(30,30)
1.4×10^{-2}	0.05301043	(30,30)
1.5×10^{-2}	0.05365790	(25,25)
1.6×10^{-2}	0.05427304	(30,30)
1.7×10^{-2}	0.05485399	(47,47)
1.8×10^{-2}	0.05539960	(47,47)
1.9×10^{-2}	0.05590895	(47,47)
2×10^{-2}	0.05638095	(25,25)
2.2×10^{-2}	0.05720679	(47,47)
2.4×10^{-2}	0.05785873	(47,47)
2.6×10^{-2}	0.05830531	(47,47)
2.8×10^{-2}	0.05849836	(47,47)
3×10^{-2}	0.05837407	(40,40)
3.5×10^{-2}	0.05632413	(47,47)
4×10^{-2}	0.05223611	(47,47)
4.5×10^{-2}	0.04850133	(47,47)
5×10^{-2}	0.04705283	(50,50)
6×10^{-2}	0.04708320	(47,47)
7×10^{-2}	0.047775/816	[12]
8×10^{-2}	0.04850180	(47,47)
9×10^{-2}	0.04911631	(47,47)
1×10^{-1}	0.049913/27	[12]

A1.2 Figures of the Energy Values

In the Figs. A1.2a–i all states of the Tables A1.2a–h with definite m and π are grouped in one figure; the values of β at which the energies have been calculated are indicated by crosses. In the Figs. A1.3a–e all states which, in the field-free limit, originate from one principal quantum number are combined.

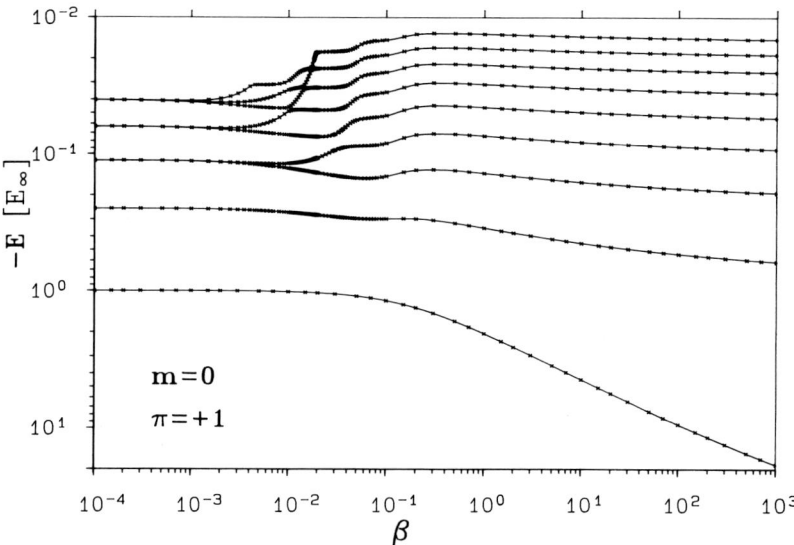

Fig. A1.2a. Energy values (in units of E_∞) of the lowest nine states in the ($m = 0$, z parity $\pi = +1$)-subspace of an electron in a static Coulomb potential as continuous functions of the magnetic field parameter $\beta = B/(4.701 \times 10^5 \text{ T})$. Crosses indicate the β-values for which the numerically accurate energy values are included in Tables A1.2a and A1.3a. Note the extreme anticrossings in the transition regime.

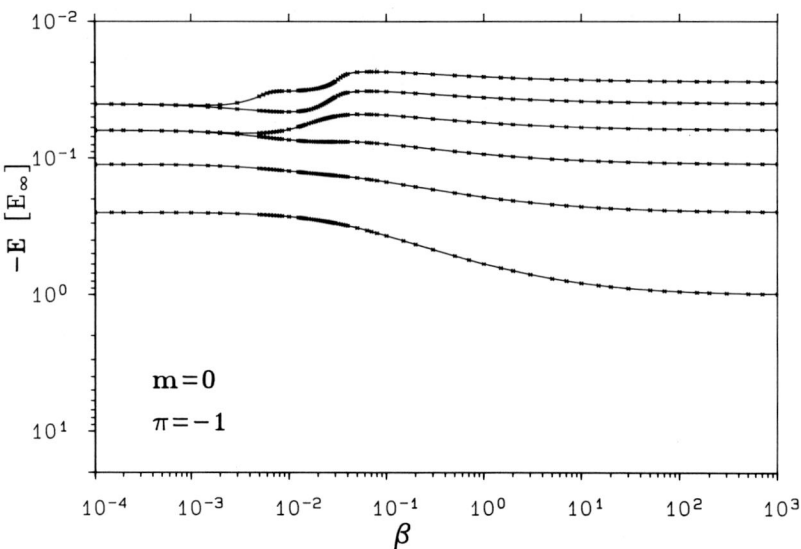

Fig. A1.2b. Energy values (in units of E_∞) of the lowest six states in the ($m = 0$, z parity $\pi = -1$)-subspace of an electron in a static Coulomb potential as continuous functions of the magnetic field parameter $\beta = B/(4.701 \times 10^5 \text{ T})$. Crosses indicate the β-values for which the numerically accurate energy values are included in Tables A1.2b and A1.3b. For $\beta = 10^3$ the energy values of these negative z parity states have reached, within the drawing accuracy, their asymptotic values, which coincide with the energies of the field-free H atom.

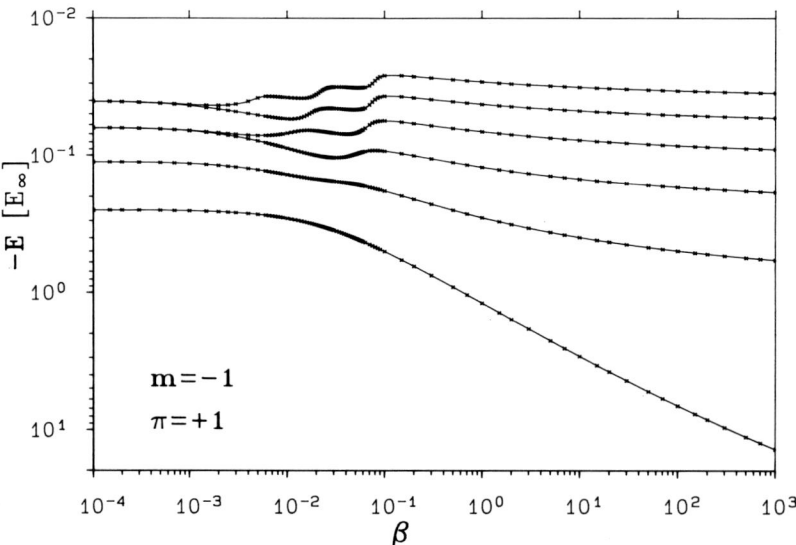

Fig. A1.2c. Energy values (in units of E_∞) of the lowest six states in the ($m = -1$, z parity $\pi = +1$)-subspace of an electron in a static Coulomb potential as continuous functions of the magnetic field parameter $\beta = B/(4.701 \times 10^5 \text{ T})$. Crosses indicate the β-values for which the numerically accurate energy values are included in Tables A1.2c and A1.3c.

A1.2 Figures of the Energy Values

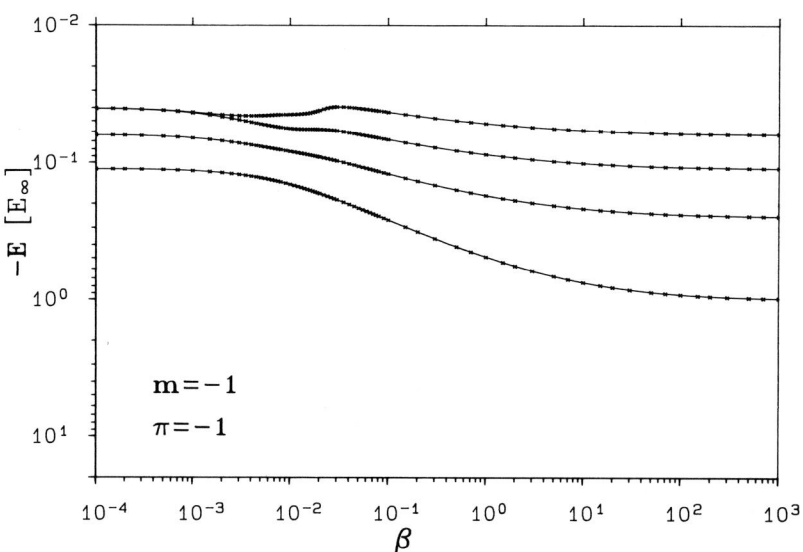

Fig. A1.2d. Energy values (in units of E_∞) of the lowest four states in the ($m = -1$, z parity $\pi = -1$)-subspace of an electron in a static Coulomb potential as continuous functions of the magnetic field parameter $\beta = B/(4.701 \times 10^5 \text{ T})$. Crosses indicate the β-values for which the numerically accurate energy values are included in Tables A1.2d and A1.3d.

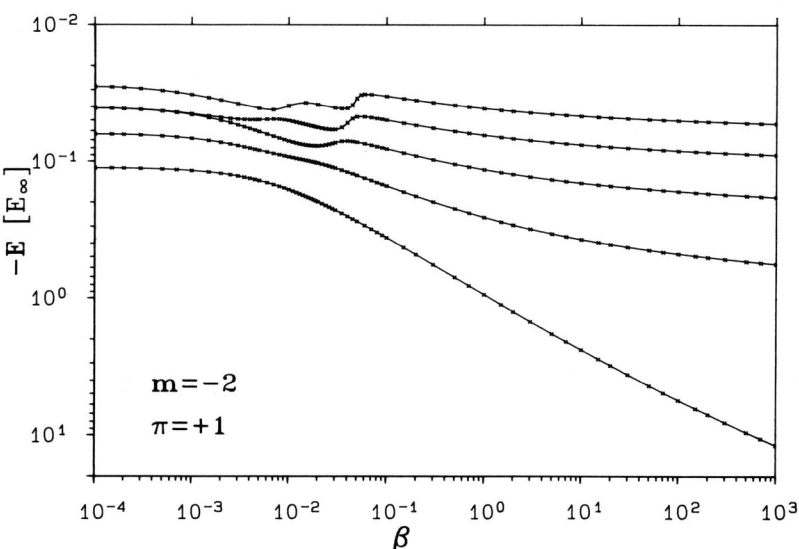

Fig. A1.2e. Energy values (in units of E_∞) of the lowest five states in the ($m = -2$, z parity $\pi = +1$)-subspace of an electron in a static Coulomb potential as continuous functions of the magnetic field parameter $\beta = B/(4.701 \times 10^5 \text{ T})$. Crosses indicate the β-values for which the numerically accurate energy values are included in Tables A1.2e and A1.3e.

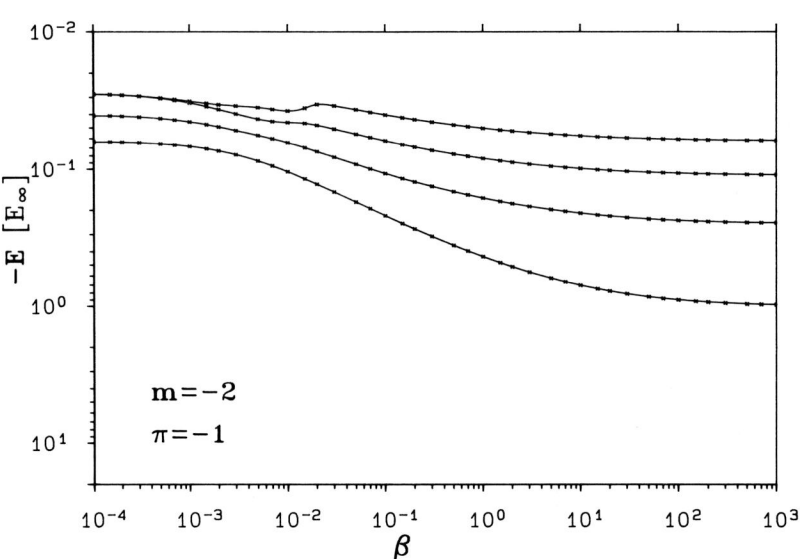

Fig. A1.2f. Energy values (in units of E_∞) of the lowest four states in the ($m = -2$, z parity $\pi = -1$)-subspace of an electron in a static Coulomb potential as continuous functions of the magnetic field parameter $\beta = B/(4.701 \times 10^5 \text{ T})$. Crosses indicate the β-values for which the numerically accurate energy values are included in Tables A1.2e and A1.2f.

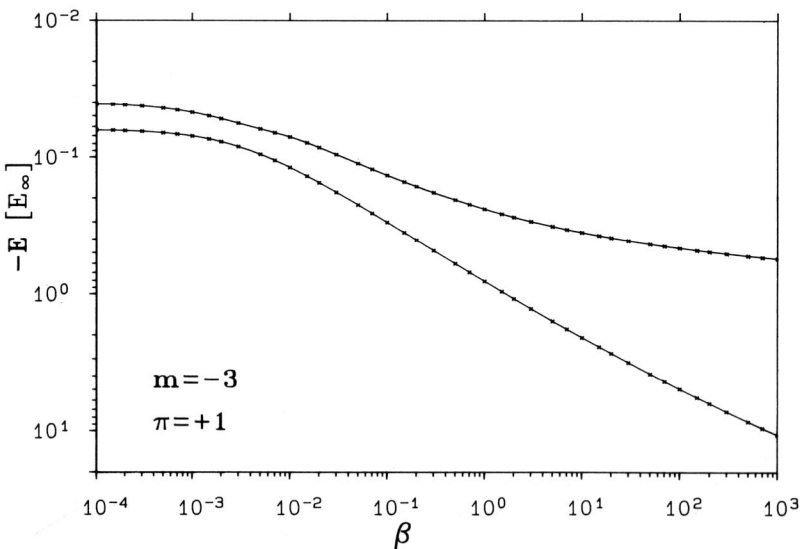

Fig. A1.2g. Energy values (in units of E_∞) of the lowest two states in the ($m = -3$, z parity $\pi = +1$)-subspace of an electron in a static Coulomb potential as continuous functions of the magnetic field parameter $\beta = B/(4.701 \times 10^5 \text{ T})$. Crosses indicate the β-values for which the numerically accurate energy values are included in Table A1.2g.

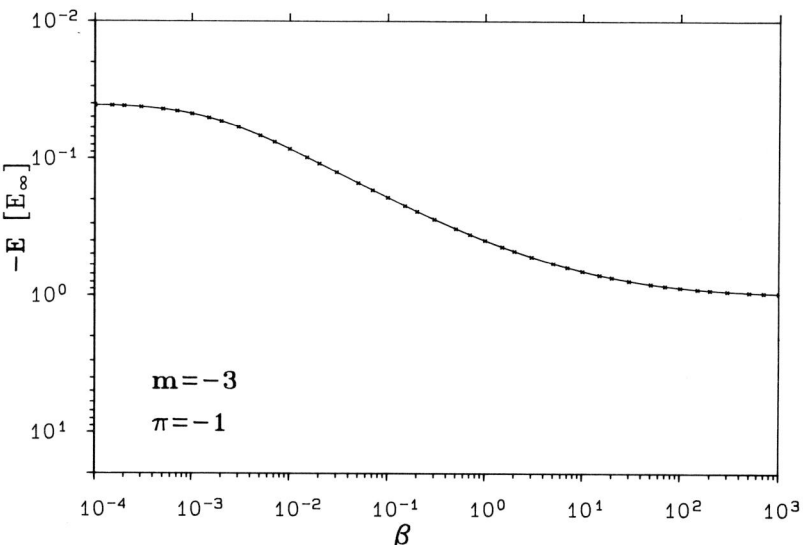

Fig. A1.2h. Energy values (in units of E_∞) of the lowest state in the ($m = -3$, z parity $\pi = -1$)-subspace of an electron in a static Coulomb potential as continuous function of the magnetic field parameter $\beta = B/(4.701 \times 10^5 \text{ T})$. Crosses indicate the β-values for which the numerically accurate energy values are included in Table A1.2g.

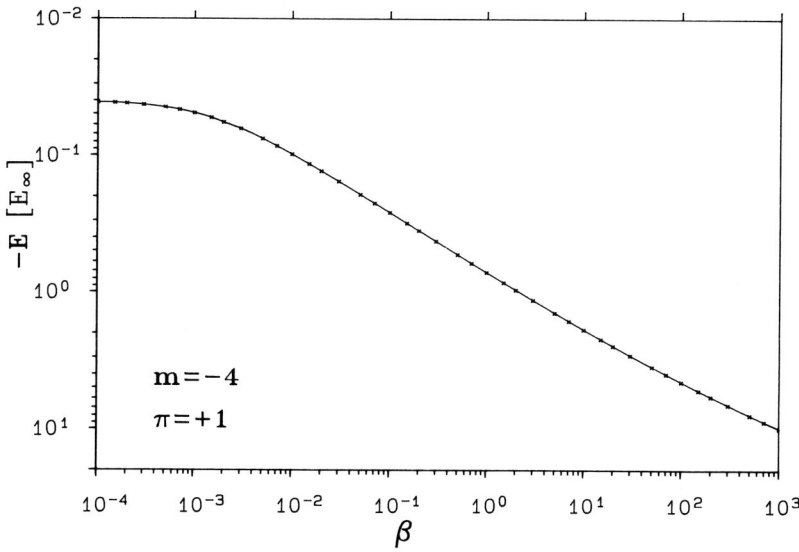

Fig. A1.2i. Energy values (in units of E_∞) of the lowest state in the ($m = -4$, z parity $\pi = +1$)-subspace of an electron in a static Coulomb potential as continuous function of the magnetic field parameter $\beta = B/(4.701 \times 10^5 \text{ T})$. Crosses indicate the β-values for which the numerically accurate energy values are included in Table A1.2h.

222   Appendix 1:   Energy Values

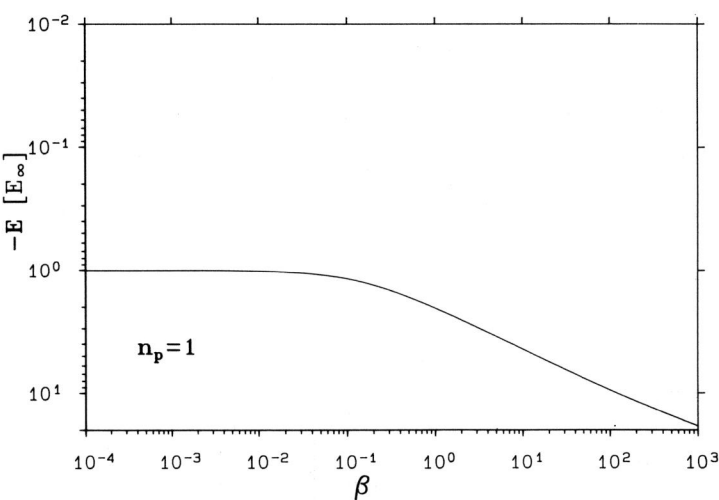

Fig. A1.3a. Energy values (in units of E_∞) of the state $1s_0/0\ 0\ 0$ of an electron in a static Coulomb potential evolving from the principal quantum number $n_p = 1$ as continuous function of the magnetic field parameter $\beta = B/(4.701 \times 10^5\ \text{T})$.

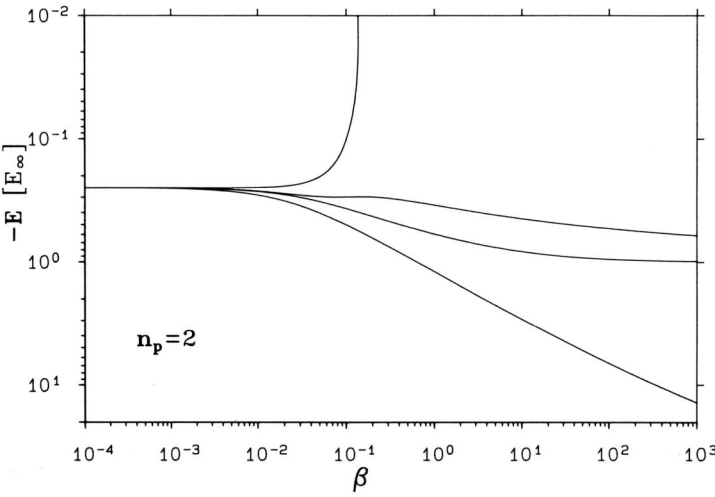

Fig. A1.3b. Energy values (in units of E_∞) of all states of an electron in a static Coulomb potential evolving from the principal quantum number $n_p = 2$ as continuous functions of the magnetic field parameter $\beta = B/(4.701 \times 10^5\ \text{T})$. Ordered according to increasing energies, the asymptotic quantum numbers of the states are in the limit $\beta \to 0$ $(n_p l m)$: $2p_{-1}, 2p_0, 2s_0, 2p_{+1}$ and in the limit $\beta \to \infty$ $(nm\nu)$: $0\ -1\ 0,\ 0\ 0\ 1,\ 0\ 0\ 2,\ 1\ 1\ 0$. Note that states with positive m are shifted to higher Landau levels as β increases.

A1.2 Figures of the Energy Values 223

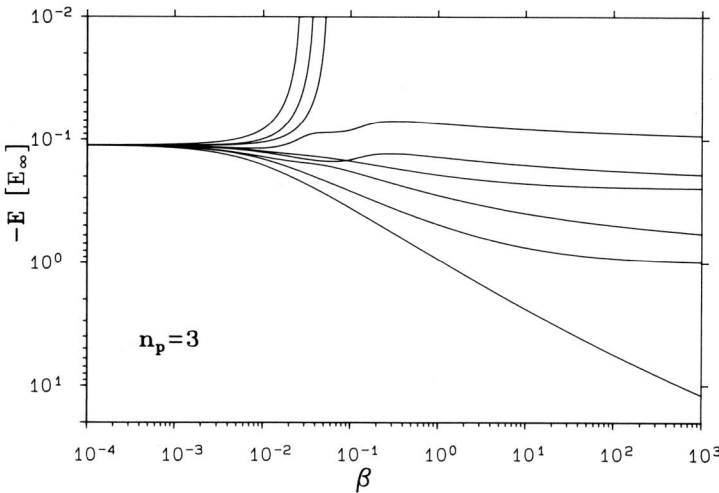

Fig. A1.3c. Energy values (in units of E_∞) of all states of an electron in a static Coulomb potential evolving from the principal quantum number $n_p = 3$ as continuous functions of the magnetic field parameter $\beta = B/(4.701 \times 10^5 \text{ T})$. Ordered according to increasing energies, the asymptotic quantum numbers of the states are in the limit $\beta \to 0$ $(n_p l m)$: $3d_{-2}$, $3d_{-1}$, $3p_{-1}$, $3d'_0$, $3p_0$, $3s'_0$, $3d_{+1}$, $3p_{+1}$, $3d_{+2}$ and in the limit $\beta \to \infty$ $(nm\nu)$: 0 -2 0, 0 -1 1, 0 -1 2, 0 0 3, 0 0 4, 0 0 6, 1 1 1, 1 1 2, 2 2 0. Note that states with positive m are shifted to higher Landau levels as β increases.

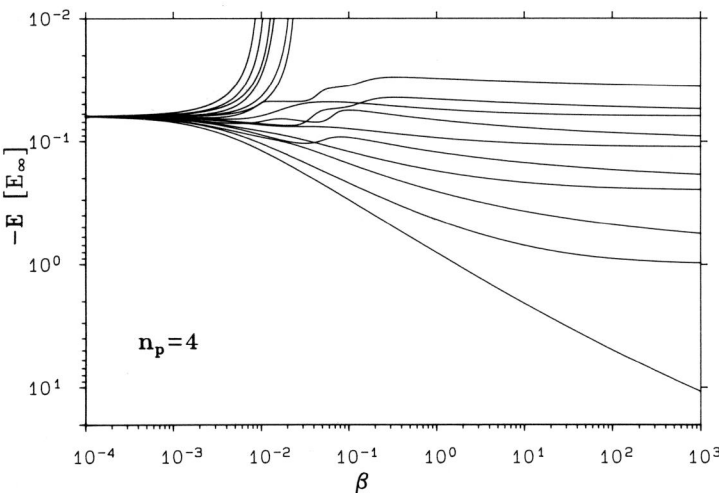

Fig. A1.3d. Energy values (in units of E_∞) of all states of an electron in a static Coulomb potential evolving from the principal quantum number $n_p = 4$ as continuous functions of the magnetic field parameter $\beta = B/(4.701 \times 10^5 \text{ T})$. Ordered according to increasing energies, the asymptotic quantum numbers of the states are in the limit $\beta \to 0$ $(n_p l m)$: $4f_{-3}$, $4f_{-2}$, $4d_{-2}$, $4f'_{-1}$, $4d_{-1}$, $4p'_{-1}$, $4f'_0$, $4d'_0$, $4p'_0$, $4s'_0$, $4f'_{+1}$, $4d_{+1}$, $4p'_{+1}$, $4f_{+2}$, $4d_{+2}$, $4f_{+3}$ and in the limit $\beta \to \infty$ $(nm\nu)$: 0 -3 0, 0 -2 1, 0 -2 2, 0 -1 3, 0 -1 4, 0 0 5, 0 -1 6, 0 0 7, 0 0 8, 0 0 10, 1 1 3, 1 1 4, 2 2 1, 2 2 2, 3 3 0. Note that states with positive m are shifted to higher Landau levels as β increases.

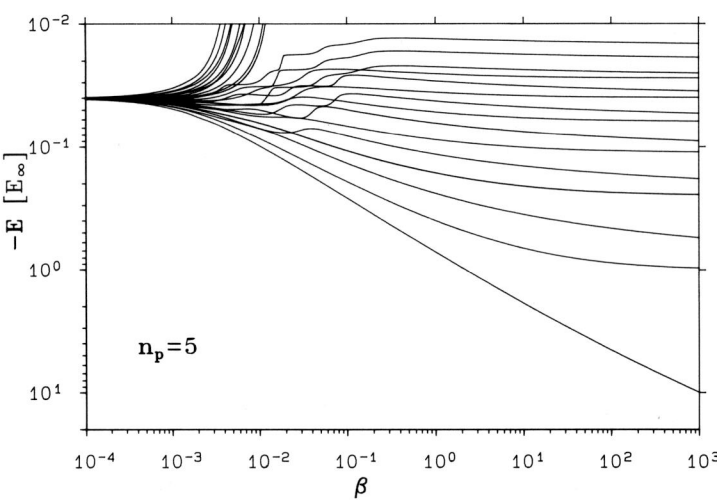

Fig. A1.3e. Energy values (in units of E_∞) of all states of an electron in a static Coulomb potential evolving from the principal quantum number $n_p = 5$ as continuous functions of the magnetic field parameter $\beta = B/(4.701 \times 10^5 \text{ T})$. Ordered according to increasing energies, the asymptotic quantum numbers of the states are in the limit $\beta \to 0$ $(n_p lm)$: $5g_{-4}$, $5g_{-3}$, $5f_{-3}$, $5g'_{-2}$, $5f_{-2}$, $5d'_{-2}$, $5g'_{-1}$, $5f'_{-1}$, $5d'_{-1}$, $5p'_{-1}$, $5g'_0$, $5f'_0$, $5d'_0$, $5p'_0$, $5s'_0$, $5g'_{+1}$, $5f'_{+1}$, $5d'_{+1}$, $5p'_{+1}$, $5g'_{+2}$, $5f_{+2}$, $5d'_{+2}$, $5g_{+3}$, $5f_{+3}$, $5g_{+4}$ and in the limit $\beta \to \infty$ $(nm\nu)$: 0 −4 0, 0 −3 1, 0 −3 2, 0 −2 3, 0 −2 4, 0 −1 5, 0 −2 6, 0 −1 7, 0 −1 8, 0 0 9, 0 −1 10, 0 0 11, 0 0 12, 0 0 14, 0 0 16, 1 1 5, 1 1 7, 1 1 8, 1 1 10, 2 2 3, 2 2 4, 2 2 6, 3 3 1, 3 3 2, 4 4 0. Note that states with positive m are shifted to higher Landau levels as β increases.

Appendix 2: Wavelengths

In Table A2.1 we have listed the relations between transitions, number of components, and the following tables and figures.

Table A2.1. Relations between transitions, number of components, tables, and figures

Transition	$n'_p \to n_p$	number of components	Figure	Stationary components	Table	Figure
Lyman α	$2 \to 1$	3	A2.1	1	A2.2a	A2.5
Lyman β	$3 \to 1$	3	A2.1	—	—	—
Lyman γ	$4 \to 1$	6	A2.1	—	—	—
Lyman δ	$5 \to 1$	6	A2.1	—	—	—
Balmer α	$3 \to 2$	15	A2.2a	5	A2.2a–b	A2.6a
Balmer β	$4 \to 2$	18	A2.2a	7	A2.2b–d	A2.6b
Balmer γ	$5 \to 2$	27	A2.2b	7	A2.2d–e	A2.6c
Balmer δ	$6 \to 2$	30	A2.2b	*	—	—
Balmer ε	$7 \to 2$	39	A2.2b	*	—	—
Paschen α	$4 \to 3$	42	A2.3	2	A2.2f	A2.7a
Paschen β	$5 \to 3$	51	A2.3	7	A2.2f–h	A2.7b
Brackett α	$5 \to 4$	90	A2.4	3	A2.2h	A2.8
$n'_p \leq 5 \to n_p \leq 5$	$n'_p \neq n_p$	261	4.2a	—	—	—
$n'_p \leq 5 \to n_p \leq 5$	$n'_p = n_p$	105	4.2b	—	—	—

* There are numerous stationary points, which, however, are not very pronounced, for details see the figures.

A2.1 Figures of the Wavelengths

In the following Figs. A2.1–A2.4 the wavelengths of all dipole transitions possible between states with (field-free) principal quantum numbers $n_p \leq 5$ are plotted in their dependence on the magnetic field strength for $10^{-4} \leq \beta \leq 10^3$.

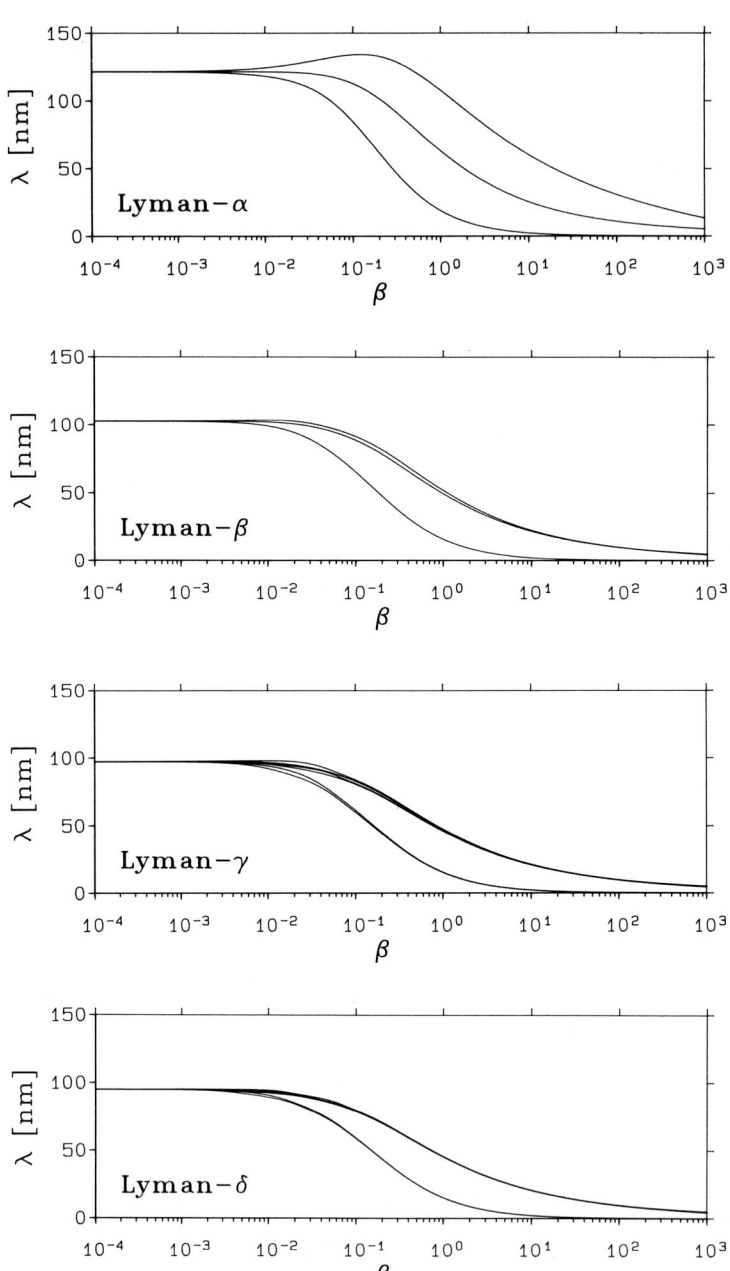

Fig. A2.1. The wavelengths of the (field-free) Lyman α, Lyman β Lyman γ and Lyman δ transitions of the hydrogen atom as continuous functions of the magnetic field strength. Note the occurrence of a *stationary* Lyman α component. The shortest wavelengths in the high-field regime correspond to cyclotron transitions of the electron. The merging of the other components into a nearly unresolved line is a consequence of the strong energetic lowering of the ground state.

A2.1 Figures of the Wavelengths 227

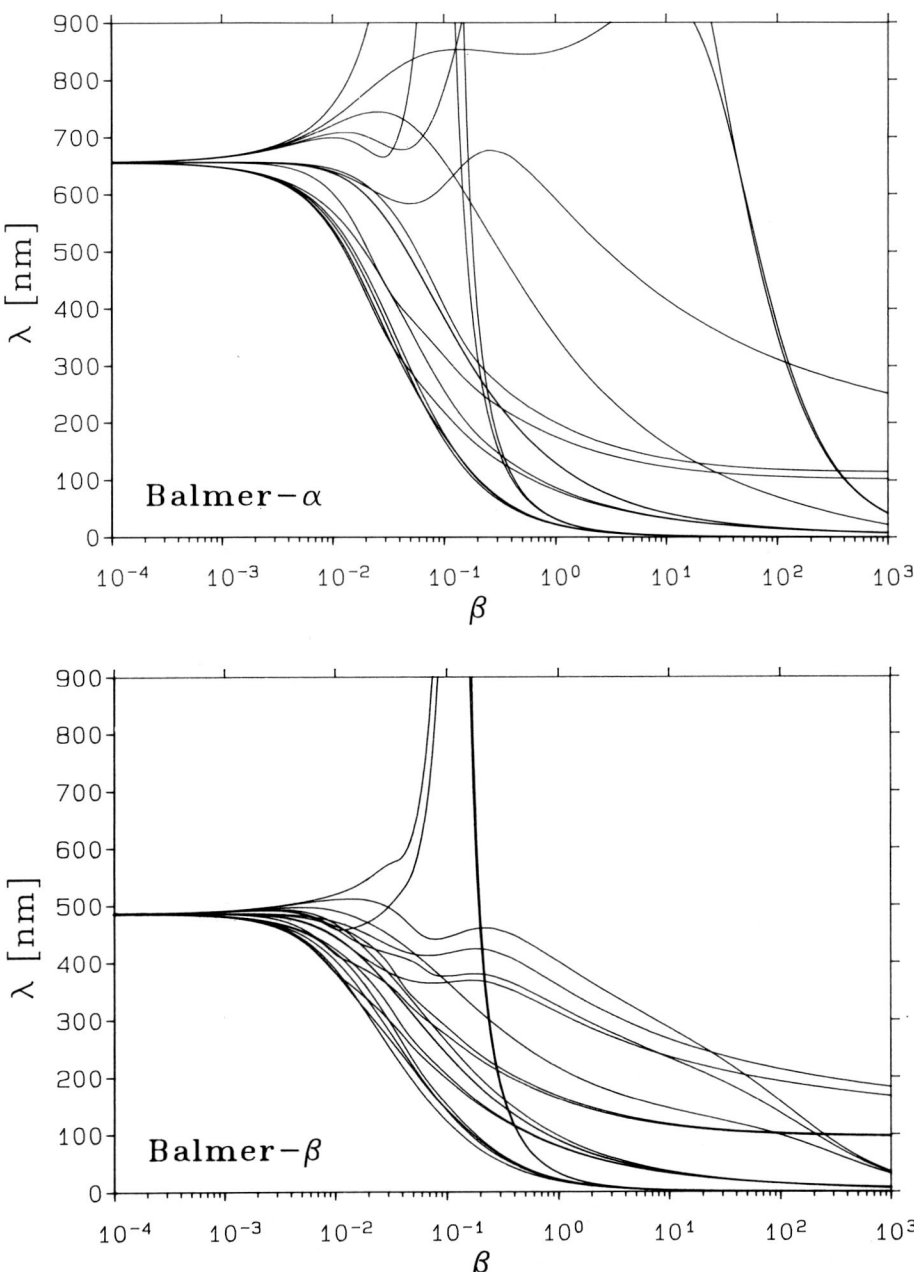

Fig. A2.2a. The wavelengths of the (field-free) Balmer α and Balmer β transitions of the hydrogen atom as continuous functions of the magnetic field strength. For small values of β the characteristic splitting into three Zeeman components is clearly recognized. Pronounced *stationary* components occur at intermediate values and the shortest wavelengths in the high-field regime correspond to cyclotron transitions of the electron.

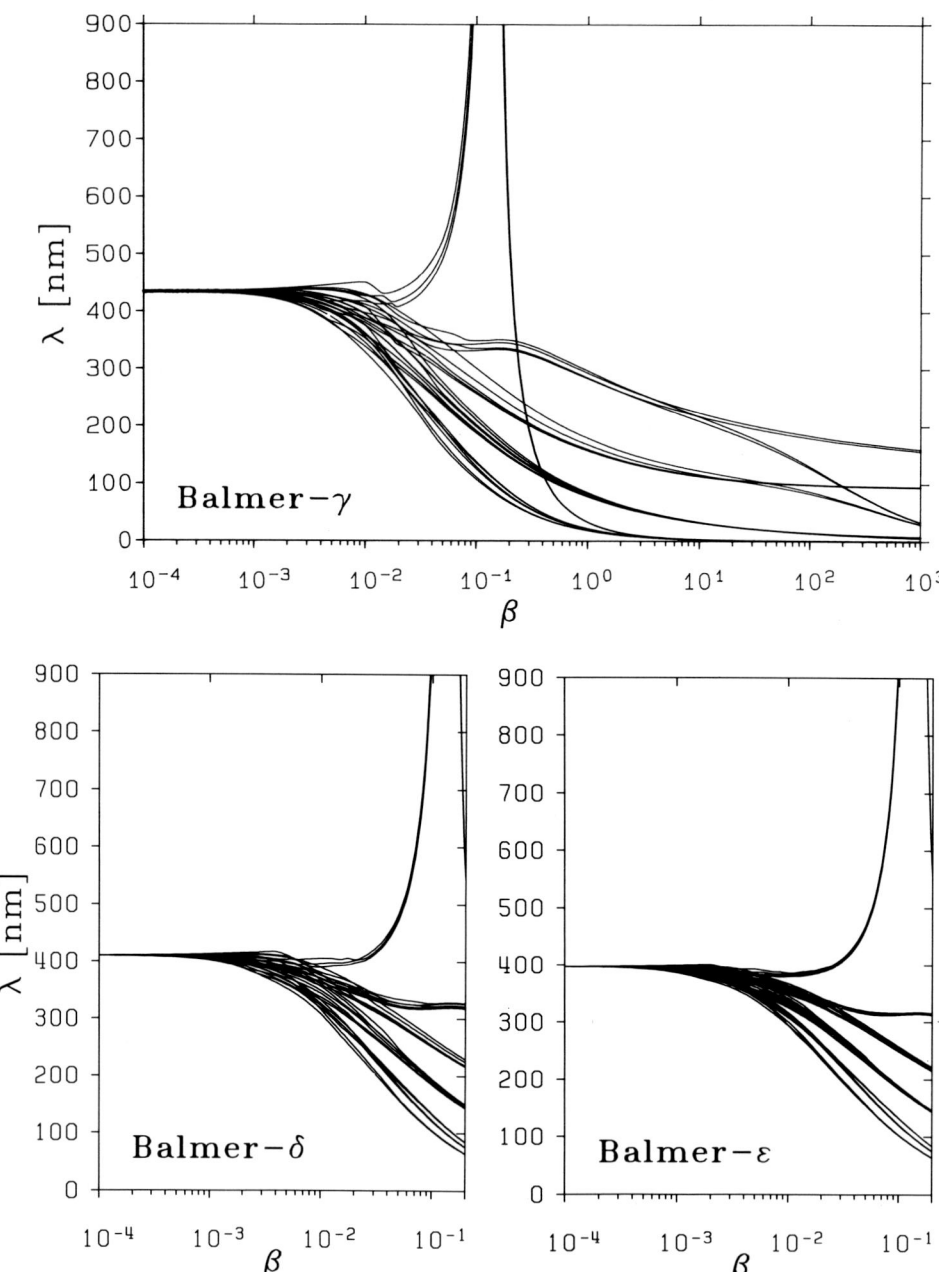

Fig. A2.2b. The wavelengths of the (field-free) Balmer γ, Balmer δ and Balmer ε transitions of the hydrogen atom as continuous functions of the magnetic field strength. Pronounced *stationary* components with wavelengths in the violet spectral range occur at intermediate values and the shortest wavelengths in the high-field regime again correspond to cyclotron transitions of the electron. Note that the Balmer δ and Balmer ε transitions are shown only in the range $10^{-4} \leq \beta \leq 10-1$.

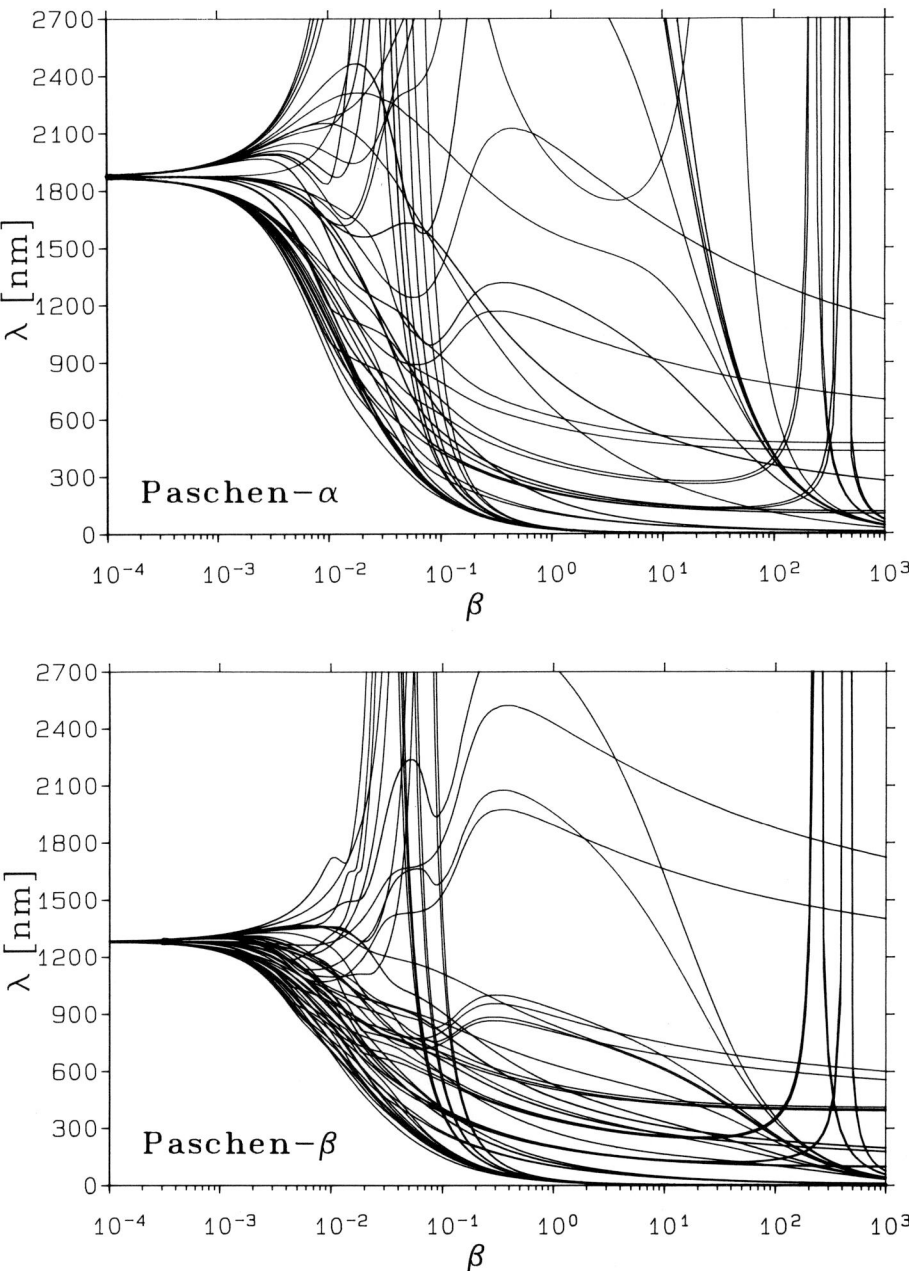

Fig. A2.3. The wavelengths of the (field-free) Paschen α and Paschen β transitions of the hydrogen atom as continuous functions of the magnetic field strength. For small values of β the characteristic splitting into three Zeeman components is clearly recognized. Pronounced *stationary* components occur at intermediate values and the shortest wavelengths in the high-field regime correspond to cyclotron transitions of the electron. Additionally structures for $\beta \geq 5$ are caused by effects of the finite proton mass.

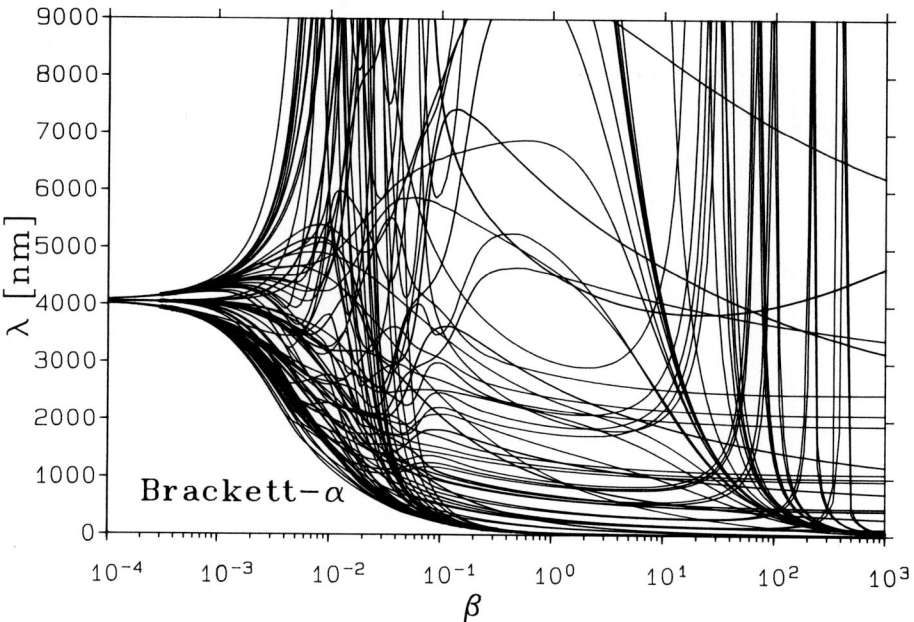

Fig. A2.4. The wavelengths of the (field-free) Brackett α transitions of the hydrogen atom as continuous functions of the magnetic field strength. For small values of β the characteristic splitting into three Zeeman components is clearly recognized. The wavelengths of nearly all *stationary* components lie in the infrared spectral range. The shortest wavelengths in the high-field regime again correspond to cyclotron transitions of the electron. Additionally structures for $\beta \geq 1$ are caused by effects of the finite proton mass.

A2.2 Tables of Stationary Wavelengths

Table A2.2. Wavelengths in nm of stationary Lyman α, Balmer α, Balmer β, Balmer γ, Paschen α, Paschen β and Brackett α transitions of the H atom. Wherever the accuracy cannot be guaranteed to within all the digits quoted, to each computed value the extrapolated upper bound is given in parentheses.

β	B in 10^2 T	$2p_{-1} \to 1s_0$	$3p_{-1} \to 2s_0$	$3p_0 \to 2s_0$	$3d_{-2} \to 2p_{-1}$
0.0010	4.696	121.891	665.487	656.262	665.643
0.0015	7.044	122.048	669.666	656.005	670.018
0.0020	9.392	122.205	673.615	655.647	674.247
0.0030	14.09	122.515	680.817	654.643	682.252
0.0050	23.48	123.118	692.441	651.597	696.434
0.0070	32.87	123.699	700.577	647.440	708.250
0.010	46.96	124.526	707.233	639.873	721.987
0.015	70.44	125.793	707.864	626.094	736.363
0.020	93.92	126.925	701.472	613.293	743.147
0.030	140.9	128.827	685.301	594.678	743.497
0.050	234.8	131.478	680.701	582.988	722.926
0.070	328.7	133.041	710.533	590.909	695.924
0.10	469.6	134.102	782.580	615.547	657.578
0.15	704.4	133.981	911.042	652.559	605.585
0.20	939.2	132.729	1015.185	670.972	565.681
0.30	1409	129.085	1150.834	674.555	508.245
0.50	2348	121.531	1282.3	648.96	437.955
0.70	3287	115.145	1346.8	622.491	394.597
			(1348.1)	(622.550)	
1.0	4696	107.598	1398.9	591.356	351.952
			(1399.8)	(591.388)	

Table A2.2. (Continued)

β	B in 10^2 T	$3d_{-1} \to 2p_0$	$3s'_0 \to 2p_{+1}$	$4p_{-1'} \to 2s_0$	$4f'_{-1} \to 2s_0$
0.0010	4.696	665.759	665.407	490.274	491.131
0.0015	7.044	670.282	669.482	491.397	493.315
0.0020	9.392	674.720	673.287	491.964	495.337
0.0030	14.09	683.330	680.068	491.587	498.914
0.0050	23.48	699.444	690.340	486.090	504.397
0.0070	32.87	714.070	696.500	476.614	508.061
0.010	46.96	733.356	699.267	459.808	511.068
0.015	70.44	759.351	692.122	437.978	511.695
0.020	93.92	779.255	678.506	427.571	508.813
0.030	140.9	806.564	664.746	416.452	496.236
0.050	234.8	834.148	795.141	401.082	459.952
0.070	328.7	845.735	1151.92	381.655	441.918
0.10	469.6	851.989	4994.21	376.278	444.612
0.15	704.4	852.983	938.272	379.97	456.18
				(380.11)	(456.56)
0.20	939.2	851.161	398.277	379.599	460.58
				(379.675)	(460.82)

Table A2.2. (Continued)

β	B in 10^2 T	$3d_{-1} \to 2p_0$	$3s'_0 \to 2p_{+1}$	$4p_{-1'} \to 2s_0$	$4f'_{-1} \to 2s_0$
0.30	1409	847.370	175.071	371.587	455.94
				(371.623)	(456.07)
0.50	2348	844.482		352.902	436.521
				(352.919)	(436.577)
0.70	3287	845.769		338.356	419.641
		(845.810)		(338.366)	(419.676)
1.0	4696	851.172		322.557	400.555
		(851.200)		(322.563)	(400.577)

Table A2.2. (Continued)

β	B in 10^2 T	$4p'_0 \to 2s_0$	$4f'_0 \to 2s_0$	$4d_{-2} \to 2p_{-1}$	$4d_{-1} \to 2p_0$
0.0010	4.696	485.552	486.070	490.441	490.786
0.0015	7.044	484.669	485.819	491.769	492.542
0.0020	9.392	483.466	485.472	492.616	493.977
0.0030	14.09	480.221	484.510	492.988	495.958
0.0050	23.48	471.275	481.674	489.482	497.016
0.0070	32.87	460.574	477.933	482.156	495.416
0.010	46.96	443.976	471.323	467.611	490.401
0.015	70.44	419.958	459.455	441.571	479.731
0.020	93.92	402.747	448.247	418.357	468.937
0.030	140.9	382.620	430.913	381.897	449.273
0.050	234.8	367.057	415.006	331.472	417.402
0.070	328.7	363.751	412.456	296.884	393.001
0.10	469.6	365.543	416.742	260.781	365.550
0.15	704.4	368.882	423.68	222.190	334.061
		(368.926)	(423.80)		
0.20	939.2	367.734	424.179	197.102	312.408
		(367.761)	(424.250)		(312.417)
0.30	1409	359.061	415.463		283.754
		(359.073)	(415.497)		(283.758)
0.50	2348	340.434	393.873		251.602
		(340.438)	(393.887)		(251.604)
0.70	3287	326.357	377.007		233.021
		(326.359)	(377.014)		(233.022)
1.0	4696	311.346	358.831		215.533
		(311.347)	(358.835)		(215.534)

Table A2.2. (Continued)

β	B in 10^2 T	$4s'_0 \to 2p_{+1}$	$5p'_{-1} \to 2s_0$	$5f'_{-1} \to 2s_0$	$5p'_0 \to 2s_0$
0.0010	4.696	490.205	435.781	437.349	432.477
0.0015	7.044	491.241	434.860	438.271	430.496
0.0020	9.392	491.690	433.017	438.820	427.935
0.0030	14.09	490.994	427.303	439.053	421.622
0.0050	23.48	484.624	412.233	437.411	407.301
0.0070	32.87	474.169	404.538	434.355	396.569
0.010	46.96	458.075	397.523	428.059	388.095

Table A2.2. (Continued)

β	B in 10^2 T	$4s'_0 \to 2p_{+1}$	$5p'_{-1} \to 2s_0$	$5f'_{-1} \to 2s_0$	$5p'_0 \to 2s_0$
0.015	70.44	461.944	386.956	410.846	376.610
0.020	93.92	471.682	374.359	391.942	366.764
0.030	140.9	497.258	356.166	376.330	350.346
0.050	234.8	559.661	345.237	365.022	334.405
0.070	328.7	706.783	340.710	356.222	331.282
0.10	469.6	1379.404	335.305	349.131	332.169
0.15	704.4	1688.990	336.756	351.110	334.051
			(336.796)	(351.179)	(334.064)
0.20	939.2	485.647	335.151	349.791	332.422
			(335.170)	(349.826)	(332.429)
0.30	1409	189.868	327.014	341.552	324.362
		(189.797)	(327.022)	(341.567)	(324.365)
0.50	2348		310.296	324.061	308.003
			(310.299)	(324.068)	(308.004)
0.70	3287		297.708	310.774	295.802
			(297.710)	(310.779)	
1.0	4696		284.197	296.473	282.830
			(284.198)	(296.475)	

Table A2.2. (Continued)

β	B in 10^2 T	$5f'_0 \to 2s_0$	$5s'_0 \to 2p_{+1}$	$5d'_0 \to 2p_{+1}$	$5g'_0 \to 2p_{+1}$
0.0010	4.696	433.678	435.700	437.255	437.904
0.0015	7.044	433.092	434.685	438.069	439.491
0.0020	9.392	432.324	432.721	438.475	440.907
0.0030	14.09	430.380	426.713	438.332	443.295
0.0050	23.48	425.575	417.754	435.688	446.775
0.0070	32.87	420.259	418.500	431.620	449.258
0.010	46.96	411.973	417.082	426.604	450.160
0.015	70.44	397.884	413.007	424.019	430.446
0.020	93.92	383.990	408.948	419.596	437.475
0.030	140.9	361.820	428.442	440.082	459.667
0.050	234.8	345.833	491.654	504.145	523.906
0.070	328.7	342.571	606.333	623.275	651.184
			(606.349)		
0.10	469.6	343.703	1051.898	1100.100	1184.452
			(1051.934)	(1100.180)	
0.15	704.4	346.006	2580.527	2363.931	2087.425
		(346.029)	(2581.842)	(2365.525)	(2089.290)
0.20	939.2	344.492	535.840	527.147	513.604
		(344.504)	(535.874)	(527.193)	(513.668)
0.30	1409	336.166	196.849	195.883	194.216
		(336.171)	(196.851)	(195.885)	(194.219)
0.50	2348	318.991	82.976	82.867	82.630
		(318.993)	(82.976)	(82.870)	(82.633)
0.70	3287	306.135	51.522	51.505	51.437
		(306.136)	(51.523)		(51.438)
1.0	4696	292.458	32.425	32.433	32.420
		(292.459)	(32.426)		(32.421)

Table A2.2. (Continued)

β	B in 10^2 T	$4p'_0 \to 3d_0$	$4p'_{-1} \to 3d'_0$	$5p'_{-1} \to 3d'_0$	$5f'_{-1} \to 3d'_0$
0.0010	4.696	1865.440	1937.127	1296.536	1310.521
0.0015	7.044	1853.096	1955.469	1288.719	1319.144
0.0020	9.392	1836.490	1965.457	1273.081	1324.577
0.0030	14.09	1792.822	1962.203	1225.995	1327.964
0.0050	23.48	1680.357	1885.227	1112.321	1316.854
0.0070	32.87	1559.224	1759.707	1061.465	1294.659
0.010	46.96	1395.227	1564.513	1020.479	1249.244
0.015	70.44	1198.631	1358.119	963.982	1127.270
0.020	93.92	1081.602	1281.404	898.604	1007.045
0.030	140.9	965.910	1215.110	813.393	926.804
0.050	234.8	893.043	1125.314	774.023	881.095
0.070	328.7	891.332	1007.104	764.624	847.442
0.10	469.6	944.169	1019.291	765.793	841.941
0.15	704.4	1062.936	1160.536	833.75	927.64
		(1063.308)	(1161.882)	(833.99)	(928.12)
0.20	939.2	1131.466	1251.889	870.94	977.23
		(1131.719)	(1252.714)	(871.07)	(977.50)
0.30	1409	1166.018	1309.401	884.523	999.61
		(1166.145)	(1309.853)	(884.586)	(999.74)
0.50	2348	1146.637	1301.610	869.0	986.4
		(1155.556)	(1313.271)	(864.0)	(979.9)
0.70	3287	1121.522	1277.284	845.8	960.5
		(1126.819)	(1284.259)	(842.8)	(956.7)
1.0	4696	1090.519	1241.854	818.9	929.8
		(1093.676)	(1246.016)	(817.1)	(927.5)

Table A2.2. (Continued)

β	B in 10^2 T	$5p'_0 \to 3d'_0$	$5f'_0 \to 3d'_0$	$5p'_{+1} \to 3d'_{+2}$	$5p'_0 \to 3s'_0$
0.0010	4.696	1267.721	1278.097	1297.362	1269.751
0.0015	7.044	1251.136	1273.318	1290.849	1255.585
0.0020	9.392	1230.133	1267.109	1277.027	1237.774
0.0030	14.09	1180.366	1251.675	1234.722	1196.161
0.0050	23.48	1077.127	1215.115	1132.956	1113.392
0.0070	32.87	1008.296	1176.989	1098.392	1069.960
0.010	46.96	960.572	1121.457	1089.302	1073.251
0.015	70.44	902.233	1034.777	1098.509	1122.406
0.020	93.92	856.052	956.169	1101.738	1204.798
0.030	140.9	783.666	843.494	1185.942	1382.425
0.050	234.8	721.618	777.026	2040.478	1433.430
0.070	328.7	718.720	774.063	18511.968	1449.893
0.10	469.6	749.633	811.054	1488.823	1530.995
0.15	704.4	817.360	892.85	468.030	1756.998
		(817.441)	(892.99)	(468.107)	(1757.370)
0.20	939.2	852.748	936.958	268.169	1898.797
		(852.795)	(937.050)	(268.181)	(1899.030)

Table A2.2. (Continued)

β	B in 10^2 T	$5p'_0 \to 3d'_0$	$5f'_0 \to 3d'_0$	$5p'_{+1} \to 3d'_{+2}$	$5p'_0 \to 3s'_0$
0.30	1409	865.385	954.836	140.335	1970.871
		(865.408)	(954.883)	(140.337)	(1970.990)
0.50	2348	851.3	940.8	69.592	1955.890
		(846.4)	(935.0)	(69.592)	(1966.486)
0.70	3287	830.6	917.5	45.599	1926.986
		(827.7)	(914.0)	(45.599)	(1933.143)
1.0	4696	807.6	891.4	29.767	1890.194
		(805.9)	(889.3)	(29.767)	(1893.940)

Table A2.2. (Continued)

β	B in 10^2 T	$5p'_{-1} \to 3s'_0$	$5s'_0 \to 4p'_{-1}$	$5d'_0 \to 4p'_{-1}$	$5g'_0 \to 4p'_{-1}$
0.0010	4.696	1298.661	3605.542	3715.553	3763.587
0.0015	7.044	1293.440	3352.631	3565.968	3663.373
0.0020	9.392	1281.267	3106.183	3430.350	3586.281
0.0030	14.09	1243.045	2669.439	3201.549	3488.421
0.0050	23.48	1151.039	2247.230	2888.343	3459.927
0.0070	32.87	1130.027	2259.285	2705.234	3592.478
0.010	46.96	1148.590	2284.857	2606.406	3838.307
0.015	70.44	1219.596	2185.594	2538.421	2793.155
0.020	93.92	1290.831	1889.586	2144.950	2718.454
0.030	140.9	1477.704	1787.545	2015.072	2512.486
0.050	234.8	1656.188	1736.642	1911.878	2242.941
0.070	328.7	1649.723	2041.404	2264.030	2705.689
			(2041.587)		
0.10	469.6	1599.997	2221.320	2476.726	2991.694
			(2221.482)	(2477.129)	
0.15	704.4	1834.594	2118.151	2328.089	2728.528
		(1835.776)	(2123.529)	(2335.065)	(2739.184)
0.20	939.2	1991.580	2030.503	2211.312	2546.167
		(1992.233)	(2033.175)	(2214.696)	(2551.154)
0.30	1409	2073.255	1935.208	2094.747	2384.949
		(2073.604)	(1936.318)	(2096.143)	(2386.885)
0.50	2348	2052.556	1842.382	1986.327	2246.600
		(2064.337)	(1842.870)	(1988.103)	(2249.649)
0.70	3287	2011.381	1793.845	1932.031	2180.490
		(2018.157)	(1794.541)	(1933.043)	(2182.249)
1.0	4696	1953.675	1754.194	1887.730	2127.713
		(1957.719)	(1754.613)	(1888.325)	(2128.766)

A2.3 Figures of Stationary Wavelengths

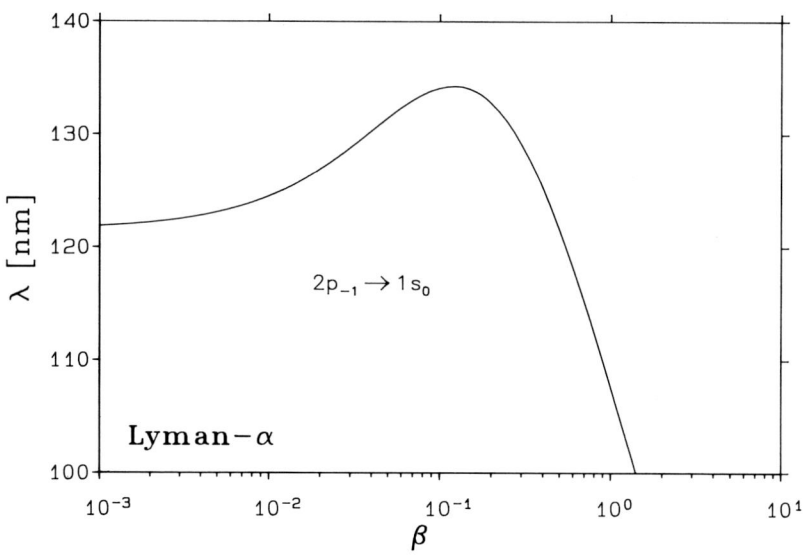

Fig. A2.5. The wavelength of the Lyman α component which becomes stationary beyond 3×10^3 T as a continuous function of the magnetic field strength in the interval 4.7×10^2 T to 4.7×10^6 T.

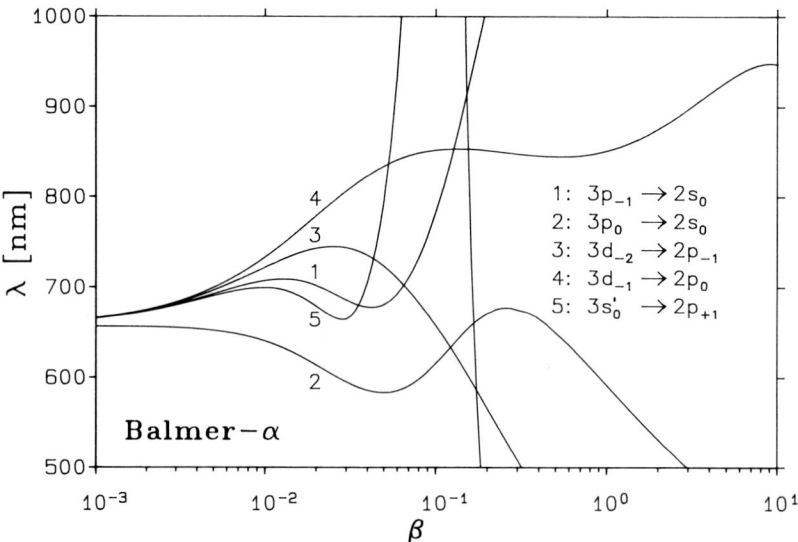

Fig. A2.6a. The wavelengths of the five stationary Balmer α components as continuous functions of the magnetic field strength in the interval 4.7×10^2 T to 4.7×10^6 T. These components are important to the correct interpretation of absorption features in the red spectral range of strongly magnetized white dwarf stars.

A2.3 Figures of Stationary Wavelengths   237

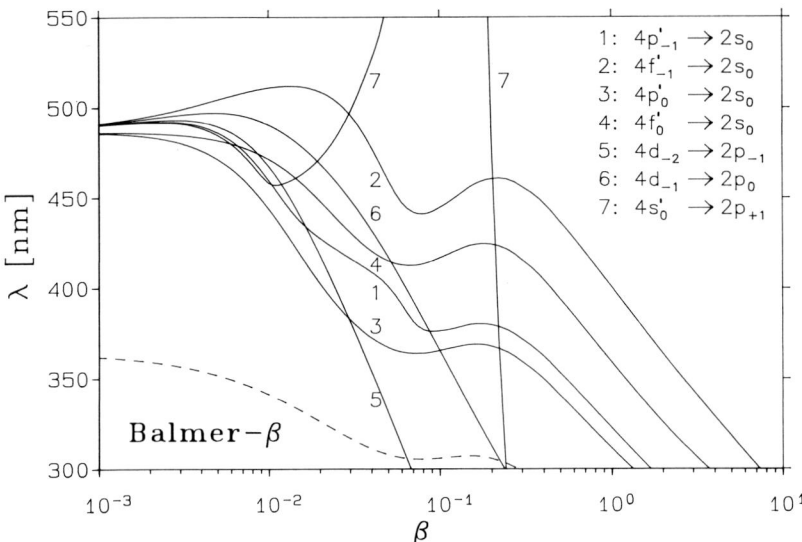

Fig. A2.6b. The wavelengths of the seven stationary Balmer β components as continuous functions of the magnetic field strength in the interval 4.7×10^2 T to 4.7×10^6 T. These components are important to the correct interpretation of absorption features in the blue spectral range of strongly magnetized white dwarf stars. The Balmer edge for transitions from $2s_0$ to the continuum is plotted as dashed curve.

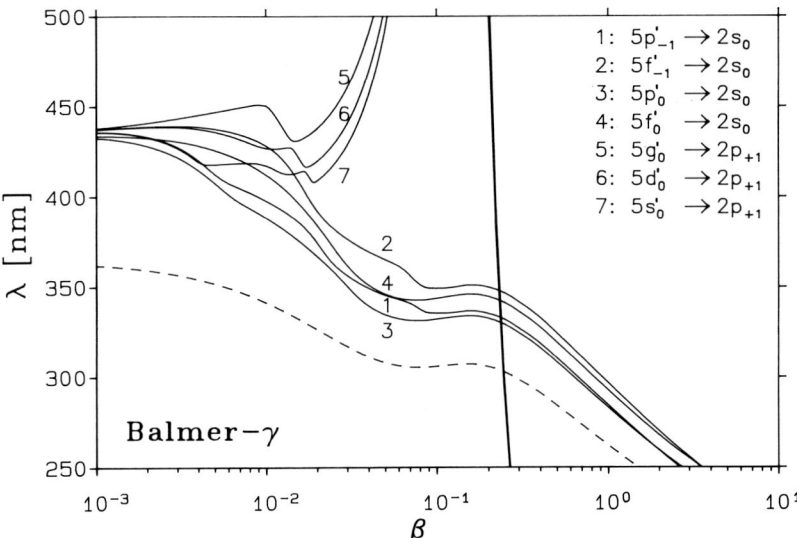

Fig. A2.6c. The wavelengths of the seven stationary Balmer γ components as continuous functions of the magnetic field strength in the interval 4.7×10^2 T to 4.7×10^6 T. These components are important to the correct interpretation of absorption features in the blue and violet spectral range of strongly magnetized white dwarf stars. The Balmer edge for transitions from $2s_0$ to the continuum is plotted as dashed curve.

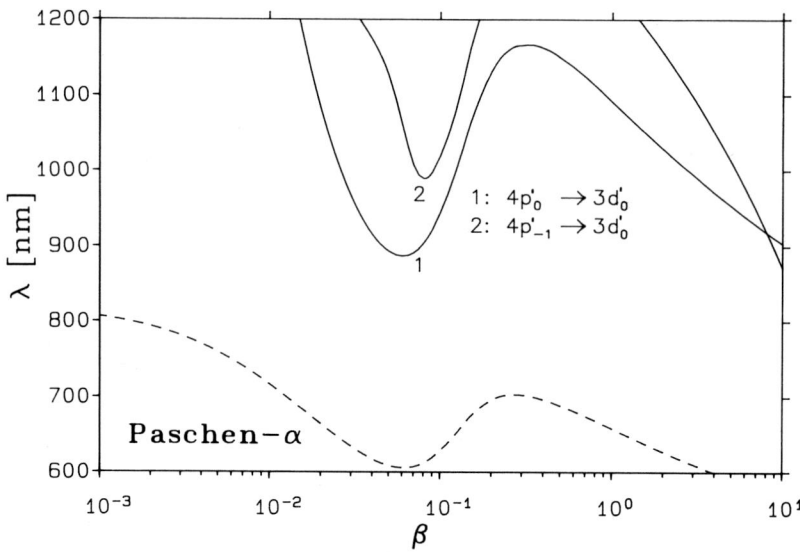

Fig. A2.7a. The wavelengths of the two Paschen α components which become stationary in the optical and the near infrared region as continuous functions of the magnetic field strength in the interval 4.7×10^2 T to 4.7×10^6 T. The Paschen edge for transitions from $3d'_0$ to the continuum is plotted as dashed curve.

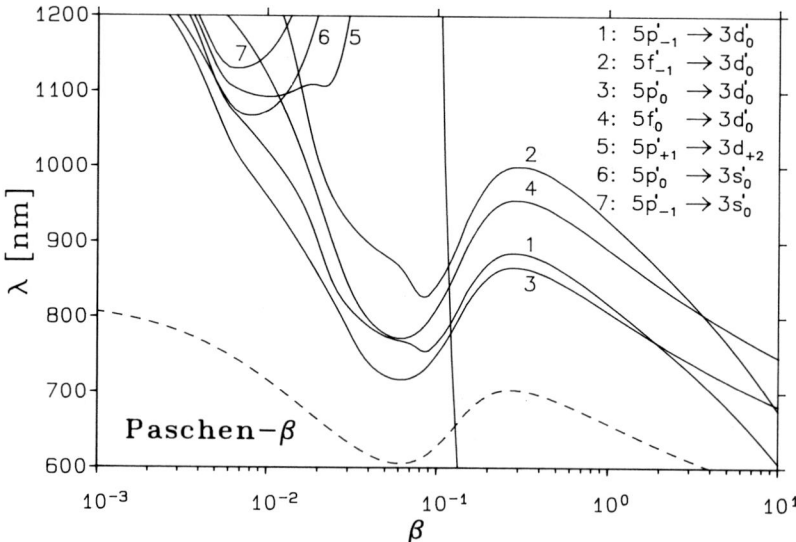

Fig. A2.7b. The wavelengths of the seven Paschen β components which become stationary in the optical and the near infrared region as continuous functions of the magnetic field strength in the interval 4.7×10^2 T to 4.7×10^6 T. The Paschen edge for transitions from $3d'_0$ to the continuum is plotted as dashed curve.

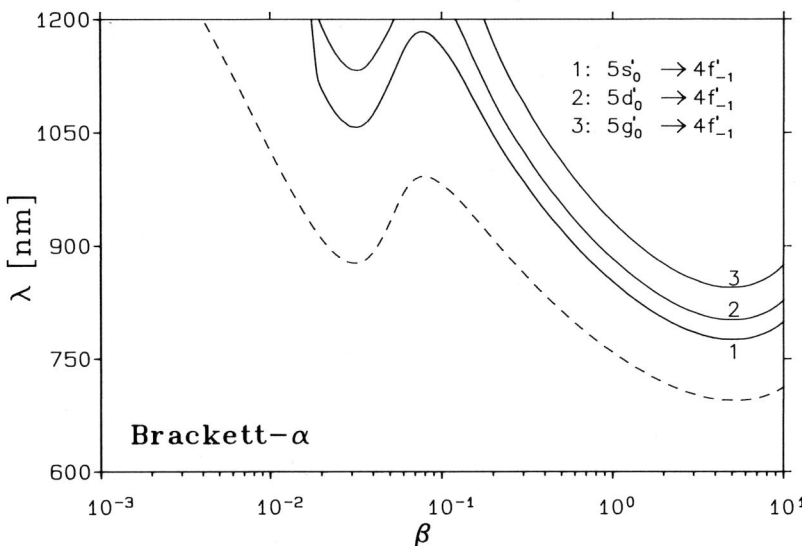

Fig. A2.8. The wavelengths of the three Brackett α components which become stationary in the optical and the near infrared region as continuous functions of the magnetic field strength in the interval 4.7×10^2 T to 4.7×10^6 T. The Brackett edge for transitions from $4f'_{-1}$ to the continuum is plotted as dashed curve. The second minima at $\beta \approx 5$ ($B \approx 2.35 \times 10^6$ T) are caused by effects of the finite proton mass.

Appendix 3: Electromagnetic Transition Probabilities

A3.1 Wavelengths, Dipole Strengths, Oscillator Strengths, and Transition Rates

In Table A3.1 we have listed the wavelengths, dipole strengths, oscillator strengths, and transition rates of all transitions possible between states with the field-free principal quantum numbers $n_p \leq 3$ in magnetic fields ranging from $\beta = 10^{-4}$ to 10^3. Initial and final states are always labelled by both their field-free quantum numbers n_p, l, m and their corresponding $\beta \to \infty$ quantum numbers n, m, ν.

Each entry consists of four numbers: the first is the wavelength of the transition in nm, the second is the dipole strength, the third is the oscillator strength, and the fourth is the transition rate in units of $w_0 \approx 8.03 \times 10^9 \text{s}^{-1}$. The values given for *the magnetic field strength, wavelengths, oscillator strengths* and *transition rates* refer to the *hydrogen atom*, i.e., use has been made of the scaling relations (2.26b) for energy differences and (2.28), (2.29) for oscillator strengths and transition rates to relate the quantities calculated for $Z = 1$ and $m_+ \to \infty$ (cf. (6.3), (6.4)) to those valid for finite proton mass. By contrast, the values given for the dipole strenghts are those computed for $m_+ \to \infty$ (in accord with the definitions (6.1), (6.2)), which, however, by virtue of the scaling law (2.27) conincide with the dipole strengths of the hydrogen atom *if* these are referred to the Bohr radius of the hydrogen atom, $a_H = \hbar/(\alpha \mu c)$, instead of $a_0 = \hbar/(\alpha m_e c)$. This makes it possible to the reader to calculate transition rates also for other mass ratios and hydrogenic ions.

A "*" following the entry of the wavelength of a transition indicates that at least one of the two states involved has come to lie in the continuum of positive energy values. States with $m < 0$ belonging to the lowest Landau level may be raised into the range of positive energies due to effects of the finiteness of the proton mass; states with $m > 0$ are raised to positive energies primarily by transverse excitation into higher Landau levels. Note that for $\beta \gg 1$ all ν-conserving $\Delta m = -1$, $\Delta n = 1$ transitions practically become cyclotron transitions, and thus possess almost the same wavelength.

Appendix 3: Electromagnetic Transition Probabilities

Table A3.1. Electromagnetic transitions between low-lying states of the hydrogen atom in magnetic fields with $10^{-4} \leq \beta \leq 10^3$. Each entry consists of four numbers: the first is the wavelength of the transition in nm, the second is the dipole strength, the third is the oscillator strength, the fourth is the transition rate.

β / B in 10^2 T	$2p_0/0\ 0\ 1$ $\leftrightarrow$ $1s_0/0\ 0\ 0$	$2p_{-1}/0\ -1\ 0$ $\leftrightarrow$ $1s_0/0\ 0\ 0$	$2p_{+1}/1\ +1\ 0$ $\leftrightarrow$ $1s_0/0\ 0\ 0$	$3p_0/0\ 0\ 3$ $\leftrightarrow$ $1s_0/0\ 0\ 0$	$3p_{-1}/0\ -1\ 2$ $\leftrightarrow$ $1s_0/0\ 0\ 0$
10^{-4} — 4.696×10^{-1}	1.216×10^3 5.549×10^{-1} 4.162×10^{-1} 7.804×10^{-2}	1.216×10^3 5.549×10^{-1} 4.161×10^{-1} 7.798×10^{-2}	1.215×10^3 5.549×10^{-1} 4.163×10^{-1} 7.810×10^{-2}	1.026×10^3 8.899×10^{-2} 7.910×10^{-2} 2.083×10^{-2}	1.026×10^3 8.899×10^{-2} 7.908×10^{-2} 2.082×10^{-2}
10^{-3} — 4.696	1.216×10^3 5.550×10^{-1} 4.162×10^{-1} 7.804×10^{-2}	1.219×10^3 5.550×10^{-1} 4.151×10^{-1} 7.743×10^{-2}	1.212×10^3 5.550×10^{-1} 4.173×10^{-1} 7.868×10^{-2}	1.026×10^3 8.909×10^{-2} 7.919×10^{-2} 2.086×10^{-2}	1.028×10^3 8.918×10^{-2} 7.910×10^{-2} 2.074×10^{-2}
7×10^{-3} — 3.287×10^1	1.215×10^3 5.567×10^{-1} 4.178×10^{-1} 7.845×10^{-2}	1.237×10^3 5.585×10^{-1} 4.117×10^{-1} 7.455×10^{-2}	1.192×10^3 5.585×10^{-1} 4.273×10^{-1} 8.337×10^{-2}	1.022×10^3 9.281×10^{-2} 8.280×10^{-2} 2.197×10^{-2}	1.034×10^3 9.673×10^{-2} 8.526×10^{-2} 2.208×10^{-2}
10^{-2} — 4.696×10^1	1.214×10^3 5.585×10^{-1} 4.194×10^{-1} 7.886×10^{-2}	1.245×10^3 5.621×10^{-1} 4.116×10^{-1} 7.355×10^{-2}	1.181×10^3 5.621×10^{-1} 4.340×10^{-1} 8.626×10^{-2}	1.019×10^3 9.544×10^{-2} 8.542×10^{-2} 2.281×10^{-2}	1.034×10^3 1.023×10^{-1} 9.021×10^{-2} 2.337×10^{-2}
2×10^{-2} — 9.392×10^1	1.210×10^3 5.680×10^{-1} 4.282×10^{-1} 8.110×10^{-2}	1.269×10^3 5.802×10^{-1} 4.168×10^{-1} 7.170×10^{-2}	1.142×10^3 5.802×10^{-1} 4.632×10^{-1} 9.838×10^{-2}	1.004×10^3 1.004×10^{-1} 9.121×10^{-2} 2.509×10^{-2}	1.025×10^3 1.156×10^{-1} 1.028×10^{-1} 2.712×10^{-2}
3×10^{-2} — 1.409×10^2	1.203×10^3 5.808×10^{-1} 4.403×10^{-1} 8.438×10^{-2}	1.288×10^3 6.033×10^{-1} 4.269×10^{-1} 7.129×10^{-2}	1.102×10^3 6.033×10^{-1} 4.993×10^{-1} 1.140×10^{-1}	9.870×10^2 9.854×10^{-2} 9.103×10^{-2} 2.589×10^{-2}	1.009×10^3 1.123×10^{-1} 1.015×10^{-1} 2.763×10^{-2}
5×10^{-2} — 2.348×10^2	1.184×10^3 6.083×10^{-1} 4.686×10^{-1} 9.265×10^{-2}	1.315×10^3 6.484×10^{-1} 4.496×10^{-1} 7.208×10^{-2}	1.021×10^3 6.484×10^{-1} 5.792×10^{-1} 1.541×10^{-1}	9.545×10^2 8.859×10^{-2} 8.463×10^{-2} 2.574×10^{-2}	9.774×10^2 8.011×10^{-2} 7.472×10^{-2} 2.168×10^{-2}
7×10^{-2} — 3.287×10^2	1.161×10^3 6.320×10^{-1} 4.963×10^{-1} 1.020×10^{-1}	1.330×10^3 6.828×10^{-1} 4.680×10^{-1} 7.326×10^{-2}	9.448×10^2 6.828×10^{-1} 6.590×10^{-1} 2.045×10^{-1}	9.246×10^2 8.014×10^{-2} 7.903×10^{-2} 2.562×10^{-2}	9.496×10^2 5.609×10^{-2} 5.385×10^{-2} 1.655×10^{-2}
10^{-1} — 4.696×10^2	1.125×10^3 6.577×10^{-1} 5.330×10^{-1} 1.167×10^{-1}	1.341×10^3 7.116×10^{-1} 4.839×10^{-1} 7.455×10^{-2}	8.446×10^2 7.116×10^{-1} 7.682×10^{-1} 2.984×10^{-1}	8.844×10^2 7.135×10^{-2} 7.355×10^{-2} 2.605×10^{-2}	9.124×10^2 3.865×10^{-2} 3.862×10^{-2} 1.285×10^{-2}
2×10^{-1} — 9.392×10^2	1.012×10^3 6.761×10^{-1} 6.089×10^{-1} 1.647×10^{-1}	1.327×10^3 6.948×10^{-1} 4.773×10^{-1} 7.507×10^{-2}	$6.135\times10^{2*}$ 6.948×10^{-1} 1.033 7.600×10^{-1}	7.814×10^2 5.703×10^{-2} 6.655×10^{-2} 3.020×10^{-2}	8.136×10^2 2.012×10^{-2} 2.255×10^{-2} 9.442×10^{-3}
3×10^{-1} — 1.409×10^3	9.232×10^2 6.533×10^{-1} 6.452×10^{-1} 2.097×10^{-1}	1.291×10^3 6.268×10^{-1} 4.427×10^{-1} 7.363×10^{-2}	$4.786\times10^{2*}$ 6.268×10^{-1} 1.194 1.444	7.098×10^2 4.951×10^{-2} 6.359×10^{-2} 3.498×10^{-2}	7.421×10^2 1.374×10^{-2} 1.689×10^{-2} 8.497×10^{-3}

A3.1 Wavelengths, Strengths, and Transition Rates

Table A3.1. (Continued)

β / B in 10^2 T	$2p_0/0\ 0\ 1$ $\leftrightarrow$ $1s_0/0\ 0\ 0$	$2p_{-1}/0\ -1\ 0$ $\leftrightarrow$ $1s_0/0\ 0\ 0$	$2p_{+1}/1\ +1\ 0$ $\leftrightarrow$ $1s_0/0\ 0\ 0$	$3p_0/0\ 0\ 3$ $\leftrightarrow$ $1s_0/0\ 0\ 0$	$3p_{-1}/0\ -1\ 2$ $\leftrightarrow$ $1s_0/0\ 0\ 0$
5×10^{-1} —— 2.348×10³	7.982×10² 5.902×10⁻¹ 6.742×10⁻¹ 2.933×10⁻¹	1.215×10³ 5.015×10⁻¹ 3.763×10⁻¹ 7.060×10⁻²	3.318×10²* 5.015×10⁻¹ 1.379 3.470	6.153×10² 4.102×10⁻² 6.079×10⁻² 4.450×10⁻²	6.455×10² 8.238×10⁻³ 1.164×10⁻² 7.738×10⁻³
7×10^{-1} —— 3.287×10³	7.145×10² 5.350×10⁻¹ 6.827×10⁻¹ 3.699×10⁻¹	1.151×10³ 4.128×10⁻¹ 3.269×10⁻¹ 6.831×10⁻²	2.541×10²* 4.128×10⁻¹ 1.482 6.359	5.541×10² 3.603×10⁻² 5.928×10⁻² 5.349×10⁻²	5.820×10² 5.748×10⁻³ 9.005×10⁻³ 7.366×10⁻³
1 —— 4.696×10³	6.292×10² 4.718×10⁻¹ 6.836×10⁻¹ 4.785×10⁻¹	1.076×10³ 3.250×10⁻¹ 2.753×10⁻¹ 6.591×10⁻²	1.883×10²* 3.250×10⁻¹ 1.574 1.230×10¹	4.927×10² 3.128×10⁻² 5.788×10⁻² 6.609×10⁻²	5.175×10² 3.851×10⁻³ 6.785×10⁻³ 7.019×10⁻³
2 —— 9.392×10³	4.824×10² 3.542×10⁻¹ 6.696×10⁻¹ 7.974×10⁻¹	9.207×10² 1.903×10⁻¹ 1.885×10⁻¹ 6.162×10⁻²	1.015×10²* 1.903×10⁻¹ 1.710 4.597×10¹	3.870×10² 2.362×10⁻² 5.564×10⁻² 1.030×10⁻¹	4.059×10² 1.672×10⁻³ 3.756×10⁻³ 6.318×10⁻³
3 —— 1.409×10⁴	4.103×10² 2.918×10⁻¹ 6.484×10⁻¹ 1.067	8.316×10² 1.351×10⁻¹ 1.481×10⁻¹ 5.935×10⁻²	6.969×10¹* 1.351×10⁻¹ 1.767 1.009×10²	3.347×10² 1.991×10⁻² 5.425×10⁻² 1.342×10⁻¹	3.502×10² 1.005×10⁻³ 2.616×10⁻³ 5.912×10⁻³
5 —— 2.348×10⁴	3.339×10² 2.252×10⁻¹ 6.149×10⁻¹ 1.528	7.261×10² 8.584×10⁻² 1.078×10⁻¹ 5.664×10⁻²	4.294×10¹* 8.584×10⁻² 1.823 2.739×10²	2.783×10² 1.598×10⁻² 5.236×10⁻² 1.873×10⁻¹	2.902×10² 5.169×10⁻⁴ 1.624×10⁻³ 5.344×10⁻³
7 —— 3.287×10⁴	2.917×10² 1.886×10⁻¹ 5.896×10⁻¹ 1.920	6.620×10² 6.308×10⁻² 8.688×10⁻² 5.493×10⁻²	3.107×10¹* 6.308×10⁻² 1.851 5.315×10²	2.465×10² 1.378×10⁻² 5.097×10⁻² 2.324×10⁻¹	2.563×10² 3.294×10⁻⁴ 1.172×10⁻³ 4.941×10⁻³
10 —— 4.696×10⁴	2.530×10² 1.555×10⁻¹ 5.604×10⁻¹ 2.426	5.992×10² 4.521×10⁻² 6.880×10⁻² 5.310×10⁻²	2.198×10¹* 4.521×10⁻² 1.875 1.076×10³	2.169×10² 1.174×10⁻² 4.935×10⁻² 2.905×10⁻¹	2.249×10² 2.025×10⁻⁴ 8.210×10⁻⁴ 4.499×10⁻³
20 —— 9.392×10⁴	1.930×10² 1.056×10⁻¹ 4.388×10⁻¹ 3.712	4.919×10² 2.336×10⁻² 4.330×10⁻² 4.958×10⁻²	1.115×10¹* 2.336×10⁻² 1.910 4.257×10³	1.699×10² 8.510×10⁻³ 4.567×10⁻² 4.386×10⁻¹	1.748×10² 7.692×10⁻⁵ 4.012×10⁻⁴ 3.639×10⁻³
30 —— 1.409×10⁵	1.655×10² 8.370×10⁻² 4.611×10⁻¹ 4.664	4.380×10² 1.578×10⁻² 3.286×10⁻² 4.747×10⁻²	7.476* 1.578×10⁻² 1.925 9.543×10³	1.477×10² 7.003×10⁻³ 4.322×10⁻² 5.490×10⁻¹	1.513×10² 4.322×10⁻⁵ 2.605×10⁻⁴ 3.155×10⁻³
50 —— 2.348×10⁵	1.371×10² 6.217×10⁻² 4.135×10⁻¹ 6.097	3.772×10² 9.591×10⁻³ 2.318×10⁻² 4.516×10⁻²	4.509* 9.591×10⁻³ 1.939 2.643×10⁴	1.243×10² 5.441×10⁻³ 3.992×10⁻² 7.164×10⁻¹	1.263×10² 2.073×10⁻⁵ 1.497×10⁻⁴ 2.598×10⁻³
10² —— 4.696×10⁵	1.073×10² 4.129×10⁻² 3.509×10⁻¹ 8.452	3.063×10² 4.853×10⁻³ 1.445×10⁻² 4.266×10⁻²	2.265* 4.853×10⁻³ 1.953 1.055×10⁵	9.901×10¹ 3.820×10⁻³ 3.518×10⁻² 9.945×10⁻¹	9.940×10¹ 7.579×10⁻⁶ 6.953×10⁻⁵ 1.950×10⁻³

Table A3.1. (Continued)

β $\\$ B in 10^2 T	$2p_0/0\ 0\ 1$ $\leftrightarrow$ $1s_0/0\ 0\ 0$	$2p_{-1}/0\ -1\ 0$ $\leftrightarrow$ $1s_0/0\ 0\ 0$	$2p_{+1}/1\ +1\ 0$ $\leftrightarrow$ $1s_0/0\ 0\ 0$	$3p_0/0\ 0\ 3$ $\leftrightarrow$ $1s_0/0\ 0\ 0$	$3p_{-1}/0\ -1\ 2$ $\leftrightarrow$ $1s_0/0\ 0\ 0$
2×10^2 $\\$ — $\\$ 9.392×10^5	8.496×10^1 $\\$ 2.732×10^{-2} $\\$ 2.933×10^{-1} $\\$ 1.126×10^1	2.459×10^2 $\\$ 2.446×10^{-3} $\\$ 9.067×10^{-3} $\\$ 4.155×10^{-2}	$1.136*$ $\\$ 2.446×10^{-3} $\\$ 1.963 $\\$ 4.218×10^5	7.959×10^1 $\\$ 2.655×10^{-3} $\\$ 3.042×10^{-2} $\\$ 1.330	7.851×10^1 $\\$ 2.754×10^{-6} $\\$ 3.198×10^{-5} $\\$ 1.437×10^{-3}
5×10^2 $\\$ — $\\$ 2.348×10^6	6.359×10^1 $\\$ 1.585×10^{-2} $\\$ 2.273×10^{-1} $\\$ 1.559×10^1	1.775×10^2 $\\$ 9.847×10^{-4} $\\$ 5.058×10^{-3} $\\$ 4.446×10^{-2}	$4.552 \times 10^{-1}*$ $\\$ 9.847×10^{-4} $\\$ 1.972 $\\$ 2.637×10^6	6.048×10^1 $\\$ 1.625×10^{-3} $\\$ 2.449×10^{-2} $\\$ 1.856	$5.751 \times 10^1*$ $\\$ 7.209×10^{-7} $\\$ 1.143×10^{-5} $\\$ 9.577×10^{-4}
10^3 $\\$ — $\\$ 4.696×10^6	5.175×10^1 $\\$ 1.056×10^{-2} $\\$ 1.861×10^{-1} $\\$ 1.925×10^1	1.325×10^2 $\\$ 4.940×10^{-4} $\\$ 3.400×10^{-3} $\\$ 5.368×10^{-2}	$2.278 \times 10^{-1}*$ $\\$ 4.940×10^{-4} $\\$ 1.977 $\\$ 1.056×10^7	4.966×10^1 $\\$ 1.119×10^{-3} $\\$ 2.054×10^{-2} $\\$ 2.308	$4.513 \times 10^1*$ $\\$ 2.622×10^{-7} $\\$ 5.298×10^{-6} $\\$ 7.209×10^{-4}

Table A3.1. (Continued)

β $\\$ B in 10^2 T	$3p_{+1}/1\ +1\ 2$ $\leftrightarrow$ $1s_0/0\ 0\ 0$	$2p_0/0\ 0\ 1$ $\leftrightarrow$ $2s_0/0\ 0\ 2$	$2p_{-1}/0\ -1\ 0$ $\leftrightarrow$ $2s_0/0\ 0\ 2$	$2p_{+1}/1\ +1\ 0$ $\leftrightarrow$ $2s_0/0\ 0\ 2$	$3p_0/0\ 0\ 3$ $\leftrightarrow$ $2s_0/0\ 0\ 2$
10^{-4} $\\$ — $\\$ 4.696×10^{-1}	1.026×10^3 $\\$ 8.899×10^{-2} $\\$ 7.912×10^{-2} $\\$ 2.084×10^{-2}	5.699×10^9 $\\$ 9.000 $\\$ 1.440×10^{-6} $\\$ 1.229×10^{-20}	4.563×10^6 $\\$ 9.000 $\\$ 1.798×10^{-3} $\\$ 2.394×10^{-11}	4.565×10^6 $\\$ 9.000 $\\$ 1.798×10^{-3} $\\$ 2.391×10^{-11}	6.565×10^3 $\\$ 3.131 $\\$ 4.348×10^{-1} $\\$ 2.797×10^{-3}
10^{-3} $\\$ — $\\$ 4.696	1.023×10^3 $\\$ 8.918×10^{-2} $\\$ 7.946×10^{-2} $\\$ 2.102×10^{-2}	5.700×10^7 $\\$ 9.000 $\\$ 1.440×10^{-4} $\\$ 1.228×10^{-14}	4.555×10^5 $\\$ 8.997 $\\$ 1.801×10^{-2} $\\$ 2.406×10^{-8}	4.573×10^5 $\\$ 8.997 $\\$ 1.794×10^{-2} $\\$ 2.377×10^{-8}	6.563×10^3 $\\$ 3.133 $\\$ 4.353×10^{-1} $\\$ 2.801×10^{-3}
7×10^{-3} $\\$ — $\\$ 3.287×10^1	1.003×10^3 $\\$ 9.673×10^{-2} $\\$ 8.796×10^{-2} $\\$ 2.425×10^{-2}	1.176×10^6 $\\$ 9.012 $\\$ 6.987×10^{-3} $\\$ 1.399×10^{-9}	6.431×10^4 $\\$ 8.844 $\\$ 1.254×10^{-1} $\\$ 8.402×10^{-6}	6.611×10^4 $\\$ 8.844 $\\$ 1.220×10^{-1} $\\$ 7.735×10^{-6}	6.474×10^3 $\\$ 3.219 $\\$ 4.532×10^{-1} $\\$ 2.997×10^{-3}
10^{-2} $\\$ — $\\$ 4.696×10^1	9.895×10^2 $\\$ 1.023×10^{-1} $\\$ 9.430×10^{-2} $\\$ 2.668×10^{-2}	5.828×10^5 $\\$ 9.024 $\\$ 1.412×10^{-2} $\\$ 1.152×10^{-8}	4.477×10^4 $\\$ 8.693 $\\$ 1.770×10^{-1} $\\$ 2.447×10^{-5}	4.654×10^4 $\\$ 8.693 $\\$ 1.703×10^{-1} $\\$ 2.179×10^{-5}	6.399×10^3 $\\$ 3.278 $\\$ 4.672×10^{-1} $\\$ 3.161×10^{-3}
2×10^{-2} $\\$ — $\\$ 9.392×10^1	9.403×10^2 $\\$ 1.156×10^{-1} $\\$ 1.121×10^{-1} $\\$ 3.511×10^{-2}	1.543×10^5 $\\$ 9.097 $\\$ 5.375×10^{-2} $\\$ 6.252×10^{-7}	2.204×10^4 $\\$ 7.955 $\\$ 3.291×10^{-1} $\\$ 1.878×10^{-4}	2.366×10^4 $\\$ 7.955 $\\$ 3.066×10^{-1} $\\$ 1.518×10^{-4}	6.133×10^3 $\\$ 3.435 $\\$ 5.107×10^{-1} $\\$ 3.762×10^{-3}
3×10^{-2} $\\$ — $\\$ 1.409×10^2	8.909×10^2 $\\$ 1.123×10^{-1} $\\$ 1.150×10^{-1} $\\$ 4.014×10^{-2}	7.409×10^4 $\\$ 9.220 $\\$ 1.135×10^{-1} $\\$ 5.727×10^{-6}	1.453×10^4 $\\$ 7.070 $\\$ 4.435×10^{-1} $\\$ 5.819×10^{-4}	1.596×10^4 $\\$ 7.070 $\\$ 4.039×10^{-1} $\\$ 4.395×10^{-4}	5.947×10^3 $\\$ 3.573 $\\$ 5.479×10^{-1} $\\$ 4.293×10^{-3}
5×10^{-2} $\\$ — $\\$ 2.348×10^2	$8.050 \times 10^{2}*$ $\\$ 8.011×10^{-2} $\\$ 9.073×10^{-2} $\\$ 3.879×10^{-2}	3.183×10^4 $\\$ 9.591 $\\$ 2.747×10^{-1} $\\$ 7.512×10^{-5}	8.650×10^3 $\\$ 5.387 $\\$ 5.679×10^{-1} $\\$ 2.103×10^{-3}	9.661×10^3 $\\$ 5.387 $\\$ 5.085×10^{-1} $\\$ 1.509×10^{-3}	5.830×10^3 $\\$ 4.126 $\\$ 6.453×10^{-1} $\\$ 5.261×10^{-3}

Table A3.1. (Continued)

β / B in 10^2 T	$3p_{+1}/1\ +1\ 2$ ↔ $1s_0/0\ 0\ 0$	$2p_0/0\ 0\ 1$ ↔ $2s_0/0\ 0\ 2$	$2p_{-1}/0\ -1\ 0$ ↔ $2s_0/0\ 0\ 2$	$2p_{+1}/1\ +1\ 0$ ↔ $2s_0/0\ 0\ 2$	$3p_0/0\ 0\ 3$ ↔ $2s_0/0\ 0\ 2$
7×10^{-2} ——— ——— 3.287×10^2	7.354×10^2 5.609×10^{-2} 6.954×10^{-2} 3.563×10^{-2}	1.953×10^4 1.004×10^1 4.685×10^{-1} 3.402×10^{-4}	6.223×10^3 4.024 5.895×10^{-1} 4.218×10^{-3}	6.846×10^3 4.024 5.359×10^{-1} 3.169×10^{-3}	5.909×10^3 5.118 7.896×10^{-1} 6.266×10^{-3}
10^{-1} ——— ——— 4.696×10^2	$6.518 \times 10^{2*}$ 3.865×10^{-2} 5.406×10^{-2} 3.526×10^{-2}	1.259×10^4 1.058×10^1 7.660×10^{-1} 1.338×10^{-3}	4.494×10^3 2.557 5.187×10^{-1} 7.117×10^{-3}	4.636×10^3 2.557 5.028×10^{-1} 6.484×10^{-3}	6.155×10^3 7.115 1.054 7.708×10^{-3}
2×10^{-1} ——— ——— 9.392×10^2	$4.749 \times 10^{2*}$ 2.012×10^{-2} 3.863×10^{-2} 4.746×10^{-2}	7.003×10^3 1.037×10^1 1.350 7.626×10^{-3}	2.651×10^3 6.574×10^{-1} 2.261×10^{-1} 8.916×10^{-3}	$2.003 \times 10^{3*}$ 6.574×10^{-1} 2.993×10^{-1} 2.066×10^{-2}	6.710×10^3 1.244×10^1 1.690 1.041×10^{-2}
3×10^{-1} ——— ——— 1.409×10^3	$3.756 \times 10^{2*}$ 1.374×10^{-2} 3.336×10^{-2} 6.552×10^{-2}	5.634×10^3 9.315 1.508 1.316×10^{-2}	2.058×10^3 2.383×10^{-1} 1.056×10^{-1} 6.908×10^{-3}	$1.207 \times 10^{3*}$ 2.383×10^{-1} 1.801×10^{-1} 3.427×10^{-2}	6.746×10^3 1.414×10^1 1.912 1.164×10^{-2}
5×10^{-1} ——— ——— 2.348×10^3	$2.674 \times 10^{2*}$ 8.238×10^{-3} 2.809×10^{-2} 1.089×10^{-1}	4.580×10^3 7.944 1.582 2.089×10^{-2}	1.542×10^3 8.129×10^{-2} 4.806×10^{-2} 5.598×10^{-3}	$6.482 \times 10^{2*}$ 8.129×10^{-2} 1.144×10^{-1} 7.541×10^{-2}	6.490×10^3 1.444×10^1 2.028 1.334×10^{-2}
7×10^{-1} ——— ——— 3.287×10^3	$2.090 \times 10^{2*}$ 5.748×10^{-3} 2.508×10^{-2} 1.592×10^{-1}	4.092×10^3 7.152 1.594 2.637×10^{-2}	1.290×10^3 4.157×10^{-2} 2.940×10^{-2} 4.898×10^{-3}	$4.363 \times 10^{2*}$ 4.157×10^{-2} 8.688×10^{-2} 1.265×10^{-1}	6.225×10^3 1.400×10^1 2.051 1.466×10^{-2}
1 ——— ——— 4.696×10^3	$1.584 \times 10^{2*}$ 3.851×10^{-3} 2.217×10^{-2} 2.450×10^{-1}	3.683×10^3 6.429 1.592 3.250×10^{-2}	1.074×10^3 2.108×10^{-2} 1.790×10^{-2} 4.302×10^{-3}	$2.898 \times 10^{2*}$ 2.108×10^{-2} 6.632×10^{-2} 2.189×10^{-1}	5.914×10^3 1.330×10^1 2.051 1.625×10^{-2}
2 ——— ——— 9.392×10^3	$8.906 \times 10^{1*}$ 1.672×10^{-3} 1.712×10^{-2} 5.980×10^{-1}	3.104×10^3 5.334 1.567 4.507×10^{-2}	7.638×10^2 6.062×10^{-3} 7.236×10^{-2} 3.437×10^{-3}	$1.341 \times 10^{2*}$ 6.062×10^{-3} 4.120×10^{-1} 6.347×10^{-1}	5.310×10^3 1.175×10^1 2.017 1.982×10^{-2}
3 ——— ——— 1.409×10^4	$6.249 \times 10^{1*}$ 1.005×10^{-3} 1.466×10^{-2} 1.041	2.858×10^3 4.850 1.548 5.250×10^{-2}	6.311×10^2 3.023×10^{-3} 4.368×10^{-3} 3.039×10^{-3}	$8.649 \times 10^{1*}$ 3.023×10^{-3} 3.188×10^{-2} 1.181	4.984×10^3 1.081×10^1 1.977 2.206×10^{-2}
5 ——— ——— 2.348×10^4	$3.944 \times 10^{1*}$ 5.169×10^{-4} 1.195×10^{-2} 2.129	2.624×10^3 4.371 1.519 6.115×10^{-2}	5.004×10^2 1.290×10^{-3} 2.350×10^{-3} 2.601×10^{-3}	$5.022 \times 10^{1*}$ 1.290×10^{-3} 2.341×10^{-2} 2.572	4.603×10^3 9.712 1.924 2.516×10^{-2}
7 ——— ——— 3.287×10^4	$2.892 \times 10^{1*}$ 3.294×10^{-4} 1.039×10^{-2} 3.441	2.505×10^3 4.119 1.500 6.622×10^{-2}	4.315×10^2 7.440×10^{-4} 1.572×10^{-3} 2.339×10^{-3}	$3.526 \times 10^{1*}$ 7.440×10^{-4} 1.924×10^{-2} 4.287	4.376×10^3 9.046 1.885 2.728×10^{-2}
10 ——— ——— 4.696×10^4	$2.072 \times 10^{1*}$ 2.025×10^{-4} 8.912×10^{-3} 5.754	2.405×10^3 3.899 1.478 7.084×10^{-2}	3.704×10^2 4.184×10^{-4} 1.030×10^{-3} 2.080×10^{-3}	$2.432 \times 10^{1*}$ 4.184×10^{-4} 1.569×10^{-2} 7.351	4.154×10^3 8.389 1.841 2.958×10^{-2}

Table A3.1. (Continued)

β / B in 10^2 T	$3p_{+1}/1\ +1\ 2$ $\leftrightarrow$ $1s_0/0\ 0\ 0$	$2p_0/0\ 0\ 1$ $\leftrightarrow$ $2s_0/0\ 0\ 2$	$2p_{-1}/0\ -1\ 0$ $\leftrightarrow$ $2s_0/0\ 0\ 2$	$2p_{+1}/1\ +1\ 0$ $\leftrightarrow$ $2s_0/0\ 0\ 2$	$3p_0/0\ 0\ 3$ $\leftrightarrow$ $2s_0/0\ 0\ 2$
20 ── 9.392×10^4	1.071×10^1* 7.692×10^{-5} 6.549×10^{-3} 1.582×10^1	2.273×10^3 3.580 1.436 7.702×10^{-2}	2.786×10^2 1.390×10^{-4} 4.547×10^{-4} 1.624×10^{-3}	1.190×10^1* 1.390×10^{-4} 1.066×10^{-2} 2.085×10^1	3.774×10^3 7.248 1.751 3.408×10^{-2}
30 ── 1.409×10^5	7.242* 4.322×10^{-5} 5.441×10^{-3} 2.875×10^1	2.228×10^3 3.445 1.410 7.874×10^{-2}	2.377×10^2 7.353×10^{-5} 2.821×10^{-4} 1.383×10^{-3}	7.858* 7.353×10^{-5} 8.532×10^{-3} 3.829×10^1	3.580×10^3 6.657 1.696 3.667×10^{-2}
50 ── 2.348×10^5	4.405* 2.073×10^{-5} 4.291×10^{-3} 6.130×10^1	2.198×10^3 3.317 1.377 7.897×10^{-2}	1.962×10^2 3.318×10^{-5} 1.543×10^{-4} 1.111×10^{-3}	4.672* 3.318×10^{-5} 6.475×10^{-3} 8.218×10^1	3.361×10^3 5.986 1.624 3.984×10^{-2}
10^2 ── 4.696×10^5	2.231* 7.579×10^{-6} 3.098×10^{-3} 1.725×10^2	2.196×10^3 3.197 1.328 7.625×10^{-2}	1.535×10^2 1.138×10^{-5} 6.760×10^{-5} 7.946×10^{-4}	2.316* 1.138×10^{-5} 4.480×10^{-3} 2.314×10^2	3.105×10^3 5.191 1.525 4.382×10^{-2}
2×10^2 ── 9.392×10^5	1.125* 2.754×10^{-6} 2.233×10^{-3} 4.892×10^2	2.231×10^3 3.120 1.275 7.099×10^{-2}	1.227×10^2 3.943×10^{-6} 2.931×10^{-5} 5.397×10^{-4}	1.152* 3.943×10^{-6} 3.122×10^{-3} 6.523×10^2	2.889×10^3 4.517 1.426 4.734×10^{-2}
5×10^2 ── 2.348×10^6	4.528×10^{-1}* 7.209×10^{-7} 1.452×10^{-3} 1.962×10^3	2.318×10^3 3.055 1.202 6.195×10^{-2}	9.501×10^1 9.846×10^{-7} 9.449×10^{-6} 2.901×10^{-4}	4.586×10^{-1}* 9.846×10^{-7} 1.958×10^{-3} 2.579×10^3	2.654×10^3 3.784 1.300 5.114×10^{-2}
10^3 ── 4.696×10^6	2.270×10^{-1}* 2.622×10^{-7} 1.053×10^{-3} 5.660×10^3	2.408×10^3 3.022 1.145 5.472×10^{-2}	8.204×10^1 3.482×10^{-7} 3.870×10^{-6} 1.594×10^{-4}	2.288×10^{-1}* 3.482×10^{-7} 1.388×10^{-3} 7.343×10^3	2.508×10^3 3.333 1.212 5.339×10^{-2}

Table A3.1. (Continued)

β / B in 10^2 T	$3p_{-1}/0\ -1\ 2$ $\leftrightarrow$ $2s_0/0\ 0\ 2$	$3p_{+1}/1\ +1\ 2$ $\leftrightarrow$ $2s_0/0\ 0\ 2$	$3d_{-1}/0\ -1\ 1$ $\leftrightarrow$ $2p_0/0\ 0\ 1$	$3d_{+1}/1\ +1\ 1$ $\leftrightarrow$ $2p_0/0\ 0\ 1$	$3d_{-1}/0\ -1\ 1$ $\leftrightarrow$ $2p_{-1}/0\ -1\ 0$
10^{-4} ── 4.696×10^{-1}	6.574×10^3 3.131 4.342×10^{-1} 2.785×10^{-3}	6.555×10^3 3.131 4.355×10^{-1} 2.809×10^{-3}	6.574×10^3 4.509 6.253×10^{-1} 4.010×10^{-3}	6.555×10^3 4.509 6.271×10^{-1} 4.044×10^{-3}	6.565×10^3 4.509 6.262×10^{-1} 4.027×10^{-3}
10^{-3} ── 4.696	6.655×10^3 3.134 4.294×10^{-1} 2.686×10^{-3}	6.466×10^3 3.134 4.419×10^{-1} 2.929×10^{-3}	6.658×10^3 4.511 6.177×10^{-1} 3.862×10^{-3}	6.469×10^3 4.511 6.358×10^{-1} 4.210×10^{-3}	6.562×10^3 4.511 6.267×10^{-1} 4.033×10^{-3}
7×10^{-3} ── 3.287×10^1	7.006×10^3 3.243 4.221×10^{-1} 2.383×10^{-3}	5.766×10^3 3.243 5.128×10^{-1} 4.274×10^{-3}	7.141×10^3 4.593 5.865×10^{-1} 3.187×10^{-3}	5.858×10^3 4.593 7.150×10^{-1} 5.774×10^{-3}	6.462×10^3 4.598 6.488×10^{-1} 4.305×10^{-3}

A3.1 Wavelengths, Strengths, and Transition Rates

Table A3.1. (Continued)

β / B in 10^2 T	$3p_{-1}/0\ -1\ 2$ $\leftrightarrow$ $2s_0/0\ 0\ 2$	$3p_{+1}/1\ +1\ 2$ $\leftrightarrow$ $2s_0/0\ 0\ 2$	$3d_{-1}/0\ -1\ 1$ $\leftrightarrow$ $2p_0/0\ 0\ 1$	$3d_{+1}/1\ +1\ 1$ $\leftrightarrow$ $2p_0/0\ 0\ 1$	$3d_{-1}/0\ -1\ 1$ $\leftrightarrow$ $2p_{-1}/0\ -1\ 0$
10^{-2} ⎯ 4.696×10^1	7.072×10^3 3.299 4.253×10^{-1} 2.356×10^{-3}	5.399×10^3 3.299 5.571×10^{-1} 5.297×10^{-3}	7.334×10^3 4.644 5.774×10^{-1} 2.975×10^{-3}	5.550×10^3 4.644 7.630×10^{-1} 6.864×10^{-3}	6.370×10^3 4.656 6.664×10^{-1} 4.550×10^{-3}
2×10^{-2} ⎯ 9.392×10^1	7.015×10^3 3.264 4.242×10^{-1} 2.389×10^{-3}	4.344×10^3 3.264 6.850×10^{-1} 1.006×10^{-2}	7.793×10^3 4.688 5.485×10^{-1} 2.503×10^{-3}	4.630×10^3 4.688 9.231×10^{-1} 1.193×10^{-2}	5.980×10^3 4.749 7.241×10^{-1} 5.611×10^{-3}
3×10^{-2} ⎯ 1.409×10^2	6.853×10^3 2.940 3.911×10^{-1} 2.308×10^{-3}	3.605×10^3 2.940 7.435×10^{-1} 1.586×10^{-2}	8.066×10^3 4.543 5.136×10^{-1} 2.188×10^{-3}	3.915×10^3 4.543 1.059 1.913×10^{-2}	5.577×10^3 4.687 7.662×10^{-1} 6.825×10^{-3}
5×10^{-2} ⎯ 2.348×10^2	6.807×10^3 2.187 2.930×10^{-1} 1.752×10^{-3}	$2.732 \times 10^3 *$ 2.187 7.298×10^{-1} 2.709×10^{-2}	8.341×10^3 4.045 4.421×10^{-1} 1.761×10^{-3}	2.950×10^3 4.045 1.251 3.981×10^{-2}	4.900×10^3 4.377 8.143×10^{-1} 9.397×10^{-3}
7×10^{-2} ⎯ 3.287×10^2	7.105×10^3 1.832 2.351×10^{-1} 1.290×10^{-3}	2.235×10^3 1.832 7.476×10^{-1} 4.149×10^{-2}	8.457×10^3 3.548 3.825×10^{-1} 1.481×10^{-3}	$2.353 \times 10^3 *$ 3.548 1.375 6.882×10^{-2}	4.391×10^3 4.038 8.385×10^{-1} 1.205×10^{-2}
10^{-1} ⎯ 4.696×10^2	7.826×10^3 1.698 1.978×10^{-1} 8.950×10^{-4}	1.767×10^3 1.698 8.762×10^{-1} 7.779×10^{-2}	8.520×10^3 2.951 3.159×10^{-1} 1.206×10^{-3}	$1.800 \times 10^3 *$ 2.951 1.495 1.279×10^{-1}	3.839×10^3 3.606 8.564×10^{-1} 1.611×10^{-2}
2×10^{-1} ⎯ 9.392×10^2	1.015×10^4 1.493 1.341×10^{-1} 3.606×10^{-4}	$1.026 \times 10^3 *$ 1.493 1.327 3.497×10^{-1}	8.512×10^3 1.854 1.986×10^{-1} 7.596×10^{-4}	$1.006 \times 10^3 *$ 1.854 1.680 4.600×10^{-1}	2.842×10^3 2.717 8.719×10^{-1} 2.992×10^{-2}
3×10^{-1} ⎯ 1.409×10^3	1.151×10^4 1.200 9.511×10^{-2} 1.990×10^{-4}	$7.135 \times 10^2 *$ 1.200 1.534 8.351×10^{-1}	8.474×10^3 1.347 1.449×10^{-1} 5.591×10^{-4}	$6.980 \times 10^2 *$ 1.347 1.759 1.000	2.345×10^3 2.242 8.716×10^{-1} 4.393×10^{-2}
5×10^{-1} ⎯ 2.348×10^3	1.282×10^4 8.190×10^{-1} 5.820×10^{-2} 9.799×10^{-5}	$4.407 \times 10^2 *$ 8.190×10^{-1} 1.694 2.417	8.445×10^3 8.704×10^{-1} 9.398×10^{-2} 3.652×10^{-4}	$4.330 \times 10^2 *$ 8.704×10^{-1} 1.833 2.709	1.823×10^3 1.728 8.639×10^{-1} 7.201×10^{-2}
7×10^{-1} ⎯ 3.287×10^3	1.347×10^4 6.148×10^{-1} 4.161×10^{-2} 6.352×10^{-5}	$3.183 \times 10^2 *$ 6.148×10^{-1} 1.761 4.818	8.458×10^3 6.435×10^{-1} 6.938×10^{-2} 2.687×10^{-4}	$3.139 \times 10^2 *$ 6.435×10^{-1} 1.869 5.258	1.540×10^3 1.445 8.555×10^{-1} 9.994×10^{-2}
1 ⎯ 4.696×10^3	1.399×10^4 4.463×10^{-1} 2.908×10^{-2} 4.117×10^{-5}	$2.245 \times 10^2 *$ 4.463×10^{-1} 1.812 9.961	8.512×10^3 4.630×10^{-1} 4.960×10^{-2} 1.897×10^{-4}	$2.222 \times 10^2 *$ 4.630×10^{-1} 1.899 1.066×10^1	1.286×10^3 1.189 8.429×10^{-1} 1.411×10^{-1}
2 ⎯ 9.392×10^3	1.458×10^4 2.331×10^{-1} 1.458×10^{-2} 1.902×10^{-5}	$1.132 \times 10^2 *$ 2.331×10^{-1} 1.878 4.060×10^1	8.755×10^3 2.397×10^{-1} 2.496×10^{-2} 9.023×10^{-5}	$1.126 \times 10^2 *$ 2.397×10^{-1} 1.940 4.239×10^1	9.081×10^2 8.068×10^{-1} 8.101×10^{-1} 2.722×10^{-1}

248   Appendix 3: Electromagnetic Transition Probabilities

Table A3.1. (Continued)

β / B in 10^2 T	$3p_{-1}/0\ -1\ 2$ $\leftrightarrow$ $2s_0/0\ 0\ 2$	$3p_{+1}/1\ +1\ 2$ $\leftrightarrow$ $2s_0/0\ 0\ 2$	$3d_{-1}/0\ -1\ 1$ $\leftrightarrow$ $2p_0/0\ 0\ 1$	$3d_{+1}/1\ +1\ 1$ $\leftrightarrow$ $2p_0/0\ 0\ 1$	$3d_{-1}/0\ -1\ 1$ $\leftrightarrow$ $2p_{-1}/0\ -1\ 0$
3 — 1.409×10^4	1.464×10^4 1.578×10^{-1} 9.829×10^{-3} 1.271×10^{-5}	$7.567\times10^{1*}$ 1.578×10^{-1} 1.902 9.203×10^{1}	8.975×10^3 1.618×10^{-1} 1.644×10^{-2} 5.655×10^{-5}	$7.542\times10^{1*}$ 1.618×10^{-1} 1.956 9.529×10^{1}	7.429×10^2 6.405×10^{-1} 7.861×10^{-1} 3.947×10^{-1}
5 — 2.348×10^4	1.423×10^4 9.601×10^{-2} 6.150×10^{-3} 8.413×10^{-6}	$4.549\times10^{1*}$ 9.601×10^{-2} 1.924 2.576×10^2	9.278×10^3 9.816×10^{-2} 9.646×10^{-3} 3.105×10^{-5}	$4.541\times10^{1*}$ 9.816×10^{-2} 1.971 2.647×10^2	5.797×10^2 4.773×10^{-1} 7.508×10^{-1} 6.192×10^{-1}
7 — 3.287×10^4	1.365×10^4 6.904×10^{-2} 4.612×10^{-3} 6.857×10^{-6}	$3.252\times10^{1*}$ 6.904×10^{-2} 1.936 5.072×10^2	9.432×10^3 7.046×10^{-2} 6.812×10^{-3} 2.121×10^{-5}	$3.249\times10^{1*}$ 7.046×10^{-2} 1.977 5.192×10^2	4.940×10^2 3.926×10^{-1} 7.246×10^{-1} 8.226×10^{-1}
10 — 4.696×10^4	1.272×10^4 4.861×10^{-2} 3.483×10^{-3} 5.962×10^{-6}	$2.278\times10^{1*}$ 4.861×10^{-2} 1.946 1.039×10^3	9.469×10^3 4.952×10^{-2} 4.768×10^{-3} 1.473×10^{-5}	$2.276\times10^{1*}$ 4.952×10^{-2} 1.983 1.061×10^3	4.185×10^2 3.188×10^{-1} 6.945×10^{-1} 1.099
20 — 9.392×10^4	1.010×10^4 2.450×10^{-2} 2.212×10^{-3} 6.015×10^{-6}	$1.140\times10^{1*}$ 2.450×10^{-2} 1.960 4.182×10^3	8.796×10^3 2.488×10^{-2} 2.578×10^{-3} 9.237×10^{-6}	$1.139\times10^{1*}$ 2.488×10^{-2} 1.991 4.248×10^3	3.065×10^2 2.117×10^{-1} 6.298×10^{-1} 1.858
30 — 1.409×10^5	8.265×10^3 1.639×10^{-2} 1.808×10^{-3} 7.335×10^{-6}	$7.599*$ 1.639×10^{-2} 1.966 9.436×10^3	7.796×10^3 1.661×10^{-2} 1.943×10^{-3} 8.860×10^{-6}	$7.599*$ 1.661×10^{-2} 1.993 9.566×10^3	2.573×10^2 1.662×10^{-1} 5.890×10^{-1} 2.466
50 — 2.348×10^5	6.010×10^3 9.868×10^{-3} 1.497×10^{-3} 1.148×10^{-5}	$4.560*$ 9.868×10^{-3} 1.973 2.629×10^4	6.064×10^3 9.981×10^{-3} 1.501×10^{-3} 1.131×10^{-5}	$4.560*$ 9.981×10^{-3} 1.996 2.659×10^4	2.080×10^2 1.222×10^{-1} 5.358×10^{-1} 3.432
10^2 — 4.696×10^5	3.538×10^3 4.951×10^{-3} 1.276×10^{-3} 2.825×10^{-5}	$2.280*$ 4.951×10^{-3} 1.979 1.055×10^5	3.689×10^3 4.996×10^{-3} 1.235×10^{-3} 2.514×10^{-5}	$2.280*$ 4.996×10^{-3} 1.997 1.065×10^5	1.580×10^2 8.018×10^{-2} 4.627×10^{-1} 5.135
2×10^2 — 9.392×10^5	1.929×10^3 2.481×10^{-3} 1.173×10^{-3} 8.738×10^{-5}	$1.140*$ 2.481×10^{-3} 1.984 4.229×10^5	2.003×10^3 2.499×10^{-3} 1.138×10^{-3} 7.863×10^{-5}	$1.140*$ 2.499×10^{-3} 1.998 4.259×10^5	1.219×10^2 5.241×10^{-2} 3.920×10^{-1} 7.310
5×10^2 — 2.348×10^6	$8.123\times10^{2*}$ 9.948×10^{-4} 1.117×10^{-3} 4.691×10^{-4}	$4.561\times10^{-1*}$ 9.948×10^{-4} 1.988 2.648×10^6	$8.291\times10^{2*}$ 9.999×10^{-4} 1.100×10^{-3} 4.432×10^{-4}	$4.561\times10^{-1*}$ 9.999×10^{-4} 1.999 2.662×10^6	$8.850\times10^{1*}$ 2.985×10^{-2} 3.075×10^{-1} 1.088×10^1
10^3 — 4.696×10^6	$4.128\times10^{2*}$ 4.980×10^{-4} 1.100×10^{-3} 1.789×10^{-3}	$2.281\times10^{-1*}$ 4.980×10^{-4} 1.991 1.061×10^7	$4.175\times10^{2*}$ 5.000×10^{-4} 1.092×10^{-3} 1.737×10^{-3}	$2.281\times10^{-1*}$ 5.000×10^{-4} 1.999 1.065×10^7	$7.057\times10^{1*}$ 1.955×10^{-2} 2.526×10^{-1} 1.405×10^1

A3.1 Wavelengths, Strengths, and Transition Rates   249

Table A3.1. (Continued)

β / B in 10^2 T	$3d_{-2}/0\ -2\ 0$ $\leftrightarrow$ $2p_{-1}/0\ -1\ 0$	$3d_{+2}/2\ +2\ 0$ $\leftrightarrow$ $2p_{+1}/1\ +1\ 0$	$3d_{-1}/0\ -1\ 1$ $\leftrightarrow$ $3p_0/0\ 0\ 3$	$3d_{+1}/1\ +1\ 1$ $\leftrightarrow$ $3p_0/0\ 0\ 3$	$3d_{-1}/0\ -1\ 1$ $\leftrightarrow$ $3p_{-1}/0\ -1\ 2$
10^{-4} — — 4.696×10^{-1}	6.574×10^3 9.017 1.251 8.019×10^{-3}	6.555×10^3 9.017 1.255 8.088×10^{-3}	4.564×10^6 2.025×10^1 4.045×10^{-3} 5.382×10^{-11}	4.564×10^6 2.025×10^1 4.045×10^{-3} 5.382×10^{-11}	1.266×10^9 2.025×10^1 1.458×10^{-5} 2.519×10^{-18}
10^{-3} — — 4.696	6.656×10^3 9.024 1.236 7.731×10^{-3}	6.468×10^3 9.024 1.272 8.427×10^{-3}	4.564×10^5 2.018×10^1 4.031×10^{-2} 5.363×10^{-8}	4.564×10^5 2.018×10^1 4.031×10^{-2} 5.364×10^{-8}	1.263×10^7 2.025×10^1 1.455×10^{-3} 2.503×10^{-12}
7×10^{-3} — — 3.287×10^1	7.083×10^3 9.259 1.192 6.585×10^{-3}	5.818×10^3 9.259 1.451 1.188×10^{-2}	6.552×10^4 1.738×10^1 2.419×10^{-1} 1.562×10^{-5}	6.488×10^4 1.738×10^1 2.443×10^{-1} 1.609×10^{-5}	2.819×10^5 2.015×10^1 6.519×10^{-2} 2.274×10^{-7}
10^{-2} — — 4.696×10^1	7.220×10^3 9.388 1.186 6.302×10^{-3}	5.485×10^3 9.388 1.561 1.437×10^{-2}	4.622×10^4 1.529×10^1 3.016×10^{-1} 3.912×10^{-5}	4.507×10^4 1.529×10^1 3.093×10^{-1} 4.218×10^{-5}	1.481×10^5 2.021×10^1 1.245×10^{-1} 1.573×10^{-6}
2×10^{-2} — — 9.392×10^1	7.431×10^3 9.429 1.157 5.804×10^{-3}	4.500×10^3 9.429 1.910 2.614×10^{-2}	2.427×10^4 9.153 3.439×10^{-1} 1.618×10^{-4}	2.153×10^4 9.153 3.875×10^{-1} 2.316×10^{-4}	4.829×10^4 2.145×10^1 4.051×10^{-1} 4.814×10^{-5}
3×10^{-2} — — 1.409×10^2	7.435×10^3 9.008 1.105 5.538×10^{-3}	$3.760\times10^{3*}$ 9.008 2.184 4.282×10^{-2}	1.734×10^4 5.451 2.867×10^{-1} 2.642×10^{-4}	1.355×10^4 5.451 3.668×10^{-1} 5.535×10^{-4}	2.822×10^4 2.360×10^1 7.626×10^{-1} 2.653×10^{-4}
5×10^{-2} — — 2.348×10^2	7.229×10^3 7.821 9.863×10^{-1} 5.230×10^{-3}	$2.798\times10^{3*}$ 7.821 2.548 9.024×10^{-2}	1.204×10^4 2.281 1.728×10^{-1} 3.303×10^{-4}	7.350×10^3 2.281 2.830×10^{-1} 1.451×10^{-3}	1.711×10^4 2.520×10^1 1.343 1.271×10^{-3}
7×10^{-2} — — 3.287×10^2	6.959×10^3 6.752 8.847×10^{-1} 5.061×10^{-3}	$2.220\times10^{3*}$ 6.752 2.774 1.560×10^{-1}	9.787×10^3 1.190 1.109×10^{-1} 3.208×10^{-4}	$4.888\times10^{3*}$ 1.190 2.220×10^{-1} 2.575×10^{-3}	1.357×10^4 2.362×10^1 1.587 2.388×10^{-3}
10^{-1} — — 4.696×10^2	6.576×10^3 5.545 7.688×10^{-1} 4.927×10^{-3}	$1.694\times10^{3*}$ 5.545 2.985 2.882×10^{-1}	8.033×10^3 5.861×10^{-1} 6.652×10^{-2} 2.856×10^{-4}	$3.187\times10^{3*}$ 5.861×10^{-1} 1.677×10^{-1} 4.573×10^{-3}	1.113×10^4 2.081×10^1 1.704 3.809×10^{-3}
2×10^{-1} — — 9.392×10^2	5.657×10^3 3.445 5.553×10^{-1} 4.808×10^{-3}	$9.494\times10^{2*}$ 3.445 3.308 1.017	5.736×10^3 1.517×10^{-1} 2.412×10^{-2} 2.032×10^{-4}	$1.424\times10^{3*}$ 1.517×10^{-1} 9.714×10^{-2} 1.327×10^{-2}	8.077×10^3 1.556×10^1 1.756 7.460×10^{-3}
3×10^{-1} — — 1.409×10^3	5.082×10^3 2.503 4.489×10^{-1} 4.816×10^{-3}	$6.616\times10^{2*}$ 2.503 3.449 2.184	4.814×10^3 7.082×10^{-2} 1.341×10^{-2} 1.604×10^{-4}	$9.034\times10^{2*}$ 7.082×10^{-2} 7.148×10^{-2} 2.427×10^{-2}	6.832×10^3 1.312×10^1 1.751 1.040×10^{-2}
5×10^{-1} — — 2.348×10^3	4.380×10^3 1.626 3.385×10^{-1} 4.890×10^{-3}	$4.133\times10^{2*}$ 1.626 3.586 5.817	3.937×10^3 2.778×10^{-2} 6.434×10^{-3} 1.150×10^{-4}	$5.162\times10^{2*}$ 2.778×10^{-2} 4.906×10^{-2} 5.101×10^{-2}	5.621×10^3 1.068×10^1 1.732 1.519×10^{-2}

250   Appendix 3: Electromagnetic Transition Probabilities

Table A3.1. (Continued)

β / B in 10^2 T	$3d_{-2}/0\ -2\ 0$ $\leftrightarrow$ $2p_{-1}/0\ -1\ 0$	$3d_{+2}/2\ +2\ 0$ $\leftrightarrow$ $2p_{+1}/1\ +1\ 0$	$3d_{-1}/0\ -1\ 1$ $\leftrightarrow$ $3p_0/0\ 0\ 3$	$3d_{+1}/1\ +1\ 1$ $\leftrightarrow$ $3p_0/0\ 0\ 3$	$3d_{-1}/0\ -1\ 1$ $\leftrightarrow$ $3p_{-1}/0\ -1\ 2$
7×10^{-1} — — 3.287×10^3	3.946×10^3 1.208 2.791×10^{-1} 4.967×10^{-3}	$3.011\times 10^{2*}$ 1.208 3.657 1.118×10^1	3.487×10^3 1.516×10^{-2} 3.964×10^{-3} 9.035×10^{-5}	$3.596\times 10^{2*}$ 1.516×10^{-2} 3.844×10^{-2} 8.237×10^{-2}	4.990×10^3 9.385 1.715 1.908×10^{-2}
1 — — 4.696×10^3	3.520×10^3 8.741×10^{-1} 2.264×10^{-1} 5.065×10^{-3}	$2.143\times 10^{2*}$ 8.741×10^{-1} 3.719 2.244×10^1	3.095×10^3 8.017×10^{-3} 2.362×10^{-3} 6.832×10^{-5}	$2.464\times 10^{2*}$ 8.017×10^{-3} 2.968×10^{-2} 1.355×10^{-1}	4.435×10^3 8.241 1.694 2.387×10^{-2}
2 — — 9.392×10^3	2.797×10^3 4.578×10^{-1} 1.493×10^{-1} 5.287×10^{-3}	$1.096\times 10^{2*}$ 4.578×10^{-1} 3.808 8.780×10^1	2.523×10^3 2.329×10^{-3} 8.415×10^{-4} 3.662×10^{-5}	$1.195\times 10^{2*}$ 2.329×10^{-3} 1.777×10^{-2} 3.448×10^{-1}	3.615×10^3 6.547 1.651 3.500×10^{-2}
3 — — 1.409×10^4	2.438×10^3 3.111×10^{-1} 1.164×10^{-1} 5.423×10^{-3}	$7.376\times 10^{1*}$ 3.111×10^{-1} 3.845 1.958×10^2	2.277×10^3 1.124×10^{-3} 4.498×10^{-4} 2.404×10^{-5}	$7.869\times 10^{1*}$ 1.124×10^{-3} 1.302×10^{-2} 5.826×10^{-1}	3.260×10^3 5.810 1.625 4.238×10^{-2}
5 — — 2.348×10^4	2.047×10^3 1.900×10^{-1} 8.463×10^{-2} 5.594×10^{-3}	$4.464\times 10^{1*}$ 1.900×10^{-1} 3.881 5.397×10^2	2.038×10^3 4.430×10^{-4} 1.982×10^{-4} 1.322×10^{-5}	$4.668\times 10^{1*}$ 4.430×10^{-4} 8.652×10^{-3} 1.100	2.910×10^3 5.083 1.593 5.211×10^{-2}
7 — — 3.287×10^4	1.824×10^3 1.370×10^{-1} 6.848×10^{-2} 5.703×10^{-3}	$3.203\times 10^{1*}$ 1.370×10^{-1} 3.900 1.054×10^3	1.917×10^3 2.375×10^{-4} 1.130×10^{-4} 8.522×10^{-6}	$3.316\times 10^{1*}$ 2.375×10^{-4} 6.530×10^{-3} 1.646	2.729×10^3 4.702 1.571 5.847×10^{-2}
10 — — 4.696×10^4	1.612×10^3 9.665×10^{-2} 5.465×10^{-2} 5.825×10^{-3}	$2.250\times 10^{1*}$ 9.665×10^{-2} 3.916 2.143×10^3	1.815×10^3 1.213×10^{-4} 6.094×10^{-5} 5.126×10^{-6}	$2.311\times 10^{1*}$ 1.213×10^{-4} 4.786×10^{-3} 2.483	2.572×10^3 4.369 1.549 6.490×10^{-2}
20 — — 9.392×10^4	1.266×10^3 4.886×10^{-2} 3.519×10^{-2} 6.086×10^{-3}	$1.131\times 10^{1*}$ 4.886×10^{-2} 3.940 8.538×10^3	1.691×10^3 3.165×10^{-5} 1.706×10^{-5} 1.653×10^{-6}	$1.149\times 10^{1*}$ 3.165×10^{-5} 2.512×10^{-3} 5.276	2.351×10^3 3.886 1.507 7.551×10^{-2}
30 — — 1.409×10^5	1.096×10^3 3.272×10^{-2} 2.722×10^{-2} 6.278×10^{-3}	$7.554*$ 3.272×10^{-2} 3.950 1.918×10^4	1.667×10^3 1.405×10^{-5} 7.684×10^{-6} 7.665×10^{-7}	$7.641*$ 1.405×10^{-5} 1.676×10^{-3} 7.954	2.264×10^3 3.681 1.482 8.009×10^{-2}
50 — — 2.348×10^5	9.099×10^2 1.972×10^{-2} 1.976×10^{-2} 6.613×10^{-3}	$4.541*$ 1.972×10^{-2} 3.959 5.321×10^4	1.702×10^3 4.898×10^{-6} 2.624×10^{-6} 2.512×10^{-7}	$4.576*$ 4.898×10^{-6} 9.759×10^{-4} 1.291×10^1	2.190×10^3 3.485 1.451 8.378×10^{-2}
10^2 — — 4.696×10^5	6.971×10^2 9.901×10^{-3} 1.295×10^{-2} 7.385×10^{-3}	$2.274*$ 9.901×10^{-3} 3.969 2.126×10^5	1.975×10^3 1.109×10^{-6} 5.118×10^{-7} 3.636×10^{-8}	$2.285*$ 1.109×10^{-6} 4.424×10^{-4} 2.349×10^1	2.142×10^3 3.303 1.406 8.494×10^{-2}
2×10^2 — — 9.392×10^5	5.200×10^2 4.964×10^{-3} 8.703×10^{-3} 8.918×10^{-3}	$1.138*$ 4.964×10^{-3} 3.975 8.500×10^5	3.389×10^3 2.351×10^{-7} 6.324×10^{-8} 1.525×10^{-9}	$1.141*$ 2.351×10^{-7} 1.878×10^{-4} 3.996×10^1	2.140×10^3 3.187 1.358 8.218×10^{-2}

A3.1 Wavelengths, Strengths, and Transition Rates

Table A3.1. (Continued)

β $\overline{B \text{ in } 10^2 \text{ T}}$	$3d_{-2}/0\ -2\ 0$ $\leftrightarrow$ $2p_{-1}/0\ -1\ 0$	$3d_{+2}/2\ +2\ 0$ $\leftrightarrow$ $2p_{+1}/1\ +1\ 0$	$3d_{-1}/0\ -1\ 1$ $\leftrightarrow$ $3p_0/0\ 0\ 3$	$3d_{+1}/1\ +1\ 1$ $\leftrightarrow$ $3p_0/0\ 0\ 3$	$3d_{-1}/0\ -1\ 1$ $\leftrightarrow$ $3p_{-1}/0\ -1\ 2$
5×10^2 — 2.348×10^6	3.309×10^2 1.990×10^{-3} 5.483×10^{-3} 1.387×10^{-2}	$4.557 \times 10^{-1}*$ 1.990×10^{-3} 3.981 5.312×10^6	$2.513 \times 10^{3}*$ 2.753×10^{-8} 9.988×10^{-9} 4.383×10^{-10}	$4.563 \times 10^{-1}*$ 2.753×10^{-8} 5.500×10^{-5} 7.320×10^1	$2.191 \times 10^{3}*$ 3.096 1.289 7.436×10^{-2}
10^3 — 4.696×10^6	2.200×10^2 9.962×10^{-4} 4.128×10^{-3} 2.363×10^{-2}	$2.280 \times 10^{-1}*$ 9.962×10^{-4} 3.984 2.124×10^7	$6.324 \times 10^{2}*$ 5.093×10^{-9} 7.344×10^{-9} 5.088×10^{-9}	$2.281 \times 10^{-1}*$ 5.093×10^{-9} 2.036×10^{-5} 1.084×10^2	$2.260 \times 10^{3}*$ 3.055 1.233 6.682×10^{-2}

Table A3.1. (Continued)

β $\overline{B \text{ in } 10^2 \text{ T}}$	$3d_{-2}/0\ -2\ 0$ $\leftrightarrow$ $3p_{-1}/0\ -1\ 2$	$3d_{+2}/2\ +2\ 0$ $\leftrightarrow$ $3p_{+1}/1\ +1\ 2$	β $\overline{B \text{ in } 10^2 \text{ T}}$	$3d_{-2}/0\ -2\ 0$ $\leftrightarrow$ $3p_{-1}/0\ -1\ 2$	$3d_{+2}/2\ +2\ 0$ $\leftrightarrow$ $3p_{+1}/1\ +1\ 2$
10^{-4} — 4.696×10^{-1}	4.556×10^6 4.050×10^1 8.105×10^{-3} 1.082×10^{-10}	4.572×10^6 4.050×10^1 8.076×10^{-3} 1.071×10^{-10}	10^{-1} — 4.696×10^2	5.045×10^3 7.481×10^{-1} 1.352×10^{-1} 1.472×10^{-3}	$4.167 \times 10^{3}*$ 7.481×10^{-1} 1.637×10^{-1} 2.613×10^{-3}
10^{-3} — 4.696	4.483×10^5 4.035×10^1 8.206×10^{-2} 1.132×10^{-7}	4.647×10^5 4.035×10^1 7.916×10^{-2} 1.016×10^{-7}	2×10^{-1} — 9.392×10^2	3.345×10^3 1.632×10^{-1} 4.448×10^{-2} 1.101×10^{-3}	$1.732 \times 10^{3}*$ 1.632×10^{-1} 8.592×10^{-2} 7.941×10^{-3}
7×10^{-3} — 3.287×10^1	5.849×10^4 3.486×10^1 5.433×10^{-1} 4.400×10^{-5}	7.364×10^4 3.486×10^1 4.316×10^{-1} 2.206×10^{-5}	3×10^{-1} — 1.409×10^3	2.659×10^3 7.416×10^{-2} 2.543×10^{-2} 9.966×10^{-4}	$1.065 \times 10^{3}*$ 7.416×10^{-2} 6.347×10^{-2} 1.549×10^{-2}
10^{-2} — 4.696×10^1	3.964×10^4 3.094×10^1 7.116×10^{-1} 1.255×10^{-4}	5.377×10^4 3.094×10^1 5.246×10^{-1} 5.028×10^{-5}	5×10^{-1} — 2.348×10^3	2.008×10^3 2.948×10^{-2} 1.339×10^{-2} 9.198×10^{-4}	$5.906 \times 10^{2}*$ 2.948×10^{-2} 4.551×10^{-2} 3.615×10^{-2}
2×10^{-2} — 9.392×10^1	1.874×10^4 1.958×10^1 9.528×10^{-1} 7.520×10^{-4}	2.917×10^4 1.958×10^1 6.119×10^{-1} 1.992×10^{-4}	7×10^{-1} — 3.287×10^3	1.677×10^3 1.655×10^{-2} 8.999×10^{-3} 8.865×10^{-4}	$4.046 \times 10^{2}*$ 1.655×10^{-2} 3.730×10^{-2} 6.312×10^{-2}
3×10^{-2} — 1.409×10^2	1.246×10^4 1.188×10^1 8.689×10^{-1} 1.550×10^{-3}	$1.952 \times 10^{4}*$ 1.188×10^1 5.549×10^{-1} 4.038×10^{-4}	1 — 4.696×10^3	1.391×10^3 9.146×10^{-3} 5.993×10^{-3} 8.577×10^{-4}	$2.730 \times 10^{2}*$ 9.146×10^{-3} 3.055×10^{-2} 1.136×10^{-1}
5×10^{-2} — 2.348×10^2	8.052×10^3 4.224 4.783×10^{-1} 2.043×10^{-3}	$1.053 \times 10^{4}*$ 4.224 3.656×10^{-1} 9.130×10^{-4}	2 — 9.392×10^3	9.801×10^2 2.995×10^{-3} 2.787×10^{-3} 8.037×10^{-4}	$1.291 \times 10^{2}*$ 2.995×10^{-3} 2.115×10^{-2} 3.515×10^{-1}
7×10^{-2} — 3.287×10^2	6.340×10^3 1.811 2.604×10^{-1} 1.795×10^{-3}	6.710×10^3 1.811 2.460×10^{-1} 1.514×10^{-3}	3 — 1.409×10^4	8.047×10^2 1.583×10^{-3} 1.794×10^{-3} 7.679×10^{-4}	$8.400 \times 10^{1}*$ 1.583×10^{-3} 1.719×10^{-2} 6.749×10^{-1}

Table A3.1. (Continued)

β / B in 10^2 T	$3d_{-2}/0\ -2\ 0$ $\leftrightarrow$ $3p_{-1}/0\ -1\ 2$	$3d_{+2}/2\ +2\ 0$ $\leftrightarrow$ $3p_{+1}/1\ +1\ 2$
5 — 2.348×10^4	6.328×10^2 7.161×10^{-4} 1.032×10^{-3} 7.142×10^{-4}	$4.919\times10^{1*}$ 7.161×10^{-4} 1.328×10^{-2} 1.520
7 — 3.287×10^4	5.428×10^2 4.263×10^{-4} 7.160×10^{-4} 6.734×10^{-4}	$3.468\times10^{1*}$ 4.263×10^{-4} 1.121×10^{-2} 2.581
10 — 4.696×10^4	4.633×10^2 2.465×10^{-4} 4.851×10^{-4} 6.260×10^{-4}	$2.400\times10^{1*}$ 2.465×10^{-4} 9.364×10^{-3} 4.504
20 — 9.392×10^4	3.451×10^2 8.541×10^{-5} 2.257×10^{-4} 5.252×10^{-4}	$1.180\times10^{1*}$ 8.541×10^{-5} 6.600×10^{-3} 1.313×10^1
30 — 1.409×10^5	2.928×10^2 4.603×10^{-5} 1.434×10^{-4} 4.636×10^{-4}	$7.809*$ 4.603×10^{-5} 5.375×10^{-3} 2.442×10^1
50 — 2.348×10^5	2.401×10^2 2.116×10^{-5} 8.037×10^{-5} 3.864×10^{-4}	$4.652*$ 2.116×10^{-5} 4.147×10^{-3} 5.310×10^1
10^2 — 4.696×10^5	1.865×10^2 7.391×10^{-6} 3.613×10^{-5} 2.878×10^{-4}	$2.310*$ 7.391×10^{-6} 2.918×10^{-3} 1.514×10^2
2×10^2 — 9.392×10^5	1.482×10^2 2.591×10^{-6} 1.594×10^{-5} 2.011×10^{-4}	$1.150*$ 2.591×10^{-6} 2.054×10^{-3} 4.307×10^2
5×10^2 — 2.348×10^6	$1.145\times10^{2*}$ 6.525×10^{-7} 5.196×10^{-6} 1.099×10^{-4}	$4.582\times10^{-1*}$ 6.525×10^{-7} 1.299×10^{-3} 1.714×10^3
10^3 — 4.696×10^6	$9.933\times10^{1*}$ 2.313×10^{-7} 2.123×10^{-6} 5.963×10^{-5}	$2.287\times10^{-1*}$ 2.313×10^{-7} 9.221×10^{-4} 4.885×10^3

A3.2 Oscillator Strengths and Transition Probabilities in Adiabatic Approximation

Table A3.2a. Oscillator strengths for $\Delta m = 0$ transitions from the tightly bound states ($\nu = 0$) of the H atom for $\beta = 50$ ($n = 0$)

initial state final state	$m = 0, \nu = 0$ $m' = 0, \nu'$	$m = -1, \nu = 0$ $m' = -1, \nu'$	$m = -2, \nu = 0$ $m' = -2, \nu'$	$m = -3, \nu = 0$ $m' = -3, \nu'$	$m = -4, \nu = 0$ $m' = -4, \nu'$
$\nu' = 1$	0.4180	0.5374	0.5975	0.6363	0.6643
$\nu' = 3$	0.0402	0.0467	0.0489	0.0498	0.0502
$\nu' = 5$	0.0114	0.0131	0.0136	0.0138	0.0138
$\nu' = 7$	0.0048	0.0054	0.0056	0.0057	0.0057
$\nu' = 9$	0.0024	0.0028	0.0029	0.0029	0.0029
$\nu' = 11$	0.0014	0.0016	0.0017	0.0017	0.0017
$\nu' = 13$	0.0009	0.0010	0.0010	0.0011	0.0011
$\nu' = 15 - \infty$	0.0027	0.0031	0.0032	0.0032	0.0032
asympt. formula	$\dfrac{0.24}{(\nu'+1)^3}$	$\dfrac{0.28}{(\nu'+1)^3}$	$\dfrac{0.29}{(\nu'+1)^3}$	$\dfrac{0.29}{(\nu'+1)^3}$	$\dfrac{0.29}{(\nu'+1)^3}$
discr. spectr.	0.4818	0.6111	0.6744	0.7145	0.7429
cont. spectr.	0.5182	0.3889	0.3256	0.2855	0.2571
total	1.0000	1.0000	1.0000	1.0000	1.0000

Table A3.2b. Oscillator strengths for $\Delta m = -1$ transitions from the tightly bound states ($\nu = 0$) of the H atom for $\beta = 50$ ($n = 0$)

initial state final state	$m = 0, \nu = 0$ $m' = -1, \nu'$	$m = -1, \nu = 0$ $m' = -2, \nu'$	$m = -2, \nu = 0$ $m' = -3, \nu'$	$m = -3, \nu = 0$ $m' = -4, \nu'$
$\nu' = 0$	2.2×10^{-2}	1.756×10^{-2}	1.534×10^{-2}	1.392×10^{-2}
$\nu' = 2$	1.5×10^{-4}	8.0×10^{-5}	5.4×10^{-5}	4.2×10^{-5}
$\nu' = 4$	3.0×10^{-5}	1.6×10^{-5}	1.1×10^{-5}	8.4×10^{-6}
$\nu' = 0 - 4$	2.20×10^{-2}	1.77×10^{-2}	1.54×10^{-2}	1.40×10^{-2}
sum rule	0	0	0	0

Table A3.2c. Oscillator strengths for $\Delta m = +1$ transitions from the tightly bound states ($\nu = 0$) of the H atom for $\beta = 50$ ($n = 0$)

initial state final state	$m = -1, \nu = 0$ $m' = 0, \nu'$	$m = -2, \nu = 0$ $m' = -1, \nu'$	$m = -3, \nu = 0$ $m' = -2, \nu'$	$m = -4, \nu = 0$ $m' = -3, \nu'$
$\nu' = 2$	1.4×10^{-4}	7.8×10^{-5}	5.3×10^{-5}	4.1×10^{-5}
$\nu' = 4$	2.7×10^{-5}	1.4×10^{-5}	1.0×10^{-5}	8.4×10^{-6}
$\nu' = 2 - 4$	1.7×10^{-4}	9.2×10^{-5}	6.3×10^{-5}	4.9×10^{-5}
sum rule	2	2	2	2

Appendix 3: Electromagnetic Transition Probabilities

Table A3.3a. Oscillator strengths for $\Delta m = 0$ transitions from the tightly bound states ($\nu = 0$) of the H atom for $\beta = 1000$ ($n = 0$)

initial state final state	$m = 0, \nu = 0$ $m' = 0, \nu'$	$m = -1, \nu = 0$ $m' = -1, \nu'$	$m = -2, \nu = 0$ $m' = -2, \nu'$	$m = -3, \nu = 0$ $m' = -3, \nu'$	$m = -4, \nu = 0$ $m' = -4, \nu'$
$\nu' = 1$	0.1865	0.2528	0.2931	0.3227	0.3462
$\nu' = 3$	0.0206	0.0268	0.0302	0.0326	0.0345
$\nu' = 5$	0.0060	0.0077	0.0087	0.0093	0.0098
$\nu' = 7$	0.0025	0.0032	0.0036	0.0039	0.0041
$\nu' = 9$	0.0013	0.0016	0.0018	0.0020	0.0021
$\nu' = 11 - \infty$	0.0026	0.0033	0.0037	0.0040	0.0042
asympt. formula	$\dfrac{1.3}{(\nu'+1)^3}$	$\dfrac{1.6}{(\nu'+1)^3}$	$\dfrac{1.8}{(\nu'+1)^3}$	$\dfrac{2.0}{(\nu'+1)^3}$	$\dfrac{2.1}{(\nu'+1)^3}$
discr. spectr.	0.2194	0.2955	0.3412	0.3705	0.4008
cont. spectr.	0.7806	0.7045	0.6588	0.6295	0.5992
total	1.0000	1.0000	1.0000	1.0000	1.0000

Table A3.3b. Oscillator strengths for $\Delta m = -1$ transitions from the tightly bound states ($\nu = 0$) of the H atom for $\beta = 1000$ ($n = 0$)

initial state final state	$m = 0, \nu = 0$ $m' = -1, \nu'$	$m = -1, \nu = 0$ $m' = -2, \nu'$	$m = -2, \nu = 0$ $m' = -3, \nu'$	$m = -3, \nu = 0$ $m' = -4, \nu'$
$\nu' = 0$	2.317×10^{-3}	1.957×10^{-3}	1.767×10^{-3}	1.642×10^{-3}
$\nu' = 2$	4.7×10^{-6}	2.7×10^{-6}	1.9×10^{-6}	1.5×10^{-6}
$\nu' = 4$	8.4×10^{-7}	4.8×10^{-7}	3.4×10^{-7}	2.6×10^{-7}
$\nu' = 0 - 4$	2.32×10^{-3}	1.96×10^{-3}	1.77×10^{-3}	1.71×10^{-3}
sum rule	0	0	0	0

Table A3.3c. Oscillator strengths for $\Delta m = +1$ transitions from the tightly bound states ($\nu = 0$) of the H atom for $\beta = 1000$ ($n = 0$)

initial state final state	$m = -1, \nu = 0$ $m' = 0, \nu'$	$m = -2, \nu = 0$ $m' = -1, \nu'$	$m = -3, \nu = 0$ $m' = -2, \nu'$	$m = -4, \nu = 0$ $m' = -3, \nu'$
$\nu' = 2$	4.5×10^{-6}	2.6×10^{-6}	1.9×10^{-6}	1.5×10^{-6}
$\nu' = 4$	7.8×10^{-7}	4.6×10^{-7}	3.3×10^{-7}	2.6×10^{-7}
$\nu' = 2 - 4$	5.3×10^{-6}	3.1×10^{-6}	2.2×10^{-6}	1.7×10^{-6}
sum rule	2	2	2	2

A3.2 Electromagnetic Transition Probabilities in Adiabatic Approximation

Table A3.4a. Transition probabilities for dipole emission of the states of the H atom with $n = 0$, $m = 0$, $\nu = 0 - 5$, in units of 10^8 per second for $\beta = 50$

initial state (m,ν)	m'	final states $\nu' = 0$	$\nu' = 1$	$\nu' = 2$	$\nu' = 3$	$\nu' = 4$	sum $\sum_{\nu'}$	$\sum_{\nu',m'}$	total lifetime in 10^{-10} s
(0,0)	0	...	...	...	...	...	0	0	∞
(0,1)	0	482.5	...	...	...	...	482.5	482.5	0.21
(0,2)	0	...	6.390	...	...	...	6.390		
	−1	0.087	...	...	...	...	0.087	6.477	15.4
(0,3)	0	56.59	...	3.18	...	...	59.77		
	−1	...	3.5×10^{-5}	...	...	...	3.5×10^{-5}	59.77	1.67
(0,4)	0	...	0.311	...	0.319	...	0.630		
	−1	0.0183	...	2.6×10^{-5}	...	...	0.018	0.648	154.3
(0,5)	0	16.698	...	0.748	...	0.308	17.75		
	−1	...	1.2×10^{-5}	...	2.8×10^{-7}	...	1.2×10^{-5}	17.75	5.63

Blank entries refer to transitions which are forbidden in emission, or not allowed by the break selection rules.

Table A3.4b. Transition probabilities for dipole emission of the states of the H atom with $n = 0$, $m = -1$, $\nu = 0 - 5$, in units of 10^8 per second for $\beta = 50$

initial state (m,ν)	m'	final states $\nu' = 0$	$\nu' = 1$	$\nu' = 2$	$\nu' = 3$	$\nu' = 4$	sum $\sum_{\nu'}$	$\sum_{\nu',m'}$	total lifetime in 10^{-10} s
(−1,0)	0	2.969	...	...	...	...	2.969	2.969	33.7
(−1,1)	0	...	1.9×10^{-5}	...	...	...	1.9×10^{-5}		
	−1	274.0	...	...	...	...	274.0	274.0	0.37
(−1,2)	0	0.200	...	2.0×10^{-5}	...	...	0.200		
	−1	...	6.741	...	...	...	6.741	6.973	14.3
	−2	0.032	...	...	...	...	0.032		
(−1,3)	0	...	3.9×10^{-5}	...	3.9×10^{-8}	...	3.9×10^{-5}		
	−1	31.42	...	2.64	...	...	34.06	34.06	2.94
	−2	...	3.6×10^{-5}	...	...	...	3.6×10^{-5}		
(−1,4)	0	0.0432	...	3.3×10^{-5}	...	1.6×10^{-7}	0.0432		
	−1	...	0.4998	...	0.3647	...	0.8645	0.9148	109.3
	−2	0.0071	...	1.5×10^{-5}	...	...	0.0071		
(−1,5)	0	...	1.4×10^{-5}	...	3.0×10^{-7}	...	1.4×10^{-5}		
	−1	9.275	...	0.585	...	0.269	10.129	10.129	9.87
	−2	...	1.3×10^{-5}	...	3.1×10^{-7}	...	1.3×10^{-5}		

Appendix 3: Electromagnetic Transition Probabilities

Table A3.4c. Transition probabilities for dipole emission of the states of the H atom with $n = 0$, $m = -2$, $\nu = 0 - 5$, in units of 10^8 per second for $\beta = 50$

initial state (m,ν)	m'	final states $\nu' = 0$	$\nu' = 1$	$\nu' = 2$	$\nu' = 3$	$\nu' = 4$	sum $\sum_{\nu'}$	total $\sum_{\nu',m'}$	lifetime in 10^{-10} s
$(-2,0)$	-1	0.3659	...	...	...	...	0.3659	0.3659	273.3
$(-2,1)$	-1	...	1.5×10^{-5}	...	...	...	1.5×10^{-5}		
	-2	197.4	...	...	...	...	197.4	197.4	0.51
$(-2,2)$	-1	0.0496	...	5.7×10^{-6}	...	...	0.0496		
	-2	...	6.623	...	...	...	6.623	6.689	15.0
	-3	0.0166	...	...	...	...	0.0166		
$(-2,3)$	-1	...	4.0×10^{-5}	...	3.3×10^{-8}	...	4.0×10^{-5}		
	-2	22.38	...	2.35	...	...	24.73	24.73	4.04
	-3	...	3.4×10^{-5}	...	...	...	3.4×10^{-5}		
$(-2,4)$	-1	0.0111	...	1.7×10^{-5}	...	5.2×10^{-8}	0.0111		
	-2	...	0.5698	...	0.3778	...	0.9476	0.9625	103.9
	-3	0.0038	...	1.0×10^{-5}	...	...	0.0038		
$(-2,5)$	-1	...	1.4×10^{-5}	...	3.2×10^{-7}	...	1.4×10^{-5}		
	-2	6.614	...	0.507	...	0.247	7.368	7.368	13.6
	-3	...	1.2×10^{-5}	...	3.0×10^{-7}	...	1.2×10^{-5}		

Table A3.4d. Transition probabilities for dipole emission of the states of the H atom with $n = 0$, $m = -3$, $\nu = 0 - 5$, in units of 10^8 per second for $\beta = 50$

initial state (m,ν)	m'	final states $\nu' = 0$	$\nu' = 1$	$\nu' = 2$	$\nu' = 3$	$\nu' = 4$	sum $\sum_{\nu'}$	total $\sum_{\nu',m'}$	lifetime in 10^{-10} s
$(-3,0)$	-2	0.1079	...	...	...	...	0.1079	0.1079	926.4
$(-3,1)$	-2	...	1.1×10^{-5}	...	...	...	1.1×10^{-5}		
	-3	155.8	...	...	...	...	155.8	155.8	0.64
$(-3,2)$	-2	0.0227	...	2.7×10^{-6}	...	...	0.0227		
	-3	...	6.420	...	...	...	6.420	6.453	15.5
	-4	0.0103	...	...	...	...	0.0103		
$(-3,3)$	-2	...	3.7×10^{-5}	...	2.7×10^{-8}	...	3.7×10^{-5}		
	-3	17.54	...	2.16	...	...	19.70	19.70	5.08
	-4	...	3.1×10^{-5}	...	...	...	3.1×10^{-5}		
$(-3,4)$	-2	0.0052	...	1.1×10^{-5}	...	2.7×10^{-8}	0.0052		
	-3	...	0.5986	...	0.3816	...	0.9802	0.9878	101.2
	-4	0.0024	...	7.7×10^{-6}	...	...	0.0024		
$(-3,5)$	-2	...	1.3×10^{-5}	...	3.2×10^{-7}	...	1.3×10^{-5}		
	-3	5.189	...	0.457	...	0.232	5.878	5.878	17.0
	-4	...	1.1×10^{-5}	...	2.8×10^{-7}	...	2.8×10^{-7}		

A3.2 Electromagnetic Transition Probabilities in Adiabatic Approximation

Table A3.5a. Transition probabilities for dipole emission of the states of the H atom with $n = 0$, $m = 0$, $\nu = 0 - 5$, in units of 10^8 per second for $\beta = 1000$

initial state (m,ν)	m'	$\nu' = 0$	$\nu' = 1$	$\nu' = 2$	$\nu' = 3$	$\nu' = 4$	sum $\sum_{\nu'}$	$\sum_{\nu',m'}$	total lifetime in 10^{-10} s
(0,0)	0	...	...	...	...	...	0	0	∞
(0,1)	0	1544	...	...	...	...	1544	1544	0.065
(0,2)	0	...	4.402	...	...	...	4.402		
	−1	0.022	...	...	...	...	0.022	4.424	22.6
(0,3)	0	185.02	...	4.29	...	...	189.3		
	−1	...	5.3×10^{-8}	...	...	...	5.3×10^{-8}	189.3	0.53
(0,4)	0	...	0.0060	...	0.1886	...	0.1946		
	−1	0.0039	...	9.0×10^{-5}	...	...	0.0039	0.1985	503.8
(0,5)	0	54.45	...	1.11	...	0.38	55.94		
	−1	...	1.9×10^{-8}	...	3.7×10^{-10}	...	1.9×10^{-8}	55.94	1.79

Table A3.5b. Transition probabilities for dipole emission of the states of the H atom with $n = 0$, $m = -1$, $\nu = 0 - 5$, in units of 10^8 per second for $\beta = 1000$

initial state (m,ν)	m'	$\nu' = 0$	$\nu' = 1$	$\nu' = 2$	$\nu' = 3$	$\nu' = 4$	sum $\sum_{\nu'}$	$\sum_{\nu',m'}$	total lifetime in 10^{-10} s
(−1,0)	0	1.358	...	...	...	...	1.358	1.358	73.6
(−1,1)	0	4.2×10^{-9}	...	...	...	...	4.2×10^{-9}		
	−1	1128	...	...	...	...	1128	1128	0.089
(−1,2)	0	0.041	...	4.1×10^{-7}	...	...	0.041		
	−1	...	5.370	...	...	...	5.370	5.420	18.5
	−2	0.0090	...	...	...	...	0.009		
(−1,3)	0	...	5.5×10^{-8}	...	7.7×10^{-10}	...	5.5×10^{-8}		
	−1	133.5	...	4.01	...	...	137.5	137.5	0.73
	−2	...	7.9×10^{-8}	...	...	...	7.9×10^{-8}		
(−1,4)	0	0.0077	...	1.0×10^{-6}	...	2.4×10^{-9}	0.0077		
	−1	...	0.0718	...	0.2380	...	0.3098	0.3192	313.3
	−2	0.0017	...	5.2×10^{-7}	...	...	0.0017		
(−1,5)	0	...	1.9×10^{-8}	...	3.8×10^{-10}	...	1.9×10^{-8}		
	−1	39.21	...	1.007	...	0.360	40.58	40.58	2.46
	−2	...	2.8×10^{-8}	...	5.5×10^{-10}	...	2.9×10^{-8}		

Table A3.5c. Transition probabilities for dipole emission of the states of the H atom with $n = 0$, $m = -2$, $\nu = 0 - 5$, in units of 10^8 per second for $\beta = 1000$

initial state (m, ν)	m'	final states $\nu' = 0$	$\nu' = 1$	$\nu' = 2$	$\nu' = 3$	$\nu' = 4$	sum $\sum_{\nu'}$	total $\sum_{\nu',m'}$	lifetime in 10^{-10} s
$(-2, 0)$	-1	0.2017	...	...	...	...	0.2017	0.2017	495.8
$(-2, 1)$	-1	...	5.3×10^{-9}	...	...	...	5.3×10^{-9}		
	-2	941.8	...	...	...	...	941.8	941.8	0.11
$(-2, 2)$	-1	0.0126	...	1.3×10^{-7}	...	...	0.0126		
	-2	...	5.862	...	...	...	5.862	5.880	17.0
	-3	0.0052	...	...	...	...	0.0052		
$(-2, 3)$	-1	...	8.2×10^{-8}	...	9.8×10^{-10}	...	8.2×10^{-12}		
	-2	110.55	...	3.84	...	...	114.39	114.39	0.87
	-3	...	9.3×10^{-8}	...	...	...	9.3×10^{-8}		
$(-2, 4)$	-1	0.0024	...	5.6×10^{-7}	...	8.0×10^{-9}	0.0024		
	-2	...	0.1325	...	0.2652	...	0.03977	0.4011	249.3
	-3	0.0010	...	3.7×10^{-7}	...	...	0.0010		
$(-2, 5)$	-1	...	2.8×10^{-8}	...	5.5×10^{-10}	...	2.9×10^{-8}		
	-2	32.46	...	0.945	...	0.350	33.75	33.75	2.96
	-3	...	3.3×10^{-8}	...	6.4×10^{-10}	...	3.4×10^{-8}		

Table A3.5d. Transition probabilities for dipole emission of the states of the H atom with $n = 0$, $m = -3$, $\nu = 0 - 5$, in units of 10^8 per second for $\beta = 1000$

initial state (m, ν)	m'	final states $\nu' = 0$	$\nu' = 1$	$\nu' = 2$	$\nu' = 3$	$\nu' = 4$	sum $\sum_{\nu'}$	total $\sum_{\nu',m'}$	lifetime in 10^{-10} s
$(-3, 0)$	-2	0.0659	...	...	...	...	0.0659	0.0659	1518
$(-3, 1)$	-2	...	5.7×10^{-9}	...	...	...	5.7×10^{-9}		
	-3	826.3	...	...	...	...	826.3	826.3	0.12
$(-3, 2)$	-2	0.0065	...	6.4×10^{-8}	...	...	0.00652		
	-3	...	6.176	...	...	...	6.176	6.186	16.2
	-4	0.0034	...	...	...	...	0.0034		
$(-3, 3)$	-2	...	9.6×10^{-8}	...	1.1×10^{-11}	...	9.6×10^{-8}		
	-3	96.46	...	3.71	...	...	100.17	100.17	1.00
	-4	...	1.0×10^{-7}	...	...	...	1.0×10^{-7}		
$(-3, 4)$	-2	0.0013	...	3.9×10^{-7}	...	4.2×10^{-10}	0.0013		
	-3	...	0.1830	...	0.2837	...	0.4667	0.4687	213.4
	-4	0.0007	...	2.9×10^{-7}	...	...	0.0007		
$(-3, 5)$	-2	...	3.3×10^{-8}	...	6.7×10^{-10}	...	3.4×10^{-8}		
	-3	28.31	...	0.900	...	0.341	29.55	29.55	3.38
	-4	...	3.7×10^{-8}	...	3.0×10^{-10}	...	3.7×10^{-8}		

A3.3 Dipole Strengths of Stationary Transitions

Table A3.6a. Dipole strengths (as defined in (6.2)) of stationary transitions of Lyman α and Balmer α

β	B in 10^2 T	$2p_{-1} \to 1s_0$	$3p_{-1} \to 2s_0$	$3p_0 \to 2s_0$	$3d_{-2} \to 2p_{-1}$	$3d_{-1} \to 2p_0$	$3s'_0 \to 2p_{+1}$
0.0001	0.4696	5.549×10^{-1}	3.131	3.131	9.017	4.509	9.780×10^{-1}
0.0010	4.696	5.550×10^{-1}	3.134	3.133	9.024	4.511	9.794×10^{-1}
0.0015	7.044						
0.0020	9.392						
0.0030	14.09						
0.0050	23.48						
0.0070	32.87	5.585×10^{-1}	3.324	3.219	9.259	4.593	
0.010	46.96	5.621×10^{-1}	3.299	3.278	9.388	4.644	1.065
0.015	70.44	5.702×10^{-1}	3.327	3.366	9.487	4.694	1.114
0.020	93.92	5.802×10^{-1}	3.264	3.435	9.429	4.688	1.148
0.030	140.9	6.033×10^{-1}	2.940	3.573	9.008	4.543	1.034
0.050	234.8	6.484×10^{-1}	2.187	4.126	7.821	4.045	4.905×10^{-1}
0.070	328.7	6.828×10^{-1}	1.832	5.118	6.752	3.548	4.220×10^{-1}
0.10	469.6	7.116×10^{-1}	1.698	7.115	5.545	2.951	4.165×10^{-1}
0.15	704.4	7.169×10^{-1}	1.625	10.36	4.250	2.282	2.413×10^{-1}
0.20	939.2	6.948×10^{-1}	1.799	12.44	3.445	1.854	9.462×10^{-1}
0.30	1409	6.268×10^{-1}	1.228	14.87	2.503	1.347	

Table A3.6b. Dipole strengths (as defined in (6.2)) of stationary transitions of Balmer β

β	B in 10^2 T	$4p'_{-1} \to 2s_0$	$4f'_{-1} \to 2s_0$	$4p'_0 \to 2s_0$	$4f'_0 \to 2s_0$	$4s'_0 \to 2p_{+1}$
0.0001	0.4696	5.148×10^{-1}	3.335×10^{-2}	3.607×10^{-1}	1.874×10^{-1}	1.678×10^{-1}
0.0010	4.696	5.190×10^{-1}	3.391×10^{-2}	3.611×10^{-1}	1.895×10^{-1}	1.696×10^{-1}
0.0015	7.044	5.239×10^{-1}	3.460×10^{-2}	3.614×10^{-1}	1.921×10^{-1}	
0.0020	9.392	5.299×10^{-1}	3.555×10^{-2}	3.615×10^{-1}	1.957×10^{-1}	
0.0030	14.09	5.432×10^{-1}	3.816×10^{-2}	3.600×10^{-1}	2.053×10^{-1}	
0.0050	23.48	5.616×10^{-1}	4.594×10^{-2}	3.443×10^{-1}	2.327×10^{-1}	
0.0070	32.87	5.532×10^{-1}	5.686×10^{-2}	3.061×10^{-1}	2.677×10^{-1}	
0.010	46.96	4.659×10^{-1}	7.916×10^{-2}	2.103×10^{-1}	3.286×10^{-1}	1.830×10^{-1}
0.015	70.44	1.548×10^{-1}	1.341×10^{-1}	3.549×10^{-2}	4.352×10^{-1}	3.171×10^{-2}
0.020	93.92	2.743×10^{-2}	2.144×10^{-1}	1.550×10^{-2}	5.256×10^{-1}	1.422×10^{-2}
0.030	140.9	3.951×10^{-4}	4.432×10^{-1}	1.893×10^{-1}	6.181×10^{-1}	2.909×10^{-4}
0.050	234.8	1.216×10^{-1}	7.412×10^{-1}	2.316×10^{-1}	6.411×10^{-1}	1.658×10^{-1}
0.070	328.7	2.909×10^{-1}	4.512×10^{-1}	2.283×10^{-1}	6.750×10^{-1}	1.032×10^{-1}
0.10	469.6	8.918×10^{-2}	2.195×10^{-1}	2.398×10^{-1}	7.608×10^{-1}	1.046×10^{-1}

Table A3.6c. Dipole strengths (as defined in (6.2)) of stationary transitions of Balmer γ and Paschen β

β	B in 10^2 T	$5p'_{-1} \to 2s_0$	$5f'_{-1} \to 2s_0$	$5p'_0 \to 2s_0$	$5f'_0 \to 2s_0$	$5p'_{-1} \to 3d'_0$
0.0001	0.4696	1.772×10^{-1}	2.252×10^{-2}	1.081×10^{-1}	9.157×10^{-2}	1.288×10^{-3}
0.0010	4.696	1.838×10^{-1}	2.381×10^{-2}	1.092×10^{-1}	9.447×10^{-2}	1.865×10^{-3}
0.0015	7.044	1.903×10^{-1}	2.533×10^{-2}	1.096×10^{-1}	9.775×10^{-2}	
0.0020	9.392	1.966×10^{-1}	2.732×10^{-2}	1.090×10^{-1}	1.020×10^{-1}	
0.0030	14.09	2.039×10^{-1}	3.249×10^{-2}	1.021×10^{-1}	1.122×10^{-1}	
0.0050	23.48	1.338×10^{-1}	4.751×10^{-2}	5.110×10^{-2}	1.367×10^{-1}	
0.0070	32.87	1.397×10^{-3}	7.173×10^{-2}	8.319×10^{-3}	1.652×10^{-1}	
0.010	46.96	1.560×10^{-2}	1.442×10^{-1}	6.328×10^{-2}	2.102×10^{-1}	3.314×10^{-1}
0.015	70.44	6.354×10^{-2}	3.313×10^{-1}	1.094×10^{-1}	2.385×10^{-1}	2.281×10^{-1}
0.020	93.92	1.765×10^{-1}	1.825×10^{-1}	1.150×10^{-1}	1.229×10^{-1}	1.009×10^{-1}
0.030	140.9	2.214×10^{-2}	1.415×10^{-2}	2.965×10^{-3}	3.012×10^{-2}	5.465×10^{-3}
0.050	234.8	9.636×10^{-6}	4.045×10^{-3}	6.480×10^{-2}	1.133×10^{-1}	2.469×10^{-2}
0.070	328.7	3.473×10^{-2}	1.311×10^{-1}	6.008×10^{-2}	1.076×10^{-1}	1.535×10^{-1}
0.10	469.6	3.199×10^{-2}	4.992×10^{-2}	5.960×10^{-2}	1.090×10^{-1}	7.488×10^{-2}

Table A3.6c. (Continued)

β	B in 10^2 T	$5s'_0 \to 2p_{+1}$	$5d'_0 \to 2p_{+1}$	$5g'_0 \to 2p_{+1}$	$5f'_{-1} \to 3d'_0$	$5p'_0 \to 3d'_0$	$5f'_0 \to 3d'_0$
0.0001	0.4696				1.750		
0.0010	4.696	6.360×10^{-2}	5.253×10^{-3}	1.466×10^{-2}	1.755		
0.0015	7.044	6.662×10^{-2}	4.681×10^{-3}	1.502×10^{-2}			
0.0020	9.392	7.001×10^{-2}	3.930×10^{-3}	1.545×10^{-2}			
0.0030	14.09	7.653×10^{-2}	2.139×10^{-3}	1.625×10^{-2}			
0.0050	23.48	1.221×10^{-2}	1.87×10^{-5}	1.612×10^{-2}			
0.0070	32.87	4.747×10^{-3}	5.078×10^{-3}	1.099×10^{-2}			
0.010	46.96	5.03×10^{-3}	2.08×10^{-2}	2.60×10^{-2}	8.616×10^{-1}	7.088×10^{-1}	1.106
0.015	70.44	1.303×10^{-2}	6.166×10^{-2}	5.635×10^{-2}	2.044×10^{-1}	3.337×10^{-1}	0.1075
0.020	93.92	4.3×10^{-3}	4.9×10^{-4}	2.0×10^{-3}	3.00×10^{-5}	4.853×10^{-2}	0.3181
0.030	140.9				5.251×10^{-2}	6.065×10^{-1}	1.759
0.050	234.8				1.176×10^{-1}	3.269×10^{-1}	0.5602
0.070	328.7				4.744×10^{-1}	2.356×10^{-1}	0.4552
0.10	469.6				1.225×10^{-1}	3.123×10^{-1}	0.6450

A3.3 Dipole Strengths of Stationary Transitions

Table A3.6d. Dipole strengths (as defined in (6.2)) of stationary Balmer β and Balmer γ transitions (π components) as a function of the magnetic field

β	B in 10^2 T	$4f_0' \to 2s_0$	$4p_0' \to 2s_0$	$5f_0' \to 2s_0$	$5p_0' \to 2s_0$
0.0010	4.696	1.90×10^{-1}	3.61×10^{-1}	9.44×10^{-2}	1.09×10^{-1}
0.0015	7.044	1.92×10^{-1}	3.61×10^{-1}	9.78×10^{-2}	1.10×10^{-1}
0.0020	9.392	1.96×10^{-1}	3.61×10^{-1}	1.02×10^{-1}	1.09×10^{-1}
0.0030	14.09	2.05×10^{-1}	3.60×10^{-1}	1.12×10^{-1}	1.02×10^{-1}
0.0050	23.48	2.33×10^{-1}	3.44×10^{-1}	1.37×10^{-1}	5.12×10^{-2}
0.0055	25.83	2.41×10^{-1}	3.37×10^{-1}	1.44×10^{-1}	2.80×10^{-2}
0.0060	28.18	2.49×10^{-1}	3.28×10^{-1}	1.51×10^{-1}	6.77×10^{-3}
0.0065	30.52	2.58×10^{-1}	3.18×10^{-1}	1.58×10^{-1}	1.86×10^{-4}
0.0070	32.87	2.68×10^{-1}	3.06×10^{-1}	1.65×10^{-1}	8.26×10^{-3}
0.0075	35.22	2.77×10^{-1}	2.93×10^{-1}	1.73×10^{-1}	2.04×10^{-2}
0.0080	37.57	2.87×10^{-1}	2.79×10^{-1}	1.80×10^{-1}	3.17×10^{-2}
0.0085	39.92	2.97×10^{-1}	2.63×10^{-1}	1.88×10^{-1}	4.12×10^{-2}
0.0100	46.96	3.29×10^{-1}	2.10×10^{-1}	2.10×10^{-1}	6.33×10^{-2}
0.0125	58.70	3.82×10^{-1}	1.15×10^{-1}	2.38×10^{-1}	9.01×10^{-2}
0.0130	61.05	3.93×10^{-1}	9.65×10^{-2}	2.41×10^{-1}	9.46×10^{-2}
0.0135	63.40	4.04×10^{-1}	7.92×10^{-2}	2.43×10^{-1}	9.88×10^{-2}
0.0140	65.74	4.14×10^{-1}	6.31×10^{-2}	2.43×10^{-1}	1.03×10^{-1}
0.0145	68.09	4.25×10^{-1}	4.85×10^{-2}	2.42×10^{-1}	1.06×10^{-1}
0.0150	70.44	4.35×10^{-1}	3.55×10^{-2}	2.39×10^{-1}	1.09×10^{-1}
0.0160	75.14	4.55×10^{-1}	1.53×10^{-2}	2.27×10^{-1}	1.15×10^{-1}
0.0170	79.83	4.74×10^{-1}	3.39×10^{-3}	2.08×10^{-1}	1.18×10^{-1}
0.0180	84.53	4.93×10^{-1}	0.34×10^{-4}	1.84×10^{-1}	1.19×10^{-1}
0.0190	89.22	5.10×10^{-1}	4.50×10^{-3}	1.54×10^{-1}	1.18×10^{-1}
0.0200	93.92	5.26×10^{-1}	1.55×10^{-2}	1.23×10^{-1}	1.15×10^{-1}
0.0220	103.3	5.54×10^{-1}	5.04×10^{-2}	6.24×10^{-2}	1.01×10^{-1}
0.0240	112.7	5.76×10^{-1}	9.21×10^{-2}	1.86×10^{-2}	7.84×10^{-2}
0.0260	122.1	5.94×10^{-1}	1.32×10^{-1}	4.50×10^{-4}	4.97×10^{-2}
0.0280	131.5	6.08×10^{-1}	1.64×10^{-1}	7.26×10^{-3}	2.14×10^{-2}
0.0300	140.9	6.18×10^{-1}	1.89×10^{-1}	3.01×10^{-2}	2.97×10^{-3}
0.0320	150.3	6.25×10^{-1}	2.07×10^{-1}	5.71×10^{-2}	1.67×10^{-3}
0.0340	159.7	6.30×10^{-1}	2.19×10^{-1}	7.97×10^{-2}	1.53×10^{-2}
0.0360	169.1	6.33×10^{-1}	2.27×10^{-1}	9.55×10^{-2}	3.36×10^{-2}
0.0380	178.5	6.35×10^{-1}	2.31×10^{-1}	1.05×10^{-1}	4.82×10^{-2}
0.0400	187.8	6.36×10^{-1}	2.33×10^{-1}	1.11×10^{-1}	5.73×10^{-2}
0.0500	234.8	6.41×10^{-1}	2.32×10^{-1}	1.13×10^{-1}	6.48×10^{-2}
0.0630	295.9	6.59×10^{-1}	2.28×10^{-1}	1.09×10^{-1}	6.12×10^{-2}
0.0700	328.7	6.75×10^{-1}	2.28×10^{-1}	1.08×10^{-1}	6.01×10^{-2}
0.0800	375.7	7.02×10^{-1}	2.31×10^{-1}	1.07×10^{-1}	5.95×10^{-2}
0.1000	469.6	7.61×10^{-1}	2.40×10^{-1}	1.09×10^{-1}	6.03×10^{-2}

Table A3.6e. Dipole strengths (as defined in (6.2)) of stationary Balmer β and Balmer γ transitions (σ components) as a function of the magnetic field

β	B in 10^2 T	$4f'_{-1} \to 2s_0$	$4p'_{-1} \to 2s_0$	$5f'_{-1} \to 2s_0$	$5p'_{-1} \to 2s_0$
0.0010	4.696	3.39×10^{-2}	5.19×10^{-1}	2.38×10^{-2}	1.84×10^{-1}
0.0015	7.044	3.46×10^{-2}	5.24×10^{-1}	2.53×10^{-2}	1.90×10^{-1}
0.0020	9.392	3.55×10^{-2}	5.30×10^{-1}	2.73×10^{-2}	1.97×10^{-1}
0.0025	11.74	3.67×10^{-2}	5.37×10^{-1}	2.97×10^{-2}	2.02×10^{-1}
0.0030	14.09	3.82×10^{-2}	5.43×10^{-1}	3.25×10^{-2}	2.04×10^{-1}
0.0040	18.78	4.16×10^{-2}	5.55×10^{-1}	3.92×10^{-2}	1.96×10^{-1}
0.0050	23.48	4.59×10^{-2}	5.62×10^{-1}	4.75×10^{-2}	1.34×10^{-1}
0.0060	28.18	5.10×10^{-2}	5.61×10^{-1}	5.81×10^{-2}	4.41×10^{-3}
0.0070	32.87	5.69×10^{-2}	5.53×10^{-1}	7.17×10^{-2}	1.40×10^{-3}
0.0080	37.57	6.35×10^{-2}	5.35×10^{-1}	8.96×10^{-2}	6.00×10^{-3}
0.0090	42.26	7.09×10^{-2}	5.07×10^{-1}	1.13×10^{-1}	1.07×10^{-2}
0.0100	46.96	7.92×10^{-2}	4.66×10^{-1}	1.44×10^{-1}	1.56×10^{-2}
0.0110	51.66	8.83×10^{-2}	4.12×10^{-1}	1.83×10^{-1}	2.12×10^{-2}
0.0120	56.35	9.83×10^{-2}	3.48×10^{-1}	2.28×10^{-1}	2.81×10^{-2}
0.0130	61.05	1.09×10^{-1}	2.78×10^{-1}	2.74×10^{-1}	3.69×10^{-2}
0.0140	65.74	1.21×10^{-1}	2.12×10^{-1}	3.11×10^{-1}	4.84×10^{-2}
0.0150	70.44	1.34×10^{-1}	1.55×10^{-1}	3.31×10^{-1}	6.35×10^{-2}
0.0160	75.14	1.48×10^{-1}	1.11×10^{-1}	3.32×10^{-1}	8.33×10^{-2}
0.0170	79.83	1.63×10^{-1}	7.84×10^{-1}	3.12×10^{-1}	1.08×10^{-1}
0.0180	84.53	1.79×10^{-1}	5.54×10^{-1}	2.76×10^{-1}	1.35×10^{-1}
0.0190	89.22	1.96×10^{-1}	3.91×10^{-1}	2.30×10^{-1}	1.61×10^{-1}
0.0200	93.92	2.14×10^{-1}	2.74×10^{-2}	1.83×10^{-1}	1.77×10^{-1}
0.0220	103.3	2.54×10^{-1}	1.30×10^{-2}	1.05×10^{-1}	1.61×10^{-1}
0.0240	112.7	2.97×10^{-1}	5.39×10^{-3}	6.03×10^{-1}	1.06×10^{-1}
0.0260	122.1	3.44×10^{-1}	1.53×10^{-3}	3.60×10^{-1}	6.08×10^{-2}
0.0280	131.5	3.93×10^{-1}	0.75×10^{-4}	2.23×10^{-2}	3.56×10^{-2}
0.0300	140.9	4.43×10^{-1}	3.95×10^{-4}	1.41×10^{-2}	2.21×10^{-2}
0.0320	150.3	4.94×10^{-1}	2.27×10^{-3}	8.96×10^{-3}	1.45×10^{-2}
0.0340	159.7	5.43×10^{-1}	5.71×10^{-3}	5.50×10^{-3}	9.84×10^{-3}
0.0360	169.1	5.89×10^{-1}	1.09×10^{-2}	3.14×10^{-3}	6.78×10^{-3}
0.0380	178.5	6.32×10^{-1}	1.80×10^{-2}	1.54×10^{-3}	4.66×10^{-3}
0.0400	187.8	6.69×10^{-1}	2.74×10^{-2}	5.39×10^{-4}	3.14×10^{-3}
0.0500	234.8	7.41×10^{-1}	1.22×10^{-1}	4.05×10^{-3}	0.10×10^{-4}
0.0600	281.8	6.21×10^{-1}	2.69×10^{-1}	3.81×10^{-2}	4.19×10^{-3}
0.0630	295.9	5.67×10^{-1}	2.97×10^{-1}	6.10×10^{-2}	8.81×10^{-3}
0.0700	328.7	4.51×10^{-1}	2.91×10^{-1}	1.31×10^{-1}	3.47×10^{-2}
0.0800	375.7	3.34×10^{-1}	1.92×10^{-1}	1.39×10^{-1}	8.98×10^{-2}
0.0900	422.6	2.64×10^{-1}	1.24×10^{-1}	7.95×10^{-2}	5.72×10^{-2}
0.1000	469.6	2.19×10^{-1}	8.92×10^{-2}	4.99×10^{-2}	3.25×10^{-2}

Appendix 4: Helium and Helium-Like Atoms

A4.1 Tables of the Energy Values

In the following tables A4.1 – A4.7, the energies (in units of $-E_Z$) of low-lying states with $M = 0, -1, -2,$ and -3 of the two-electron systems Fe^{24+} ($Z = 26$), Si^{12+} ($Z = 14$), O^{6+} ($Z = 8$), He ($Z = 2$), and H^- ($Z = 1$) resulting from single- and multi-configuration calculations are listed for the magnetic field strength parameter $10^{-4} \leq \beta_Z \leq 10^3$.

Table A4.1a. Total single-configuration Hartree-Fock energies (in units of $-E_Z$) of the four lowest $M = 0$ states in the range $10^{-4} \leq \beta_Z \leq 10^3$ for Fe^{24+} ($Z = 26$). Above the underlined number or above the first gap the spherical ansatz was used, below the cylindrical one. Double lines indicate the crossing of two states. The quantum numbers $1s_0$ or $m = 0, \nu = 0$ of the other electron are omitted for the sake of brevity.

	$M = 0$			
β_Z	$2p_0/0\ 1$	$2s_0/0\ 2$	$3p_0/0\ 3$	$\begin{array}{c}3s_0\\3d_0\end{array}/0\ 4$
1×10^{-4}	1.233240	1.236078	1.103608	1.104395
1.5×10^{-4}	1.233439	1.236278	1.103807	1.104593
2×10^{-4}	1.233639	1.236478	1.104005	1.104790
3×10^{-4}	1.234038	1.236876	1.104401	1.105183
5×10^{-4}	1.234836	1.237671	1.105189	1.105959
7×10^{-4}	1.235632	1.238464	1.105970	1.106724
1×10^{-3}	1.236825	1.239648	1.107129	1.107849
1.5×10^{-3}	1.238806	1.241609	1.109031	1.109665
2×10^{-3}	1.240781	1.243554	1.110893	1.111408
3×10^{-3}	1.244707	1.247398	1.114501	<u>1.114680</u>
5×10^{-3}	1.252471	1.254901	1.121264	1.121110
7×10^{-3}	1.260118	1.262158	1.127455	1.127590
1×10^{-2}	1.271373	1.272594	1.135764	1.136499
1.5×10^{-2}	1.289565	1.288830	1.147385	1.149517
2×10^{-2}	1.307080	1.303709	1.156753	1.160700
3×10^{-2}	1.340228	1.329831	1.170505	1.179004
5×10^{-2}	1.400051	1.370227	1.185038	1.204875
7×10^{-2}	1.452941	1.398627	1.188830	1.221575
1×10^{-1}	1.522363	1.424986		1.235441
1.5×10^{-1}	1.618597	1.438652		1.238507
2×10^{-1}	1.697063			
3×10^{-1}	1.817005			
5×10^{-1}	<u>1.963311</u>	1.643156	1.530050	1.480786
7×10^{-1}	2.080797	1.857436	1.734683	1.682927
1	2.357844	2.112849	1.979660	1.925456
2	2.991028	2.705089	2.551153	2.492874
5	4.057023	3.723804	3.541790	3.479778
1×10^1	5.074973	4.714647	4.511069	4.447617
2×10^1	6.313945	5.935712	5.710421	5.646716
5×10^1	8.359299	7.971556	7.717632	7.655182
1×10^2	10.27113	9.885381	9.610112	9.549565
2×10^2	12.54762	12.16975	11.87370	11.81558
5×10^2	16.19450	15.83279	15.51050	15.45602
1×10^3	19.49248	19.14530	18.80430	18.75267

Table A4.1b. Total single-configuration Hartree-Fock energies (in units of $-E_Z$) of the five lowest $M = -1$ states in the range $10^{-4} \leq \beta_Z \leq 10^3$ for Fe^{24+} ($Z = 26$). For the meaning of the lines see Table A4.1a.

	$M = -1$				
β_Z	$2p_{-1}/-10$	$3d_{-1}/-11$	$3p_{-1}/-12$	$4d_{-1}/-13$	$\begin{array}{c}4p_{-1}\\4f_{-1}\end{array}/-14$
1×10^{-4}	1.233439	1.103349	1.103807	1.058393	1.058579
1.5×10^{-4}	1.233739	1.103648	1.104105	1.058689	1.058873
2×10^{-4}	1.234039	1.103947	1.104402	1.058984	1.059163
3×10^{-4}	1.234637	1.104543	1.104994	1.059568	1.059738
5×10^{-4}	1.235833	1.105730	1.106169	1.060718	1.060855
7×10^{-4}	1.237026	1.106911	1.107332	1.061843	<u>1.061932</u>
1×10^{-3}	1.238812	1.108670	1.109052	1.063485	1.063576
1.5×10^{-3}	1.241778	1.111571	1.111858	1.066100	1.066318
2×10^{-3}	1.244730	1.114432	1.114587	1.068568	1.068960
3×10^{-3}	1.250592	1.120038	1.119820	1.073096	1.073965
5×10^{-3}	1.262154	1.130800	1.129431	1.080732	1.082985
7×10^{-3}	1.273500	1.141001	1.138004	1.086869	1.090925
1×10^{-2}	1.290125	1.155371	1.149198	1.094034	1.101309
1.5×10^{-2}	1.316831	1.177250	1.164340	1.102311	1.115751
2×10^{-2}	1.342373	1.197095	1.176196	1.107583	1.127745
3×10^{-2}	1.390401	1.232368	1.193188	1.112549	1.146998
5×10^{-2}	1.476779	1.291439	1.210865		1.174195
7×10^{-2}	1.553640	1.340842	1.215705		1.192332
1×10^{-1}	1.656296	1.403273			1.208826
1.5×10^{-1}	1.803880	1.486558			1.217332
2×10^{-1}	1.930817	1.551922			
3×10^{-1}	2.142909	<u>1.646463</u>			
5×10^{-1}	2.463514	1.749408	1.594581	1.510799	1.467585
7×10^{-1}	2.701274	1.982693	1.808547	1.716222	1.670059
1	<u>2.965275</u>	2.259447	2.063867	1.962138	1.912967
2	3.844768	2.895296	2.656552	2.535684	2.425007
5	5.480077	3.971097	3.676852	3.529287	3.469143
1×10^1	7.101377	5.000093	4.669404	4.500813	4.437782
2×10^1	9.127898	6.251628	5.892443	5.702264	5.637656
5×10^1	12.55574	8.313237	7.931114	7.711816	7.647074
1×10^2	15.81871	10.23556	9.847147	9.605704	9.542123
2×10^2	19.75189	12.52037	12.13370	11.87037	11.80875
5×10^2	26.12550	16.17499	15.79961	15.50813	15.44994
1×10^3	31.94415	19.47683	19.11425	18.80238	18.74710

Table A4.1c. Total single-configuration Hartree-Fock energies (in units of $-E_Z$) of the five lowest $M = -2$ states in the range $10^{-4} \leq \beta_Z \leq 10^3$ for Fe^{24+} ($Z = 26$). For the meaning of the lines see Table A4.1a. In brackets the differences in per cent with respect to the two-configuration calculations are given.

	$M = -2$				
β_Z	$3d_{-2}/\!-\!2\ 0$	$4f_{-2}/\!-\!2\ 1$	$4d_{-2}/\!-\!2\ 2$	$5f_{-2}/\!-\!2\ 3$	$5g_{-2}/\!-\!2\ 4$
1×10^{-4}	1.103549	1.058582	1.058592	1.037774	1.037777
1.5×10^{-4}	1.103947	1.058979	1.058986	1.038164	1.038170
2×10^{-4}	1.104345	1.059375	1.059378	1.038550	1.038561
3×10^{-4}	1.105139	1.060161	1.060154	1.039309	1.039334
5×10^{-4}	1.106720	1.061720	1.061679	1.040780	1.040849
7×10^{-4}	1.108292	1.063257	1.063168	1.042188	1.042322
1×10^{-3}	1.110631	1.065525	1.065331	1.044186	1.044456
1.5×10^{-3}	1.114484	1.069204	1.068759	1.047233	1.047820
2×10^{-3}	1.118278	1.072760	1.071976	1.049963	1.050965
3×10^{-3}	1.125695	1.079533	1.077829	1.053882	1.056691
5×10^{-3}	1.139880	1.091899	1.087621	1.061720	1.066442
7×10^{-3}	1.153277	1.103014	1.095500	1.066833	1.074626
1×10^{-2}	1.172109 (.005)	1.117965	1.104834	1.072214	1.084991
1.5×10^{-2}	1.200820	1.139720	1.116122	1.077453	1.099040
2×10^{-2}	1.227008 (.03)	1.158815	1.123995	1.079795	1.110519
3×10^{-2}	1.274125	1.191922	1.133505		1.128746
5×10^{-2}	1.355157 (.19)	1.246117	1.138294		1.154236
7×10^{-2}	1.425356	1.290763			1.171079
1×10^{-1}	1.517830 (.55)	1.346526			1.186166
1.5×10^{-1}	1.649477	1.419731			1.193233
2×10^{-1}	1.761955 (1.3)	1.475938			
3×10^{-1}	1.948585	1.553959			
5×10^{-1}	2.227084 (4.5)	1.701154	1.571092	1.499918	1.460712
7×10^{-1}	2.429693	1.932795	1.784708	1.705625	1.663331
1	2.687204	2.208288	2.039790	1.951920	1.906412
2	3.547819	2.843385	2.632372	2.526394	
5	5.067033	3.922064	3.653159	3.521533	
1×10^1	6.577203	4.956040	4.646432	4.494346	4.432563
2×10^1	8.469501	6.214324	5.870373	5.697095	5.632837
5×10^1	11.68024	8.285966	7.910398	7.708242	7.642764
1×10^2	14.74627	10.21548	9.827541	9.603140	
2×10^2	18.45269	12.50639	12.11526	11.86860	
5×10^2	24.47945	16.16695	15.78268		
1×10^3	30.00008	19.47179	19.09842		

Table A4.1d. Total single-configuration Hartree-Fock energies (in units of $-E_Z$) of the three lowest $M = -3$ states in the range $10^{-4} \leq \beta_Z \leq 10^3$ for Fe^{24+} ($Z = 26$). For the meaning of the lines see Table A4.1a. The thick line in the last column designates the crossing of the intruder $2p_{-1}3d_{-2}/-1\ 0\ -2\ 0$ with the state $1s_0 5f_{-3}/0\ 0\ -3\ 2$. Above the thick line the energies of the state $0\ 0\ -3\ 2$ are given, below the thick line those of the state $-1\ 0\ -2\ 0$.

	$M = -3$		
β_Z	$4f_{-3}/-3\ 0$	$5g_{-3}/-3\ 1$	$5f_{-3}/\begin{smallmatrix}0\ 0\ -3\ 2\\ -1\ 0\ -2\ 0\end{smallmatrix}$
1×10^{-4}	1.058781	1.037976	1.037972
1.5×10^{-4}	1.059277	1.038468	1.038458
2×10^{-4}	1.059771	1.038956	1.038939
3×10^{-4}	1.060754	1.039924	1.039885
5×10^{-4}	1.062298	1.041820	1.041714
7×10^{-4}	1.064615	1.043667	1.043459
1×10^{-3}	1.067439	1.046344	1.045929
1.5×10^{-3}	1.072014	1.050578	1.049682
2×10^{-3}	1.076431	1.054555	1.053040
3×10^{-3}	1.084827	1.061859	1.058805
5×10^{-3}	1.100157	1.074558	1.067712
7×10^{-3}	1.113989	1.085538	1.074406
1×10^{-2}	1.132745	1.099947	1.081929
1.5×10^{-2}	1.160438	1.120518	1.090410
2×10^{-2}	1.185184	1.138365	1.095758
3×10^{-2}	1.229127	1.169070	1.100811
5×10^{-2}	1.304030	1.218977	
7×10^{-2}	1.368717	1.259857	
1×10^{-1}	1.453866	1.310611	
1.5×10^{-1}	1.575026	1.376530	
2×10^{-1}	1.678433	1.426278	
3×10^{-1}	1.849552	1.492868	
5×10^{-1}	2.102977	1.543333	1.553098
7×10^{-1}	2.284914		1.769231
1	2.563040	2.174590	2.045680
2	3.385420	2.808229	2.730066
5	4.838622		3.955656
1×10^1	6.284834		5.191084
2×10^1	8.099054	6.186388	6.757937
5×10^1	11.18190		9.453750
1×10^2	14.13051		12.06235
2×10^2	17.70032		15.25098
5×10^2	23.51568		20.50035
1×10^3	28.85272		25.36418

Table A4.2a. Total single-configuration Hartree-Fock energies (in units of $-E_Z$) of the four lowest $M=0$ states in the range $10^{-4} \leq \beta_Z \leq 10^3$ for Si^{12+} ($Z=14$). Above the underlined number or above the first gap the spherical ansatz was used, below the cylindrical one. Double lines indicate the crossing of two states. The quantum numbers $1s_0$ or $m=0, \nu=0$ of the other electron are omitted for the sake of brevity.

	$M=0$			
β_Z	$2p_0$/0 1	$2s_0$/0 2	$3p_0$/0 3	$3s_0$/$3d_0$/0 4
1×10^{-4}	1.218862	1.224020	1.097027	1.098448
1.5×10^{-4}	1.219062	1.224220	1.097226	1.098646
2×10^{-4}	1.219262	1.224419	1.097424	1.098843
3×10^{-4}	1.219661	1.224818	1.097820	1.099235
5×10^{-4}	1.220458	1.225612	1.098606	1.100010
7×10^{-4}	1.221255	1.226405	1.099386	1.100773
1×10^{-3}	1.222447	1.227588	1.100544	1.101894
1.5×10^{-3}	1.224427	1.229548	1.102439	1.103701
2×10^{-3}	1.226400	1.231491	1.104292	1.105432
3×10^{-3}	1.230323	1.235330	1.107877	1.108670
5×10^{-3}	1.238075	1.242815	1.114568	<u>1.114290</u>
7×10^{-3}	1.245705	1.250046	1.120658	1.120475
1×10^{-2}	1.256923	1.260428	1.128779	1.129208
1.5×10^{-2}	1.275031	1.276541	1.140028	1.141884
2×10^{-2}	1.292435	1.291264	1.149000	1.152701
3×10^{-2}	1.325301	1.317011	1.161792	1.170277
5×10^{-2}	1.384439	1.356556	1.175135	1.194818
7×10^{-2}	1.436596	1.384092	1.177787	1.210358
1×10^{-1}	1.504946	1.409218		1.222716
1.5×10^{-1}	1.599537	1.421025		1.223680
2×10^{-1}	1.676529			
3×10^{-1}	1.793882			
5×10^{-1}	<u>1.935893</u>	1.629064	1.520903	1.474087
7×10^{-1}	2.058344	1.842257	1.725038	1.675919
1	2.332992	2.096510	1.969503	1.918130
2	2.961218	2.868507	2.540056	2.484967
5	4.020369	3.702364	3.529582	3.471176
1×10^1	5.033256	4.691171	4.498132	4.438508
2×10^1	6.267564	5.910327	5.696860	5.637180
5×10^1	8.307710	7.943838	7.703399	7.645144
1×10^2	10.21645	9.856032	9.595506	9.539187
2×10^2	12.49058	12.13887	11.85880	11.80489
5×10^2	16.13529	15.80004	15.49533	15.44496
1×10^3	19.43219	19.11124	18.78892	18.74136

Table A4.2b. Total single-configuration Hartree-Fock energies (in units of $-E_Z$) of the five lowest $M = -1$ states in the range $10^{-4} \leq \beta_Z \leq 10^3$ for Si^{12+} ($Z = 14$). For the meaning of the lines see Table A4.2a.

	$M = -1$				
β_Z	$2p_{-1}/-1\ 0$	$3d_{-1}/-1\ 1$	$3p_{-1}/-1\ 2$	$4d_{-1}/-1\ 3$	$\begin{array}{c}4p_{-1}\\4f_{-1}\end{array}/-1\ 4$
1×10^{-4}	1.219062	1.096438	1.097226	1.054505	1.054827
1.5×10^{-4}	1.219362	1.096737	1.097524	1.054801	1.055120
2×10^{-4}	1.219661	1.097035	1.097821	1.055095	1.055410
3×10^{-4}	1.220260	1.097631	1.098413	1.055678	1.055983
5×10^{-4}	1.221455	1.098817	1.099586	1.056825	1.057095
7×10^{-4}	1.222648	1.099997	1.100746	1.057945	1.058163
1×10^{-3}	1.224433	1.101753	1.102462	1.059575	<u>1.059686</u>
1.5×10^{-3}	1.227397	1.104647	1.105255	1.062163	1.062391
2×10^{-3}	1.230346	1.107498	1.107968	1.064597	1.065009
3×10^{-3}	1.236202	1.113077	1.113155	1.069032	1.069950
5×10^{-3}	1.247740	1.123758	1.122630	1.076431	1.078801
7×10^{-3}	1.259053	1.133850	1.131025	1.082302	1.086545
1×10^{-2}	1.275609	1.148018	1.141908	1.089070	1.096623
1.5×10^{-2}	1.302157	1.169514	1.156491	1.096733	1.110572
2×10^{-2}	1.327503	1.188955	1.167794	1.101460	1.122107
3×10^{-2}	1.375069	1.223438	1.183776	1.105487	1.140541
5×10^{-2}	1.460437	1.281091	1.199797		1.166375
7×10^{-2}	1.536315	1.329269	1.203304		1.183375
1×10^{-1}	1.637624	1.390118			1.198422
1.5×10^{-1}	1.783258	1.471208			1.204927
2×10^{-1}	1.908505	1.534723			
3×10^{-1}	2.117696	<u>1.626174</u>			
5×10^{-1}	2.433566	1.731164	1.582676	1.502496	1.461593
7×10^{-1}	2.667383	1.962276	1.795594	1.707363	1.663740
1	<u>2.926330</u>	2.236545	2.049775	1.952693	1.906307
2	3.801037	2.867077	2.640198	2.525125	2.420917
5	5.415593	3.935110	3.657498	3.517344	3.461042
1×10^1	7.015637	4.958019	4.647840	4.487921	4.429142
2×10^1	9.014852	6.203669	5.868754	5.688528	5.628515
5×10^1	12.39510	8.258392	7.904777	7.697242	7.637345
1×10^2	15.61152	10.17560	9.818934	9.590594	9.531993
2×10^2	19.48740	12.45808	12.10375	11.85483	11.79826
5×10^2	25.76614	16.10981	15.76749	15.49223	15.43902
1×10^3	31.49662	19.41033	19.08059	18.78633	18.73588

Table A4.2c. Total single-configuration Hartree-Fock energies (in units of $-E_Z$) of the five lowest $M = -2$ states in the range $10^{-4} \leq \beta_Z \leq 10^3$ for Si^{12+} ($Z = 14$). For the meaning of the lines see Table A4.2a.

	$M = -2$				
β_Z	$3d_{-2}/-2\ 0$	$4f_{-2}/-2\ 1$	$4d_{-2}/-2\ 2$	$5f_{-2}/-2\ 3$	$\begin{matrix}5d_{-2}\\5g_{-2}\end{matrix}/-2\ 4$
1×10^{-4}	1.096637	1.054688	1.054704	1.035281	<u>1.035287</u>
1.5×10^{-4}	1.097034	1.055084	1.055097	1.035670	<u>1.035677</u>
2×10^{-4}	1.097433	1.055479	1.055489	1.036055	1.036067
3×10^{-4}	1.098227	1.056265	1.056263	1.036812	1.036838
5×10^{-4}	1.099807	1.057821	1.057783	1.038273	1.038347
7×10^{-4}	1.101376	1.059354	1.059264	1.039668	1.039811
1×10^{-3}	1.103712	1.061612	1.061411	1.041639	1.041927
1.5×10^{-3}	1.107554	1.065269	1.064800	1.044626	1.045250
2×10^{-3}	1.111333	1.068796	1.067966	1.047283	1.048344
3×10^{-3}	1.118711	1.075492	1.073691	1.051792	1.053950
5×10^{-3}	1.132780	1.087659	1.083172	1.058537	1.063437
7×10^{-3}	1.146025	1.098551	1.090718	1.063335	1.071363
1×10^{-2}	1.164589	1.113161	1.099585	1.068293	1.081368
1.5×10^{-2}	1.192812	1.134375	1.110170	1.072927	1.094886
2×10^{-2}	1.218508	1.152973	1.117429	1.074748	1.105899
3×10^{-2}	1.264693	1.185201	1.125893		1.123324
5×10^{-2}	1.344102	1.237943	1.129023		1.147514
7×10^{-2}	1.412926	1.281382			1.163290
1×10^{-1}	1.503647	1.335612			1.177024
1.5×10^{-1}	1.632887	1.406704			1.182224
2×10^{-1}	1.743353	1.461133			
3×10^{-1}	1.926642	<u>1.536184</u>			
5×10^{-1}	2.199889	1.685059	1.560308	1.492252	1.455087
7×10^{-1}	<u>2.398244</u>	1.914693	1.772917	1.697399	1.657371
1	2.655944	2.187873	2.026894	1.943099	1.900101
2	3.505599	2.817945	2.617263	2.516433	
5	5.004853	3.889093	3.635062	3.510125	
1×10^1	6.494583	4.916991	4.626091	4.481983	4.424223
2×10^1	8.360620	6.169224	5.847868	5.683829	5.623978
5×10^1	11.52555	8.233522	7.885193	7.693971	7.633288
1×10^2	14.54674	10.15837	9.800404	9.588265	
2×10^2	18.19787	12.44557	12.08630	11.85325	
5×10^2	24.13293	16.10270	15.75147		
1×10^3	29.56817	19.40591	19.06563		

Table A4.2d. Total single-configuration Hartree-Fock energies (in units of $-E_Z$) of the three lowest $M = -3$ states in the range $10^{-4} \leq \beta_Z \leq 10^3$ for Si^{12+} ($Z = 14$). For the meaning of the lines see Table A4.2a. The thick line in the last column designates the crossing of the intruder $2p_{-1}3d_{-2}/-1\ 0\ -2\ 0$ with the state $1s_0 5f_{-3}/0\ 0\ -3\ 2$. Above the thick line the energies of the state $0\ 0\ -3\ 2$ are given, below the thick line those of the state $-1\ 0\ -2\ 0$.

	$M = -3$		
β_Z	$4f_{-3}/-3\ 0$	$5g_{-3}/-3\ 1$	$5f_{-3}/\begin{smallmatrix}0\ 0\ -3\ 2\\-1\ 0\ -2\ 0\end{smallmatrix}$
1×10^{-4}	1.054887	1.035483	1.035478
1.5×10^{-4}	1.055382	1.035974	1.035964
2×10^{-4}	1.055876	1.036462	1.036444
3×10^{-4}	1.056857	1.037427	1.037386
5×10^{-4}	1.058797	1.039316	1.039202
7×10^{-4}	1.060709	1.041152	1.040930
1×10^{-3}	1.063521	1.043808	1.043365
1.5×10^{-3}	1.068066	1.047993	1.047042
2×10^{-3}	1.072444	1.051910	1.050310
3×10^{-3}	1.080743	1.059078	1.055881
5×10^{-3}	1.095831	1.071485	1.064404
7×10^{-3}	1.109402	1.082185	1.070749
1×10^{-2}	1.127768	1.096209	1.077811
1.5×10^{-2}	1.154856	1.116214	1.085634
2×10^{-2}	1.179056	1.133567	1.090418
3×10^{-2}	1.222039	1.163425	1.094513
5×10^{-2}	1.295366	1.211962	
7×10^{-2}	1.358752	1.251719	
1×10^{-1}	1.442253	1.301051	
1.5×10^{-1}	1.561148	1.365009	
2×10^{-1}	1.662655	1.413107	
3×10^{-1}	1.830610	1.476934	
5×10^{-1}	2.079058	1.523031	1.542427
7×10^{-1}	2.256968		1.758290
1	2.534676	2.156067	2.017574
2	3.347034	2.784966	2.692070
5	4.781916		3.899617
1×10^1	6.209288		5.116526
2×10^1	7.999207	6.143844	6.659527
5×10^1	11.03945		9.313594
1×10^2	13.94614		11.88118
2×10^2	17.46404		15.01906
5×10^2	23.19283		20.18387
1×10^3	28.44884		24.96859

Table A4.3a. Total single-configuration Hartree-Fock energies (in units of $-E_Z$) of the four lowest $M=0$ states in the range $10^{-4} \leq \beta_Z \leq 10^3$ for O^{6+} ($Z=8$). Above the underlined number or above the first gap the spherical ansatz was used, below the cylindrical one. Double lines indicate the crossing of two states. The quantum numbers $1s_0$ or $m=0, \nu=0$ of the other electron are omitted for the sake of brevity.

	\multicolumn{4}{c}{$M=0$}			
β_Z	$2p_0$/0 1	$2s_0$/0 2	$3p_0$/0 3	$3s_0$/0 4, $3d_0$
1×10^{-4}	1.196180	1.204862	1.086719	1.089085
1.5×10^{-4}	1.196380	1.205062	1.086918	1.089283
2×10^{-4}	1.196580	1.205261	1.087116	1.089480
3×10^{-4}	1.196979	1.205659	1.087512	1.089871
5×10^{-4}	1.197776	1.206454	1.088297	1.090644
7×10^{-4}	1.198572	1.207246	1.089074	1.091404
1×10^{-3}	1.199764	1.208428	1.090227	1.092518
1.5×10^{-3}	1.201742	1.210385	1.092111	1.094309
2×10^{-3}	1.203713	1.212325	1.093949	1.096018
3×10^{-3}	1.207629	1.216154	1.097489	1.099193
5×10^{-3}	1.215359	1.223607	1.104046	<u>1.104630</u>
7×10^{-3}	1.222957	1.230792	1.109952	1.109375
1×10^{-2}	1.234107	1.241079	1.117734	1.117795
1.5×10^{-2}	1.252058	1.256975	1.128332	1.129871
2×10^{-2}	1.269258	1.271426	1.136620	1.140059
3×10^{-2}	1.301618	1.296527	1.148275	1.156404
5×10^{-2}	1.359553	1.334630	1.159166	1.178732
7×10^{-2}	1.410453	1.360724	1.159950	1.192356
1×10^{-1}	1.476990	1.383812		1.202238
1.5×10^{-1}	1.568845	1.392566		
2×10^{-1}	1.643402			
3×10^{-1}	1.756507			
5×10^{-1}	<u>1.891505</u>	1.606597	1.506433	1.463520
7×10^{-1}	2.022203	1.818087	1.709797	1.664874
1	2.293023	2.070527	1.953470	1.906597
2	2.913373	2.657031	2.522572	2.472538
5	3.961738	3.668751	3.510387	3.457680
1×10^1	4.966735	4.654105	4.477821	4.424329
2×10^1	6.193857	5.870301	5.675598	5.622352
5×10^1	8.226119	7.900194	7.681141	7.629549
1×10^2	10.13028	9.809856	9.572682	9.523073
2×10^2	12.40098	12.09032	11.83556	11.78830
5×10^2	16.04260	15.74859	15.47171	15.42781
1×10^3	19.33800	19.05774	18.76518	18.72356

Table A4.3b. Total single-configuration Hartree-Fock energies (in units of $-E_Z$) of the five lowest $M = -1$ states in the range $10^{-4} \leq \beta_Z \leq 10^3$ for O^{6+} ($Z = 8$). For the meaning of the lines see Table A4.3a.

	$M = -1$				
β_Z	$2p_{-1}/-1\ 0$	$3d_{-1}/-1\ 1$	$3p_{-1}/-1\ 2$	$4d_{-1}/-1\ 3$	$\begin{array}{c}4p_{-1}\\4f_{-1}\end{array}/-14$
1×10^{-4}	1.196380	1.085712	1.086918	1.048472	1.048967
1.5×10^{-4}	1.196680	1.086011	1.087216	1.048767	1.049259
2×10^{-4}	1.196979	1.086309	1.087513	1.049061	1.049548
3×10^{-4}	1.197578	1.086904	1.088103	1.049642	1.050117
5×10^{-4}	1.198773	1.088089	1.089274	1.050781	1.051219
7×10^{-4}	1.199965	1.089266	1.090430	1.051891	1.052273
1×10^{-3}	1.201749	1.091017	1.092136	1.053501	<u>1.053764</u>
1.5×10^{-3}	1.204709	1.093898	1.094907	1.056040	1.056291
2×10^{-3}	1.207654	1.096732	1.097588	1.058408	1.058865
3×10^{-3}	1.213495	1.102261	1.102690	1.062676	1.063900
5×10^{-3}	1.224991	1.112793	1.111917	1.069656	1.072240
7×10^{-3}	1.236240	1.122686	1.119993	1.075071	1.079645
1×10^{-2}	1.252668	1.136496	1.130328	1.081168	1.089204
1.5×10^{-2}	1.278928	1.157322	1.143947	1.087814	1.102325
2×10^{-2}	1.303920	1.176073	1.154317	1.091643	1.113100
3×10^{-2}	1.350668	1.209228	1.168617	1.094133	1.130187
5×10^{-2}	1.434293	1.264524	1.181917		1.153788
7×10^{-2}	1.508509	1.310682	1.183252		1.168932
1×10^{-1}	1.607568	1.368934			1.181618
1.5×10^{-1}	1.749980	1.446430			
2×10^{-1}	1.872454	1.506924			
3×10^{-1}	2.076904	<u>1.593336</u>			
5×10^{-1}	2.385057	1.701879	1.563737	1.489429	1.452150
7×10^{-1}	2.612456	1.929541	1.775019	1.693443	1.653791
1	<u>2.863180</u>	2.199878	2.027425	1.937876	1.895834
2	3.730118	2.822045	2.614334	2.508609	2.414566
5	5.311076	3.877995	3.626997	3.498734	3.448435
1×10^1	6.876725	4.891570	4.613931	4.467887	4.415718
2×10^1	8.831785	6.128336	5.831577	5.667246	5.614344
5×10^1	12.13512	8.172894	7.863527	7.674544	7.622283
1×10^2	15.27639	10.08496	9.774799	9.567120	9.516326
2×10^2	19.05982	12.36199	12.05689	11.83079	11.78204
5×10^2	25.18561	16.00980	15.71730	15.46769	15.42214
1×10^3	30.77404	19.30860	19.02805	18.76157	18.71856

Table A4.3c. Total single-configuration Hartree-Fock energies (in units of $-E_Z$) of the five lowest $M = -2$ states in the range $10^{-4} \leq \beta_Z \leq 10^3$ for O^{6+} ($Z = 8$). For the meaning of the lines see Table A4.3a.

β_Z	$3d_{-2}/-2\ 0$	$4f_{-2}/-2\ 1$	$4d_{-2}/-2\ 2$	$5f_{-2}/-2\ 3$	$\genfrac{}{}{0pt}{}{5d_{-2}}{5g_{-2}}/-2\ 4$
1×10^{-4}	1.085912	1.048649	1.048670	1.031415	<u>1.031424</u>
1.5×10^{-4}	1.086301	1.049045	1.049063	1.031803	1.031811
2×10^{-4}	1.086707	1.049439	1.049453	1.032186	1.032199
3×10^{-4}	1.087500	1.050224	1.050225	1.032937	1.032967
5×10^{-4}	1.089077	1.051773	1.051735	1.034382	1.034465
7×10^{-4}	1.090643	1.053298	1.053200	1.035752	1.035912
1×10^{-3}	1.092970	1.055539	1.055317	1.037673	1.037994
1.5×10^{-3}	1.096793	1.059154	1.058635	1.040551	1.041243
2×10^{-3}	1.100546	1.062627	1.061707	1.043079	1.044244
3×10^{-3}	1.107851	1.069183	1.067202	1.047303	1.049640
5×10^{-3}	1.121708	1.081001	1.076142	1.053477	1.058678
7×10^{-3}	1.134679	1.091508	1.083131	1.057749	1.066171
1×10^{-2}	1.152772	1.105540	1.091210	1.062016	1.075580
1.5×10^{-2}	1.180154	1.125851	1.100634	1.065660	1.088223
2×10^{-2}	1.205017	1.143629	1.106886	1.066629	1.098473
3×10^{-2}	1.249644	1.174414	1.113639		1.114587
5×10^{-2}	1.326369	1.224778	1.114057		1.136660
7×10^{-2}	1.392931	1.266248			1.150696
1×10^{-1}	1.480777	1.317978			1.162229
1.5×10^{-1}	1.606080	1.385628			1.164394
2×10^{-1}	1.713262	1.437162			
3×10^{-1}	1.891105	<u>1.507381</u>			
5×10^{-1}	2.155803	1.659246	1.543152	1.480180	1.446268
7×10^{-1}	<u>2.347237</u>	1.885695	1.754183	1.684467	1.648039
1	2.605299	2.155213	2.006437	1.929254	1.890235
2	3.437222	2.777374	2.593353	2.500838	
5	4.904205	3.836780	3.606523	3.492332	
1×10^{1}	6.360907	4.855316	4.594197	4.462616	4.411267
2×10^{1}	8.184533	6.098356	5.812551	5.663100	5.610237
5×10^{1}	11.27553	8.151713	7.845693	7.671741	7.618609
1×10^{2}	14.22440	10.06977	9.757926	9.565151	
2×10^{2}	17.78643	12.35169	12.04100	11.82947	
5×10^{2}	23.57379	16.00406	15.70272		
1×10^{3}	28.87156	19.30507	19.01444		

Table A4.3d. Total single-configuration Hartree-Fock energies (in units of $-E_Z$) of the three lowest $M = -3$ states in the range $10^{-4} \leq \beta_Z \leq 10^3$ for O^{6+} ($Z = 8$). For the meaning of the lines see Table A4.3a. The thick line in the last column designates the crossing of the intruder $2p_{-1}3d_{-2}/-1\ 0\ -2\ 0$ with the state $1s_0 5f_{-3}/0\ 0\ -3\ 2$. Above the thick line the energies of the state $0\ 0\ -3\ 2$ are given, below the thick line those of the state $-1\ 0\ -2\ 0$.

	$M = -3$		
β_Z	$4f_{-3}/-3\ 0$	$5g_{-3}/-3\ 1$	$5f_{-3}/\begin{smallmatrix}0\ 0\ -3\ 2\\-1\ 0\ -2\ 0\end{smallmatrix}$
1×10^{-4}	1.048848	1.031617	1.031612
1.5×10^{-4}	1.049342	1.032807	1.032096
2×10^{-4}	1.049835	1.032594	1.032573
3×10^{-4}	1.050814	1.033555	1.033508
5×10^{-4}	1.052747	1.035430	1.035302
7×10^{-4}	1.054647	1.037246	1.036997
1×10^{-3}	1.057436	1.039861	1.039368
1.5×10^{-3}	1.061928	1.043959	1.042909
2×10^{-3}	1.066237	1.047769	1.046020
3×10^{-3}	1.074360	1.054699	1.051257
5×10^{-3}	1.089026	1.066612	1.059136
7×10^{-3}	1.102148	1.076844	1.064905
1×10^{-2}	1.119859	1.090228	1.071208
1.5×10^{-2}	1.145942	1.109305	1.077958
2×10^{-2}	1.169240	1.125850	1.081821
3×10^{-2}	1.210649	1.154323	1.084354
5×10^{-2}	1.281402	1.200630	
7×10^{-2}	1.342665	1.238558	
1×10^{-1}	1.423480	1.285576	
1.5×10^{-1}	1.538686	1.346346	
2×10^{-1}	1.637102	1.391759	
3×10^{-1}	1.799907	1.451093	
5×10^{-1}	2.040259	1.490088	
7×10^{-1}	2.211623		1.740894
1	2.488748	2.126422	1.992983
2	3.284899	2.747843	2.630497
5	4.690171		3.808849
1×10^1	6.087114		4.995810
2×10^1	7.837802	6.076884	6.500266
5×10^1	10.80931		9.086920
1×10^2	13.64841		11.58831
2×10^2	17.08266		14.64436
5×10^2	22.67204		19.67293
1×10^3	27.79762		24.33022

Table A4.4a. Total single-configuration Hartree-Fock energies (in units of $-E_Z$) of the four lowest $M = 0$ states in the range $10^{-4} \leq \beta_Z \leq 10^3$ for He ($Z = 2$). Above the underlined number or above the first gap the spherical ansatz was used, below the cylindrical one. Double lines indicate the crossing of two states. The quantum numbers $1s_0$ or $m = 0, \nu = 0$ of the other electron are omitted for the sake of brevity.

	$M = 0$			
β_Z	$2p_0/0\ 1$	$2s_0/0\ 2$	$3p_0/0\ 3$	$3s_0/3d_0\ /0\ 4$
1×10^{-4}	1.066118	1.087525	1.029192	1.034639
1.5×10^{-4}	1.066318	1.087724	1.029389	1.034834
2×10^{-4}	1.066517	1.087923	1.029584	1.035028
3×10^{-4}	1.066915	1.088320	1.029971	1.035409
5×10^{-4}	1.067708	1.089110	1.030728	1.036151
7×10^{-4}	1.068497	1.089895	1.031464	1.036863
1×10^{-3}	1.069674	1.091064	1.032529	1.037877
1.5×10^{-3}	1.071619	1.092987	1.034204	1.039430
2×10^{-3}	1.073542	1.094880	1.035758	1.040818
3×10^{-3}	1.077324	1.098574	1.038543	1.043149
5×10^{-3}	1.084638	1.105611	1.043051	1.046343
7×10^{-3}	1.091640	1.112200	1.046503	<u>1.048056</u>
1×10^{-2}	1.101621	1.121320	1.050323	1.051754
1.5×10^{-2}	1.117091	1.134752	1.054344	1.058052
2×10^{-2}	1.131403	1.146350	1.056496	1.062731
3×10^{-2}	1.157473	1.165284	1.057378	1.069034
5×10^{-2}	1.202796	1.191055		1.074354
7×10^{-2}	1.242104	1.205693		
1×10^{-1}	1.293285	1.213276		
1.5×10^{-1}	1.363529			
2×10^{-1}	1.419698			
3×10^{-1}	1.501664			
5×10^{-1}	<u>1.586401</u>	1.466975	1.420246	1.401688
7×10^{-1}	1.783255	1.669039	1.619575	1.600590
1	2.030449	1.911572	1.859132	1.839816
2	2.603700	2.479282	2.420778	2.401215
5	3.591614	3.467173	3.400032	3.380970
1×10^1	4.556043	4.436167	4.362121	4.343978
2×10^1	5.749506	5.636690	5.555612	5.538689
5×10^1	7.749139	7.647293	7.557075	7.541956
1×10^2	9.636980	9.543374	9.446581	9.432803
2×10^2	11.89702	11.81108	11.70812	11.69558
5×10^2	15.53066	15.45368	15.34327	15.33018
1×10^3	18.82296	18.75185	18.63635	18.62414

Table A4.4b. Total single-configuration Hartree-Fock energies (in units of $-E_Z$) of the five lowest $M = -1$ states in the range $10^{-4} \leq \beta_Z \leq 10^3$ for He ($Z = 2$). For the meaning of the lines see Table A4.4a.

β_Z	$2p_{-1}/-1\ 0$	$3d_{-1}/-1\ 1$	$3p_{-1}/-1\ 2$	$4d_{-1}/-1\ 3$	$\begin{array}{c}4p_{-1}\\4f_{-1}\end{array}/-14$
1×10^{-4}	1.066318	1.028383	1.029390	1.016218	1.016643
1.5×10^{-4}	1.066617	1.028680	1.029683	1.016504	1.016920
2×10^{-4}	1.066915	1.028975	1.029974	1.016784	1.017189
3×10^{-4}	1.067511	1.029560	1.030547	1.017327	1.017699
5×10^{-4}	1.068697	1.030714	1.031662	1.018347	<u>1.018618</u>
7×10^{-4}	1.069876	1.031845	1.032736	1.019283	1.019458
1×10^{-3}	1.071632	1.033500	1.034270	1.020547	1.020895
1.5×10^{-3}	1.074525	1.036153	1.036638	1.022334	1.023065
2×10^{-3}	1.077376	1.038681	1.038788	1.023804	1.025004
3×10^{-3}	1.082954	1.043415	1.042536	1.026060	1.028369
5×10^{-3}	1.093650	1.051864	1.048376	1.028869	1.033754
7×10^{-3}	1.103798	1.059343	1.052717	1.030350	1.038038
1×10^{-2}	1.118164	1.069341	1.057446	1.031127	1.043220
1.5×10^{-2}	1.140341	1.083892	1.062418		1.049836
2×10^{-2}	1.160889	1.096730	1.065152		1.054875
3×10^{-2}	1.198643	1.119198	1.066592		1.062051
5×10^{-2}	1.265760	1.156543			1.069565
7×10^{-2}	1.325790	1.187735			1.071479
1×10^{-1}	1.406878	1.227014			
1.5×10^{-1}	1.524887	1.278548			
2×10^{-1}	1.627019	1.317387			
3×10^{-1}	1.797380	<u>1.367915</u>			
5×10^{-1}	2.050773	1.513048	1.447610	1.411384	1.397173
7×10^{-1}	<u>2.232898</u>	1.720237	1.649877	1.611378	1.596216
1	2.467513	1.967755	1.892643	1.851177	1.835585
2	3.239397	2.543299	2.460815	2.413333	2.378957
5	4.589940	3.536891	3.449255	3.393334	3.377294
1×10^{1}	5.920577	4.506673	4.418600	4.356025	4.340493
2×10^{1}	7.574957	5.705596	5.619428	5.550116	5.535373
5×10^{1}	10.35671	7.711842	7.630406	7.552321	7.538837
1×10^{2}	12.99041	9.603891	9.526776	9.442323	9.429820
2×10^{2}	16.15163	11.86749	11.79479	11.70430	11.69273
5×10^{2}	21.25212	15.50499	15.43785	15.33993	15.32948
1×10^{3}	25.89174	18.79972	19.73641	18.63332	18.62362

Table A4.4c. Total single-configuration Hartree-Fock energies (in units of $-E_Z$) of the five lowest $M = -2$ states in the range $10^{-4} \leq \beta_Z \leq 10^3$ for He ($Z = 2$). For the meaning of the lines see Table A4.4a. In brackets the differences in per cent with respect to the two-configuration calculations are given.

	$M = -2$				
β_Z	$3d_{-2}/-2\ 0$	$4f_{-2}/-2\ 1$	$4d_{-2}/-2\ 2$	$5f_{-2}/-2\ 3$	$5g_{-2}/-2\ 4$
1×10^{-4}	1.028882	1.016415	1.016413	1.010770	1.010780
1.5×10^{-4}	1.028976	1.016803	1.016791	1.011133	1.011156
2×10^{-4}	1.029369	1.017187	1.017161	1.011481	1.011522
3×10^{-4}	1.030147	1.017939	1.017876	1.012137	1.012226
5×10^{-4}	1.031678	1.019389	1.019209	1.013297	1.013532
7×10^{-4}	1.033176	1.020770	1.020423	1.014287	1.014718
1×10^{-3}	1.035361	1.022723	1.022048	1.015526	1.016316
1.5×10^{-3}	1.038850	1.025714	1.024334	1.017122	1.018619
2×10^{-3}	1.042166	1.028440	1.026221	1.018322	1.020602
3×10^{-3}	1.048362	1.033313	1.029173	1.019974	1.023940
5×10^{-3}	1.059447	1.041580	1.033105	1.021583	1.029139
7×10^{-3}	1.069347	1.048656	1.035538	1.021932	1.033214
1×10^{-2}	1.082761 (.036)	1.057937	1.037575		1.038109
1.5×10^{-2}	1.102693	1.071255	1.038442		1.044334
2×10^{-2}	1.120695 (.12)	1.082901			1.049070
3×10^{-2}	1.153132	1.103133			1.055817
5×10^{-2}	1.209702 (.46)	1.136422			1.062885
7×10^{-2}	1.259648	1.163879			1.064660
1×10^{-1}	1.326604 (1.0)	1.197894			
1.5×10^{-1}	1.423571	1.241210			
2×10^{-1}	1.507284 (2.1)	1.272324			
3×10^{-1}	1.646441	1.308503			
5×10^{-1}	1.850782	1.493702	1.438048	1.408373	1.394764
7×10^{-1}	1.993526 (6.2)	1.701195	1.640345	1.608109	1.593875
1	2.257230	1.949328	1.883178	1.848132	1.833318
2	2.968309	2.526865	2.451562	2.410773	
5	4.216099	3.524231	3.440393	3.391465	
1×10^1	5.449238	4.497132	4.410084	4.354651	4.338712
2×10^1	6.986673	5.698889	5.611287	5.549167	5.533728
5×10^1	9.580601	7.708030	7.622788	7.551786	7.537361
1×10^2	12.04487	9.601520	9.519571		
2×10^2	15.01167	11.86608	11.78801	11.70410	
5×10^2	19.81549	15.50431	15.43163		
1×10^3	24.20044	18.79933	18.73061		

Table A4.4d. Total single-configuration Hartree-Fock energies (in units of $-E_Z$) of the three lowest $M = -3$ states in the range $10^{-4} \leq \beta_Z \leq 10^3$ for He ($Z = 2$). For the meaning of the lines see Table A4.4a. The thick line in the last column designates the crossing of the intruder $2p_{-1}3d_{-2}/-1\ 0\ -2\ 0$ with the state $1s_0 5f_{-3}/0\ 0\ -3\ 2$. Above the thick line the energies of the state $0\ 0\ -3\ 2$ are given, below the thick line those of the state $-1\ 0\ -2\ 0$.

	$M = -3$		
β_Z	$4f_{-3}/-3\ 0$	$5g_{-3}/-3\ 1$	$5f_{-3}/\begin{smallmatrix}0\ 0\ -3\ 2\\-1\ 0\ -2\ 0\end{smallmatrix}$
1×10^{-4}	1.016612	1.010976	1.010960
1.5×10^{-4}	1.017096	1.011446	1.011411
2×10^{-4}	1.017574	1.011905	1.011842
3×10^{-4}	1.018511	1.012789	1.012652
5×10^{-4}	1.020312	1.014435	1.014080
7×10^{-4}	1.022024	1.015943	1.015299
1×10^{-3}	1.024444	1.017998	1.016836
1.5×10^{-3}	1.028153	1.021023	1.018864
2×10^{-3}	1.031547	1.023699	1.020445
3×10^{-3}	1.037670	1.028375	1.022801
5×10^{-3}	1.048255	1.036152	1.025659
7×10^{-3}	1.057527	1.042739	1.027155
1×10^{-2}	1.069981	1.051334	1.027947
1.5×10^{-2}	1.088411	1.063627	
2×10^{-2}	1.105040	1.074356	
3×10^{-2}	1.135021	1.092964	
5×10^{-2}	1.187352	1.123486	
7×10^{-2}	1.233536	1.148524	
1×10^{-1}	1.295329	1.179261	
1.5×10^{-1}	1.384478	1.217633	
2×10^{-1}	1.461085	1.244194	
3×10^{-1}	1.587627	1.271992	
5×10^{-1}	1.771071		
7×10^{-1}	1.896485		1.634840
1	2.174382	1.938918	1.877701
2	2.860499	2.517499	2.446193
5	4.065392		3.435240
1×10^1	5.257071		4.567643
2×10^1	6.743890	5.695484	5.410624
5×10^1	9.254668		7.541388
1×10^2	11.64215		9.596888
2×10^2	14.51896		12.10357
5×10^2	19.81203		16.22089
1×10^3	23.44321		20.02917

Table A4.5. Total energies of the three DTB states with $M = -1, -2$, and -3 for H^- ($Z = 1$) in the range $10^{-4} \leq \beta_Z \leq 10^3$ resulting from one-configuration calculations. Above the underlined number or above the first gap the spherical ansatz was used, below the cylindrical one. The quantum numbers $1s_0$ or $m = 0, \nu = 0$ of the other electron are omitted for the sake of brevity. In brackets the differences in per cent with respect to the two-configuration calculations are given.

	$M = -1$	$M = -2$	$M = -3$
β_Z	$2p_{-1}/-1\ 0$	$3d_{-2}/-2\ 0$	$4f_{-3}/-3\ 0$
1×10^{-4}	1.0001528	1.0001519	1.0001515
1.5×10^{-4}	1.0002291	1.0002278	1.0002272
2×10^{-4}	1.0003055	1.0003038	1.0003029
3×10^{-4}	1.0004582	1.0004556	1.0004542
5×10^{-4}	1.0007636	1.0007591	1.0007569
7×10^{-4}	1.0010688	1.0010625	1.0010593
1×10^{-3}	1.0015266	1.0015173	1.0015127
1.5×10^{-3}	1.0022893	1.0022744	1.0022676
2×10^{-3}	1.0030466	1.0030305	1.0030214
3×10^{-3}	1.0045766	1.0045398	1.0045262
5×10^{-3}	1.0076282	1.0075467	1.0075236
7×10^{-3}	1.0106858	1.0105380	1.0105051
1×10^{-2}	1.0152899 (.232)	1.0149964	1.0149474
1.5×10^{-2}	1.0230269 (.338)	1.0223534	1.0222718
2×10^{-2}	1.0308585 (.438)	1.0296218	1.0294976
3×10^{-2}	1.0468294 (.617)	1.0439077	1.0436572
5×10^{-2}	1.0798648 (.909)	1.0715523	1.0708418
7×10^{-2}	1.1137596 (1.14)	1.0980490	1.0965804
1×10^{-1}	1.1645216 (1.43)	1.1357480	1.1326499
1.5×10^{-1}	1.2448496 (1.87)	1.1933107	1.1866111
2×10^{-1}	1.3174515 (2.29)	1.2446382	1.2338000
3×10^{-1}	1.4401505 (3.14)	1.3308791	1.3115099
5×10^{-1}	1.6188115 (4.80)	<u>1.4534927</u>	1.4185237
7×10^{-1}	<u>1.7397581</u> (6.39)	1.587055	
1	1.995964	1.828723	
2	2.607165	2.395885	
5	3.669310	3.389708	
1×10^1	4.709163	4.365121	
2×10^1	5.995322	5.576044	
5×10^1	8.145986	7.971556	
1×10^2	10.17240	9.530427	
2×10^2	12.59597	11.83428	
5×10^2	16.49214	15.54685	15.37259
1×10^3	20.02577	18.92268	18.70656

Table A4.6. Total energies (in units of $-E_Z$) of selected states in the range $10^{-4} \leq \beta_Z \leq 3$ for Fe^{24+} ($Z = 26$) resulting from multi-configuration calculations (for $2p_{-1}/-1\,0$ three configurations were used, for all other states two). The underlined digits are those, up to which the results of one- and multi-configuration calculations are identical. If no digit is underlined, they differ in all digits.

	$M = 0$		$M = -1$		$M = -2$		$M = -3$
β_Z	$2p_0$ 0 1	$2s_0$ 0 2	$2p_{-1}$ -1 0	$3d_{-1}$ -1 1	$3d_{-2}$ -2 0	$4f_{-2}$ -2 1	$4f_{-3}$ -3 0
1×10^{-4}	1.23324$\underline{0}$	1.236078	1.23343$\underline{9}$	1.10334$\underline{9}$	1.10354$\underline{9}$	1.058582	1.05878$\underline{1}$
1.5×10^{-4}	1.23343$\underline{9}$	1.23627$\underline{8}$	1.23373$\underline{9}$	1.10364$\underline{8}$	1.10394$\underline{7}$	1.05897$\underline{9}$	1.05927$\underline{7}$
2×10^{-4}	1.23363$\underline{9}$	1.23647$\underline{8}$	1.23403$\underline{9}$	1.10394$\underline{7}$	1.10434$\underline{5}$	1.05937$\underline{5}$	1.05977$\underline{1}$
3×10^{-4}	1.23403$\underline{8}$	1.23687$\underline{6}$	1.23463$\underline{7}$	1.10454$\underline{3}$	1.10513$\underline{9}$	1.06016$\underline{1}$	1.06075$\underline{4}$
5×10^{-4}	1.23483$\underline{6}$	1.23767$\underline{1}$	1.23583$\underline{3}$	1.10573$\underline{0}$	1.10672$\underline{0}$	1.06172$\underline{0}$	1.06269$\underline{8}$
7×10^{-4}	1.23563$\underline{2}$	1.23846$\underline{3}$	1.23702$\underline{6}$	1.10691$\underline{1}$	1.10829$\underline{2}$	1.06325$\underline{7}$	1.06461$\underline{5}$
1×10^{-3}	1.23682$\underline{5}$	1.23964$\underline{8}$	1.23881$\underline{2}$	1.10867$\underline{0}$	1.11063$\underline{1}$	1.06552$\underline{5}$	1.06744$\underline{0}$
1.5×10^{-3}	1.23880$\underline{6}$	1.24160$\underline{9}$	1.24177$\underline{8}$	1.11157$\underline{1}$	1.11448$\underline{4}$	1.06920$\underline{5}$	1.07201$\underline{5}$
2×10^{-3}	1.24078$\underline{1}$	1.24355$\underline{4}$	1.24473$\underline{0}$	1.11443$\underline{2}$	1.11827$\underline{8}$	1.07276$\underline{4}$	1.07643$\underline{2}$
3×10^{-3}	1.24470$\underline{1}$	1.24739$\underline{8}$	1.25059$\underline{2}$	1.12003$\underline{9}$	1.12569$\underline{6}$	1.07954$\underline{8}$	1.08483$\underline{3}$
5×10^{-3}	1.25247$\underline{1}$	1.25490$\underline{2}$	1.26215$\underline{4}$	1.13081$\underline{1}$	1.13988$\underline{5}$	1.09198$\underline{0}$	1.10018$\underline{9}$
7×10^{-3}	1.26012$\underline{0}$	1.26216$\underline{3}$	1.27350$\underline{1}$	1.14103$\underline{9}$	1.15329$\underline{4}$	1.10323$\underline{1}$	1.11407$\underline{2}$
1×10^{-2}	1.2713$\underline{7}$8	1.2726$\underline{1}$3	1.29012$\underline{9}$	1.1554$\underline{9}$4	1.1721$\underline{6}$3	1.1185$\underline{1}$5	1.1329$\underline{5}$0
1.5×10^{-2}	1.2895$\underline{9}$2	1.2889$\underline{1}$8	1.3168$\underline{4}$7	1.1777$\underline{6}$1	1.2009$\underline{9}$4	1.14$\underline{1}$113	1.1609$\underline{4}$5
2×10^{-2}	1.307$\underline{1}$56	1.3039$\underline{6}$2	1.342$\underline{4}$18	1.1980$\underline{1}$1	1.227374	1.16$\underline{1}$318	1.18608$\underline{7}$
3×10^{-2}	1.340$\underline{5}$31	1.3308$\underline{5}$3	1.390$\underline{5}$68	1.234$\underline{7}$91	1.275054	1.1$\underline{9}$7184	1.23$\underline{1}$012
5×10^{-2}	1.4$\underline{0}$1497	1.37$\underline{5}$284	1.47$\underline{7}$493	1.298$\underline{3}$49	1.3$\underline{5}$7730	1.2$\underline{5}$8167	1.30$\underline{8}$361
7×10^{-2}	1.4$\underline{5}$6484	1.4$\underline{1}$1655	1.5$\underline{5}$5294	1.3$\underline{5}$3491	1.4$\underline{3}$0028	1.3$\underline{1}$0550	1.3$\underline{7}$5872
1×10^{-1}	1.5$\underline{3}$0582	1.4$\underline{5}$7157	1.6$\underline{5}$9976	1.4$\underline{2}$5936	1.5$\underline{2}$6186	1.3$\underline{7}$8916	1.4$\underline{6}$5678
1.5×10^{-1}	1.6$\underline{3}$7600	1.$\underline{5}$17780	1.8$\underline{1}$2191	1.5$\underline{2}$8125	1.6$\underline{6}$4869	1.4$\underline{7}$4657	1.5$\underline{9}$5268
2×10^{-1}	1.7$\underline{2}$9391	1.$\underline{5}$64699	1.9$\underline{4}$4919	1.6$\underline{1}$4049	1.7$\underline{8}$5086	1.5$\underline{5}$4303	1.7$\underline{0}$7617
3×10^{-1}	1.$\underline{8}$80682	1.$\underline{6}$27973	2.$\underline{1}$71002	1.$\underline{7}$52569	1.$\underline{9}$88468	1.$\underline{6}$81834	1.$\underline{8}$97525
5×10^{-1}	2.100010	1.$\underline{6}$71837	2.$\underline{5}$25682	1.$\underline{9}$45562	2.$\underline{3}$03309	1.$\underline{8}$54616	2.$\underline{1}$90509
7×10^{-1}	2.250831		2.$\underline{8}$02481	2.070307	2.$\underline{5}$44347	1.$\underline{9}$60911	2.$\underline{4}$13442
1	2.$\underline{3}$99976		3.130699	2.$\underline{1}$80299	2.$\underline{8}$23075	2.$\underline{0}$45166	2.$\underline{6}$68969
1.5	2.519848		3.538497	2.$\underline{2}$39217	3.154546		2.$\underline{9}$67885
2	2.$\underline{5}$43339		3.$\underline{8}$39886		3.$\underline{3}$84041		
3			4.254694		3.662525		

Table A4.7. Total energies (in units of $-E_Z$) of selected states in the range $10^{-4} \leq \beta_Z \leq 3$ for He ($Z = 2$) resulting from multi-configuration calculations (for $2p_{-1}/-1\,0$ three configurations were used, for all other states two). The underlined digits are those, up to which the results of one- and multi-configuration calculations are identical. If no digit is underlined, they differ in all digits.

	$M = 0$		$M = -1$		$M = -2$		$M = -3$
β_Z	$2p_0$ 0 1	$2s_0$ 0 2	$2p_{-1}$ $-1\,0$	$3d_{-1}$ $-1\,1$	$3d_{-2}$ $-2\,0$	$4f_{-2}$ $-2\,1$	$4f_{-3}$ $-3\,0$
1×10^{-4}	1.06611$\underline{8}$	1.08752$\underline{5}$	1.06631$\underline{8}$	1.02838$\underline{3}$	1.02858$\underline{2}$	1.01641$\underline{5}$	1.01661$\underline{2}$
1.5×10^{-4}	1.06631$\underline{8}$	1.08772$\underline{4}$	1.06661$\underline{7}$	1.02868$\underline{0}$	1.02897$\underline{6}$	1.01680$\underline{3}$	1.01709$\underline{6}$
2×10^{-4}	1.06651$\underline{7}$	1.08792$\underline{3}$	1.06691$\underline{5}$	1.02897$\underline{5}$	1.02936$\underline{9}$	1.01718$\underline{7}$	1.01757$\underline{4}$
3×10^{-4}	1.06691$\underline{5}$	1.08832$\underline{0}$	1.06751$\underline{1}$	1.02956$\underline{0}$	1.03014$\underline{7}$	1.01793$\underline{9}$	1.01851$\underline{1}$
5×10^{-4}	1.06770$\underline{8}$	1.08911$\underline{0}$	1.06869$\underline{7}$	1.03071$\underline{4}$	1.03167$\underline{8}$	1.01039$\underline{0}$	1.02031$\underline{2}$
7×10^{-4}	1.06849$\underline{7}$	1.08989$\underline{5}$	1.06987$\underline{6}$	1.03184$\underline{5}$	1.03317$\underline{6}$	1.02077$\underline{5}$	1.02202$\underline{5}$
1×10^{-3}	1.06967$\underline{4}$	1.08752$\underline{5}$	1.07163$\underline{2}$	1.03350$\underline{1}$	1.03536$\underline{1}$	1.02273$\underline{1}$	1.02444$\underline{8}$
1.5×10^{-3}	1.07161$\underline{9}$	1.08772$\underline{4}$	1.07452$\underline{5}$	1.03615$\underline{7}$	1.03885$\underline{2}$	1.02574$\underline{4}$	1.02816$\underline{5}$
2×10^{-3}	1.07354$\underline{3}$	1.08792$\underline{3}$	1.07737$\underline{6}$	1.03869$\underline{4}$	1.04217$\underline{2}$	1.02850$\underline{8}$	1.03157$\underline{3}$
3×10^{-3}	1.07732$\underline{6}$	1.08832$\underline{0}$	1.08295$\underline{5}$	1.04346$\underline{1}$	1.04838$\underline{0}$	1.03350$\underline{4}$	1.03774$\underline{1}$
5×10^{-3}	1.08465$\underline{0}$	1.08911$\underline{0}$	1.09365$\underline{7}$	1.05206$\underline{6}$	1.05952$\underline{8}$	1.04216$\underline{0}$	1.04846$\underline{5}$
7×10^{-3}	1.09167$\underline{9}$	1.08989$\underline{5}$	1.10382$\underline{0}$	1.05981$\underline{0}$	1.06952$\underline{9}$	1.04975$\underline{3}$	1.05792$\underline{0}$
1×10^{-2}	1.10174$\underline{8}$	1.12145$\underline{2}$	1.11823$\underline{1}$	1.07036$\underline{8}$	1.08314$\underline{9}$	1.05996$\underline{1}$	1.07070$\underline{6}$
1.5×10^{-2}	1.11751$\underline{2}$	1.13524$\underline{8}$	1.14054$\underline{4}$	1.08615$\underline{4}$	1.10353$\underline{2}$	1.07506$\underline{7}$	1.08978$\underline{3}$
2×10^{-2}	1.13229$\underline{8}$	1.14752$\underline{8}$	1.16129$\underline{8}$	1.10047$\underline{8}$	1.12207$\underline{6}$	1.08868$\underline{7}$	1.10713$\underline{5}$
3×10^{-2}	1.15976$\underline{1}$	1.16888$\underline{1}$	1.19963$\underline{1}$	1.12636$\underline{8}$	1.15572$\underline{2}$	1.11319$\underline{0}$	1.13869$\underline{8}$
5×10^{-2}	1.20905$\underline{3}$	1.20364$\underline{5}$	1.26835$\underline{8}$	1.17153$\underline{0}$	1.21525$\underline{2}$	1.15517$\underline{2}$	1.19449$\underline{8}$
7×10^{-2}	1.25326$\underline{9}$	1.23204$\underline{1}$	1.33038$\underline{8}$	1.21120$\underline{7}$	1.26838$\underline{9}$	1.19288$\underline{6}$	1.24437$\underline{2}$
1×10^{-1}	1.31284$\underline{3}$	1.26699$\underline{9}$	1.41496$\underline{2}$	1.26387$\underline{5}$	1.34041$\underline{6}$	1.24196$\underline{4}$	1.31192$\underline{3}$
1.5×10^{-1}	1.39879$\underline{9}$	1.31094$\underline{1}$	1.53971$\underline{3}$	1.33869$\underline{4}$	1.44626$\underline{9}$	1.30872$\underline{9}$	1.41100$\underline{5}$
2×10^{-1}	1.47213$\underline{1}$	1.34161$\underline{7}$	1.64946$\underline{0}$	1.40155$\underline{1}$	1.53920$\underline{8}$	1.36766$\underline{8}$	1.49779$\underline{1}$
3×10^{-1}	1.59130$\underline{1}$	1.37440$\underline{7}$	1.83694$\underline{9}$	1.50148$\underline{2}$	1.69747$\underline{9}$	1.45837$\underline{5}$	1.64513$\underline{5}$
5×10^{-1}	$\underline{1}$.757406		$\underline{2}$.129597	$\underline{1}$.63345$\underline{1}$	$\underline{1}$.94170$\underline{4}$	$\underline{1}$.57191$\underline{2}$	$\underline{1}$.87120$\underline{7}$
7×10^{-1}	$\underline{1}$.863526		$\underline{2}$.355134	$\underline{1}$.708622	$\underline{2}$.125664	$\underline{1}$.630168	$\underline{2}$.040050
1	$\underline{1}$.954228		$\underline{2}$.617895	$\underline{1}$.756067	$\underline{2}$.332528	$\underline{1}$.653932	$\underline{2}$.227566
1.5	$\underline{1}$.994483		$\underline{2}$.934609		$\underline{2}$.565106		$\underline{2}$.433057
2			$\underline{3}$.158623		$\underline{2}$.711170		$\underline{2}$.555858
3			3.442903		$\underline{2}$.849574		$\underline{2}$.654298

A4.2 Wavelengths, Dipole Strengths, Oscillator Strengths, and Transition Rates

In the following tables A4.8 – A4.10, wavelengths (in Å), dipole strengths d (in units of a_Z^2), oscillator strengths f, transition probabilities w (in units of $w_Z = Z^4 8.02 \times 10^9 \, \text{s}^{-1}$) and transition energies $\hbar\omega$ (in units of E_Z) for $\Delta M = 0, 1$ transitions of two-electron systems with $Z = 2, 8, 14, 26, \infty$ calculated from Hartree-Fock wave functions in adiabatic approximation are listed for the magnetic field strength parameter $0.1 \leq \beta_Z \leq 10^3$.

Table A4.8a. Wavelengths λ and oscillator strengths f of some transitions $n'l'm' \leftrightarrow nlm$ in the range $5 \times 10^{-5} \leq \beta_z \leq 10^{-2}$ for He with $m', m \leq 0$. For each pair the wavelength (in Å) is given in the first line, the strength in the second line. The wavelenghts in brackets have been determined by correcting for the zero field discrepancy (cf. Sect 9.4.4).

$\beta_{Z=2}$	$1s_0 3d_{m'} \leftrightarrow 1s_0 2p_m$				
B in 10^2 T	$1s_0 3d_0$ $\leftrightarrow$ $1s_0 2p_0$	$1s_0 3d_{-1}$ $\leftrightarrow$ $1s_0 2p_{-1}$	$1s_0 3d_{-1}$ $\leftrightarrow$ $1s_0 2p_0$	$1s_0 3d_0$ $\leftrightarrow$ $1s_0 2p_{-1}$	$1s_0 3d_{-2}$ $\leftrightarrow$ $1s_0 2p_{-1}$
5×10^{-5} 0.94	6005.8 0.74813 (5877.3)	6005.8 0.56109 (5877.3)	6021.6 0.55962 (5892.5)	5990.0 0.18752 (5862.2)	6021.6 1.1193 (5892.5)
1×10^{-4} 1.88	6005.6 0.74821 (5877.1)	6005.5 0.56115 (5877.1)	6037.3 0.55822 (5904.5)	5974.1 0.18803 (5847.0)	6037.1 1.1165 (5907.4)
3×10^{-4} 5.64	6003.1 0.74906 (5874.7)	6003.0 0.56171 (5874.6)	6098.8 0.55311 (5966.3)	5910.2 0.19013 (5785.8)	6097.3 1.1068 (5965.0)
5×10^{-4} 9.40	5998.1 0.75075 (5870.0)	5997.9 0.56282 (5869.7)	6158.3 0.54878 (6023.3)	5845.8 0.19236 (5724.0)	6154.1 1.0992 (6019.3)
1×10^{-3} 18.8	5975.2 0.75841 (5848.0)	5974.4 0.56780 (5847.3)	6297.8 0.54101 (6156.7)	5683.3 0.19847 (5568.1)	6280.9 1.0880 (6140.6)
5×10^{-3} 94.0	5435.5 0.90459 (5330.1)	5452.0 0.64177 (5346.0)	6951.3 0.54149 (6779.8)	4473.6 0.25494 (4402.0)	6660.7 1.1067 (6503.1)
1×10^{-2} 188	4568.5 1.0742 (4493.8)	4666.2 0.68943 (4588.3)	7057.5 0.53410 (6880.8)	3430.5 0.30295 (3388.2)	6435.0 1.0646 (6287.7)

Table A4.8b. Wavelengths λ and oscillator strengths f of some transitions $n'l'm' \leftrightarrow nlm$ in the range $5 \times 10^{-5} \leq \beta_z \leq 10^{-2}$ for He with $m', m \leq 0$. For each pair the wavelength (in Å) is given in the first line, the strength in the second line.

$\beta_{Z=2}$	$1s_04d_{m'} \leftrightarrow 1s_02p_m$				
B in 10^2 T	$1s_04d_0$ $\leftrightarrow$ $1s_02p_0$	$1s_04d_{-1}$ $\leftrightarrow$ $1s_02p_{-1}$	$1s_04d_{-1}$ $\leftrightarrow$ $1s_02p_0$	$1s_04d_0$ $\leftrightarrow$ $1s_02p_{-1}$	$1s_04d_{-2}$ $\leftrightarrow$ $1s_02p_{-1}$
5×10^{-5} 0.94	4548.1 0.14782	4548.0 0.11087	4557.1 0.11065	4539.0 0.03703	4557.0 0.22133
1×10^{-4} 1.88	4547.4 0.14792	4547.3 0.11096	4565.5 0.11052	4529.4 0.03713	4565.0 0.22116
3×10^{-4} 5.64	4540.8 0.14898	4539.6 0.11193	4594.2 0.11057	4487.5 0.03770	4589.9 0.22217
5×10^{-4} 9.40	4527.9 0.15098	4524.6 0.11374	4615.3 0.11142	4440.5 0.03852	4603.5 0.22543
1×10^{-3} 18.8	4470.9 0.15884	4459.5 0.12068	4637.3 0.11569	4305.5 0.04137	4594.5 0.23841
5×10^{-3} 94.0	3580.1 0.19309	3516.7 0.15417	4085.0 0.11880	3135.9 0.06080	3762.8 0.24138
13×10^{-2} 188	2668.0 0.15513	2617.5 0.14485	3231.8 0.07711	2235.0 0.06561	2826.9 0.16689

Table A4.8c. Wavelengths λ and oscillator strengths f of some transitions $n'l'm' \leftrightarrow nlm$ in the range $5 \times 10^{-5} \leq \beta_z \leq 10^{-2}$ for He with $m', m \leq 0$. For each pair the wavelength (in Å) is given in the first line, the strength in the second line.

$\beta_{Z=2}$	$1s_05d_{m'} \leftrightarrow 1s_02p_m$				
B in 10^2 T	$1s_05d_0$ $\leftrightarrow$ $1s_02p_0$	$1s_05d_{-1}$ $\leftrightarrow$ $1s_02p_{-1}$	$1s_05d_{-1}$ $\leftrightarrow$ $1s_02p_0$	$1s_05d_0$ $\leftrightarrow$ $1s_02p_{-1}$	$1s_05d_{-2}$ $\leftrightarrow$ $1s_02p_{-1}$
5×10^{-5} 0.94	4088.5 0.05627	4088.4 0.04221	4059.7 0.04213	4081.1 0.01409	4095.4 0.08432
1×10^{-4} 1.88	4087.1 0.05645	4086.7 0.04237	4101.4 0.04222	4072.5 0.01416	4100.3 0.08465
3×10^{-4} 5.64	4072.6 0.05824	4069.7 0.04397	4113.5 0.04349	4029.7 0.01472	4104.0 0.08860
5×10^{-4} 9.40	4045.3 0.06133	4037.8 0.04667	4109.9 0.04580	3975.5 0.01562	4085.8 0.09506
1×10^{-3} 18.8	3936.6 0.07087	3913.6 0.05462	4049.8 0.05257	3807.7 0.01839	3976.7 0.11142
5×10^{-3} 94.0	2796.1 0.08347	2712.4 0.06552	3038.5 0.05246	2517.6 0.02581	2735.4 0.10074
13×10^{-2} 188	1937.8 0.05679	1868.2 0.04951	2161.4 0.03152	1698.8 0.02224	1959.2 0.05995

Table A4.8d. Wavelengths λ and oscillator strengths f of some transitions $n'l'm' \leftrightarrow nlm$ in the range $5 \times 10^{-5} \leq \beta_z \leq 10^{-2}$ for He with $m', m \leq 0$. For each pair the wavelength (in Å) is given in the first line, the strength in the second line.

$\beta_{Z=2}$	$1s_03s_0 \leftrightarrow 1s_02p_m$		$1s_04s_0 \leftrightarrow 1s_02p_m$		$1s_05s_0 \leftrightarrow 1s_02p_m$	
B in 10^2 T	$1s_03s_0$ $\leftrightarrow$ $1s_02p_0$	$1s_03s_0$ $\leftrightarrow$ $1s_02p_{-1}$	$1s_04s_0$ $\leftrightarrow$ $1s_02p_0$	$1s_04s_0$ $\leftrightarrow$ $1s_02p_{-1}$	$1s_05s_0$ $\leftrightarrow$ $1s_02p_0$	$1s_05s_0$ $\leftrightarrow$ $1s_02p_{-1}$
5×10^{-5} 0.94	7237.6 0.21667	7214.7 0.21667	4795.8 0.03224	4785.7 0.03231	4185.1 0.01147	4177.4 0.01149
1×10^{-4} 1.88	7237.0 0.21602	7191.5 0.21737	4794.9 0.03226	4774.8 0.03240	4183.2 0.01151	4167.9 0.01155
3×10^{-4} 5.64	7231.1 0.21628	7096.8 0.22019	4785.0 0.3247	4725.8 0.03289	4163.7 0.01189	4118.8 0.01203
5×10^{-4} 9.40	7219.2 0.21681	6999.8 0.22308	4765.6 0.03287	4669.0 0.03359	4127.0 0.01254	4054.3 0.01278
1×10^{-3} 18.8	7164.7 0.21920	6749.2 0.23055	4680.7 0.03440	4499.6 0.03594	3981.7 0.01450	3849.9 0.01507
5×10^{-3} 94.0	5949.0 0.26925	4815.7 0.27974	3424.0 0.03728	3015.6 0.04872	2566.7 0.01637	2330.1 0.02015
13×10^{-2} 188	4302.3 0.33643	3278.2 0.28433	2306.2 0.02350	1975.4 0.04641	1640.5 0.01105	1465.8 0.01620

Table A4.9a. Dipole strength d (in units of a_Z^2), oscillator strength f, transition probability w (in units of $w_Z = Z^4 8.02 \times 10^9 \, \text{s}^{-1}$) and transition energy $\hbar\omega$ (in units of E_Z) of the $\Delta M = 0$ transition $0\ 0\ -1\ 1 \leftrightarrow 0\ 0\ -1\ 0$ for various values of β_Z and of the nuclear charge Z calculated from Hartree-Fock wave functions in adiabatic approximation. In each block d is given in the first line, f in the second, w in the third, and $\hbar\omega$ in the last line.

β_Z	$Z = 2$	$Z = 8$	$Z = 14$	$Z = 26$	$Z = \infty$
			$0\ 0\ -1\ 1 \leftrightarrow 0\ 0\ -1\ 0$		
0.1	5.797 0.8745 0.006635 0.1509	5.465 0.9370 0.009183 0.1715	5.405 0.9385 0.009434 0.1737	5.368 0.9390 0.009577 0.1749	
0.5	2.130 0.7531 0.03138 0.3536	2.0225 0.8752 0.05463 0.4327	1.997 0.8828 0.0570 0.4420	1.981 0.8868 0.05921 0.4476	
5	0.47092 0.4959 0.1833 1.053	0.50091 0.7178 0.4914 1.433	0.49738 0.7364 0.5380 1.480	0.4948 0.7452 0.5633 1.506	
10	0.28996 0.4100 0.2732 1.414	0.32821 0.6516 0.8559 1.985	0.32730 0.6735 0.9504 2.058	0.32636 0.6858 1.009 2.101	0.32499 0.6992 1.079 2.151
50	0.090177 0.2385 0.5561 2.645	0.12048 0.4774 2.498 3.962	0.12193 0.5044 2.877 4.137	0.12258 0.5200 3.120 4.243	0.12314 0.5374 3.412 4.364
100	0.054294 0.1839 0.7029 3.387	0.077528 0.4025 3.616 5.191	0.07906 0.4297 4.231 5.435	0.079831 0.4457 4.631 5.583	0.080579 0.4637 5.118 5.754
250	0.028052 0.1293 0.9159 4.610				
500	0.017253 0.09916 1.091 5.747	0.027678 0.2540 7.128 9.176	0.02871 0.2772 8.617 9.656	0.029286 0.2914 9.618 9.951	0.029905 0.3078 10.87 10.29
1000	0.010774 0.07638 1.280 7.092	0.017864 0.2048 8.975 11.47	0.018643 0.2253 10.97 12.09	0.019090 0.2380 12.33 12.47	0.019580 0.2528 14.05 12.91

Table A4.9b. Dipole strength d (in units of a_Z^2), oscillator strength f, transition probability w (in units of $w_Z = Z^4 8.02 \times 10^9\,\mathrm{s}^{-1}$) and transition energy $\hbar\omega$ (in units of E_Z) of the $\Delta M = 0$ transition $0\,0\,-2\,1) \leftrightarrow (0\,0\,-2\,0$ for various values of β_Z and of the nuclear charge Z calculated from Hartree-Fock wave functions in adiabatic approximation. In each block d is given in the first line, f in the second, w in the third, and $\hbar\omega$ in the last line.

β_Z	$0\,0\,-2\,1 \leftrightarrow 0\,0\,-2\,0$				
	$Z = 2$	$Z = 8$	$Z = 14$	$Z = 26$	$Z = \infty$
0.1	10.94 0.9104 0.002103 0.08325	0.329 0.9530 0.004159 0.1144	8.078 0.9555 0.004455 0.1183	7.934 0.9568 0.004638 0.1206	
0.5	3.898 0.8233 0.01224 0.2112	2.998 0.9070 0.02767 0.3025	2.908 0.9130 0.03000 0.3140	2.856 0.9163 0.03145 0.3209	
1	2.4960 0.7685 0.02429 0.03079	1.94866 0.8771 0.05922 0.4501	1.89111 0.8852 0.06464 0.4681	1.85763 0.8896 0.06802 0.4789	
5	0.86370 0.5976 0.09535 0.6919	0.72528 0.7742 0.2940 1.067	0.70702 0.7889 0.3276 1.116	0.69615 0.7971 0.3483 1.145	
10	0.53673 0.5110 0.1544 0.9521	0.47404 0.7137 0.5393 1.506	0.46381 0.7317 0.6070 1.578	0.45759 0.7418 0.6499 1.621	0.45044 0.7528 0.7010 1.671
50	0.16992 0.3182 0.3719 1.873	0.17400 0.5435 1.768 3.124	0.17250 0.5674 2.051 3.292	0.17145 0.5819 2.235 3.394	0.17011 0.5975 2.457 3.512
100	0.10238 0.2502 0.4978 2.443	0.11199 0.4653 2.677 4.155	0.11183 0.4908 3.150 4.388	0.11161 0.5057 3.460 4.531	0.11125 0.5224 3.839 4.696
250	0.052494 0.1786 0.6896 3.403	0.062148 0.3670 4.267 5.906	0.062688 0.3927 5.137 6.265	0.062934 0.4070 5.676 6.468	

Table A4.9c. Dipole strength d (in units of a_Z^2), oscillator strength f, transition probability w (in units of $w_Z = Z^4 8.02 \times 10^9 \, \text{s}^{-1}$) and transition energy $\hbar\omega$ (in units of E_Z) of the $\Delta M = 0$ transition $0\,0\,-3\,1 \leftrightarrow 0\,0\,-3\,0$ for various values of β_Z and of the nuclear charge Z calculated from Hartree-Fock wave functions in adiabatic approximation. In each block d is given in the first line, f in the second, w in the third, and $\hbar\omega$ in the last line.

		$0\,0\,-3\,1 \leftrightarrow 0\,0\,-3\,0$				
β_Z	$Z = 2$	$Z = 8$	$Z = 14$	$Z = 26$	$Z = \infty$
0.1	15.62 0.9374 0.1125×10^{-2} 0.6001×10^{-1}	10.83 0.9629 0.2536×10^{-2} 0.8889×10^{-1}	10.42 0.9647 0.2755×10^{-2} 0.9256×10^{-1}	10.19 0.9657 0.2891×10^{-2} 0.9477×10^{-1}	
0.5	5.450 0.8657 0.7283×10^{-2} 0.1589	3.838 0.9246 0.1788×10^{-1} 0.2409	3.696 0.9291 0.1957×10^{-1} 0.2514	3.615 0.9316 0.2063×10^{-1} 0.2577	
1	3.475 0.8183 0.1512×10^{-1} 0.2355	2.481 0.8988 0.3933×10^{-1} 0.3623	2.391 0.9051 0.4325×10^{-1} 0.3786	2.339 0.9085 0.4570×10^{-1} 0.3884	
5	1.206 0.6614 0.6629 0.5483	0.9155 0.8071 0.2091 0.8816	0.8857 0.8192 0.2336 0.9249	0.8684 0.8260 0.2491 0.9511	
10	0.7540 0.5770 0.1126 0.7651	0.5976 0.7513 0.3958 1.257	0.5800 0.7666 0.4464 1.322	0.5696 0.7752 0.4785 1.361	0.5580 0.7845 0.5168 1.406
50	0.2424 0.3752 0.2998 1.548	0.2197 0.5872 1.398 2.673	0.2157 0.6094 1.621 2.825	0.2133 0.6221 1.765 2.917	0.2104 0.6363 1.939 3.024
100	0.1465 0.2991 0.4154 2.041	0.1416 0.5082 2.183 3.590	0.1400 0.5321 2.564 3.802	0.1389 0.5461 2.813 3.931	0.1376 0.5617 3.118 4.081

Table A4.9d. Dipole strength d (in units of a_Z^2), oscillator strength f, transition probability w (in units of $w_Z = Z^4 8.02 \times 10^9 \text{ s}^{-1}$) and transition energy $\hbar\omega$ (in units of E_Z) of the $\Delta M = 1$ transition $0\,0\,-2\,0 \leftrightarrow 0\,0\,-1\,0$ for various values of β_Z and of the nuclear charge Z calculated from Hartree-Fock wave functions in adiabatic approximation. In each block d is given in the first line, f in the second, w in the third, and $\hbar\omega$ in the last line.

	$0\,0\,-2\,0 \leftrightarrow 0\,0\,-1\,0$				
β_Z	$Z = 2$	$Z = 8$	$Z = 14$	$Z = 26$	$Z = \infty$
0.1	9.686 0.8245 0.1992×10^{-2} 0.3513×10^{-1}	9.833 0.8853 0.2392×10^{-2} 0.9003×10^{-1}	9.846 0.8920 0.2441×10^{-2} 0.9060×10^{-1}	9.853 0.8960 0.2470×10^{-2} 0.9094×10^{-1}	
0.5	1.951 0.3154 0.2749×10^{-2} 0.1617	1.974 0.3411 0.3398×10^{-2} 0.1729	1.9755 0.3440 0.3478×10^{-2} 0.1742	1.976 0.3457 0.3526×10^{-2} 0.1749	
1	0.9781 0.2057 0.303×10^{-2} 0.2103	0.9882 0.2223 0.380×10^{-2} 0.2260	0.9890 0.2253 0.390×10^{-2} 0.2278	0.9894 0.2265 0.396×10^{-2} 0.2289	
5	0.1967 0.7351×10^{-1} 0.342×10^{-2} 0.3738	0.1982 0.8064×10^{-1} 0.445×10^{-2} 0.4069	0.1983 0.8145×10^{-1} 0.458×10^{-2} 0.4107	0.1984 0.8194×10^{-1} 0.466×10^{-2} 0.4130	
10	0.9852×10^{-1} 0.4643×10^{-1} 0.344×10^{-2} 0.4713	0.9920×10^{-1} 0.5117×10^{-1} 0.454×10^{-2} 0.5158	0.9925×10^{-1} 0.5171×10^{-1} 0.468×10^{-2} 0.5211	0.9928×10^{-1} 0.5204×10^{-1} 0.477×10^{-2} 0.5242	0.9931×10^{-1} 0.5241×10^{-1} 0.487×10^{-2} 0.528
50	0.1978×10^{-1} 0.1535×10^{-1} 0.308×10^{-2} 0.7761	0.1988×10^{-1} 0.1709×10^{-1} 0.421×10^{-2} 0.7431	0.1989×10^{-1} 0.1729×10^{-1} 0.436×10^{-2} 0.8696	0.1989×10^{-1} 0.1742×10^{-1} 0.445×10^{-2} 0.8755	0.1990×10^{-1} 0.1756×10^{-1} 0.456×10^{-2} 0.882
100	0.9902×10^{-2} 0.9362×10^{-2} 0.279×10^{-2} 0.9455	0.9948×10^{-2} 0.1047×10^{-1} 0.386×10^{-2} 1.052	0.9951×10^{-2} 0.1060×10^{-1} 0.400×10^{-2} 1.065	0.9953×10^{-2} 0.1067×10^{-1} 0.409×10^{-2} 1.072	0.9955×10^{-2} 0.1076×10^{-1} 0.419×10^{-2} 1.081
250	0.3967×10^{-2} 0.4791×10^{-2} 0.233×10^{-2} 1.2008				
500	0.1985×10^{-3} 0.2852×10^{-2} 0.196×10^{-2} 1.436	0.1992×10^{-2} 0.3211×10^{-2} 0.278×10^{-1} 1.612	0.1993×10^{-2} 0.3255×10^{-2} 0.289×10^{-2} 1.633	0.1993×10^{-2} 0.3281×10^{-2} 0.296×10^{-2} 1.646	0.1994×10^{-2} 0.3311×10^{-2} 0.304×10^{-2} 1.661
1000	0.9936×10^{-3} 0.1680×10^{-2} 0.160×10^{-2} 1.691	0.9967×10^{-3} 0.1986×10^{-2} 0.229×10^{-2} 1.902	0.9969×10^{-3} 0.1922×10^{-2} 0.238×10^{-2} 1.928	0.9970×10^{-3} 0.1938×10^{-2} 0.244×10^{-2} 1.946	0.9972×10^{-3} 0.1957×10^{-2} 0.251×10^{-2} 1.962

Table A4.9e. Dipole strength d (in units of a_Z^2), oscillator strength f, transition probability w (in units of $w_Z = Z^4 8.02 \times 10^9$ s^{-1}) and transition energy $\hbar\omega$ (in units of E_Z) of the $\Delta M = 1$ transition $0\,0\,-3\,0 \leftrightarrow 0\,0\,-2\,0$ for various values of β_Z and of the nuclear charge Z calculated from Hartree-Fock wave functions in adiabatic approximation. In each block d is given in the first line, f in the second, w in the third, and $\hbar\omega$ in the last line.

			$0\,0\,-3\,0 \leftrightarrow 0\,0\,-2\,0$		
β_Z	$Z = 2$	$Z = 8$	$Z = 14$	$Z = 26$	$Z = \infty$
0.1	1.483×10 0.4871 0.1719×10^{-3} 0.3264×10^{-1}	1.4897×10 0.6746 0.4611×10^{-3} 0.4528×10^{-1}	1.4902×10 0.7010 0.5171×10^{-3} 0.4704×10^{-1}	1.4905×10 0.7173 0.5537×10^{-3} 0.4812×10^{-1}	
0.5	2.974 0.1878 0.2496×10^{-3} 0.6315×10^{-1}	2.983 0.2639 0.6880×10^{-3} 0.8844×10^{-1}	2.9840 0.2745 0.7740×10^{-3} 0.9198×10^{-1}	2.9844 0.2810 0.8302×10^{-3} 0.9415×10^{-1}	
1	1.488 0.1233 0.2821×10^{-3} 0.8285×10^{-1}	1.4925 0.1739 0.7976×10^{-3} 0.1165	1.4928 0.1810 0.8874×10^{-3} 0.1213	1.4929 0.1854 0.9526×10^{-3} 0.1242	
5	0.2982 0.4495×10^{-1} 0.3403×10^{-3} 0.1507	0.29882 0.6396×10^{-1} 0.9766×10^{-3} 0.2140	0.29887 0.6663×10^{-1} 0.1104×10^{-2} 0.2229	0.29889 0.6827×10^{-1} 0.1187×10^{-2} 0.2284	
10	0.1492 0.2867×10^{-1} 0.3530×10^{-3} 0.1922	0.14948 0.4093×10^{-1} 0.1023×10^{-2} 0.2738	0.14949 0.4265×10^{-1} 0.1157×10^{-2} 0.2853	0.14951 0.4371×10^{-1} 0.1245×10^{-3} 0.2924	0.14952 0.4495×10^{-1} 0.1354×10^{-2} 0.3006
50	0.2988×10^{-1} 0.9739×10^{-2} 0.3449×10^{-3} 0.3259	0.29921×10^{-1} 0.1395×10^{-1} 0.1011×10^{-2} 0.4662	0.29924×10^{-1} 0.1456×10^{-1} 0.1146×10^{-2} 0.4861	0.29925×10^{-1} 0.1491×10^{-1} 0.1234×10^{-2} 0.4983	0.29927×10^{-1} 0.1534×10^{-1} 0.1344×10^{-2} 0.5126
100	0.1495×10^{-1} 0.6020×10^{-2} 0.3254×10^{-3} 0.4027	0.14965×10^{-1} 0.8620×10^{-2} 0.9532×10^{-3} 0.5760	0.14966×10^{-1} 0.8989×10^{-2} 0.1081×10^{-2} 0.6006	0.14967×10^{-1} 0.9216×10^{-2} 0.1165×10^{-2} 0.6158	0.14968×10^{-1} 0.9482×10^{-2} 0.1268×10^{-2} 0.6335
500	0.2992×10^{-2} 0.1895×10^{-2} 0.2535×10^{-3} 0.6335	0.29948×10^{-2} 0.2701×10^{-2} 0.7320×10^{-3} 0.9018	0.29950×10^{-2} 0.2816×10^{-2} 0.8295×10^{-3} 0.9401	0.29951×10^{-2} 0.2887×10^{-2} 0.8937×10^{-3} 0.9638	0.29953×10^{-2} 0.2970×10^{-2} 0.9730×10^{-3}
1000	0.1496×10^{-2} 0.1133×10^{-2} 0.2166×10^{-3} 0.7572	0.149772×10^{-2} 0.1608×10^{-2} 0.6184×10^{-3} 1.074	0.149781×10^{-2} 0.1677×10^{-2} 0.7002×10^{-3} 1.119	0.149785×10^{-2} 0.1719×10^{-2} 0.7541×10^{-3} 1.147	0.149797×10^{-2} 0.1768×10^{-2} 0.8207×10^{-3}

Table A4.9f. Dipole strength d (in units of a_Z^2), oscillator strength f, transition probability w (in units of $w_Z = Z^4 8.02 \times 10^9 \text{ s}^{-1}$) and transition energy $\hbar\omega$ (in units of E_Z) of the $\Delta M = 1$ transition 00 02 $\leftrightarrow$ 00 $-$1 0 for various values of β_Z and of the nuclear charge Z calculated from Hartree-Fock wave functions in adiabatic approximation. In each block d is given in the first line, f in the second, w in the third, and $\hbar\omega$ in the last line.

		0 0 0 2 $\leftrightarrow$ 0 0 $-$1 0				
β_Z	$Z = 2$	$Z = 8$	$Z = 14$	$Z = 26$	$Z = \infty$	
0.1	0.1769 0.3032×10⁻¹ 0.2968×10⁻³ 0.1714	0.1987 0.4075×10⁻¹ 0.5709×10⁻³ 0.2050	0.2001 0.4177×10⁻¹ 0.6068×10⁻³ 0.2087	0.2008 0.4236×10⁻¹ 0.6282×10⁻³ 0.2109		
0.5	0.2087×10⁻¹ 0.8342×10⁻² 0.4441×10⁻³ 0.3996	0.2386×10⁻¹ 0.1260×10⁻¹ 0.1171×10⁻² 0.5280	0.2395×10⁻¹ 0.1303×10⁻¹ 0.1286×10⁻² 0.5441	0.2399×10⁻¹ 0.1328×10⁻¹ 0.1358×10⁻² 0.5538		
1	0.7976×10⁻² 0.4434×10⁻² 0.457×10⁻³ 0.5559	0.9199×10⁻² 0.6998×10⁻² 0.135×10⁻² 0.7608	0.9230×10⁻² 0.7266×10⁻² 0.150×10⁻² 0.7872	0.9240×10⁻² 0.7422×10⁻² 0.160×10⁻² 0.8032		
5	0.8081×10⁻³ 0.9073×10⁻³ 0.381×10⁻³ 1.123	0.9293×10⁻³ 0.1526×10⁻² 0.137×10⁻² 1.643	0.9305×10⁻³ 0.1594×10⁻² 0.156×10⁻² 1.713	0.9301×10⁻³ 0.1633×10⁻² 0.168×10⁻² 1.756		
10	0.2980×10⁻³ 0.4423×10⁻³ 0.325×10⁻³ 1.484	0.3374×10⁻³ 0.7499×10⁻³ 0.124×10⁻² 2.223	0.3372×10⁻³ 0.7837×10⁻³ 0.141×10⁻² 2.324	0.3367×10⁻³ 0.8035×10⁻³ 0.153×10⁻² 2.387	0.3358×10⁻³ 0.8256×10⁻³ 0.166×10⁻² 2.459	
50	0.2924×10⁻⁴ 0.7922×10⁻⁴ 0.194×10⁻³ 2.709	0.3109×10⁻⁴ 0.1317×10⁻³ 0.787×10⁻³ 4.235	0.3085×10⁻⁴ 0.1373×10⁻³ 0.907×10⁻³ 4.451	0.3068×10⁻⁴ 0.1406×10⁻³ 0.985×10⁻³ 4.584	0.3045×10⁻⁴ 0.1443×10⁻³ 0.108×10⁻² 4.739	
100	0.1077×10⁻⁴ 0.3712×10⁻⁴ 0.147×10⁻³ 3.447	0.1108×10⁻⁴ 0.6055×10⁻⁴ 0.603×10⁻³ 5.467	0.1095×10⁻⁴ 0.6301×10⁻⁴ 0.696×10⁻³ 5.755	0.1086×10⁻⁴ 0.6443×10⁻⁴ 0.756×10⁻³ 5.933	0.1075×10⁻⁴ 0.6601×10⁻⁴ 0.830×10⁻³ 6.141	
250	0.2883×10⁻⁵ 0.1345×10⁻⁴ 0.976×10⁻⁴ 4.665					
500	0.1066×10⁻⁵ 0.6183×10⁻⁵ 0.693×10⁻⁴ 5.798	0.1014×10⁻⁵ 0.9569×10⁻⁵ 0.284×10⁻³ 9.437	0.9927×10⁻⁶ 0.9893×10⁻⁵ 0.328×10⁻³ 9.966	0.9792×10⁻⁶ 0.1008×10⁻⁴ 0.356×10⁻³ 1.029×10	0.9631×10⁻⁶ 0.1028×10⁻⁴ 0.391×10⁻³ 1.067×10	
1000	0.3954×10⁻⁶ 0.2823×10⁻⁵ 0.480×10⁻⁴ 7.140	0.3646×10⁻⁶ 0.4271×10⁻⁵ 0.195×10⁻³ 1.172×10	0.3555×10⁻⁶ 0.4403×10⁻⁵ 0.225×10⁻³ 1.239×10	0.3499×10⁻⁶ 0.4478×10⁻⁵ 0.245×10⁻³ 1.280×10	0.3432×10⁻⁶ 0.4558×10⁻⁵ 0.268×10⁻³ 1.328×10	

Table A4.10. Dipole strength d (in units of a_Z^2), oscillator strength f, transition probability w (in units of $w_Z = Z^4 8.02 \times 10^9$ s^{-1}) and transition energy $\hbar\omega$ (in units of E_Z) of the $\Delta M = 1$ transition $0\ 0\ -1\ 1 \leftrightarrow 0\ 0\ 0\ 1$ for He ($Z = 2$) in the range $10 \leq \beta_{Z=2} \leq 1000$ calculated from Hartree-Fock wave functions in adiabatic approximation. In each block d is given in the first line, f in the second, w in the third, and $\hbar\omega$ in the last line.

$\beta_{Z=2}$	$0\ 0\ -1\ 1 \leftrightarrow 0\ 0\ 0\ 1$
B in 10^8 T	$Z = 2$
10 0.188	0.48377×10^{-1} 0.23882×10^{-2} 0.1940×10^{-5} 0.4937×10^{-1}
50 0.940	0.98080×10^{-2} 0.3661×10^{-3} 0.1701×10^{-6} 0.3733×10^{-1}
100 1.88	0.49228×10^{-2} 0.16297×10^{-3} 0.5954×10^{-7} 0.3311×10^{-1}
250 4.70	0.19765×10^{-2} 0.5634×10^{-4} 0.1526×10^{-7} 0.2850×10^{-1}
500 9.40	0.99034×10^{-3} 0.2541×10^{-4} 0.5575×10^{-8} 0.2566×10^{-1}
1000 18.8	0.49599×10^{-3} 0.1153×10^{-4} 0.2079×10^{-8} 0.2325×10^{-1}

References

Chapter 1

Aldrich, C., and Greene, R.L. (1979): Hydrogen-like systems in arbitrary magnetic fields. *Phys. Stat. Sol. b* **93**, 343.

Alijah, A., Hinze, J., and Broad, J.T. (1990): Photoionisation of hydrogen in a strong magnetic field. *J. Phys. B: At. Mol. Opt. Phys.* **23**, 45.

Angel, J.R.P. (1978): Magnetic white dwarfs. *Ann. Rev. Astron. Astrophys.* **16**, 487.

Angel, J.R.P., Borra, E.F., and Landstreet, J.D. (1981): The magnetic fields of white dwarfs. *Astrophys. J. Suppl. Ser.* **45**, 457.

Angel, J.R.P., Liebert, J., and Stockman, H.S. (1985): The optical spectrum of hydrogen at 160–350 million gauss in the white dwarf Grw+70°8247. *Astrophys. J.* **292**, 260.

Avron, J.E., Herbst, I.W., and Simon B. (1978): Separation of center of mass in homogeneous magnetic fields. *Ann. Phys. (N.Y.)* **114**, 431.

Baye, D. (1982): An approximate constant of motion for the problem of an atomic ion in a homogeneous magnetic field. *J. Phys. B: At. Mol. Phys.* **15**, L795.

Baye, D. (1983): Separation of centre-of-mass motion for a charged two-body system in a homogeneous magnetic field. *J. Phys. A: Math. Gen.* **16**, 3207.

Baye, D., and Vincke, M. (1984): A simple variational basis for the study of hydrogen atoms in strong magnetic fields. *J. Phys. B: At. Mol. Phys.* **17**, L631.

Baye, D., and Vincke, M. (1986): Centre-of-mass energy of hydrogenic ions in a magnetic field. *J. Phys. B: At. Mol. Phys.* **19**, 4051.

Bender, C.M., Mlodinow, L.D., and Papanicolaou, N. (1982): Semiclassical perturbation theory for the hydrogen atom in a uniform magnetic field. *Phys. Rev. A* **25**, 1305.

Brandi, H.S., Santos, R.R., and Miranda, L.C.M. (1976): Hydrogen atoms in strong magnetic fields: oscillator strengths. *Lett. Nuovo Cimento* **16**, 187.

Burdyuzha, V.V., and Pavlov-Verevkin, V.B. (1981): The spectrum of the hydrogen atom and hydrogenlike ions in the magnetic field of a neutron star. *Sov. Astron.* **25**, 187.

Cabib, D., Fabri, E., and Fiorio, G. (1971): The ground state of the exciton in a magnetic field. *Solid State Commun.* **9**, 1517.

Cizek, J., and Vrscay, E.R. (1982): Large order perturbation theory in the context of atomic and molecular physics – interdisciplinary aspects. *Int. J. Quantum Chem.* **21**, 27.

Clark, C.W., and Taylor, K.T. (1982): The quadratic Zeeman effect in hydrogen Rydberg series: application of Sturmian functions. *J. Phys. B: At. Mol. Phys.* **15**, 1175.

Cohen, M., and Herman, G. (1981): The hydrogen atom in strong magnetic fields. *J. Phys. B: At. Mol. Phys.* **14**, 2761.

Cohen, M. H., Putney, A., and Goodrich, R. W. (1993): The strong magnetic field in G227–35. *Astrophys. J.* **405** , L67.

Delande, D., and Gay, J.C. (1981): On a possible dynamical symmetry in diamagnetism. *Phys. Lett.* **82A**, 393.

Delande, D., and Gay, J.C. (1984): Group theory applied to the hydrogen atom in a strong magnetic field. Derivation of the effective diamagnetic Hamiltonian. *J. Phys. B: At. Mol. Phys.* **17**, L335.

Delande, D., Bommier, A., and Gay, J.C. (1991): Positive-energy spectrum of the hydrogen atom in a magnetic field. *Phys. Rev. Lett.* **66**, 141.

Doman, B.G.S. (1980): Relativistic energy levels of hydrogen in strong magnetic fields. *J. Phys. B: At. Mol. Phys.* **13**, 3335.
Forster, H., Strupat, W., Rösner, W., Wunner, G., Ruder, H., and Herold, H., (1984): Hydrogen atoms in arbitrary magnetic fields: II. Bound-bound transitions. *J. Phys. B: At. Mol. Phys.* **17**, 1301.
Friedrich, H. (1982): Bound-state spectrum of the hydrogen atom in strong magnetic fields. *Phys. Rev. A* **26**, 1827.
Friedrich, H., and Chu, M. (1983): Autoionizing states of the hydrogen atom in strong magnetic fields. *Phys. Rev. A* **28**, 1423.
Friedrich, H., and Wintgen, D. (1989): The hydrogen atom in a uniform magnetic field - an example of chaos. *Phys. Rep.* **183**, 37.
Garstang, R.H., and Kemic, S.B. (1974): Hydrogen and helium spectra in large magnetic fields. *Astrophys. Space Sci.* **31**, 103.
Garstang, R.H. (1977): Atoms in high magnetic fields. *Rep. Prog. Phys.* **40**, 105.
Garstang, R.H. (1982): High magnetic field spectroscopy in astrophysics. *J. Phys. (Paris), Colloq. C2* **43**, 19.
Garton, W.R.S., and Tomkins, F.S. (1969): Diamagnetic Zeeman effect and magnetic configuration mixing in long spectral series of Ba I. *Astrophys. J.* **158**, 839.
Gay, J.C., Delande, D., Biraben, F., and Penent, F. (1983): Diamagnetism of the hydrogen atom – an elementary derivation of the adiabatic invariant. *J. Phys. B: At. Mol. Phys.* **16**, L693.
Greene, C.H. (1983): Atomic photoionization in a strong magnetic field. *Phys. Rev. A* **28**, 2209.
Greenstein, J.L., Henry, R.J.W., and O'Connell, R.F. (1985): Further identifications of hydrogen in Grw+70°8247. *Astrophys. J.* **289**, L25.
Gutzwiller, M. C. (1990): *Chaos in Classical and Quantum Mechanics.* Springer-Verlag New York, Berlin, Heidelberg.
Haake, F. (1991): *Quantum Signatures of Chaos.* Springer-Verlag Berlin, Heidelberg, New York.
Handy, C.R., Bessis, D., Sigismondi, G., and Morley, T.D. (1988): Rapidly converging bounds for the ground-state energy of hydrogenic atoms in superstrong magnetic fields. *Phys. Rev. Lett.* **60**, 253.
Hasegawa, H., and Howard, R.E. (1961): Optical absorption spectrum of hydrogenic atoms in a strong magnetic field. *J. Phys. Chem. Solids* **21**, 179.
Hasegawa, H., Robnik, M., and Wunner, G. (1989): Classical and quantal chaos in the diamagnetic Kepler problem. *Prog. Theor. Phys. Suppl.* **98**, 198.
Helfand, D.J., and Huang, J.-H. (eds.), (1987): *The Origin and Evolution of Neutron Stars,* Reidel Publishing Company, Dordrecht.
Henry, R.J.W., and O'Connell, R.F. (1985): Hydrogen spectrum in magnetic white dwarfs: Hα, Hβ, and Hγ transitions. *Publ. Astron. Soc. Pac.*, **97**, 333.
Henry, R.J.W., O'Connell, R.F., Smith, E.R., Chanmugam, G., and Rajagopal, A.K. (1974): Energy spectrum of H$^-$ in a strong magnetic field. *Phys. Rev. D* **9**, 329.
Herold, H., Ruder, H., and Wunner, G. (1981): The two-body problem in the presence of a homogeneous magnetic field. *J. Phys. B: At. Mol. Phys.* **14**, 751.
Holle, A., Main, J., Wiebusch, G., Rottke, H., and Welge, K.H. (1988): Quasi-Landau spectrum of the chaotic diamagnetic hydrogen atom. *Phys. Rev. Lett.* **61**, 161.
Hylton, D.J., and Rau, A.R.P. (1980): Longitudinally excited states of hydrogen in intense magnetic fields. *Phys. Rev. A* **22**, 321.
Iu, C., Welch, G.R., Kash, M.M., Kleppner, D., Delande, D., and Gay, J.C. (1991): Diamagnetic Rydberg atom: Confrontation of calculated and observed spectra. *Phys. Rev. Lett.* **66**, 145.
Johnson, B.R., Hirschfelder, J.O., and Yang, K.-H. (1983): Interaction of atoms, molecules, and ions with constant electric and magnetic fields. *Rev. Mod. Phys.* **55**, 109.

Jones, P.B. (1985a): Density functional calculations of the ground-state energies of atoms and infinite linear molecules in very strong magnetic fields. *Mon. Not. R. astr. Soc.* **216**, 503.

Jones, P.B. (1985b): Density-functional calculations of the cohesive energy of condensed matter in very strong magnetic fields. *Phys. Rev. Lett.* **55**, 1338.

Jones, P.B. (1986): Properties of condensed matter in very strong magnetic fields. *Mon. Not. R. astr. Soc.* **218**, 477.

Kara, S.M., and McDowell, M.R.C. (1980): Energy levels and bound-bound transitions of hydrogen atoms in strong magnetic fields. *J. Phys. B: At. Mol. Phys.* **13**, 1337.

Kara, S.M., and McDowell, M.R.C. (1981): The photoionization of atomic hydrogen and hydrogenic ions in intense magnetic fields. *J. Phys. B: At. Mol. Phys.* **14**, 1719.

Kaschiev, M.S., Vinitsky, S.I., and Vukajlović, F.R. (1980): Hydrogen atom H and H_2^+ molecule in strong magnetic fields. *Phys. Rev. A* **22**, 557.

Kemic, S.B. (1973), *PhD thesis*, Univ. of Colorado at Boulder.

Kemic, S.B. (1974a): Hydrogen and helium features in magnetic white dwarfs. *Astrophys. J.* **193**, 213.

Kemic, S.B. (1974b): Wavelengths and strengths of hydrogen and helium transitions in large magnetic fields. *JILA Report* **113**, Univ. of Colorado.

Kemp, J.C., Swedlund, J.B., Landstreet, J.D., and Angel, J.R.P. (1970): Discovery of circularly polarized light from a white dwarf. *Astrophys. J.* **161**, L77.

Kössl, D., Wolff, R.G., Müller, E., and Hillebrandt, W. (1988): Density functional calculations in strong magnetic fields: The ground state properties of atoms. *Astron. Astrophys.* **205**, 347.

Larsen, D.M. (1979): Variational studies of bound states of the H^- ion in a magnetic field. *Phys. Rev. B* **20**, 5217.

Latter, W.B., Schmidt, G.D., and Green, R.F. (1987) The rotationally modulated Zeeman spectrum at nearly 10^9 Gauss of the white dwarf PG 1031+234. *Astrophys. J.* **320**, 308.

Le Guillou, J.C., and Zinn-Justin, J. (1983): The hydrogen atom in strong magnetic fields: summation of the weak field series expansion. *Ann. Phys. (N.Y.)* **147**, 57.

Lindgren, K.A.U., and Virtamo, J.T. (1979): Relativistic hydrogen atom in a strong magnetic field. *J. Phys. B: At. Mol. Phys.* **12**, 3465.

Liu, C.-R., and Starace, A.F. (1987): Atomic hydrogen in a uniform magnetic field: Low-lying energy levels for fields above 10^9 G. *Phys. Rev. A* **35**, 647.

Loudon, R. (1959): One-dimensional hydrogen atom. *Am. J. Phys.* **27**, 649.

Mega, C., Herold, H., Rösner, W., Ruder, H., and Wunner, G. (1984): Approximate continuum wave functions for the strongly magnetized hydrogenic problem. *Phys. Rev. A* **30**, 1507.

de Melo, L.C., Das, T.K., Ferreira, R.C., Miranda, L.C.M., and Brandi, H.S. (1978): The H_2^+ molecule in strong magnetic fields, studied by the method of linear combinations of orbitals. *Phys. Rev. A* **18**, 12.

Miller, M.C., and Neuhauser, D. (1991): Atoms in very strong magnetic fields. *Mon. Not. R. astr. Soc.* **253**, 107.

Monteiro, T.S., and Taylor, K.T. (1990): The H_2 molecule in a magnetic field. *J. Phys. B: At. Mol. Opt. Phys.* **23**, 427.

Mueller, R.O., Rau, A.R.P., and Spruch, L. (1975): Lowest energy levels of H^-, He, and Li^+ in intense magnetic fields. *Phys. Rev. A* **11**, 789.

Müller, E. (1984): Variational calculation of iron and helium atoms and molecular chains in superstrong magnetic fields. *Astron. Astrophys.* **130**, 415.

Nagase, F. (1989): Accretion-Powered X-Ray Pulsars. *Publ. Astron. Soc. Japan* **41**, 1.

Neuhauser, D., Langanke, K., and Koonin, S.E. (1986): Hartree-Fock calculations of atoms and molecular chains in strong magnetic fields. *Phys. Rev. A* **33**, 2084.

O'Connell, R.F. (1979): Effect of the proton mass on the spectrum of the hydrogen atom in a strong magnetic field. *Phys. Lett.* **70A**, 389.

O'Mahony, P.F. (1989): Quasi-Landau modulations in nonhydrogenic systems in a magnetic field. *Phys. Rev. Lett.* **63**, 2653.

O'Mahony, P.F., and Mota-Furtado, F. (1991): Continuum spectrum of an atom or molecule in a magnetic field. *Phys. Rev. Lett.* **67**, 2283.

O'Mahony, P.F., and Taylor, K.T. (1986a): Quadratic Zeeman effect for nonhydrogenic systems: application to the Sr and Ba atoms. *Phys. Rev. Lett.* **57**, 2931.

O'Mahony, P.F., and Taylor, K.T. (1986b): The quadratic Zeeman effect in caesium: departures from hydrogenic behaviour. *J. Phys. B: At. Mol. Phys.* **19**, L65.

Ozaki, J., and Tomishima, Y. (1981): Energies of the H_2^+ ion in strong magnetic fields. *Phys. Lett.* **82A**, 449.

Park, C.-H. and Starace, A.F. (1984): H^- and He in a uniform magnetic field: Ground-state wave functions, energies, and binding energies for fields below 10^9 G. *Phys. Rev. A* **29**, 442.

Pavlov-Verevkin, V.B., and Zhilinskii, B.I. (1980a): The hydrogen atom in a superstrong magnetic field. *Phys. Lett.* **75A**, 279.

Pavlov-Verevkin, V.B., and Zhilinskii, B.I. (1980b): Neutral hydrogen-like system in a magnetic field. *Phys. Lett.* **78A**, 244.

Peek, J.M. and Katriel, J. (1980): Hydrogen molecular ion in a high magnetic field. *Phys. Rev. A* **21**, 413.

Potekhin, A. Yu., and Pavlov, G. G. (1993): Photoionization of the hydrogen atom in strong magnetic fields. *Astrophys. J.* **407**, 330.

Praddaude, H.C. (1972): Energy levels of hydrogenlike atoms in a magnetic field. *Phys. Rev. A* **6**, 1321.

Pringle, J.E., and Wade, R.A. (eds.), (1985): *Interacting binary stars*, Cambridge University Press.

Pröschel, P., Rösner, W., Wunner, G., Ruder, H., and Herold, H. (1982): Hartree-Fock calculations for atoms in strong magnetic fields. I: energy levels of two-electron systems. *J. Phys. B: At. Mol. Phys.* **15**, 1959.

Rafelski, J., and Müller, B. (1976): Magnetic Splitting of Quasimolecular Electronic States in Strong Fields. *Phys. Rev. Lett.* **36**, L517.

Rech, P.C., Gallas, M.R., and Gallas, J.A.C. (1986): Zeeman diamagnetism in hydrogen at arbitrary field strengths. *J. Phys. B: At. Mol. Phys.* **19**, L215.

Rösner, W., Herold, H., Ruder, H., and Wunner, G. (1983): Approximate solution of the strongly magnetized hydrogenic problem with the use of an asymptotic property. *Phys. Rev. A* **28**, 2071.

Rösner, W., Wunner, G., Herold, H., and Ruder, H. (1984): Hydrogen atoms in arbitrary magnetic fields: I. Energy levels and wavefunctions. *J. Phys. B: At. Mol. Phys.* **17**, 29.

Rosen, G. (1986): Rigorous analytical lower bound on the ground-state energies of hydrogenic atoms in high magnetic fields. *Phys. Rev. A* **34**, 1556.

Ruder, H., Wunner, G., Herold, H., and Reinecke, M. (1981a): On the validity of perturbation calculations of energy levels and transitions of hydrogen atoms in strong magnetic fields. *J. Phys. B: At. Mol. Phys.* **14**, L45.

Ruder, H., Wunner, G., Herold, H., and Trümper, J. (1981b): Iron lines in superstrong magnetic fields. *Phys. Rev. Lett.* **46**, 1700.

Schiff, L.I., and Snyder, H. (1939): Theory of the quadratic Zeeman effect. *Phys. Rev.* **55**, 59.

Schmidt, G.D., West, S.C., Liebert, J., Green, R.F., and Stockman, H.S. (1986b): The new magnetic white dwarf PG 1031+234: polarization and field structure at more than 500 million gauss. *Astrophys. J.* **309**, 218.

Schmitt, W., Herold, H., Ruder, H., and Wunner, G. (1981): The photoionization of the hydrogen atom in strong magnetic fields. *Astron. Astrophys.* **94**, 194.

Schwope, A. D., Beuermann, K., Jordan, S., and Thomas, H.-C. (1993): Cyclotron and Zeeman spectroscopy of MR Serpentis in low and high states of accretion. *Astron. Astrophys.* **278**, 487.

Shapiro, S.L., and Teukolsky, S.A. (1983): *Black Holes, White Dwarfs, and Neutron Stars*, Wiley and Sons, New York.

Shertzer, J. (1989): Finite-element analysis of hydrogen in superstrong magnetic fields. *Phys. Rev. A* **39**, 3833.

Shertzer, J., Ram-Mohan, L.R., and Dossa, D. (1989): Finite-element calculation of low-lying states of hydrogen in a superstrong magnetic field. *Phys. Rev. A* **40**, 4777.
Simola, J., and Virtamo, J. (1978): Energy levels of hydrogen atoms in a strong magnetic field. *J. Phys. B: At. Mol. Phys.* **11**, 3309.
Smith, E.R., Henry, R.J.W., Surmelian G.L., O'Connell, R.F., and Rajagopal, A.K. (1972): Energy spectrum of the hydrogen atom in a strong magnetic field. *Phys. Rev. D* **6**, 3700.
Smith, E.R., Henry, R.J.W., Surmelian G.L., and O'Connell, R.F. (1973a): Hydrogen atom in a strong magnetic field: bound-bound transitions. *Astrophys. J.* **179**, 659.
Smith, E.R., Henry, R.J.W., Surmelian G.L., and O'Connell, R.F. (1973b): Erratum. *Astrophys. J.* **182**, 651.
Starace, A.F., and Webster, G.L. (1979): Atomic hydrogen in a uniform magnetic field: Low-lying energy levels for fields below 10^9 G. *Phys. Rev. A* **19**, 1629.
Surmelian, G.L., Henry, R.J.W., and O'Connell, R.F. (1974): Energy spectrum of He I and H^- in a strong magnetic field. *Phys. Lett.* **49A**, 431.
Thurner, G., Körbel, H., Braun, M., Herold, H., Ruder, H., and Wunner, G. (1993): Hartree-Fock calculations for excited states of two-electron systems in strong magnetic fields. *J. Phys. B: At. Mol. Opt. Phys.* **26**, 4719.
Trümper, J., Pietsch, W., Reppin, C., Sacco, B., Kendziorra, E., and Staubert, R. (1977): Evidence for strong cyclotron emission in the hard x-ray spectrum of Her X-1. *Ann. N.Y. Acad. Sci.* **302**, 538.
Trümper, J., Pietsch, W., Reppin, C., Voges, W., Staubert, R., and Kendziorra, E. (1978): Evidence for strong cyclotron line emission in the hard X-ray spectrum of Hercules X-1. *Astrophys. J.* **219**, L105.
Trümper, J., Lewin, W.H.G., and Brinkmann, W. (eds.), (1986): *The Evolution of Galactic X-Ray Binaries*, Reidel Publishing Company, Dordrecht.
Vincke, M., and Baye, D. (1989): Variational study of H^- and He in strong magnetic fields. *J. Phys. B: At. Mol. Opt. Phys.* **22**, 2089.
Ventura, J., Herold, H., Ruder, H., and Geyer, F. (1992): Photoabsorption in magnetic neutron star atmospheres. *Astron. Astrophys.* **261**, 235.
Watanabe, S. and Komine, H. (1991): Adiabatic-expansion method applied to diamagnetic Rydberg atoms. *Phys. Rev. Lett.* **67**, 3227.
Wickramasinghe, D.T. and Cropper, M. (1988): Spectropolarimetry of the magnetic white dwarf PG1015+014: evidence for a 100-MG field. *Mon. Not. R. astr. Soc.* **235**, 1451.
Wickramasinghe, D.T. and Ferrario, L. (1988): A centered dipole model for the high field magnetic white dwarf Grw+70°8247. *Astrophys. J.* **327**, 222.
Wille, U. (1987a): Vibrational and rotational properties of the H_2^+ molecular ion in a strong magnetic field. *J. Phys. B: At. Mol. Phys.* **20**, L417.
Wille, U. (1987b): Resonant charge transfer in slow H^+-H collisions in the presence of a strong magnetic field. *Phys. Lett. A* **125**, 52.
Williams, A.C., Darbro, W., Weisskopf, M.C., and Elsner, R.F. (1985): Hydrogen-like atoms on the surface of neutron stars – intense magnetic field effects. *Astrophys. J.* **289**, 782.
Wintgen, D., and Friedrich, H. (1986a): Matching the low-field region and the high-field region for the hydrogen atom in a uniform magnetic field. *J. Phys. B: At. Mol. Phys.* **19**, 991.
Wintgen, D., and Friedrich, H. (1986b): Approximate separability for the hydrogen atom in a uniform magnetic field. *J. Phys. B: At. Mol. Phys.* **19**, 1261.
Wintgen, D., and Friedrich, H. (1986c): Regularity and irregularity in spectra of the magnetized hydrogen atom. *Phys. Rev. Lett.* **57**, 571.
Wunner, G., Ruder, H., and Herold, H. (1980): Comment on the effect of the proton mass on the spectrum of the hydrogen atom in a very strong magnetic field. *Phys. Lett.* **79A**, 159.
Wunner, G., and Ruder, H. (1980a): Electromagnetic transitions for the hydrogen atom in strong magnetic fields. *Astrophys. J.* **242**, 828.

Wunner, G., and Ruder, H. (1981): Hydrogen atom in strong magnetic fields: polynomial approximations for the magnetic-field dependence of the energy values. *Astron. Astrophys.* **95**, 204.
Wunner, G., Ruder, H., and Herold, H. (1981a): Energy levels, electromagnetic transitions and annihilation of positronium in strong magnetic fields. *J. Phys. B: At. Mol. Phys.* **14**, 765.
Wunner, G., Ruder, H., and Herold, H. (1981b): Energy levels and oscillator strengths for the two-body problem in magnetic fields. *Astrophys. J.* **247**, 374.
Wunner, G., Ruder, H., and Herold, H. (1981c): Quality of one-configuration Hartree-Fock-type calculations for the H atom in arbitrary magnetic fields. *Phys. Lett.* **85A**, 430.
Wunner, G., and Ruder, H. (1982): Energy levels and electromagnetic transitions of atoms in superstrong magnetic fields. *J. Phys. (Paris), Colloq. C2* **43**, 137.
Wunner, G., Rösner, W., Ruder, H., and Herold, H. (1982a): Energy values and sum rules for hydrogenic atoms in magnetic fields of arbitrary strength using numerical wave functions: comparison with variational results. *Astrophys. J.* **262**, 407.
Wunner, G., Ruder, H., Schmitt, W., Herold, H., and McDowell, M.R.C. (1982b): Rigorous and approximate scaling laws for the photoionization cross-section of hydrogenic ions in magnetic fields. *Mon. Not. R. astr. Soc.* **198**, 769.
Wunner, G., Herold, H., and Ruder, H. (1982c): Energy values for the H_2^+ ion in superstrong magnetic fields using the adiabatic approximation. *Phys. Lett.* **88A**, 344.
Wunner, G., Ruder, H., Herold, H., and Schmitt, W. (1983b): Cross sections for photoionization and photo-recombination of hydrogenic atoms in strong magnetic fields. *Astron. Astrophys.* **117**, 156.
Wunner, G., Rösner, W., Herold, H., and Ruder, H. (1985a): The importance of spin-orbit coupling in the magnetized hydrogen problem. *J. Phys. B: At. Mol. Phys.* **18**, L179.
Wunner, G., Rösner, W., Herold, H., and Ruder, H. (1985b): Stationary hydrogen lines in white dwarf magnetic fields and the spectrum of the magnetic degenerate Grw+70°8247. *Astron. Astrophys.* **149**, 102.
Wunner, G., Geyer, F., and Ruder, H. (1987): Atomic data relevant to line formation in strongly magnetized white dwarf stars. *Astrophys. Space Sci.* **131**, 595.
Wunner, G., Schweizer, W., and Ruder, H. (1989): Hydrogen Atoms in Strong Magnetic Fields - in the Laboratory and in the Cosmos. *The Hydrogen Atom*, G.F. Bassani, M. Inguscio, T.W. Hänsch (eds.), Springer-Verlag Berlin, Heidelberg 1989, 300.
Zimmerman, M.L., Kash, M.M., and Kleppner, D. (1980): Evidence of an approximate symmetry for hydrogen in a uniform magnetic field. *Phys. Rev. Lett.* **45**, 1092.

Chapter 2

Avron, J.E., Herbst, I.W., and Simon B. (1978): Separation of center of mass in homogeneous magnetic fields. *Ann. Phys. (N.Y.)* **114**, 431.
Blumberg, W.A.M., Itano, W.M., and Larson, D.J. (1979): Theory of the photodetachment of negative ions in magnetic field. *Phys. Rev. D* **19**, 139.
Canuto, V., and Ventura, J. (1977): Quantizing magnetic fields in astrophysics. *Fundam. Cosmic Phys.* **2**, 203.
Herold, H., Ruder, H., and Wunner, G. (1981): The two-body problem in the presence of a homogeneous magnetic field. *J. Phys. B: At. Mol. Phys.* **14**, 751.
Johnson, B.R., Hirschfelder, J.O., and Yang, K.-H. (1983): Interaction of atoms, molecules, and ions with constant electric and magnetic fields. *Rev. Mod. Phys.* **55**, 109.
O'Connell, R.F. (1979): Effect of the proton mass on the spectrum of the hydrogen atom in a strong magnetic field. *Phys. Lett.* **70A**, 389.
Pavlov-Verevkin, V.B., and Zhilinskii, B.I. (1980a): The hydrogen atom in a superstrong magnetic field. *Phys. Lett.* **75A**, 279.
Pavlov-Verevkin, V.B., and Zhilinskii, B.I. (1980b): Neutral hydrogen-like system in a magnetic field. *Phys. Lett.* **78A**, 244.

Surmelian, G.L., and O'Connell, R.F. (1974): Energy spectrum of hydrogen-like atoms in a strong magnetic field. *Astrophys. J.* **190**, 741.
Wunner, G., Ruder, H., and Herold, H. (1980): Comment on the effect of the proton mass on the spectrum of the hydrogen atom in a very strong magnetic field. *Phys. Lett.* **79A**, 159.
Wunner, G., Ruder, H., and Herold, H. (1981a): Energy levels, electromagnetic transitions and annihilation of positronium in strong magnetic fields. *J. Phys. B: At. Mol. Phys.* **14**, 765.
Wunner, G., Ruder, H., and Herold, H. (1981b): Energy levels and oscillator strengths for the two-body problem in magnetic fields. *Astrophys. J.* **247**, 374.

Chapter 3

Abramowitz, M., and Stegun I.A. (1972): *Handbook of mathematical functions*, Dover Publications, New York.
Canuto, V., and Ventura, J. (1977): Quantizing magnetic fields in astrophysics. *Fundam. Cosmic Phys.* **2**, 203.
Clark, C.W., and Taylor, K.T. (1980): The quadratic Zeeman effect in hydrogen Rydberg series. *J. Phys. B: At. Mol. Phys.* **13**, L737.
Clark, C.W., and Taylor, K.T. (1982): The quadratic Zeeman effect in hydrogen Rydberg series: application of Sturmian functions. *J. Phys. B: At. Mol. Phys.* **15**, 1175.
Davidson, E. (1975): The iterative calculation of a few of the lowest eigenvalues and corresponding eigenvectors of large real-symmetric matrices. *J. Comput. Phys.* **17**, 87.
Delande, D., and Gay, J.C. (1984): Group theory applied to the hydrogen atom in a strong magnetic field. Derivation of the effective diamagnetic Hamiltonian. *J. Phys. B: At. Mol. Phys.* **17**, L335.
Edmonds, A.R. (1973): Studies of the quadratic Zeeman effect. I. Application of the sturmian functions. *J. Phys. B: At. Mol. Phys.* **6**, 1603.
Englefield, M.J. (1971): *Group Theory and the Coulomb Problem*, Wiley, New York.
Ericsson, T., and Ruhe, A. (1980): The spectral transformation Lanczos method for the numerical solution of large sparse generalized symmetric eigenvalue problems. *Math. Comput.* **35**, 1251.
Friedrich, H. (1982): Bound-state spectrum of the hydrogen atom in strong magnetic fields. *Phys. Rev. A* **26**, 1827.
Froese-Fischer, C. (1977): *The Hartree-Fock Method for Atoms: A Numerical Approach*, Wiley, New York.
Froese-Fischer, C. (1978): A general multi-configuration Hartree-Fock program. *Comput. Phys. Commun.* **14**, 145.
Geyer, F. (1987): Berechnung der Photoionisation des Positroniums in Pulsarmagnetosphären und von Wasserstoffenergiewerten in Weißen-Zwerg-Magnetfeldern. *PhD thesis*, Univ. Tübingen.
Herrick, D.R. (1982): Symmetry of the quadratic Zeeman effect for hydrogen. *Phys. Rev. A* **26**, 323.
Pröschel, P. (1982): Hartree-Fock-Rechnungen in extrem starken magnetischen Feldern. *PhD thesis*, Univ. Erlangen.
Pröschel, P., Rösner, W., Wunner, G., Ruder, H., and Herold, H. (1982): Hartree-Fock calculations for atoms in strong magnetic fields. I: energy levels of two-electron systems. *J. Phys. B: At. Mol. Phys.* **15**, 1959.
Rösner, W., Herold, H., Ruder, H., and Wunner, G. (1983): Approximate solution of the strongly magnetized hydrogenic problem with the use of an asymptotic property. *Phys. Rev. A* **28**, 2071.
Schiff, L.I., and Snyder, H. (1939): Theory of the quadratic Zeeman effect. *Phys. Rev.* **55**, 59.
Simola, J., and Virtamo, J. (1978): Energy levels of hydrogen atoms in a strong magnetic field. *J. Phys. B: At. Mol. Phys.* **11**, 3309.

Wintgen, D. (1985): Das Wasserstoffatom im starken Magnetfeld. *PhD thesis*, TU München.
Wintgen, D., and Friedrich, H. (1986a): Matching the low-field region and the high-field region for the hydrogen atom in a uniform magnetic field. *J. Phys. B: At. Mol. Phys.* **19**, 991.
Wintgen, D., and Friedrich, H. (1986b): Approximate separability for the hydrogen atom in a uniform magnetic field. *J. Phys. B: At. Mol. Phys.* **19**, 1261.
Wunner, G., Kost, M., and Ruder, H. (1986a): "Circular" states of Rydberg atoms in strong magnetic fields. *Phys. Rev. A* **33**, 1444.
Wunner, G. (1986): Note on the usefulness of perturbation theory in calculating energy levels and transitions of hydrogen Rydberg atoms in strong magnetic fields. *J. Phys. B: At. Mol. Phys.* **19**, 1623.
Zeller, G. (1990): Berechnung von Rydbergzuständen des Wasserstoffatoms in starken Magnetfeldern und Anwendungen in der Quantenchaologie. *PhD thesis*, Univ. Tübingen.

Chapter 4

Daugherty, J.K., and Ventura, J. (1977): Cyclotron lines in the Her X-1 spectrum: structure and higher harmonics. *Astron. Astrophys.* **61**, 723.
Delande, D., and Gay, J.C. (1981): On a possible dynamical symmetry in diamagnetism. *Phys. Lett.* **82A**, 393.
Finkelnburg, W. (1967): *Einführung in die Atomphysik*, Springer-Verlag Berlin, Heidelberg, New York.
Garstang, R.H., and Kemic, S.B. (1974): Hydrogen and helium spectra in large magnetic fields. *Astrophys. Space Sci.* **31**, 103.
Garstang, R.H. (1977): Atoms in high magnetic fields. *Rep. Prog. Phys.* **40**, 105.
Herold, H., Ruder, H., and Wunner, G. (1982a): Comment on the cyclotron emission rates in superstrong magnetic fields. *Phys. Lett.* **91**, 272.
Lindgren, K.A.U., and Virtamo, J.T. (1979): Relativistic hydrogen atom in a strong magnetic field. *J. Phys. B: At. Mol. Phys.* **12**, 3465.
Loudon, R. (1959): One-dimensional hydrogen atom. *Am. J. Phys.* **27**, 649.
Rösner, W., Herold, H., Ruder, H., and Wunner, G. (1983): Approximate solution of the strongly magnetized hydrogenic problem with the use of an asymptotic property. *Phys. Rev. A* **28**, 2071.
Ruder, H., Ertl, T., Geyer, F., Herold, H., and Kraus, U. (1989): Line-of-sight integration: a powerful tool for visualization of three-dimensional scalar fields. *Comput. & Graphics* **13**, 223.
White, H.E. (1931): Pictorial representations of the electron cloud for hydrogen-like atoms. *Phys. Rev* **37**, 1416.
Wunner, G., Ruder, H., and Herold, H. (1980): Comment on the effect of the proton mass on the spectrum of the hydrogen atom in a very strong magnetic field. *Phys. Lett.* **79A**, 159.
Zimmerman, M.L., Kash, M.M., and Kleppner, D. (1980): Evidence of an approximate symmetry for hydrogen in a uniform magnetic field. *Phys. Rev. Lett.* **45**, 1092.

Chapter 5

Friedrich, H. (1982): Bound-state spectrum of the hydrogen atom in strong magnetic fields. *Phys. Rev. A* **26**, 1827.
Loudon, R. (1959): One-dimensional hydrogen atom. *Am. J. Phys.* **27**, 649.

Chapter 6

Bethe, H.A. (1933): *Quantenmechanik der Ein- und Zweielektronenproblem*, Handbuch der Physik, Vol. **24**, eds. H. Geiger and K. Scheel, Springer-Verlag, Berlin.
Bethe, H.A., and Salpeter, E.E. (1957): *Quantum Mechanics of One- and Two-Electron Atoms*, Handbuch der Physik, Vol. **35**, ed. S. Flügge, Springer-Verlag, Berlin, Göttingen, Heidelberg; also appeared as a book in Springer-Verlag, Berlin, Göttingen, Heidelberg (1957).
Canuto, V., and Ventura, J. (1977): Quantizing magnetic fields in astrophysics. *Fundam. Cosmic Phys.* **2**, 203.
Hasegawa, H., and Howard, R.E. (1961): Optical Absorption Spectrum of Hydrogenic Atoms in a Strong Magnetic Field. *J. Phys. Chem. Solids* **21**, 179.
Heitler, W. (1953): *The Quantum Theory of Radiation*, Oxford Clarendon Press.
Loudon, R. (1959): One-dimensional hydrogen atom. *Am. J. Phys.* **27**, 649.
Messiah, A. (1973): *Quantum Mechanics I and II*, North-Holland Publ. Comp., Amsterdam.
Wunner, G., and Ruder, H. (1980a): Electromagnetic transitions for the hydrogen atom in strong magnetic fields. *Astrophys. J.* **242**, 828.
Wunner, G., and Ruder, H. (1980b): Lyman- and Balmer-like transitions for the hydrogen atom in strong magnetic fields. *Astron. Astrophys.* **89**, 241.
Wunner, G., Herold, H., and Ruder, H. (1983a): Radiative and thermal widths of Landau-excited hydrogen atoms in very strong magnetic fields. *J. Phys. B: At. Mol. Phys.* **16**, 2937.

Chapter 7

Achilleos, N., Remillard, R.A., and Wickramasinghe, D.T. (1991): Serendipitous identifications of two magnetic white dwarfs. *Mon. Not. R. astr. Soc.* **253**, 522.
Angel, J.R.P. (1978): Magnetic white dwarfs. *Ann. Rev. Astron. Astrophys.* **16**, 487.
Angel, J.R.P., Liebert, J., and Stockman, H.S. (1985): The optical spectrum of hydrogen at 160–350 million gauss in the white dwarf Grw+70°8247. *Astrophys. J.* **292**, 260.
Bergeron, P., Ruiz, M-T., and Leggett, S.K. (1992): Discovery of two cool magnetic white dwarfs. *Astrophys. J.* **400**, 315.
Bergeron, P., Ruiz, M-T., and Leggett, S.K. (1993): G42-46: an unresolved double degenerate binary containing a magnetic DA component. Astrophys. J. **407**, 733.
Chanmugam, G. (1992): Magnetic fields of degenerate stars. *Ann. Rev. Astron. Astrophys.* **30**, 143.
Cohen, M. H., Putney, A., and Goodrich, R. W. (1993): The strong magnetic field in G227-35. *Astrophys. J.* **405**, L67.
Friedrich, S. (1993): Theorie und Beobachtung von Spektren magnetischer Weißer Zwergsterne . *PhD thesis*, Univ. Tübingen.
Greenstein, J.L., and Matthews, M.S. (1957): Studies of the white dwarfs. I. Broad features in white dwarf spectra. *Astrophys. J.* **126**, 14.
Greenstein, J.L., Henry, R.J.W., and O'Connell, R.F. (1985): Further identifications of hydrogen in Grw+70°8247. *Astrophys. J.* **289**, L25.
Henry, R.J.W., and O'Connell, R.F. (1985): Hydrogen spectrum in magnetic white dwarfs: Hα, Hβ, and Hγ transitions. *Publ. Astron. Soc. Pac.*, **97**, 333.
Kemic, S.B. (1974a): Hydrogen and helium features in magnetic white dwarfs. *Astrophys. J.* **193**, 213.
Kemp, J.C., Swedlund, J.B., Landstreet, J.D., and Angel, J.R.P. (1970): Discovery of circularly polarized light from a white dwarf. *Astrophys. J.* **161**, L77.
Koester, D. and Chanmugam, G. (1990): Physics of white dwarf stars. *Rep. Prog. Phys.* **53**, 837.
Landstreet, J.D. (1992): Magnetic fields at the surfaces of stars. *Astron. Astrophys. Rev.* **4**, 35.

Latter, W. B., Schmidt, G. D., and Green, R. F. (1987): The rotationally modulated Zeeman spectrum at nearly 10^9 Gauss of the white dwarf PG 1031+234. *Astrophys. J.* **320**, 308.

Liebert, J., Bergeron, P., Schmidt, G. D., and Saffer, R. A. (1993): Discovery of a magnetic/nonmagnetic double-degenerate binary system. *Astrophys. J.* **418**, 426.

Liebert, J., Schmidt, G. D., Lesser, M., Stepanian, J. A., Lipovetsky, V. A., Chaffee, F. H., Foltz, C. B., and Bergeron, P. (1994): Discovery of a dwarf carbon star with a white dwarf companion and of a highly magnetic degenerate star. *Astrophys. J.* **421**, 733.

Minkowski, R. (1938): *Ann. Rept. Dir. Mt. Wilson Obs.* 28.

Östreicher, R., Seifert, W., Friedrich, S., Ruder, H., Schaich, M., Wolf, and D., Wunner, G. (1992): Hydrogen in the strong magnetic field of the white dwarf PG 1031+234. *Astron. Astrophys.* **257**, 353.

Schmidt, G.D., Stockman, H.S., and Grandi, S.A. (1986a): The optical continua of magnetic variables. *Astrophys. J.* **300**, 804.

Schmidt, G.D., West, S.C., Liebert, J., Green, R.F., and Stockman, H.S. (1986b): The new magnetic white dwarf PG 1031+234: polarization and field structure at more than 500 million gauss. *Astrophys. J.* **309**, 218.

Schmidt, G.D., Stockman, H.S., and Smith, P.S. (1992): Discovery of a sub-megagauss magnetic white dwarf through spectropolarimetry. *Astrophys. J. Lett.* **398**, L57.

Schwope, A. D., Beuermann, K., Jordan, S., and Thomas, H.-C., (1993) Cyclotron and Zeeman spectroscopy of MR Serpentis in low and high states of accretion. *Astron. Astrophys.* **278**, 487.

Wickramasinghe, D.T. and Cropper, M. (1988): Spectropolarimetry of the magnetic white dwarf PG1015+014: evidence for a 100-MG field. *Mon. Not. R. astr. Soc.* **235**, 1451.

Wickramasinghe, D.T. and Ferrario, L. (1988): A centered dipole model for the high field magnetic white dwarf Grw+70°8247. *Astrophys. J.* **327**, 222.

Wunner, G., Rösner, W., Herold, H., and Ruder, H. (1985b): Stationary hydrogen lines in white dwarf magnetic fields and the spectrum of the magnetic degenerate Grw+70°8247. *Astron. Astrophys.* **149**, 102.

Zimmerman, M.L., Kash, M.M., and Kleppner, D. (1980): Evidence of an approximate symmetry for hydrogen in a uniform magnetic field. *Phys. Rev. Lett.* **45**, 1092.

Chapter 8

Bethe, H.A., and Salpeter, E.E. (1957): *Quantum Mechanics of One- and Two-Electron Atoms*, Handbuch der Physik, Vol. **35**, ed. S. Flügge, Springer-Verlag, Berlin, Göttingen, Heidelberg; also appeared as a book in Springer-Verlag, Berlin, Göttingen, Heidelberg (1957).

Constantinescu, D.H. (1972): Electron self-energy in a magnetic field. *Nucl. Phys.* **B44**, 288.

Daugherty, J.K., and Ventura, J. (1977): Cyclotron lines in the Her X-1 spectrum: structure and higher harmonics. *Astron. Astrophys.* **61**, 723.

Daugherty, J.K., and Ventura, J. (1978): Absorption of radiation by electrons in intense magnetic fields. *Phys. Rev. D* **18**, 1053.

Friedrich, H., and Chu, M. (1983): Autoionizing states of the hydrogen atom in strong magnetic fields. *Phys. Rev. A* **28**, 1423.

Geprägs, R., Riffert, H., Herold, H., Ruder, H., Wunner, G. (1994): Electron self-energy in a homogeneous magnetic field. *Phys. Rev. D* **49**, 5582.

Greene, C.H. (1983): Atomic photoionization in a strong magnetic field. *Phys. Rev. A* **28**, 2209.

Herold, H., Ruder, H., and Wunner, G. (1981): The two-body problem in the presence of a homogeneous magnetic field. *J. Phys. B: At. Mol. Phys.* **14**, 751.

Herold, H., Ruder, H., and Wunner, G. (1982): Cyclotron emission in strongly magnetized plasmas. *Astron. Astrophys.* **115**, 90.

Lindgren, K.A.U., and Virtamo, J.T. (1979): Relativistic hydrogen atom in a strong magnetic field. *J. Phys. B: At. Mol. Phys.* **12**, 3465.

Messiah, A. (1973): *Quantum Mechanics I and II*, North-Holland Publ. Comp., Amsterdam.
Simola, J., and Virtamo, J. (1978): Energy levels of hydrogen atoms in a strong magnetic field. *J. Phys. B: At. Mol. Phys.* **11**, 3309.
Wunner, G., Ruder, H., and Herold, H. (1980): Comment on the effect of the proton mass on the spectrum of the hydrogen atom in a very strong magnetic field. *Phys. Lett.* **79A**, 159.
Wunner, G., Rösner, W., Herold, H., and Ruder, H. (1985a): The importance of spin-orbit coupling in the magnetized hydrogen problem. *J. Phys. B: At. Mol. Phys.* **18**, L179.

Chapter 9

Canuto, V., and Ventura, J. (1977): Quantizing Magnetic Fields in Astrophysics. *Fundam. Cosmic Phys.* **2**, 203.
Drake, G.W.F. (1993): High-precision calculations for the Rydberg states of helium. In *Long-Range Casimir Forces: Theory and Recent Experiments on Atomic Systems*, Levin, F.S., and Micha, D.A. (eds.), Plenum Press, New York, 107
Fenimore, E.E., Conner, J.P., Epstein, R.I., Klebesadel, R.W., Laros, J.G., Yoshida, A., Fujii, M., Hayashida, K., Itoh, M., Murakami, T., Nishimura, J., Yamagami, T., Kondo, I., and Kawai, N. (1988): Interpretations of multiple absorption features in a gamma-ray burst spectrum. *Astrophys. J.* **335**: L71.
Froese-Fischer, C. (1977): *The Hartree-Fock Method for Atoms: A Numerical Approach*, Wiley, New York.
Froese-Fischer, C. (1978): A general multi-configuration Hartree-Fock program. *Comput. Phys. Commun.* **14**, 145.
Garstang, R.H., and Kemic, S.B. (1974): Hydrogen and helium spectra in large magnetic fields. *Astrophys. Space Sci.* **31**, 103.
Kemic, S.B. (1974b): Wavelengths and strengths of hydrogen and helium transitions in large magnetic fields. *JILA Report* **113**, Univ. of Colorado.
Krivec, R., Haftel, M.I., and Mandelzweig, V.B. (1991): Precise nonvariational calculation of excited states of helium with the correlation-function hyperspherical-harmonic method. *Phys. Rev. A* **44**, 7158.
Larsen, D.M. (1979): Variational studies of bound states of the H^- ion in a magnetic field. *Phys. Rev. B* **20**, 5217.
Miller, M.C., and Neuhauser, D. (1991): Atoms in very strong magnetic fields. *Mon. Not. R. astr. Soc.* **253**, 107.
Mueller, R.O., Rau, A.R.P., and Spruch, L. (1975): Lowest energy levels of H^-, He, and Li^+ in intense magnetic fields. *Phys. Rev. A* **11**, 789.
Pekeris, C.L. (1958): Ground state of two-electron atoms. *Phys. Rev.* **112**, 1649.
Pröschel, P., Rösner, W., Wunner, G., Ruder, H., and Herold, H. (1982): Hartree-Fock calculations for atoms in strong magnetic fields. I: energy levels of two-electron systems. *J. Phys. B: At. Mol. Phys.* **15**, 1959.
Schmidt, G.D., Latter, W.B., and Foltz, C.B. (1990): On the spectrum and field strength of the magnetic white dwarf GD 229. *Astrophys. J.* **350**, 758.
Simola, J., and Virtamo, J. (1978): Energy levels of hydrogen atoms in a strong magnetic field. *J. Phys. B: At. Mol. Phys.* **11**, 3309.
Ventura, J. (1989): Radiation from cooling neutron stars. *Timing Neutron Stars*, H. Ögelman, E.P.J. van den Heuvel (eds.), Kluwer Academic Publishers, Dordrecht 1989, 491.
Vincke, M., and Baye, D. (1989): Variational study of H^- and He in strong magnetic fields. *J. Phys. B: At. Mol. Opt. Phys.* **22**, 2089.

Chapter 10

Bayfield, J. E., and Koch, P. M. (1974): Multiphoton ionization of highly excited hydrogen atoms. *Phys. Rev Lett.* **33**, 258.
Berry, M. (1987): Quantum Chaology - The Bakerian Lecture 1987. *Proc. R. Soc. London A* **413**, 183.
Berry, M. (1989): Quantum chaology, not quantum chaos. *Phys. Scri.* **40**, 335.
Blümel, R., and Smilansky, U. (1984): Quantum mechanical suppression of classical stochasticity in the dynamics of periodically perturbed surface-state electrons. *Phys. Rev. Lett.* **52**, 137.
Blümel, R., and Smilansky, U. (1987): Microwave ionization of highly excited hydrogen atoms. *Z. Phys. D: At. Mol. Clusters* **6**, 83.
Bohigas, O., Giannoni, M. J., and Schmit, C. (1984a): Characterization of chaotic quantum spectra and universality of level fluctuation laws. *Phys. Rev. Lett.* **52**, 1.
Bohigas, O., Giannoni, M. J., and Schmit, C. (1984b): Spectral properties of the Laplacian and random matrix theories. *J. Phys. Lett.* **45**, L1015.
Casati, G., Chirikov, B. V., Izrailev, F. M., and Ford, J., in Casati, G., and Ford, J. (eds) (1979): Stochastic Behaviour of Classical and Quantum Hamiltonian Systems. *Lecture Notes in Physics*, Vol. **93**, Springer-Verlag Berlin, Heidelberg, New York, p. 334.
Casati, G., Chirikov, B. V., Guarneri, I., and Shepelyansky, D. L., (1986): Dynamical stability of quantum "chaotic" motion in a hydrogen atom. *Phys. Rev. Lett.* **56**, 2437.
Casati, G., Guarneri, I., and Shepelyansky, D. L. (1988): Hydrogen atom in momochromatic field: Chaos and dynamical photonic localization. *IEEE J. Quantum Electron.* **24**, 1420.
Clark, C.W., and Taylor, K.T. (1982): The quadratic Zeeman effect in hydrogen Rydberg series: application of Sturmian functions. *J. Phys. B: At. Mol. Phys.* **15**, 1175.
Delande, D., and Gay, J.C. (1986): Quantum chaos and statistical properties of energy levels: numerical study of the hydrogen atom in a magnetic field. *Phys. Rev. Lett.* **57**, 2006.
Delande, D., Bommier, A., and Gay, J.C. (1991): Positive-energy spectrum of the hydrogen atom in a magnetic field. *Phys. Rev. Lett.* **66**, 141.
Friedrich, H., and Wintgen, D. (1989): The hydrogen atom in a uniform magnetic field - an example of chaos. *Phys. Rep.* **183**, 37.
Galvez, E.J., Sauer, B.E., Moorman, L., Koch, P.M., and Richards, D. (1988): Microwave ionization of H atoms: breakdown of classical dynamics for high frequencies. *Phys. Rev. Lett.* **61**, 2011.
Gay, J.-C. (ed.) (1990): From Regular to Irregular Atomic Physics. *Comments At. Mol. Phys* **25**, 1; also appeared as a book (1990), Gordon & Breach, Science Publishers S. A.
Gutzwiller, M. C. (1980): Classical quantization of a Hamiltonian with ergodic behavior. *Phys. Rev. Lett.* **45**, 150.
Gutzwiller, M. C. (1990): Chaos in Classical and Quantum Mechanics. Springer-Verlag New York, Berlin, Heidelberg.
Haake, F. (1991): Quantum Signatures of Chaos. Springer-Verlag Berlin, Heidelberg, New York.
Hasegawa, H., Robnik, M., and Wunner, G. (1989): Classical and Quantal Chaos in the Diamagnetic Kepler Problem. *Prog. Theor. Phys. Suppl.* **98**, 198.
Heller, E. J. (1984): Bound-state eigenfunctions of classically chaotic Hamiltonian systems: scars of periodic orbits. *Phys. Rev. Lett.* **53**, 1515.
Holle, A., Wiebusch, G., Main, J., Hager, B., Rottke, H., and Welge, K.H. (1986): Diamagnetism of the hydrogen atom in the quasi-Landau regime. *Phys. Rev. Lett.* **56**, 2594.
Holle, A., Wiebusch, G., Main, J., Welge, K.H., Zeller, G., Wunner, G., Ertl, T., and Ruder, H. (1987): Hydrogenic Rydberg atoms in strong magnetic fields: theoretical and experimental spectra in the transition region from regularity to irregularity. *Z. Phys. D: At. Mol. Clusters* **5**, 279.
Holle, A., Main, J., Wiebusch, G., Rottke, H., and Welge, K.H. (1988): Quasi-Landau spectrum of the chaotic diamagnetic hydrogen atom. *Phys. Rev. Lett.* **61**, 161.

Iu, C., Welch, G.R., Kash, M.M., Kleppner, D., Delande, D., and Gay, J.C. (1991): Diamagnetic Rydberg atom: Confrontation of calculated and observed spectra. *Phys. Rev. Lett.* **66**, 145.
Jensen, R.V., Susskind, S.M., and Sanders, M.M. (1991): Chaotic ionization of highly excited hydrogen atoms: Comparison of classical and quantum theory with experiment. *Phys. Rep.* **201**, 1.
Leopold, J. G., and Percival, I. C. (1979): Ionisation of highly excited atoms by electric fields: III. Microwave ionisation and excitation. *J. Phys. B: At. Mol. Phys.* **12**, 709.
Main, J., Wiebusch, G., Holle, A., and Welge, K.H. (1986): New quasi-Landau structure of highly excited atoms: The hydrogen atom. *Phys. Rev. Lett.* **57**, 2789.
Reinhardt, W. P. (1982): Complex coordinates in the theory of atomic and molecular structure and dynamics. *Ann. Rev. Phys. Chem.* **33**, 223.
Schweizer, W., Niemeier, R., Friedrich, H, Wunner, G., and Ruder, H. (1988): Liapunov Exponents for Classical Orbits of the Hydrogen Atom in a Magnetic Field. *Phys. Rev. A*, **38**, 1724.
Schweizer, W., Niemeier, R., Wunner, G., and Ruder, H. (1993): The diamagnetic Kepler problem: Bifurcations and catastrophes. *Z. Phys. D: At. Mol. Clusters*, **25**, 95.
Solov'ev, E. A. (1982a): Approximate motion integral for a hydrogen atom in a magnetic field. *Sov. Phys. JETP Lett.* **34**, 265.
Solov'ev, E. A. (1982b): The hydrogen atom in a weak magnetic field. *Sov. Phys. JETP* **55**, 1017.
Takahashi, K. (1989a): Similarity and essential difference between coarse-grained classical and quantum mechanics. *J. Phys. Soc. Jpn.* **58**, 3514.
Takahashi, K. (1989b): Distribution functions in classical and quantum mechanics. *Prog. Theor. Phys. Suppl.* **98**, 109.
van Leeuwen, K. A. H., v. Oppen, G., Renwick, S., Bowlin, J. B., Koch, P. M., Jensen, R. V., Rath, O., Richards, D., and Leopold, J. G. (1985): Microwave ionization of hydrogen atoms: Experiment versus classical dynamics. *Phys. Rev. Lett.* **55**, 2231.
Wintgen, D., and Friedrich, H. (1986c): Regularity and irregularity in spectra of the magnetized hydrogen atom. *Phys. Rev. Lett.* **57**, 571.
Wintgen, D., and Friedrich, H. (1987): Classical and quantum-mechanical transition between regularity and irregularity in a hamiltonian system. *Phys. Rev. A* **35**, 1464.
Wunner, G., Woelk, U., Zech, I., Zeller, G., Ertl, T., Geyer, F., Schweizer, W., and Ruder, H. (1986b): Rydberg atoms in uniform magnetic fields: uncovering the transition from regularity to irregularity in a quantum system. *Phys. Rev. Lett.* **57**, 3261.
Zeller, G. (1990): Berechnung von Rydbergzuständen des Wasserstoffatoms in starken Magnetfeldern und Anwendungen in der Quantenchaologie. *PhD thesis*, Univ. Tübingen.

Subject Index

adiabatic approximation 7–10, 24, 26, 77, 122, 133
–, energies 57–68
–, electromagnetic transitions 78–87, 253–258
anticrossings 35, 39, 47–48, 93–94, 110–112, 126–127, 154, 217
asymptotic property of effective potentials 57–58

Balmer α 40, 90–92, 101, 108, 225, 227, 236
Balmer β 40, 90–94, 101, 108, 225, 227, 237
Balmer γ 40, 90–94, 225, 228, 237
Balmer δ 40, 225, 228
Balmer ε 40, 225, 228
Balmer jump 93
Balmer transitions 40, 90–93
– to Rydberg states 71, 157–164
billiard, rectangular 166, 171
–, Sinai 166, 171
–, stadium 166, 171, 178
Brackett α 40, 91–93, 225, 230, 239
Brackett transitions 40, 90–91

centre-of-mass coordinates 15–16
chaos, classical 153, 165–167
–, quantum 6, 10, 153, 164–181
complex-rotation method 107, 161
configuration mixing 135, 137, 139
correspondence diagrams, helium 123–133
correspondence diagrams, hydrogen 7, 26–28, 109–110
–, with spin-orbit coupling 9, 110–113
correspondence, low-field high-field states 26–27
coupled-channels-like method 31
cyclotron energy 3, 11, 17
–, electron 3, 11, 118
–, proton 39, 43, 75–77, 121
cyclotron frequency 16
–, electron 11
–, proton 11, 21

cyclotron line 4–5
cyclotron transition rates 120
cyclotron transitions, electron 3, 41–43, 75, 121, 226–230
–, proton 41–43
cylindrical expansion 25, 29, 36, 46, 75

Δ_3-statistics 170–171, 174–175
density functional method 10, 183
deterministic 165
deuterium 156–164
diamagnetic Kepler problem 172, 179
–, approximate constant of motion 154
–, level statistics 174
–, n_p-mixing regime 154
–, quasi-integrability 154
–, scaling property 172
dipole matrix elements 69–70, 73–75, 94, 146
dipole strengths 20–21, 69–71, 78–85, 118–120, 143–144
–, tables of 241–252, 286–292
distribution function, phase space 177
–, Wigner 177, 179–181
distribution, Poisson 170–171, 174
distribution, spatial probability 45–56, 154, 178
distribution, Wigner 170–171, 174
double tightly bound (DTB) states 126, 128, 137–140, 280

energies, tables of 185–216, 263–282
–, figures of 217–224
effective potentials 25–26, 29, 84, 115, 118–119
–, asymptotic property 57–58
expansion, cylindrical 25, 29, 36, 46, 75
–, spherical 25, 29, 36, 46, 75

fine structure constant 11
fine structure splitting 109

gauge 14
gauge transformation 13

308   Subject Index

Gaussian orthogonal ensemble (GOE)   170–171, 175
Gutzwiller formula   175

harmonic interaction   17–18
Hartree-Fock method   7, 9, 29–31, 124, 129–136, 141–142
hydrogen-like states   38, 45, 57, 77–78, 85–87, 116–118, 121

kicked rotator   169

Lamb shift, magnetic   9, 120
Lanczos method   32
Landau channels   31, 134
Landau energy   15, 31
Landau ground state   9, 75, 115
Landau levels   38–39, 75, 118, 120–122, 222–224
–, widths   120–122
Landau quantum number   17, 25–26, 78, 118, 124
Landau resonances   8
Landau states   3, 16, 25–26, 114–116, 120–122
level statistics   170
–, H atom in magnetic fields   174
–, rectangular billiard   171
–, Sinai billiard   171
–, stadium billiard   171
Liapunov exponent   165–166, 175, 179
Liapunov-stable   178–179
Liapunov-unstable   175, 178–179, 181
line-of-sight integration   50–56
lithium   161, 165
Lyman α 40, 90–91, 225–226, 236
Lyman β 40, 225–226
Lyman γ 40, 225–226
Lyman δ 40, 225–226
Lyman transitions   40, 90–93

magnetic field parameter   11
magnetic field, reference strength   11
magnetic Lamb shift   9, 120
microwave ionization   167–169
momentum, generalized   13–15, 73, 113, 116
–, kinetic   14, 18–19
–, law of conservation   14, 73
– space   24, 177
–, transverse   19, 113, 116
motional electric field   14, 113, 156

N-body problem   13–14
nearest-neighbour distribution   170–171

neutron stars   1–3
–, magnetic fields   1–3, 6, 11, 24, 43, 120
–, Hercules X-1   3–5
non-crossing rule   26, 109–110, 124, 126, 128
nonintegrability   24, 166

oscillator strengths   8, 10, 20–21, 69, 79–81, 84-85, 120, 134, 143–152, 155–156
–, tables of   241–252, 283–292

Paschen α 40, 91–93, 225, 229, 238
Paschen β 40, 91–93, 225, 229, 238
Paschen transitions   40, 90
Paschen-Back regime   109, 113
periodic orbits   175, 177–181
phase space distribution functions   177
Poincaré surface of section   172–173, 177–181
Poisson distribution   170–171, 174
probability, transition   6, 8, 10, 20–21, 40, 69–75, 81, 84–86, 97, 107, 120–121, 134, 143–144, 146–152, 155, 161
proton cyclotron energy   39, 43, 75–77, 121
proton cyclotron frequency   11, 21
proton cyclotron transition   41–43
proton mass effects   8, 21, 23, 36, 39, 41–43, 70, 72–76, 89, 92–93, 113–116, 121 229–230, 239
pulsars, radio   2–3
pulsars, X-ray   2–3
–, Hercules X-1   3–5

quantum chaology   167
quantum chaos   6, 10, 153, 164–181
quantum excess   57, 64–65, 67–68
quantum number, angular momentum   16, 123
–, Landau   17, 25, 78, 118, 124
–, magnetic   16, 21, 25–26, 57, 107, 123–124, 126–128
–, principal   24, 26, 31, 33, 36–42, 51, 81, 123, 153, 155, 167–168

random matrix theory   170
rate, transition   6, 8, 10, 20–21, 40, 69–75, 81, 84–86, 97, 107, 120–121, 134, 143–144, 146–152, 155, 161
rectangular billiard   166
–, level statistics   171
relativistic effects   7, 9, 43, 109
resonances in chaotic spectra   175
Runge-Lenz vector   154
Rydberg atoms   8, 10

Rydberg energy 11
Rydberg series 38
Rydberg states 6, 33, 153, 156–165, 178–179
–, deuterium 156–164
–, hydrogen 155
–, level scheme 155
–, lithium 165
–, microwave ionization 167
–, transitions 71, 155–165

scaled energy 172
scaling laws 19–21, 23, 36, 39, 70, 74, 78, 89, 113, 126, 156
–, mass 19–21
–, energy 19–21
–, nuclear charge 21
–, wave functions 19–21
scars of wave functions 177–178
selection rules 79, 82, 135–136
semiclassical quantization 175, 179
Sinai billiard 166
–, level statistics 171
single tightly bound (STB) states 126, 140
spatial probability distribution 45–56, 154, 178
spherical expansion 25, 29, 36, 46, 75
spin-orbit coupling 9, 109–113, 123–124
–, Hamiltonian 111
spectral rigidity 170–171, 174–175
stadium billiard 166
–, level statistics 171
–, scars 178
states, hydrogen-like 38, 45, 57, 77–78, 85–87, 116–118, 121
states, Landau 3, 16, 25–26, 114–116, 120–122
states, Rydberg 6, 33, 153, 156–165, 178–179
states, tightly bound 38, 43, 45–46, 57, 61, 64, 67, 77, 82–84, 87, 91, 114, 118, 121, 146, 187
–, double (DTB) 126, 128, 137–140, 280
–, single (STB) 126, 140
stationary lines 10, 40–41, 44, 89–108
–, figures of wavelengths 236–239
–, table of extrema 90, 91
–, tables of dipole strengths 259–262
–, tables of wavelengths 231–235
statistics, Δ_3 170–171, 174–175
statistics, energy levels 169
statistics, nearest-neighbour 170–171

strengths, dipole 20–21, 69–71, 78–85, 118–120, 143–144
–, oscillator 8, 10, 20–21, 69, 79–81, 84–85, 120, 134, 143–152, 155–156
Sturmian functions 8, 32
sum rules 75, 79–80, 82, 84–85, 120
suppression of classical chaos 169, 180

transition probabilities (rates) 6, 8, 10, 20–21, 40, 69–75, 81, 84–86, 97, 107, 120–121, 134, 143–144, 146–152, 155, 161
–, tables of 241–252, 286–292
two-body problem 14–19, 72, 113
two-electron systems 10, 124–126, 136–152, 263

wavelengths, helium atom 143–145
–, tables of 283–285
wavelengths, hydrogen atom 39–44
–, figures of 225–230
–, stationary lines 89–94
–, tables of 241–252
Wigner distribution 170–171, 174
Wigner distribution function 177, 179–181
white dwarf magnetic fields 1–2, 7, 11, 24, 95–108
white dwarfs 1–2, 6–7, 10
–, G227–35 101
–, GD229 123
–, Grw+70°8247 95–99
–, LB 11146 105
–, MR Serpentis 104
–, PG 0945+245 105
–, PG 1015+014 101–104
–, PG 1031+234 99–100
–, SBS 1349+5434 100–101
white dwarfs, spectra 10, 183–184
–, G227–35 103
–, Grw+70°8247 96–98, 108
–, LB 11146 105
–, MR Serpentis 104
–, PG 0945+245 105
–, PG 1015+014 102
–, PG 1031+234 99
–, SBS 1349+5434 100
white dwarfs, table of 106–107

Zeeman components 40, 143, 227, 229–230
Zeeman effect 69, 109
–, linear 2, 24, 38, 40, 187
–, quadratic 2, 24, 38, 40, 91, 95, 187
z-parity 23, 25, 31, 40, 79, 82, 85, 109, 124, 127, 136

Printing: Mercedesdruck, Berlin
Binding: Buchbinderei Lüderitz & Bauer, Berlin

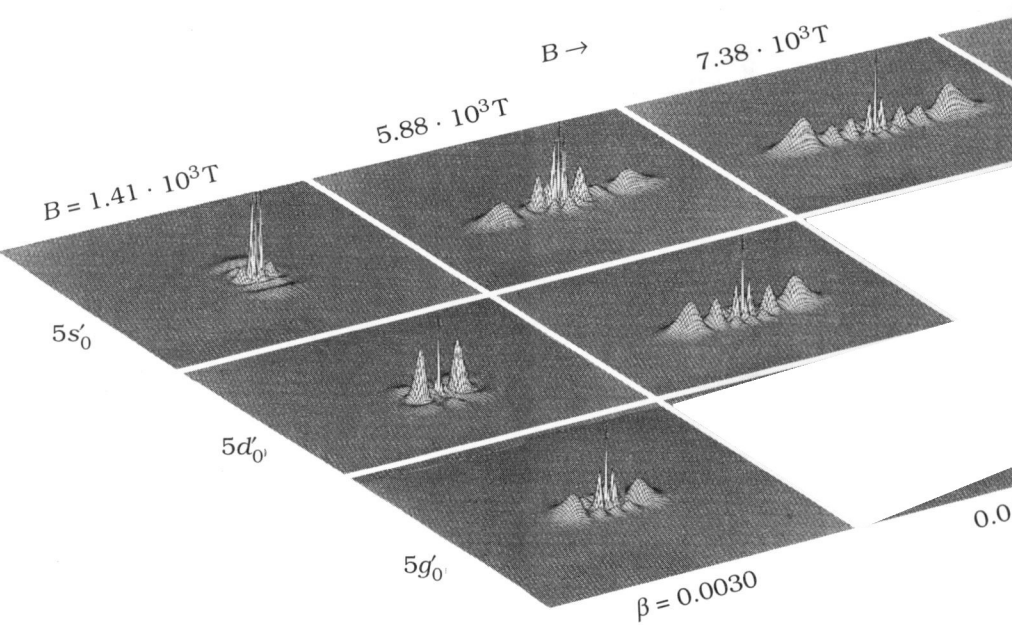